Hydrogen Bonding in Organic Synthesis

Edited by
Petri M. Pihko

Further Reading

Dodziuk, H. (ed.)

Strained Hydrocarbons

Beyond the van't Hoff and Le Bel Hypothesis

2009
Hardcover
ISBN: 978-3-527-31767-7

Yamamoto, H., Ishihara, K. (eds.)

Acid Catalysis in Modern Organic Synthesis

2008
Hardcover
ISBN: 978-3-527-31724-0

Dalko, P. I. (ed.)

Enantioselective Organocatalysis

Reactions and Experimental Procedures

2007
Hardcover
ISBN: 978-3-527-31522-2

de Vries, J. G., Elsevier, C. J. (eds.)

The Handbook of Homogeneous Hydrogenation

2007
Hardcover
ISBN: 978-3-527-31161-3

Nolan, S. P. (ed.)

N-Heterocyclic Carbenes in Synthesis

2006
Hardcover
ISBN: 978-3-527-31400-3

Tietze, L. F., Brasche, G., Gericke, K. M.

Domino Reactions in Organic Synthesis

Hardcover
ISBN: 978-3-527-29060-4

Hydrogen Bonding in Organic Synthesis

Edited by
Petri M. Pihko

WILEY-VCH Verlag GmbH & Co. KGaA

The Editor

Prof. Petri M. Pihko
University of Jyväskylä
Department of Chemistry
40014 Jyväskylä
Finland

■ All books published by Wiley-VCH are carefully produced. Nevertheless, authors, editors, and publisher do not warrant the information contained in these books, including this book, to be free of errors. Readers are advised to keep in mind that statements, data, illustrations, procedural details or other items may inadvertently be inaccurate.

Library of Congress Card No.: applied for

British Library Cataloguing-in-Publication Data
A catalogue record for this book is available from the British Library.

Bibliographic information published by the Deutsche Nationalbibliothek
The Deutsche Nationalbibliothek lists this publication in the Deutsche Nationalbibliografie; detailed bibliographic data are available on the Internet at http://dnb.d-nb.de.

© 2009 WILEY-VCH Verlag GmbH & Co. KGaA, Weinheim

All rights reserved (including those of translation into other languages). No part of this book may be reproduced in any form – by photoprinting, microfilm, or any other means – nor transmitted or translated into a machine language without written permission from the publishers. Registered names, trademarks, etc. used in this book, even when not specifically marked as such, are not to be considered unprotected by law.

Cover   Schulz Grafik-Design, Fußgönheim
Typesetting   SNP Best-set Typesetter Ltd., Hong Kong
Printing   STRAUSS Gmbh, Mörlenbach
Binding   Litges & Dopf GmbH, Heppenheim

Printed in the Federal Republic of Germany
Printed on acid-free paper

ISBN: 978-3-527-31895-7

Contents

Preface *IX*
List of Contributors *XI*

1 **Introduction** *1*
  Petri Pihko
1.1 Introduction *1*
1.2 Hydrogen Bonding in Organic Synthesis *3*
  References *4*

2 **Hydrogen-Bond Catalysis or Brønsted-Acid Catalysis? General Considerations** *5*
  Takahiko Akiyama
2.1 Introduction *5*
2.2 What is the Hydrogen Bond? *6*
2.3 Hydrogen-Bond Catalysis or Brønsted-Acid Catalysis *7*
2.4 Brønsted-Acid Catalysis *9*
2.5 Hydrogen-Bond Catalysis *11*
  References *13*

3 **Computational Studies of Organocatalytic Processes Based on Hydrogen Bonding** *15*
  Albrecht Berkessel and Kerstin Etzenbach-Effers
3.1 Introduction *15*
3.1.1 Catalytic Functions of Hydrogen Bonds *18*
3.2 Dynamic Kinetic Resolution (DKR) of Azlactones–Thioureas Can Act as Oxyanion Holes Comparable to Serine Hydrolases *19*
3.2.1 The Calculated Reaction Path of the Alcoholytic Ring Opening of Azlactones *19*
3.2.2 How Hydrogen Bonds Determine the Enantioselectivity of the Alcoholytic Azlactone Opening *23*
3.3 On the Bifunctionality of Chiral Thiourea–*Tert*-Amine-Based Organocatalysts: Competing Routes to C–C Bond Formation in a Michael Addition *25*

Hydrogen Bonding in Organic Synthesis. Edited by Petri M. Pihko
Copyright © 2009 WILEY-VCH Verlag GmbH & Co. KGaA, Weinheim
ISBN: 978-3-527-31895-7

3.4 Dramatic Acceleration of Olefin Epoxidation in Fluorinated Alcohols: Activation of Hydrogen Peroxide by Multiple Hydrogen Bond Networks 29
3.4.1 Hydrogen Bond Donor Features of HFIP 30
3.4.2 The Catalytic Activity of HFIP in the Epoxidation Reaction 30
3.5 TADDOL-Promoted Enantioselective Hetero-Diels–Alder Reaction of Danishefsky's Diene with Benzaldehyde – Another Example for Catalysis by Cooperative Hydrogen Bonding 37
3.6 Epilog 40
References 41

4 Oxyanion Holes and Their Mimics 43
Petri Pihko, Sanna Rapakko, and Rik K. Wierenga
4.1 Introduction 43
4.1.1 What are Oxyanion Holes? 44
4.1.2 Contributions of Oxyanion Holes to Catalysis 44
4.1.3 Properties of Hydrogen Bonds of Oxyanion Holes 47
4.2 A More Detailed Description of the Two Classes of Oxyanion Holes in Enzymes 49
4.2.1 A Historical Perspective 49
4.2.2 Oxyanion Holes with Tetrahedral Intermediates 52
4.2.3 Oxyanion Holes with Enolate Intermediates 56
4.2.3.1 Examples of Enolate Oxyanion Holes 58
4.3 Oxyanion Hole Mimics 61
4.3.1 Mimics of Enzymatic Oxyanion Holes and Similar Systems 61
4.3.2 Utilization of Oxyanion Holes in Enzymes for Other Reactions 64
4.4 Concluding Remarks 67
Acknowledgments 67
References 67

5 Brønsted Acids, H-Bond Donors, and Combined Acid Systems in Asymmetric Catalysis 73
Hisashi Yamamoto and Joshua N. Payette
5.1 Introduction 73
5.2 Brønsted Acid (Phosphoric Acid and Derivatives) 75
5.2.1 Binapthylphosphoric Acids 75
5.2.1.1 Mannich Reaction 75
5.2.1.2 Hydrophosphonylation 78
5.2.1.3 Friedel–Crafts 79
5.2.1.4 Diels–Alder 83
5.2.1.5 Miscellaneous Reactions 85
5.2.1.6 Nonimine Electrophiles 89
5.2.1.7 Transfer Hydrogenation 89
5.2.2 Nonbinol-Based Phosphoric Acids 91
5.2.3 N-Triflyl Phosphoramide 95

5.2.4	Asymmetric Counteranion-Directed Catalysis	*98*
5.3	N—H Hydrogen Bond Catalysts	*99*
5.3.1	Guanidine Organic Base	*99*
5.3.2	Ammonium Salt Catalysis	*106*
5.3.3	Chiral Tetraaminophosphonium Salt	*109*
5.4	Combined Acid Catalysis	*109*
5.4.1	Brønsted-Acid-Assisted Brønsted Acid Catalysis	*110*
5.4.1.1	Diol Activation of Carbonyl Electrophiles	*111*
5.4.1.2	Diol Activation of Other Electrophiles	*116*
5.4.1.3	Miscellaneous BBA and Related Systems	*120*
5.4.2	Lewis-Acid-Assisted Brønsted Acid Catalysis	*122*
5.4.3	Brønsted-Acid-Assisted Lewis Acid Catalysis (Cationic Oxazaborolidine)	*126*
5.4.3.1	Diels–Alder Reactions	*126*
5.4.3.2	Miscellaneous Reactions	*132*
	References	*136*

6 **(Thio)urea Organocatalysts** *141*
Mike Kotke and Peter R. Schreiner

6.1	Introduction and Background	*141*
6.2	Synthetic Applications of Hydrogen-Bonding (Thio)urea Organocatalysts	*149*
6.2.1	Nonstereoselective (Thio)urea Organocatalysts	*149*
6.2.1.1	Privileged Hydrogen-Bonding N,N′-bis-[3,5-(Trifluoromethyl)phenyl]thiourea	*149*
6.2.1.2	Miscellaneous Nonstereoselective (Thio)urea Organocatalysts	*174*
6.2.2	Stereoselective (Thio)urea Organocatalysts	*185*
6.2.2.1	(Thio)ureas Derived From Trans-1,2-Diaminocyclohexane and Related Chiral Primary Diamines	*185*
6.2.2.2	(Thio)ureas Derived from Cinchona Alkaloids	*253*
6.2.2.3	(Thio)urea Catalysts Derived from Chiral Amino Alcohols	*288*
6.2.2.4	Binaphthyl-Based (Thio)urea Derivatives	*296*
6.2.2.5	Guanidine-Based Thiourea Derivatives	*307*
6.2.2.6	Saccharide-Based (Thio)urea Derivatives	*315*
6.2.2.7	Miscellaneous Stereoselective (Thio)urea Derivatives	*324*
6.3	Summary and Outlook	*330*
	Acknowledgment	*332*
	Abbreviations and Acronyms	*333*
	References	*336*
	Appendix: Structure Index	*345*

7 **Highlights of Hydrogen Bonding in Total Synthesis** *353*
Mitsuru Shoji and Yujiro Hayashi

7.1	Introduction	*353*
7.2	Intramolecular Hydrogen Bonding in Total Syntheses	*353*

7.2.1	Thermodynamic Control of Stereochemistry	353
7.2.1.1	Pinnatoxin A	353
7.2.1.2	Azaspiracid-1	355
7.2.2	Kinetic Control Stereochemistry	355
7.2.2.1	Pancratistatin	355
7.2.2.2	Tunicamycins	357
7.2.2.3	Callystatin	358
7.2.2.4	Resorcylides	359
7.2.2.5	Strychnofoline	361
7.2.2.6	Asialo GM_1	361
7.2.3	Activation/Deactivation of Reactions	362
7.2.3.1	Rishirilide B	362
7.2.3.2	2-Desoxystemodione	363
7.2.3.3	Leucascandrolide A	363
7.2.3.4	Azaspirene	364
7.3	Intermolecular Hydrogen Bondings in Total Syntheses	365
7.3.1	Henbest Epoxidation	365
7.3.2	Epoxyquinols	366
7.3.3	Epoxide-Opening Cascades	367
7.4	Conclusions	369
	References	369

Index 373

Preface

The purpose of this book is to provide the reader with an overview of how hydrogen bonding can contribute to the advancement of the practice of organic synthesis. The field has grown explosively in recent years, as evidenced by the number of highlights and reviews devoted to hydrogen bonding in the service of of organic synthesis. Advances in small-molecule catalysis, computational and experimental studies of hydrogen bonding catalysis, and structural characterization of enzymes with hydrogen bonding at the core of their catalytic activity have all contributed to the advances in the field. It is nearly impossible for practitioners of organic synthesis to keep abreast of all these developments, and I hope that covering most of these aspects within the framework of a single textbook would assist the synthetic community in assessing the current power as well as future potential of the field.

The field is covered from seven different angles. The first two introductory chapters, Chapter 1 (Petri Pihko) and Chapter 2 (Takahiko Akiyama), illustrate the importance of hydrogen bonding in chemistry and chemical catalysis.

The details of how hydrogen bonding contributes to catalysis are illustrated in the following chapters. Chapter 3 by Albrecht Berkessel and Kerstin Etzenbach-Effers describes computational studies of hydrogen bonding catalysts, an essential feature in analyzing the contributions of hydrogen bonding to catalysis. In Chapter 4, Pihko, Rapakko, and Wierenga provide a general overview of hydrogen bonding in enzymatic catalysis, and they goes deeper into the structural features of oxyanion holes, the powerhouses of many hydrogen bonding enzymes. The idea behind this chapter is to present an overview of the catalytic machineries of enzymes and to provide a contrast to the present status of development of small-molecule hydrogen bonding catalysts.

The small-molecule catalysts are covered in Chapters 5 and 6. In Chapter 5, Joshua Payette and Hisashi Yamamoto discuss the importance of polar Bronsted-acid-type catalysts as well as cooperative effects in hydrogen bonding catalysis. Chapter 6 by Mike Kotke and Peter Schreiner is then devoted to the single most popular small-molecule catalyst types, the thiourea catalysts. Chapter 6, the longest of all chapters, also provides an excellent overview of the history and development of the field of small-molecule hydrogen bond catalysis.

Finally, the applications of hydrogen bonding in total synthesis of complex molecules are illustrated in the concluding Chapter 7 by Mitsuru Shoji and Yujiro

Hayashi. Although the applications of hydrogen bonding catalysts in natural product synthesis are still in their infancy, hydrogen bonding has been used many times as a driving force for desired selectivity in total synthesis.

In summary, I hope that this textbook will both stimulate fruitful research in the field and also encourage practitioners of organic synthesis to use hydrogen bonding creatively as a tool to solve synthetic challenges.

Jyväskylä, Finland *Petri Pihko*

List of Contributors

Takahiko Akiyama
Gakushuin University
Faculty of Science
Department of Chemistry
1-5-1 Mejiro, Toshima-ku
171-8588 Tokyo
Japan

Albrecht Berkessel
Universität Köln
Institut für Organische Chemie
Greinstr. 4
50939 Köln
Germany

Kerstin Etzenbach-Effers
Universität Köln
Institut für Organische Chemie
Greinstr. 4
50939 Köln
Germany

Yujiro Hayashi
Tokyo University of Science
Department of Industrial Chemistry
Faculty of Engineering
Kagurazaka
Shinjuku-ku
Tokyo 162-8601
Japan

Mike Kotke
Justus-Liebig University Giessen
Institute of Organic Chemistry
Heinrich-Buff-Ring 58
Giessen 35392
Germany

Joshua N. Payette
The University of Chicago
Department of Chemistry
5735 S. Ellis Ave. (GHJ 409)
Chicago, IL 60637
USA

Petri Pihko
University of Jyväskylä
Department of Chemistry
P.O.B. 35
40014 Jyväskylä
Finland

Sanna Rapakko
University of Oulu
Department of Biochemistry
Biocenter Oulu
P.O.B. 3000
90014 Oulu
Finland

Peter R. Schreiner
Justus-Liebig University Giessen
Institute of Organic Chemistry
Heinrich-Buff-Ring 58
Giessen 35392
Germany

Mitsuru Shoji
Tohoku University
Department of Chemistry
Graduate School of Science
Aramaki
Aoba-ku
Sendai 980-8578
Japan

Rik K. Wierenga
University of Oulu
Department of Biochemistry
Biocenter Oulu
P.O.B. 3000
90014 Oulu
Finland

Hisashi Yamamoto
The University of Chicago
Department of Chemistry
5735 S. Ellis Ave. (GHJ 409)
Chicago, IL 60637
USA

1
Introduction
Petri Pihko

1.1
Introduction

The purpose of this book is to provide the reader with an overview of how hydrogen bonding can contribute to the advancement of the practice of organic synthesis.

Hydrogen bonds form typically between polar or polarized X–H bonds and electronegative acceptor atoms [1–4]. The resulting weak bond, X–H······A, is called the hydrogen bond, and it possesses a significant electrostatic character. Consistent with this, bond strengths of hydrogen bonds in the gas phase are significantly larger with charged partners than with neutral partners. Typical strengths of hydrogen bonds are indicated in Table 1.1.

In fact, a vivid demonstration of the power of hydrogen bonds is provided by the behavior of sulfuric acid. As every student of chemistry knows, accidental addition of water to concentrated sulfuric acid can lead to a very exothermic reaction that causes the water to boil and may splash concentrated acid everywhere. For this reason, students are always taught to add sulfuric acid *cautiously*, with stirring, to water – never the other way round! When mixed with water, sulfuric acid dissociates rapidly to generate strongly solvated hydrogen bonded ions. Especially the H_3O^+ ion is very strongly hydrogen bonded to water and its solvation shell in water extends beyond its three closest neighbors, giving a solvation energy of >40 kcal/mol. Although there is an entropic cost in orienting the water molecules toward the newly generated H_3O^+ and HSO_4^- ions, the strong hydrogen bonds that are formed can more than compensate for this and are largely responsible for the heat that is generated.[1)]

Even between neutral molecules, hydrogen bonds are in fact quite strong forces. They are indeed strong enough to maintain strength in a variety of structures. These include ice and a vast range of other crystalline structures – in crystals,

1) The enthalpy of dilution of sulfuric acid is ca. 880 kJ/mol at infinite dilution (N. N. Greenwood and A. Earnshaw (1984) *Chemistry of the Elements*. Pergamon Press, Oxford, p. 837). This value compares favorably with the calculated enthalpy of hydration of H^+ (–1150 kJ/mol, see Table 1) if one assumes that the first proton of H_2SO_4 dissociates completely.

Table 1.1 Calculated and experimental hydrogen bond strengths.

Bond type	Calculated strength in the gas phase (kJ/mol)	Experimental strength in the gas phase (kJ/mol)	Calculated distance $d_{H\cdots B}$ (Å)	Calculation method/notes
H-O-H⋯O(H)-H	−20.6 [5]	−22.7 + −2.9 [6]	1.86	MP2
H-O-H⋯O(H)⁻	−108.4 [7]	−111.3 + −4.2 [8]	1.30	BLAP3 Sadlej
[H₂O⋯H⋯OH₂]⁺	−104.3 [9]	−132.3 [10]	1.20	C_2 symmetric (Zundel cation).
H₂OH⁺⋯3OH₂ (first solvation shell for H₃O⁺)	−290.22 [11]	−287.7 [12]	N/A	Eigen cation MP2
H⁺(H₂O)ₙ	−1150.1 [13]			Commentary on values [14]
H-O-H⋯O=CH(O⁻)	−71.4 [15]		1.67	MP2/6–31++G** −76.0 kJ/mol for bidentate binding
CH₃NH₃⁺⋯OH₂	−71.0 [16]	−70.6 [17]	1.72	B3LYP/6–31+G(d)

hydrogen bonds are a very powerful directing force that keeps the molecules together. The key structures of life would be impossible without hydrogen bonding: the delicate folds of proteins, the paired and folded forms nucleic acids, DNA and different RNAs, and the fibers of cellulose are all largely dependent on hydrogen bonds for their structure.

The strength of hydrogen bond is also strongly dependent on the solvent. In polar solvents, especially solvents capable of strong intermolecular hydrogen bonds such as water, hydrogen bonds between two nonwater molecules must be relatively strong in order to compete with hydrogen bonds provided by water. Experimentally, it has been established with careful site-directed mutagenesis studies of enzymes that reasonable net binding energies, in the range of 13–20 kJ/mol, are only observed when one of the components is charged [18].

1.2 Hydrogen Bonding in Organic Synthesis

Hydrogen bonds can be used in two different ways to assist in organic synthesis. First, hydrogen bonds could be used to stabilize desired structures or intermediates. This is a *thermodynamic* method of using hydrogen bonds as an assisting force in organic synthesis. As an example, Nicolaou and co-workers used an intramolecular hydrogen bond that can be used to stabilize an otherwise unattainable thermodynamically unstable nonanomeric spiroketal structure (Scheme 1.1) [19]. These methods have been used extensively in total synthesis, and they will be reviewed in Chapter 7 by Shoji and Hayashi.

A second method to utilize hydrogen bonding in organic synthesis is to use hydrogen bonds as an assisting force in catalysis. The catalysts affect reaction rates, and therefore this is a *kinetic* way of using hydrogen bonding.

In order to accelerate reactions, a catalyst should bind the transition states more strongly than starting materials. This means that typically hydrogen bonding in catalysis functions best if partial or full negative charges are generated in the substrate during the reaction. For example, addition of nucleophiles to carbonyl groups generates negatively charged tetrahedral intermediates with a charge

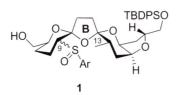

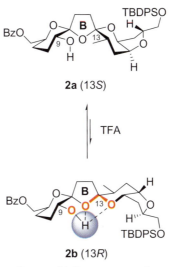

Scheme 1.1 Using hydrogen bonding as a thermodynamic force to stabilize an otherwise unattainable structure.

largely residing on the oxygen atom. Such intermediates (and transition states leading to them) can be stabilized by hydrogen bonding [20]. These strategies are used by numerous enzymes and also by small-molecule catalysts and their importance in organic synthesis lies in the mildness of the conditions as well as the immense potential for selective catalysis.

These catalysts, their structures, modes of action, and uses, are discussed in the rest of the book. Both synthetic small-molecule catalysts as well as some of Nature's finest enzymes are discussed and the role of hydrogen bonding in catalysis is described in detail.

References

1 Jeffrey, G.A. (1997) *An Introduction of Hydrogen Bonding*, Oxford University Press, Oxford.
2 Pimentel, G.C. and McLellan, A.L. (1960) *The Hydrogen Bond*, Freeman, San Francisco.
3 Steiner, T. (2002) *Angew. Chem. Int. Ed.*, **41**, 48.
4 Hine, J. (1971) *J. Am. Chem. Soc.*, **94**, 5766.
5 Raub, S. and Marian, C.M. (2007) *J. Comput. Chem.*, **28**, 1503–1515.
6 Curtiss, L.A., Frurip, D.J. and Blander, M. (1979) *J. Chem. Phys.*, **71**, 2703.
7 Wei, D., Proynov, E.I., Milet, A. and Salahub, D.R. (2000) *J. Phys. Chem. A*, **104**, 2384–2395.
8 Pudzianowsk, A.T. (1995) *J. Chem. Phys.*, **102**, 8029.
9 Xie, Y., Remington, R.B. and Schaefer, H.F. III (1994) *J. Phys. Chem.*, **101**, 4878–4884.
10 Grimsrud, E.P. and Kebarle, P. (1973) *J. Am. Chem. Soc.*, **95**, 7939.
11 Gresh, N., Leboeuf, M. and Salahub, D. (1994) Modeling the Hydrogen Bond, ACS Symposium Series 569, American Chemical Society, Washington, DC, pp. 82–112.
12 Kebarle, P. (1977) *Ann. Rev. Phys. Chem.*, **28**, 455.
13 Tissandier, M.D., Cowen, K.A., Feng, W.Y., Gundlach, E., Cohen, M.H., Earhart, A.D., Coe, J.V. and Tuttle, T.R. Jr. (1998) *J. Phys. Chem. A*, **102**, 7787–7794.
14 Camaioni, D.M. and Schwerdtfeger, C.A. (2005) *J. Phys. Chem. A*, **109**, 10795–10797.
15 Pan, Y. and McAllister, M.A. (1997) *J. Am. Chem. Soc.*, **119**, 7561–7566.
16 Kim, K.Y., Cho, U.-I. and Boo, D.W. (2001) *Bull. Korean Chem. Soc.*, **22**, 597–604.
17 Meot-Ner, M. (Mautner) (1984) *J. Am. Chem. Soc.*, **106**, 1265.
18 (a) Fersht, A.R., Shi, J.-P., Knill-Jones, J., Lowe, D.M., Wilkinson, A.J., Blow, D.M., Brick, P., Carter, P., Waye, M.M.Y. and Winter, G. (1985) *Nature*, **314**, 235–238.
(b) Fersht, A.R. (1999) *Structure and Mechanism in Protein Science*, W.H. Freeman and Company, New York.
19 Nicolaou, K.C., Qian, W., Bernal, F., Uesaka, N., Pihko, P.M. and Hinrichs, J. (2001) *Angew. Chem. Int. Ed.*, **40**, 4068–4071.
20 (a) Jencks, W.P. (1976) *Acc. Chem. Res.*, **9**, 425.
(b) Jencks, W.P. (1980) *Acc. Chem. Res.*, **13**, 161.

2
Hydrogen-Bond Catalysis or Brønsted-Acid Catalysis? General Considerations

Takahiko Akiyama

2.1
Introduction

The last few years have witnessed major advances in the use of small organic molecules as organic acid catalysts in asymmetric catalysis [1]. Selected examples of such organic acid catalysts include urea and thiourea [2], TADDOL [3], BINOL [4], and phosphoric acid derived from BINOL [5] (Figure 2.1).

Those catalysts activate carbonyl compounds and imines electrophilically, and thereby decreasing the LUMO level. The highly enantiofacial attack of carbon nucleophile on the activated carbonyl compounds and imines has been achieved by means of the chiral-acid catalysts and corresponding alcohols and amines have been obtained respectively with excellent optical purity. The proton or hydrogen bond plays a crucial role in accelerating the reaction. In this regards, those catalysts are called "hydrogen-bond catalyst," "chiral Brønsted-acid catalyst," or "chiral proton catalyst." A number of publications on the catalysts family have appeared and several review articles on the topic have been published. The titles of the selected review articles are shown below.

1. Metal-free organocatalysis through explicit hydrogen-bonding interactions. (P. R. Schreiner, 2003) [1a].
2. Activation of carbonyl compounds by double hydrogen bonding: an emerging tool in asymmetric catalysis (P. M. Pihko, 2004) [1b].
3. Protonated chiral catalysts: versatile tools for asymmetric synthesis (C. Bolm, 2005) [1c].
4. Designer acids: combined acid catalysis for asymmetric synthesis (H. Yamamoto, 2005) [1d].
5. Recent breakthroughs in enantioselective Brønsted acid and Brønsted base catalysis (P. M. Pihko, 2005) [1e].
6. Recognition and activation by ureas and thioureas: stereoselective reactions using ureas and thioureas as hydrogen-bonding donors (Y. Takemoto, 2005) [2].
7. Asymmetric catalysis by chiral hydrogen-bond donors (E. N. Jacobsen, 2006) [1f].

Hydrogen Bonding in Organic Synthesis. Edited by Petri M. Pihko
Copyright © 2009 WILEY-VCH Verlag GmbH & Co. KGaA, Weinheim
ISBN: 978-3-527-31895-7

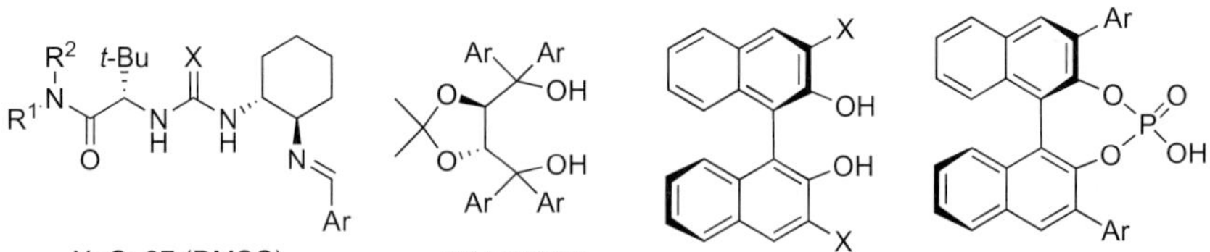

Figure 2.1 Chiral organic acid family, pK_a value in parenthesis.

8. Chiral phosphoric acids: powerful organocatalysts for asymmetric addition reactions to imines (S. J. Connon, 2006) [5a].
9. Recent progress in chiral Brønsted-acid catalysis. (T. Akiyama, 2006) [1g].
10. Small-molecule H-bond donors in asymmetric catalysis. (E. N. Jacobsen, 2007) [1h].
11. Stronger Brønsted acids. (T. Akiyama, 2007) [5b].

The difference between hydrogen-bond catalysis and Brønsted-acid catalysis is not always clear in the literatures. In this chapter, the differences and similarities of the hydrogen-bond catalysis and Brønsted-acid catalysis will be addressed.

2.2
What is the Hydrogen Bond?

The hydrogen bond has been known for more than 100 years but still a topic of vital scientific research due to its importance in not only organic chemistry but also in inorganic chemistry, supramolecular chemistry, biochemistry, and several other fields [6]. The hydrogen bond is X—H······A, where the acceptor A is an electronegative atom [7]. The following definition was proposed by Steiner [8]. An X—H······A interaction is called a "hydrogen bond," if (1) it constitutes a local bond, and (2) X—H acts as a proton donor to A. In the hydrogen bond X—H······A, the group X—H is called the hydrogen-bond donor and A is called the hydrogen-bond acceptor. In short, they are called proton donor and proton acceptor, respectively.

In ice, the O—H distance is 0.97 Å, while the H······O distance is 1.79 Å [9]. Hydrogen bonds encompass a broad range of bonding interactions, from electrostatic to covalent. Although hydrogen bonds exist with a continuum of strengths, they are classified into three categories according to strength by Jeffrey [10]. He called hydrogen bonds "moderate" if they resemble those between water molecules or in carbohydrates (one could also call them "normal") (Table 2.1).

There is a continuous transition to covalent bonding structure **2**, in which no distinction can be made between X—H and H—A if X=A. This situation is considered to be that of a hydrogen atom forming two covalent bonds having bond order $s = 1/2$. In the case of a stronger acid, proton transfer occurs to give **3**. There is

Table 2.1 Properties of hydrogen bonds.

	Strong	Moderate	Weak
Type of bonding	Mostly covalent	Mostly electrostatic	Electrostatic
Length of H-bond (Å)	1.2–1.5	1.5–2.2	2.2–3.2
Bond angles (°)	175–180	130–180	90–150
Bond energy (kcal/mol)	14–40	4–15	<4

$$X-H\cdots A \rightleftharpoons X-H-A \rightleftharpoons X\cdots H-A \rightleftharpoons X^{\ominus}\ H-A^{\oplus}$$
$$\quad\quad 1 \quad\quad\quad\quad 2 \quad\quad\quad\quad 3 \quad\quad\quad\quad 4$$

Scheme 2.1 Proton transfer reaction.

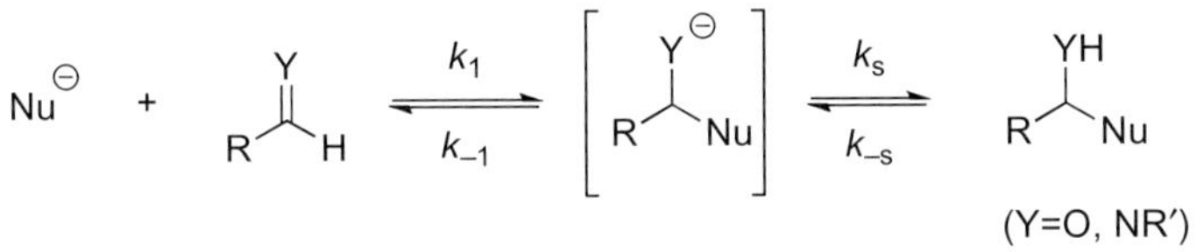

Scheme 2.2 Nucleophilic addition of strong nucleophile.

also a gradual transition from hydrogen bonding to purely ionic interactions 4. In this case the hydrogen bond may be regarded as an incipient proton-transfer reaction (Scheme 2.1).

2.3
Hydrogen-Bond Catalysis or Brønsted-Acid Catalysis

Nucleophilic addition to carbonyl compounds or imines are classified based on the nucleophilicity of the nucleophile and/or electrophile.

The addition of strong nucleophile to reactive electrophile gives a relatively stable anionic addition intermediate with a lifetime long enough to diffuse through the solution and abstract a proton before it reverts to reactants ($k_s > k_{-1}$) (Scheme 2.2).

When the nucleophile is weaker and the electrophile is more stable, the intermediate reverts to reactant faster than it abstracts a proton ($k_s < k_{-1}$) (Scheme 2.3) [11].

In order to accelerate the reaction, acid catalyst is employed. In the presence of acid catalyst, carbonyl compounds preassociate with the acid followed by proton

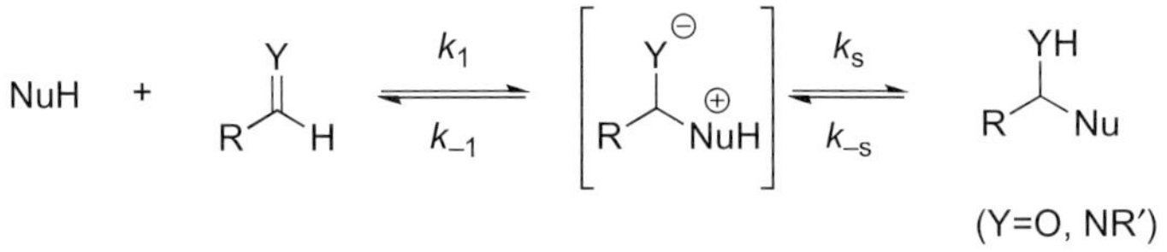

Scheme 2.3 Nucleophilic addition of weaker nucleophile.

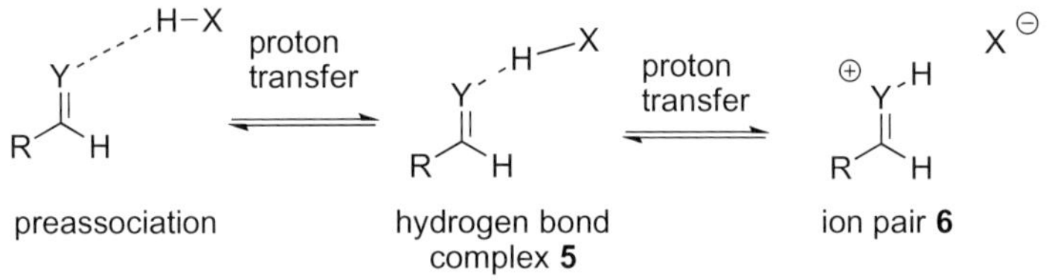

Scheme 2.4 Activation of carbonyl compound by Brønsted acid.

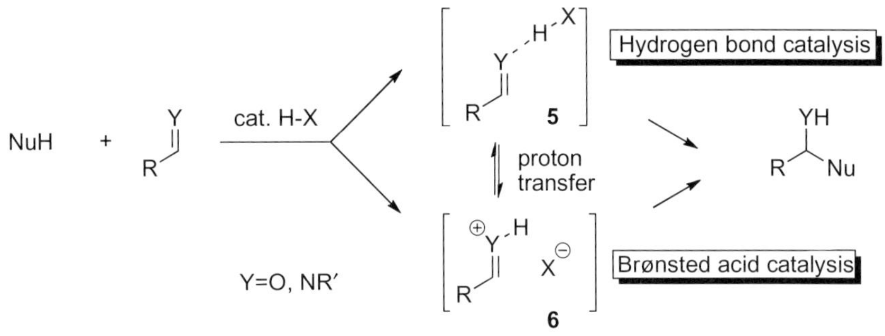

Scheme 2.5 Hydrogen-bond catalysis versus Brønsted-acid catalysis.

transfer to afford a hydrogen-bond complex **5**. Further proton transfer resulted in the formation of an ion pair **6** (Scheme 2.4).

When a nucleophile approaches a C=O or a C=N bond, there is a transfer of electron density to the oxygen or nitrogen. The oxygen or nitrogen then becomes more basic as the reaction progresses. If there is a H—Y bond that is preassociated to the C=O or C=N, the addition of the nucleophile will cause an increase in the electron density at oxygen, and this will perturb the H—Y bond.

The hydrogen-bond complex **5** and ion pair **6** are activated form of the carbonyl compounds. The nucleophilic addition of carbon nucleophile to carbonyl compounds and imines may be accelerated by acid catalysis. Nucleophilic attack to carbonyl compounds or imine took place either by way of **5** or **6** to furnish addition product. If HX activates carbonyl compound by forming hydrogen-bond complex **5** and nucleophilic addition takes place to give an adduct, the reaction is a hydrogen-bond catalyzed reaction (Scheme 2.5). In contrast, when ion pair **6** is formed and nucleophilic addition occurs, the reaction is a Brønsted-acid-catalyzed reaction.

The mechanism is closely related to general acid catalysis in which a proton transfers to the transition state in the rate-determining step, and to specific acid catalysis in which an electrophile is protonated prior to nucleophilic attack.

Because there is an equilibrium between hydrogen-bond complex **5** and ion pair **6**, it is not always easy to differentiate between hydrogen-bond catalysis and Brønsted-acid catalysis explicitly.

Use of a stronger acid in combination of imines, bearing basic element, will result in the formation of an iminium salt intermediate. Phosphoric acid is classified as a Brønsted-acid catalyst in the Mannich-type reaction (*vide infra*). In contrast, use of a neutral acid in combination of aldehyde or ketone, bearing less basic element will result in the formation of a hydrogen-bond complex. Thiourea and TADDOL are classified as hydrogen-bond catalyst. Investigation of the transition state structure by means of the theoretical study will be useful to differentiate between hydrogen-bond catalysis and Brønsted-acid catalysis.

2.4
Brønsted-Acid Catalysis

Examples of the Brønsted-acid catalysts and hydrogen-bond catalysts are shown in Figure 2.1. We have recently reported the Mannich-type reaction of ketene silyl acetals with aldimines derived from aromatic aldehyde catalyzed by chiral phosphoric acid **7** (Figure 2.2, Scheme 2.6) [12]. The corresponding β-amino esters were obtained with high *syn*-diastereoselectivities and excellent enantioselectivities.

Based on the experimental results showing that the presence of an *N-o*-hydroxyphenyl group is essential for attaining excellent enantioselectivity, we

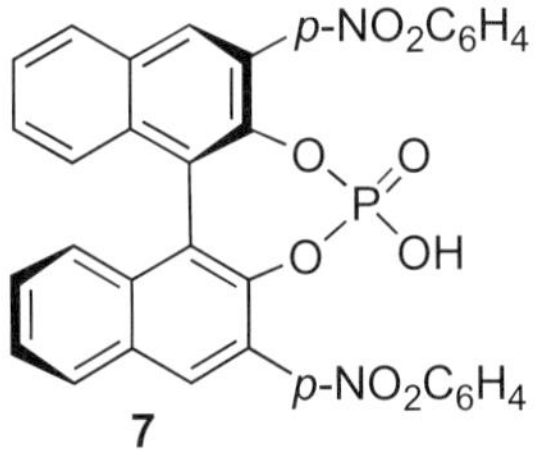

Figure 2.2 A chiral phosphoric acid.

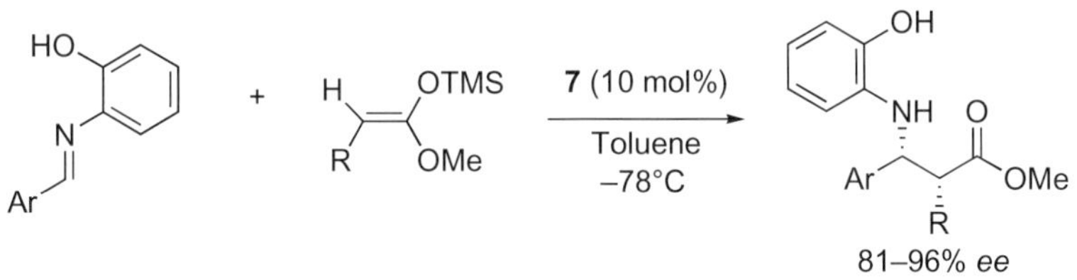

Scheme 2.6 Mannich-type reaction catalyzed by a phosphoric acid.

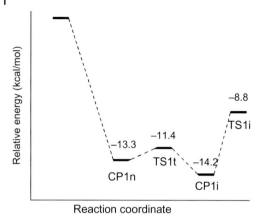

Figure 2.3 Energy profile of the pathway relative energy (kcal/mol).

hypothesized that the reaction proceeded through a nine-membered cyclic transition state, wherein phosphoryl oxygen forms a hydrogen bond with the o-OH group. In order to clarify the reaction mechanism, we calculated the transition state of the Mannich-type reaction employing a phosphoric acid derived from biphenol in place of binaphthol as a model [13]. Geometries were fully optimized and characterized by frequency calculation using hybrid density functional theory (BHandHLYP) with the 6–31G* basis set (Figure 2.3). The phosphoric acid and an aldimine form nine-membered cyclic structure **CP1n**, wherein phosphoric acid forms a hydrogen bond with nitrogen of the aldimine and at the same time phosphoryl oxygen forms a hydrogen bond with the OH group (Figure 2.4). Proton transfer resulted in the formation of the zwitterionic complex of iminium salt **CP1i**, followed by nucleophilic attack of ketene silyl acetal. Selected bond lengths are shown in Figure 2.4.

As shown in Figure 2.3, the activation energy of protonation (**TS1t**) is only 1.9 kcal/mol and the proton-transfer reaction exhibits a very flat energy profile.

In the above example, although the hydrogen-bond complex was formed initially, the nucleophilic addition reaction was supposed to proceed via protonated form **CP1i**, which clearly indicated that the phosphoric acid worked as a Brønsted-acid catalyst.

Gridnev and Terada calculated the optimized structure of the complex derived from phosphoric acid and N-acyl imine [14], wherein the hydrogen-bond complex was observed by density functional theory (DFT) calculations.

The combination of a stronger Brønsted acid with an electrophile bearing a basic electrophile would result in a Brønsted-acid-catalyzed reaction. Use of aldehyde in place of aldimine with neutral acid would lead to a hydrogen-bond-catalyzed reaction.

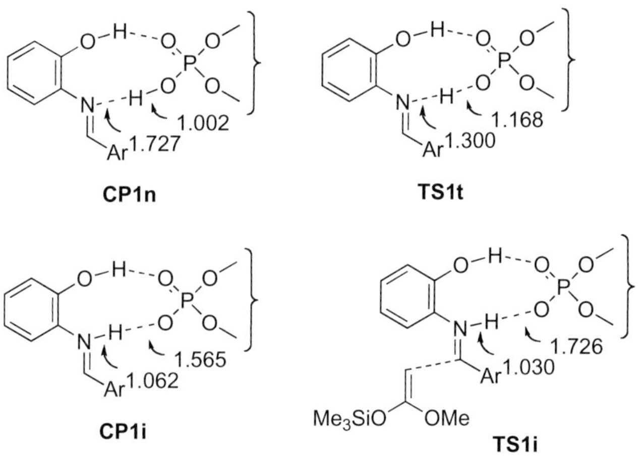

Figure 2.4 Structure of the complex and transition state. Bond length is shown in Å.

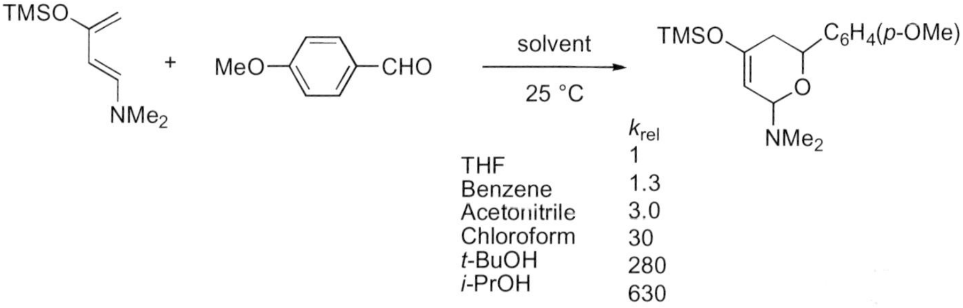

Scheme 2.7

solvent	k_{rel}
THF	1
Benzene	1.3
Acetonitrile	3.0
Chloroform	30
t-BuOH	280
i-PrOH	630

2.5 Hydrogen-Bond Catalysis

Because a comprehensive discussion of the transition state of hydrogen-bond catalysis will be presented by Berkessel in Chapter 3, the hydrogen bond catalyzed hetero Diels–Alder reaction of butadiene with carbonyl compounds will be discussed briefly here. Huang and Rawal reported that the hetero Diels–Alder reaction of aminodiene with aldehyde exhibited significant solvent effects (Scheme 2.7) [15]. The reaction in $CHCl_3$ was accelerated 30 times in comparison with that in THF, while that in i-PrOH was accelerated 630 times. They proposed that the Diels–Alder reaction was promoted by the hydrogen-bond activation of aldehyde. This finding resulted in the development of TADDOL catalyst [3].

Domingo and Andrés reported a DFT study of the hetero Diels–Alder reaction of aminodiene with acetone in $CHCl_3$ (Scheme 2.8) [16]. The transition state structure clearly shows that OH distance is 1.837 Å (Figure 2.5), which is in the range

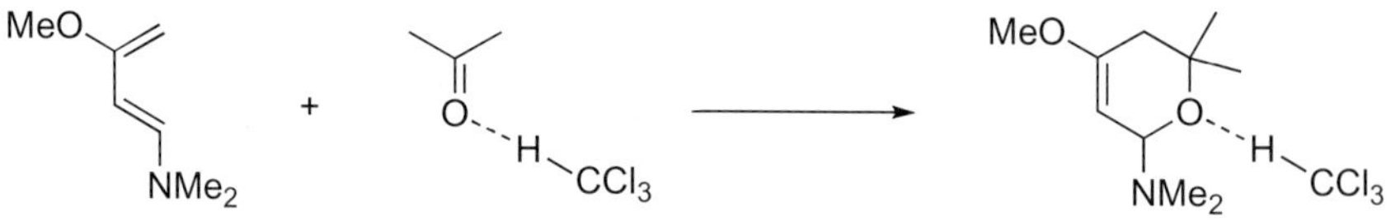

Scheme 2.8

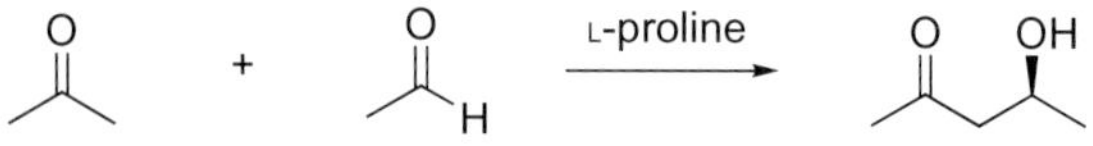

Scheme 2.9

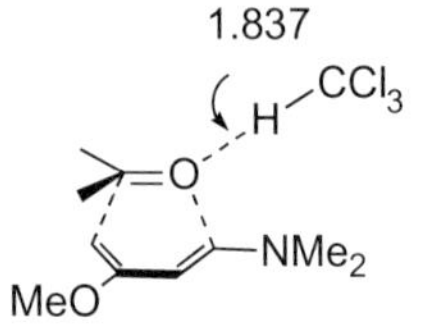

Figure 2.5 Transition state of the hetero Diels–Alder reaction.

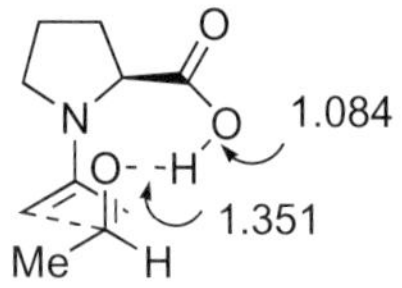

Figure 2.6 Transition state of the aldol reaction.

of hydrogen-bond catalysis. This reaction involves hydrogen-bond catalysis without the eventual proton transfer step.

Proline-catalyzed reactions, which have been attracting the attention of synthetic organic chemists, are another examples of hydrogen-bond catalysis [17]. Proline-catalyzed intermolecular aldol reaction of acetone with acetaldehyde is shown in Scheme 2.9. Houk and co-workers applied DFT calculation to rationalize the stereoselectivities [18]. Proline reacts with acetone to generate enamine intermediate *in situ* and the carboxyl group activates the carbonyl group of aldehyde (Figure 2.6). Proline is a bifunctional catalyst bearing an amino group and a hydrogen-bond donor. The OH bond length is 1.084 Å, which is in the range of hydrogen-bond catalysis.

In addition to the activation of carbonyl compounds and imines, Schreiner studied on thiourea-catalyzed acetalization reaction, in which ortho esters were activated by hydrogen bond [19]. Jacobsen has utilized the hydrogen-bond catalysis in reactions with acyliminium ions, wherein hydrogen bond activates the acyliminium salt through complexation with chloride [20].

In summary, the differences between Brønsted-acid catalysis and hydrogen-bond catalysis were discussed. Because there is a gradual transition from hydrogen-bond catalysis to Brønsted-acid catalysis, it is not always easy to differentiate the two modes of catalysis. However, the combination of a stronger acid and an imine will be a Brønsted-acid-catalyzed reaction, while the combination of a neutral acid and an aldehyde will be a hydrogen-bond-catalyzed reaction.

References

1. (a) Schreiner, P.R. (2003) *Chem. Soc. Rev.*, **32**, 289.
   (b) Pihko, P.M. (2004) *Angew. Chem. Int. Ed.*, **43**, 2062.
   (c) Bolm, C., Rantanen, T., Schiffers, I. and Zani, L. (2005) *Angew. Chem. Int. Ed.*, **44**, 1758.
   (d) Yamamoto, H. and Futatsugi, K. (2005) *Angew. Chem. Int. Ed.*, **44**, 1924.
   (e) Pihko, P.M. (2005) *Lett. Org. Chem.*, **2**, 398–403.
   (f) Taylor, M.S. and Jacobsen, E.N. (2006) *Angew. Chem. Int. Ed.*, **45**, 1520.
   (g) Akiyama, T., Itoh, J. and Fuchibe, K. (2006) *Adv. Synth. Catal.*, **348**, 999.
   (h) Doyle, G. and Jacobsen, E.N. (2007) *Chem. Rev.*, **107**, 5713.
2. (a) Takemoto, Y. (2005) *Org. Biomol. Chem.*, **3**, 4299.
   (b) Miyabe, H. and Takemoto, y. (2008) *Bull. Chem. Soc. Jpn.*, **81**, 785.
3. (a) Huang, Y., Unni, A.K., Thadani, A.N. and Rawal, V.H. (2003) *Nature*, **424**, 146.
   (b) Thadani, A.N., Stankovic, A.R. and Rawal, V.H. (2004) *Proc. Natl. Acad. Sci. USA*, **101**, 5846.
4. McDougal, N.T. and Schaus, S.E. (2003) *J. Am. Chem. Soc.*, **125**, 12094.
5. (a) Connon, S.J. (2006) *Angew. Chem. Int. Ed.*, **45**, 3909.
   (b) Akiyama, T. (2007) *Chem. Rev.*, **107**, 5744.
   (c) Terada, M. (2008) *Chem. Commun.*, 4097.
6. (a) Calhorda, M.J. (2000) *Chem. Commun.*, 801.
   (b) Mautner, M.M.-N. (2005) *Chem. Rev.*, **105**, 213.
7. Hine, J. (1971) *J. Am. Chem. Soc.*, **94**, 5766.
8. Steiner, T. (2002) *Angew. Chem. Int. Ed.*, **41**, 48.
9. Pimentel, G.C. and McLellan, A.L. (1960) *The Hydrogen Bond*, Freeman, San Francisco.
10. Jeffrey, G.A. (1997) *An Introduction of Hydrogen Bonding*, Oxford University Press, Oxford.
11. (a) Jencks, W.P. (1976) *Acc. Chem. Res.*, **9**, 425.
    (b) Jencks, W.P. (1980) *Acc. Chem. Res.*, **13**, 161.
    (c) Scheiner, S. (1994) *Acc. Chem. Res.*, **27**, 402.
12. Akiyama, T., Itoh, J., Yokota, K. and Fuchibe, K. (2004) *Angew. Chem. Int. Ed.*, **43**, 1566.
13. Yamanaka, M., Itoh, J., Fuchibe, K. and Akiyama, T. (2007) *J. Am. Chem. Soc.*, **129**, 6756.
14. Gridnev, I.D., Kouchi, M., Sorimachi, K. and Terada, M. (2007) *Tetrahedron Lett.*, **48**, 497.
15. Huang, Y. and Rawal, V.H. (2002) *J. Am. Chem. Soc.*, **124**, 9662.
16. Domingo, L.R. and Andrés, J. (2003) *J. Org. Chem.*, **68**, 8662.
17. For pioneering works see: (a) Eder, U., Sauer, G. and Wiechert, R. (1971) *Angew. Chem. Int. Ed. Engl.*, **10**, 496.
    (b) Hajos, Z.G. and Parrish, D.R. (1974) *J. Org. Chem.*, **39**, 1615.
    For recent examples see: (c) List, B., Lerner, R.A. and Barbas, C.F. III. (2000) *J. Am. Chem. Soc.*, **122**, 2395.
    For reviews see: (d) Mukherjee, S., Yang, J.W., Hoffmann, S. and List, B. (2007) *Chem. Rev.*, **107**, 5471.
    (e) Melchiorre, P., Marigo, M., Carlone, A. and Bartoli, G. (2008) *Angew. Chem. Int. Ed.*, **47**, 6138.
18. (a) Bahmanyar, S., Houk, K.N., Martin, H.J. and List, B. (2003) *J. Am. Chem. Soc.*, **125**, 2475.

(b) Allemann, C., Gordillo, R., Clemente, F.R., Cheong, P.H.-Y. and Houk, K.N. (2004) *Acc. Chem. Res.*, **37**, 558.

19 (a) Kotke, M. and Schreiner, P.R. (2006) *Tetrahedron*, **62**, 434.

See also (b) Kotke, M. and Schreiner, P.R. (2007) *Synthesis*, 779.

20 Reisman, S.E., Doyle, A.G. and Jacobsen, E.N. (2008) *J. Am. Chem. Soc.*, **130**, 7198.

3
Computational Studies of Organocatalytic Processes Based on Hydrogen Bonding

Albrecht Berkessel and Kerstin Etzenbach-Effers

3.1
Introduction

Organocatalysis has been a rapidly growing area of research over the last decade [1]. On a mechanistic basis, the vast array of organocatalytic transformations can be divided into the two subgroups "covalent organocatalysis" and "noncovalent organocatalysis." In the former case, a covalent intermediate is formed between the substrate(s) and the catalyst within the catalytic cycle. Typical examples are proline-catalyzed aldol reactions that proceed via enamine intermediates [2], or cycloadditions, conjugate additions, etc., which proceed via iminium ions derived from enal substrates and amine catalysts [3]. In contrast, noncovalent organocatalysis relies solely on noncovalent interactions such as hydrogen bonding or the formation of ion pairs. Organocatalysis had its roots in "covalent" processes, such as the proline-catalyzed Hajos–Parrish–Eder–Sauer–Wiechert aldol condensation [4]. However, the importance of hydrogen bonding for (stereo)selective organocatalysis has also been recognized early, and the recent past has seen tremendous development in this area as well [1a, 5].

Hydrogen bonding to substrates such as carbonyl compounds, imines, etc., results in electrophilic activation toward nucleophilic attack (Scheme 3.1). Thus, hydrogen bonding represents a third mode of electrophilic activation, besides substrate coordination to, for example, a metal-based Lewis acid or iminium ion formation (Scheme 3.1). Typical hydrogen bond donors such as (thio)ureas are therefore often referred to as "pseudo-Lewis-acids."

Substrate activation by hydrogen bonding is related to, but different from Brønsted-acid catalysis [1, 5c]. In the latter case, proton transfer from the catalyst to the substrate(s) occurs. The terms "specific Brønsted-acid catalysis" and "general Brønsted-acid catalysis" are used, depending on whether proton transfer occurs to the substrate in its ground state, or to the transition state. In specific Brønsted-acid catalysis, the substrate electrophile is reversibly protonated in a pre-equilibrium

Hydrogen Bonding in Organic Synthesis. Edited by Petri M. Pihko
Copyright © 2009 WILEY-VCH Verlag GmbH & Co. KGaA, Weinheim
ISBN: 978-3-527-31895-7

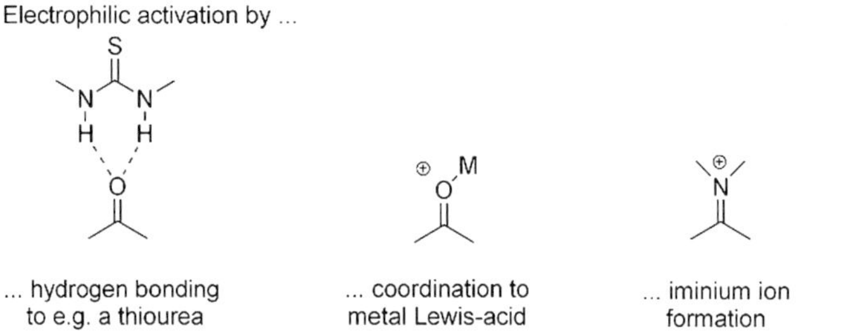

Scheme 3.1 Three modes of carbonyl activation toward nucleophilic attack.

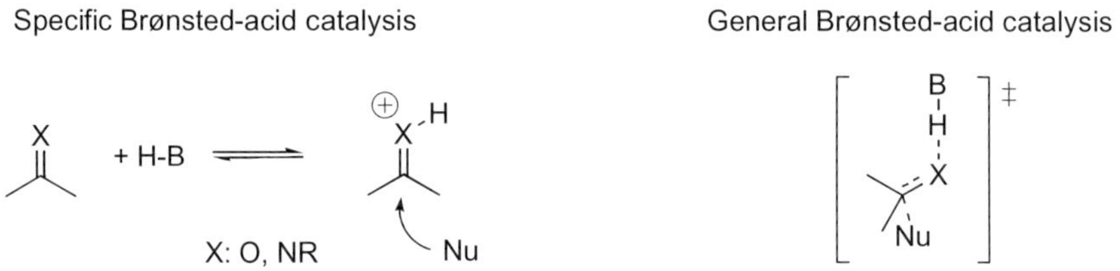

Scheme 3.2 Specific and general Brønsted-acid catalysis.

step, prior to the nucleophilic attack (Scheme 3.2). In general acid catalysis, however, the proton is (partially or fully) transferred in the transition state of the rate-determining step (Scheme 3.2). Clearly, the formation of a hydrogen bond precedes proton transfer.

Consequently, the processes most relevant to the topic of this chapter, that is, "hydrogen bonds in organocatalytic transition states," are (i) transition state stabilization by pure hydrogen bonding (without full proton transfer), and (ii) general Brønsted-acid/Brønsted-based catalyzed reactions which are initiated by hydrogen bonding but move further to distinct proton transfer.

At this point of the introduction, seminal contributions to the development and understanding of organocatalysis by hydrogen bonding by Peter R. Schreiner and co-workers need to be acknowledged. Their contribution cited in reference [6] illustrate and highlight the concepts of electrophilic (i.e., Lewis-acid like) substrate activation by hydrogen bonding [6a, b], as well as oxyanion stabilization by hydrogen bonding to organocatalysts [6c, d]. For a detailed discussion, please refer to the corresponding chapter of this book. Furthermore, please note that hydrogen bonding as the basis of (mostly biologic) catalysis has been discussed and analyzed, although not by computational means, as early as the 1970s and 1980s by Jencks and Hine [7].

Up to now, only few organocatalytic reactions of the above types have been investigated with post-Hartree–Fock methods.[1] Potential reasons are computational costs, spatial and conformational flexibility (*ab-initio* methods do not necessarily find the absolute minimum, but the minimum closest to a given starting structure – which might turn out to be a relative minimum only), and the problem of properly treating solvent effects. Nevertheless, some examples for quantum mechanically analyzed reaction mechanisms exist and will be discussed in this chapter. They allow a detailed insight at atomic level into organocatalyst function, and provide an especially detailed view on the significance of hydrogen bonding. In the majority of current theoretical publications dealing with organocatalysis, Becke's [8a, b] three parameter hybrid functional B3 and the Lee, Yang, and Parr (LYP) correlation functional [8c] are used in combination with standard split valence basis sets (e.g., 6-31G). In most cases, polarization functions that allow a greater flexibility of angle are added (for example, [d,p] means additional d-functions for second-row atoms, and additional p-functions for hydrogen atoms) [9]. In some cases, diffuse functions (abbreviated with +) are used as well, which allow an increased distance between nucleus and electron (one plus sign indicates additional diffuse functions for nonhydrogen atoms only; two plus signs indicate additional diffuse functions for hydrogen as well). They are recommended for negatively charged molecules or for the description of lone pair effects [9]. In this contribution, we focus solely on small metal- free organocatalysts (including catalytically active solvents). We also exclude covalently catalyzed reactions, for example, proline-catalyzed aldol reactions, although this reaction is well investigated at DFT level [10–17], and although a hydrogen bond is involved (the carboxyl group of the proline catalyst activates the electrophile toward the attack by the enamine by hydrogen bonding).

Transition states are clearly the most interesting stages of a reaction path. Nevertheless, we also consider starting complexes and intermediates, provided that they contribute useful information about the mode of operation of hydrogen bond mediated catalysis.

---

1) The Hartree–Fock theory neglects correlations between electrons. This means that one single electron is only subjected to an average potential by the other electrons of a system. This leads for example, to errors in bond lengths and angles and dissociation energies. Therefore, more exact methods, the so-called post-Hartree–Fock methods were developed which are either based on perturbation theory (e.g., second-order Møller–Plesset perturbation theory, MP2), or on the variational principle (e.g., configuration interaction, CI or coupled cluster methods CC). Compared to the Hartree–Fock method, these techniques are very time consuming. An alternative approach to the electronic structure is density functional theory (DFT) methods in which the electron density distribution rather than the many electron wavefunction plays a central role. Difficulties in expressing the exchange part of the energy can be relieved by including a component of the exact exchange energy calculated from Hartree–Fock theory: Functionals of this type are known as hybrid functionals. Widely used for DFT calculations is the hybrid functional B3LYP: a correlation functional developed by Becke combined with an exchange term from Lee *et al.* [8]. It provides in many cases access to qualitatively good results at computational costs comparable to Hartree–Fock methods.

3.1.1
Catalytic Functions of Hydrogen Bonds

Hydrogen bonds can preorganize the spatial arrangement of the reactants In cases where hydrogen bond donor/acceptor functions are attached to a (chiral) scaffold, they can steer the assembly of a well-defined catalyst–substrate complex. The positions of hydrogen bond donors and acceptors determine the stereoselectivity of the reaction.

Hydrogen bonds can activate the reactants by polarization The binding of substrates via hydrogen bonds (either as hydrogen bond acceptor or as donor) is necessarily associated with changes in electron densities. In catalytic systems, the resulting polarization leads to an activation of the reactants.

Hydrogen bonds can stabilize the charges of transition states and intermediates Hydrogen bonds are flexible with regard to bond length and angle. This feature is of utmost importance when charge separation occurs along the reaction pathway, and in particular in the transition state(s): Hydrogen bonds have the ability to, for example, contract and to thus stabilize developing (negative) charges. On the other hand, when the product stage is approached, the hydrogen bonds can expand again, and the product–catalyst complex can dissociate.

In hydrogen bond catalyzed reactions we find basically three different tasks that hydrogen bonds can perform:

1. There are hydrogen bonds which just stabilize charge in a transition state or intermediate. In these cases, the proton is shared between the donor and the acceptor during the transition state, and remains attached to the hydrogen bond donor afterward.

2. In some transition states, however, a second type of hydrogen bond can be encountered, which is shorter and leads to a real proton transfer from the donor to the acceptor. By some authors this phenomenon is termed a low-barrier hydrogen bond (LBHB) [18]. In particular, the lifetimes and the binding energies of LBHBs still appear to be controversially discussed [19]. Apolar organic solvents as reaction media are reminiscent of hydrophobic binding pockets of enzymes. In such surroundings, hydrogen bonds between hetero atoms with matched pK_s values can be very short and strong [18].

3. A third class, the so-called cooperative hydrogen bonds, play another important role. The latter are typically *intra*molecular hydrogen bonds which can tune the *inter*molecular hydrogen bonding to, for example, a substrate with regard to acidity (Brønsted-acid-assisted Brønsted-acid catalysis (BBA)) [20] and they are often observed in diols as for example, TADDOLs ($\alpha,\alpha,\alpha',\alpha'$-tetraaryl-1,3-dioxolan-4,5-dimethanol) [21] or BINOL (1,1'-bi-2-naphtol) [22].

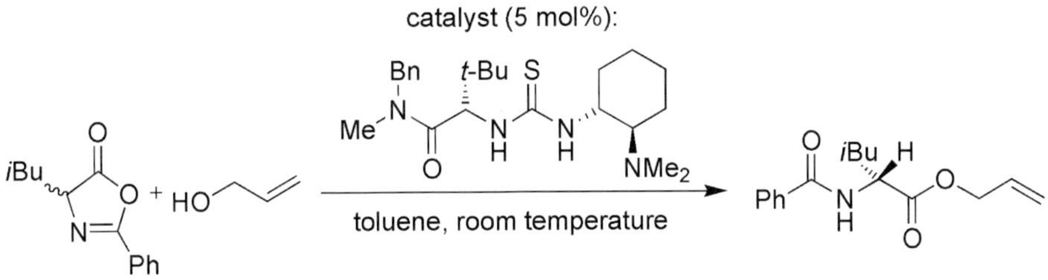

Scheme 3.3 Example for the dynamic kinetic resolution of azlactones.

3.2
Dynamic Kinetic Resolution (DKR) of Azlactones–Thioureas Can Act as Oxyanion Holes Comparable to Serine Hydrolases

Our group recently reported that bifunctional (thio)urea–*tert*-amine organocatalysts catalyze the alcoholytic dynamic kinetic resolution of azlactones (Scheme 3.3). The method affords highly enantio-enriched *N*-protected α-amino acid esters [23, 24].[2] We chose this transformation for a detailed computational study as the catalysis (both in terms of rate and stereoselectivity) is solely affected by hydrogen bonding: activation of the azlactone clearly hinges on H-bonding to the catalyst's thiourea moiety, whereas the binding/activation of the alcohol nucleophile occurs at the Brønsted-basic *tert* amine.

This reaction encompasses a number of interesting features (general Brønsted acid/Brønsted-based catalysis, bifunctional catalysis, enantioselective organocatalysis, very short hydrogen bonds, similarity to serine protease mechanism, oxyanion hole), and we were able to obtain a complete set of DFT-based data for the entire reaction path, from the starting catalyst–substrate complex to the product complex.

3.2.1
The Calculated Reaction Path of the Alcoholytic Ring Opening of Azlactones

For the calculations, we used a simplified model system in which all substituents were replaced by methyl groups (Scheme 3.4). Experimentally, the methyl substituted catalyst and methanol as nucleophile are active, but the enantiomeric excesses obtained fall below those obtained with the *tert*-leucine amide-derived catalyst in combination with allyl alcohol (Scheme 3.3).

The first step of the catalytic process is the hydrogen bond directed assembly and orientation of the reactants. In this example, the azlactone and methanol form

---

2) The formation of a ternary complex is entropically disfavored relative to binary ones. However, kinetic and spectroscopic investigations (A. Berkessel, F. Cleemann, S. Mukherjee, K. Etzenbach-Effers, N. Schlörer, unpublished) gave no indication of, for example, a ping–pong mechanism involving covalent intermediates.

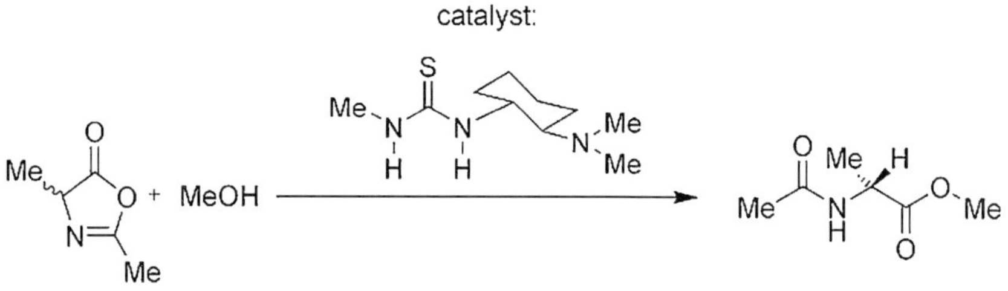

Scheme 3.4 Model system for the DFT calculations of the alcoholytic ring opening of azlactones.

a ternary *starting complex* with the organocatalyst (Figure 3.1).[3] The pseudo-Lewis acidic thiourea forms two bifurcated, nearly symmetric hydrogen bonds (2.147 Å, <(O,H,N) = 155.5° and 2.146 Å, <(O,H,N) = 155.8°) to the carbonyl oxygen atom of the azlactone, whereas the basic tertiary amino group binds the proton of the methanolic hydroxy function (1.918 Å, <(O,H,N) = 166.5°). The position of these two groups is defined by the chiral scaffold of (1R, 2R)-cyclohexane-1,2-diamine (DACH).

As exemplified for the (R)-azlactone, in principle two modes of binding are possible with this hydrogen bonding pattern. The orientation of the azlactone in Figure 3.1 (*starting complex*) leads to an attack to the *re*-side of the azlactone's carbonyl group. A 180° turn would result in a *si*-side attack, but this arrangement is disfavored because of nonbonding interactions between the methyl group at the azlactone's center of chirality and the methyl group of the incoming alcohol nucleophile. An energetically preferred arrangement for the (R)-azlactone results when the alcohol is located at the *Re*-site of the carbonyl group, preorganized for the subsequent nucleophilic attack.

Once the reactants are bound to the catalyst (*starting complex*), polarization and activation by three hydrogen bonds takes place. This process is evidenced by the change of the natural charges of the *free* azlactone and methanol molecules compared to their charges in the *starting complex*. The negative natural bond order (NBO) [25] charge of the carbonyl oxygen atom rises due to the bifurcated hydrogen bonds donated by the thiourea moiety (−0.068 e). As a consequence, the positive NBO charge of the carbonyl carbon atom increases (+0.047 e). Simultaneously, the electron density at the oxygen atom of the methanol molecule is increased (−0.057 e), due to the hydrogen bond between its hydroxy function and the tertiary amine moiety of the catalyst. In summary, the catalytic system is now perfectly orientated and activated by three hydrogen bonds for the following nucleophilic attack.

In the first transition state **TS1** (Figure 3.1) the hydrogen bonds decrease the activation energy by stabilizing the increasing charges at the participating oxygen atoms. One of the bifurcated hydrogen bonds to the carbonyl oxygen atom is

---

3) The formation of a ternary complex is entropically disfavored relative to binary ones. However, kinetic and spectroscopic investigations (A. Berkessel, F. Cleemann, S. Mukherjee, K. Etzenbach-Effers, N. Schlörer, unpublished) gave no indication of, for example, a ping–pong mechanism involving covalent intermediates.

3.2 Dynamic Kinetic Resolution (DKR) of Azlactones–Thioureas Can Act

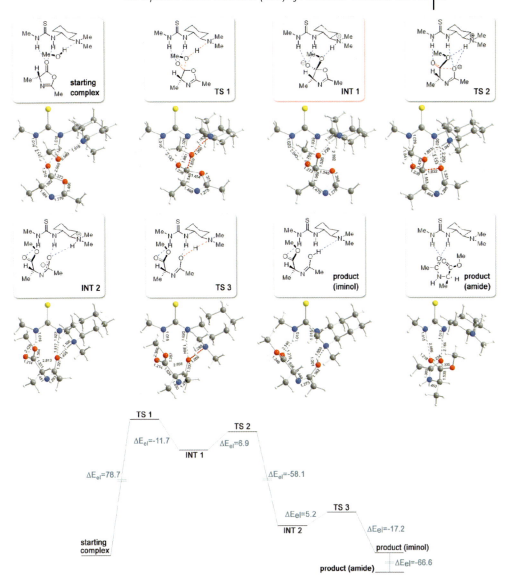

Figure 3.1 The reaction path of the alcoholytic ring opening of azlactones: geometries and relative electronic energies (kJ/mol) of the stationary points (B3LYP/6-311++G(d,p)// B3LYP/6-31++G(d,p), gas phase).

significantly shortened to 1.861 Å (−0.285 Å, <(O,H,N) = 157.2°, adjacent to the cyclohexane ring). The negative charge of the attacking hydroxyl oxygen atom is stabilized by an even stronger contraction (−0.739 Å to 1.183, <(O,H,N) = 166.8°) of the hydrogen bond to the catalyst's tertiary amine. Here we see an example for a special type of hydrogen bond, as during nucleophilic attack, the proton is

Table 3.1 NBO charges of the stationary points (black: natural charge, red: change to the previous stationary point (B3LYP/6-31++G(d,p)).

	O1	C2	N3	O4	N1'	N2'	N3'
Starting complex	−0.540	+0.571	−0.472	−0.614	−0.660	−0.659	−0.570
TS1	−0.590	+0.563	−0.504	−0.736	−0.655	−0.685	−0.538
	−0.050	−0.008	−0.032	−0.122	+0.005	−0.026	+0.032
INT1	−0.628	+0.564	−0.528	−0.834	−0.657	−0.699	−0.517
	−0.038	+0.001	−0.024	−0.098	−0.002	−0.014	+0.021
TS2	−0.735	+0.572	−0.567	−0.737	−0.656	−0.692	−0.520
	−0.107	+0.008	−0.039	+0.097	+0.001	+0.007	−0.003
INT2	−0.895	+0.614	−0.601	−0.629	−0.652	−0.683	−0.530
	−0.160	+0.042	−0.034	+0.108	+0.004	+0.009	−0.010
TS3	−0.860	+0.620	−0.578	−0.624	−0.656	−0.673	−0.552
	+0.035	+0.006	+0.023	+0.005	−0.004	+0.010	−0.022
Product (iminol)	−0.776	+0.609	−0.553	−0.632	−0.659	−0.661	−0.577
	+0.084	−0.011	+0.025	−0.008	−0.003	+0.012	−0.025
Product (amide)	−0.671	+0.701	−0.654	−0.645	−0.668	−0.648	−0.555
	+0.105	+0.092	−0.101	−0.013	−0.009	+0.013	+0.022

transferred along a nearly linear (<(O,H,N) = 166.8°) hydrogen bond from the donor alcohol to the acceptor amine ("low-barrier hydrogen bond" with an O—H—N-distance of 1.360 Å [O—H] + 1.183 Å [H—N] = 2.543 Å [O—N]).

From the first transition state (**TS1**, Figure 3.1), the reaction path leads to the tetrahedral intermediate 1 (**INT1**). In the latter, the proton transfer from methanol to the tertiary amine function is completed (from 1.183 to 1.059 Å), and the negative charge at the former carbonyl oxygen atom reaches its maximum. This charge is compensated by a further shortening of the bifurcated hydrogen bonds to 2.040 Å (−0.103 Å) and 1.765 Å (−0.096 Å) (Figure 3.1). The thiourea moiety thus forms an "oxyanion hole" similar to the amide groups of the serine protease backbone [26].

In the following transition state **TS2**, the opening of the azlactone ring takes place. The bond between the carbonyl carbon and ether oxygen atoms is stretched from 1.545 to 1.832 Å. Negative charge is transferred from the carbonyl to the ether oxygen atom in transition state 2 (**TS2**) (change in natural charge −0.102 e; see Table 3.1 for a summary), and one of the bifurcated hydrogen bonds from the carbonyl oxygen to the thiourea moiety is cleaved. Two new bifurcated hydrogen bonds (2.133 and 2.290 Å) starting to the (former) azlactone ether oxygen atom are formed to stabilize the newly developing negative charge. As the ring opening proceeds, the negative charge at the (former) azlactone ether oxygen atom raises to its maximum (change in natural charge −0.162 e), and the intermediate 2 (**INT2**) is reached. In this intermediate, the catalyst's protonated tertiary amine and the NH-group adjacent to the cyclohexane ring together form a charge-stabilizing "oxyanion hole" (length of the hydrogen bonds 1.455 and 1.855 Å).

In the third transition state (**TS3**), the neutral catalyst is recovered by transferring the proton back from the catalyst to the substrate. In other words, the (former) azlactone ether oxygen atom deprotonates the tertiary ammonium ion. For proton transfer, again a "low-barrier hydrogen bond" is formed (N–O distance 2.479 Å, <(O,H,N) = 166.2°). In the product complex, the catalyst is neutral and the N-acylamino acid ester is bound in its iminol form to the catalyst (*Product* [**iminol**]). Finally, an additional 66.6 kJ/mol are gained by the subsequent iminol-amide tautomerization (*Product* [**amide**]).

Clearly, the strength of hydrogen bonds depends on the reaction medium: In practice, the nonpolar solvent toluene is routinely used. It can be considered to mimic a hydrophobic binding pocket of an enzyme and clearly supports the formation of moderate (1.5–2.2 Å) and even strong (1.2–1.5 Å) hydrogen bonds [27].

3.2.2
How Hydrogen Bonds Determine the Enantioselectivity of the Alcoholytic Azlactone Opening

In order to explain the enantioselectivity of the alcoholytic azlactone opening, we calculated the four possible ternary starting complexes (catalyst–azlactone–methanol) **re(R)**, **re(S)**, **si(R)** and **si(S)** (Figure 3.2), together with the first (and rate-determining) transition states. In the complexes **re(R)** and **si(S)**, the methyl group bound to the azlactone's center of chirality and the methyl group of the attacking methanol are located on opposite sides of the azlactone ring. As a consequence, there is no significant interaction between them. However, in the complexes **re(S)** and **si(R)**, where both methyl groups show significant steric interaction, there is pronounced nonbonding interaction between them. This fact is reflected in the activation energies, with one exception: the activation energy of **si(S)ts** is remarkably higher than that of **re(R)ts**, although the steric interaction of the methyl groups is comparable. This effect is due to unfavorable charge separation in the transition state: As the carbonyl oxygen atom develops a partial negative charge during the nucleophilic attack of the alcohol nucleophile, the charge separation is larger for **si(S)ts** (dipole moment of **re(R)ts**: 5.66 D, dipole moment of **si(S)ts**: 6.08 D). Additionally, in **re(R)ts**, a lone pair of the lactone oxygen atom points in the direction of the developing positive charge at the tertiary amine function of the catalyst. Overall, in **re(R)ts**, the negative charge is distributed and stabilized on the azlactone oxygen atoms more effectively than in **si(S)ts**.

In summary, the hydrogen bond pattern of the catalyst disfavors some principally possible arrangements due to steric interactions, and others due to a lack of charge distribution and charge stabilization. In this example, **re(R)ts** remains as the only favored transition state (see activation energies in Figure 3.3).

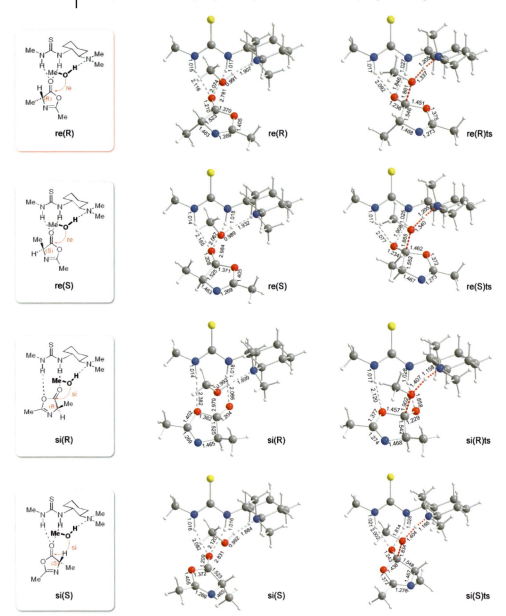

Figure 3.2 Four possible ternary (*R/S*)-azlactone–methanol–catalyst complexes optimized with B3LYP/6-31+G(d).

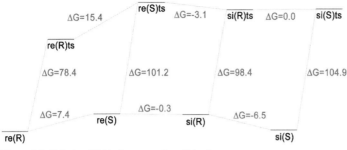

Figure 3.3 Relative Gibb's free energies of the four ternary azlactone–methanol–catalyst complexes and the corresponding transition states at 298 K, gas phase (B3LYP/6-311++G(d,p)// B3LYP/6-31+G(d)).

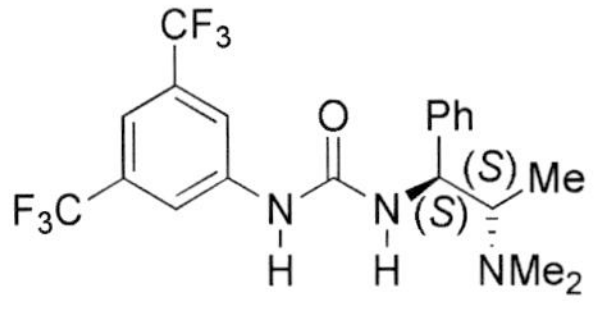

Scheme 3.5 Pseudoephedrine-derived catalyst which favors the ring opening of (S)-azlactones.

Clearly, upon using the enantiomeric catalyst [(S,S) instead of (R,R)] the opposite enantioselectivity of the overall process results. However, this effect is also seen with catalysts that are of analogous configuration, but not derived from *trans*-1,2-diaminocyclohexane (DACH). For example, the pseudoephedrine-derived catalyst shown in Scheme 3.5, having (S)-configuration at the centers of chirality, shows some preference for the (S)-azlactone in alcoholytic ring opening [28].

3.3
On the Bifunctionality of Chiral Thiourea–*Tert*-Amine-Based Organocatalysts: Competing Routes to C—C Bond Formation in a Michael Addition

Takemoto *et al.* were the first to report that bifunctional organocatalysts of the thiourea–*tert*-amine type efficiently promote certain Michael reactions, for example, the addition of β-dicarbonyl compounds to nitro olefins [29–31].

Pápai *et al.* selected as model reaction the addition of 2,4-pentanedione (acetylacetone) to *trans*-(R)-nitrostyrene, catalyzed by the bifunctional thiourea catalyst as shown in Scheme 3.6 [32]. The analogous Michael addition involving dimethyl malonate and nitroethylene as substrates, and a simplified catalyst was calculated at the same level of theory by Liu *et al.* [33]. Himo *et al.* performed a density functional study on the related *cinchona*-thiourea catalyzed Henry-reaction between nitromethane and benzaldehyde [34].

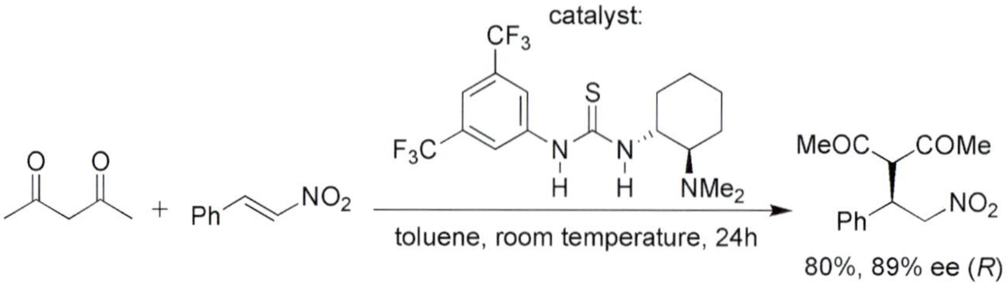

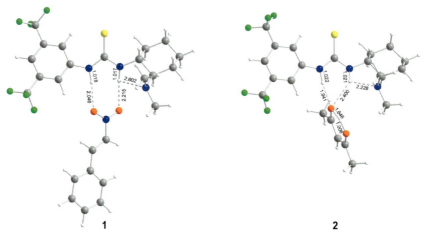

Scheme 3.6 Enantioselective Michael- ddition of acetylacetone to nitrostyrene catalyzed by a bifunctional thiourea catalyst.

Figure 3.4 Optimized structures (B3LYP/6-31G(d)) of the most stable catalyst–substrate adducts. Bond distances characteristic for hydrogen bonds are given in Å.

As shown by Takemoto and co-workers, the nitro-Michael reaction shown in Scheme 3.6 proceeds efficiently (within 1 h) at room temperature, producing the Michael adduct in good yield (80%) and with high enantiomeric excess (89% *ee*, with (*R*)-configuration of the major enantiomer) [30]. The theoretical analysis by Pápai *et al.* revealed that both the nitroolefin (2.05 and 2.21 Å) (Figure 3.4, **1**) and the enol form of acetylacetone (1.94 and 2.40 Å) (Figure 3.4, **2**) can form two hydrogen bonds with the thiourea moiety of the catalyst. A proton transfer from the coordinated enol to the amino function of the catalyst can easily take place, as the transition state related to this process (**TS2-3′**) (Figure 3.5) represents only a relatively small energy barrier (6.6 kcal/mol) with respect to adduct **2** (see Figure 3.4, **2**) and the resulting ion pair (**3′**) is predicted to be only 2.2 kcal/mol (gas phase) above **2** (0.7 kcal/mol in toluene) (see Figure 3.5, **3′**).

3.3 On the Bifunctionality of Chiral Thiourea–Tert-Amine-Based Organocatalysts | 27

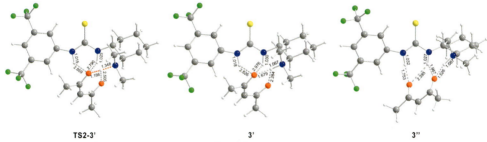

Figure 3.5 Optimized structures (B3LYP/6-31G(d)) of the stationary points located for the protonation between the thiourea-derived catalyst and the enol form of acetylacetone. Bond distances characteristic for hydrogen bonds are given in Å, broken or formed bonds in red.

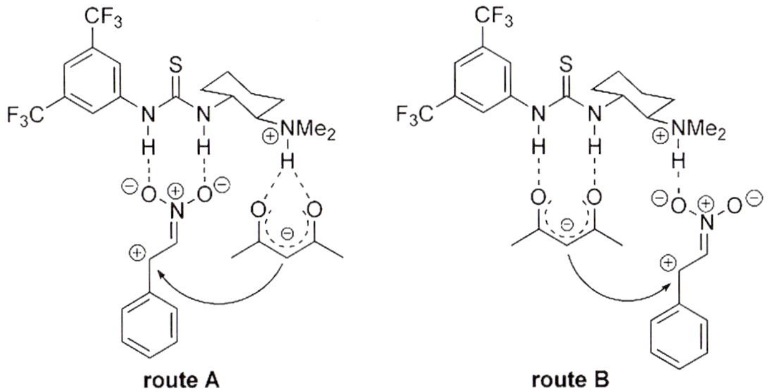

Scheme 3.7 Two alternative reaction routes for the organocatalytic Michael addition of acetylacetone to nitrostyrene.

The enolate anion in complex **3′** is stabilized by three N–H···O bonds that involve the protonated amine moiety (1.68 and 2.28 Å) and one of the N–H groups (1.80 Å) of the thiourea. In complex **3″**, the enolate is tilted from its original position to maximize the number of N–H···O bonds. In this arrangement, all three N–H units are involved in the hydrogen bond network of the thiourea.

Two distinct reaction pathways can be envisioned for the C–C bond formation step of this catalytic process (see Scheme 3.7). According to the mechanism proposed by Takemoto et al. [30], the nitroolefin interacts with the thiourea moiety of complex **3′** (Scheme 3.7, route A), forming a ternary complex, wherein both substrates are activated, and C–C bond formation can occur to produce the nitronate form of the adduct. Alternatively, the facile interconversion between **3′** and **3″** may allow an interaction of the nitroolefin with the cationic ammonium group of the protonated catalyst (Scheme 3.7, route B). In both cases, ternary complexes result

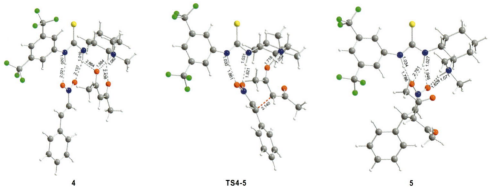

4   **TS4-5**   **5**

Figure 3.6 Optimized structures (B3LYP/6-31G(d)) of the stationary points located along route A. Lengths of hydrogen bonds are given in Å, bonds broken or formed are indicated in red.

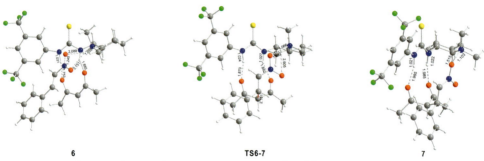

6   **TS6-7**   **7**

Figure 3.7 Optimized structures (B3LYP/6-31G(d)) of the stationary points located along route B. Lengths of hydrogen bonds are given in Å, bonds broken or formed are indicated in red.

that are the precursor for the C–C coupling step. On both routes, the hydrogen bonds to the nucleophilically attacked nitrostyrene are contracted to compensate the development of negative charge (route A: hydrogen bonds to the thiourea functionality: −0.160 and −0.316 Å (Figure 3.6), route B: hydrogen bonds to the protonated amino group: −0.437 and −0.546 Å (Figure 3.7)). Simultaneously, the hydrogen bonds to the nucleophile are stretched (route A: hydrogen bonds to the protonated amino group: +0.134 and +0.180 Å (Figure 3.6), route B: hydrogen bonds to the thiourea functionality: +0.180 and −0.003 Å (Figure 3.7)).

The C–C bond forming step also accounts for the enantioselectivity of the overall process. In the transition states affording the (R)-product (**TS 4–5** (Figure 3.6), **TS 6–7** (Figure 3.7)), the substrates are aligned in a staggered conformation

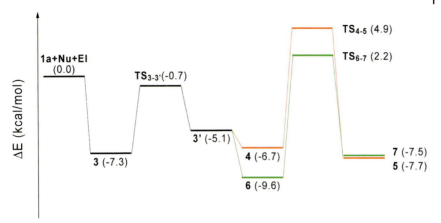

Figure 3.8 Organocatalytic Michael addition: energy profiles of paths A and B, both leading to (R)-configured product, as obtained from gas-phase calculations (B3LYP/6-311G(d,p)// B3LYP/6-31G(d)).

along the forming C—C bond, thus minimizing nonbonding interactions. Such favorable orientation cannot be adopted in the transition states leading to the (S)-configured product: the electrophilic β-carbon atom of the Michael acceptor (nitrostyrene) is displaced from its ideal position when the nucleophile attacks its si-face (Figure 3.8). C—C bond formation can only take place with a compromise of either the hydrogen bonding catalyst–substrate interactions, or the staggered geometry of the reacting molecules. These results underline the importance of the relative spatial arrangement of the hydrogen bond donor and acceptor in a bifunctional catalyst. To obtain best asymmetric induction, it should ideally be compatible only with the transition state geometry leading to the desired product stereoisomer.

3.4
Dramatic Acceleration of Olefin Epoxidation in Fluorinated Alcohols: Activation of Hydrogen Peroxide by Multiple Hydrogen Bond Networks

As a third example for an organocatalytic reaction, based on multiple hydrogen bonding and mechanistically investigated by DFT, we selected olefin epoxidation with hydrogen peroxide in fluorinated alcohol solvents, such as 1,1,1,3,3,3-hexafluoro-2-propanol (HFIP) (Scheme 3.8). Here we encounter a new type of catalytic hydrogen bond: the cooperative hydrogen bond.

In this example the solvent—a fluorinated alcohol—forms higher order aggregates and activates H_2O_2 for the epoxidation of electron rich olefins. HFIP accelerates this oxidation reaction up to 100 000-fold (relative to that in 1,4-dioxane as

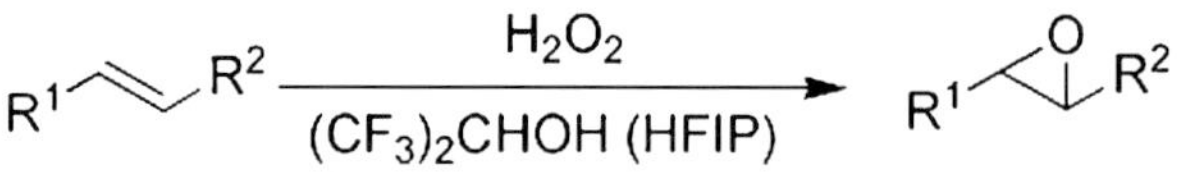

$$R^1\diagdown\!\!\!\diagup R^2 \xrightarrow[\text{(CF}_3\text{)}_2\text{CHOH (HFIP)}]{H_2O_2} R^1\diagdown\!\!\overset{O}{\triangle}\!\!\diagup R^2$$

Scheme 3.8 Epoxidation of alkenes with hydrogen peroxide in HFIP as solvent.

solvent). Which hydrogen bond network involving H_2O_2, olefin, and fluorinated alcohol gives rise to such spectacular accelerations?

3.4.1
Hydrogen Bond Donor Features of HFIP

For understanding the catalytic properties of HFIP, it is necessary to take a closer look at the hydrogen bond donor properties of HFIP, and the factors they are influenced by [35]. The hydrogen bond donor ability of fluorinated alcohols, and in particular HFIP, is mainly dependent on two parameters: (i) the conformation of the alcohol monomer along the C–O bond [35, 36] and (ii) the cooperative aggregation to hydrogen bonded alcohol clusters [35].

In a polarizable environment, the absolute minimum structure of HFIP carries the OH synclinal (sc) or almost synperiplanar (sp) to the adjacent CH (Figure 3.9) [35]. On the basis of quantum-chemical considerations as well as single-crystal X-ray structures in which HFIP acts as hydrogen bond donor, HFIP always takes on such a sc or even sp conformation. In this conformation, the hydrogen bond donor ability of HFIP is significantly increased (Figures 3.10 and 3.11) [35].

Furthermore, the hydrogen bond donor ability of an HFIP hydroxyl group is greatly enhanced upon coordination of a second or even third molecule of HFIP (Figures 3.12 and 3.13). Aggregation beyond the trimer has no significant additional effect [35, 37].

Therefore, the following mechanistic investigation of the epoxidation of olefins with hydrogen peroxide is constrained to reaction pathways which (i) involve HFIP in an sc or even sp conformation and (ii) to hydrogen-bonded HFIP aggregates comprising up to four alcohol monomers.

3.4.2
The Catalytic Activity of HFIP in the Epoxidation Reaction

Kinetic investigations of the epoxidation of Z-cyclooctene by aqueous H_2O_2 in HFIP show that the reaction follows a first-order dependence with respect to the substrate olefin as well as to the oxidant, suggesting a monomolecular participation of these components in the rate-determining step [37]. On the other hand, a rate order of 2–3 with respect to the concentration of HFIP is observed for several cosolvents. The large negative $\Delta S^\ddagger$ of -39 cal/mol K points to a highly ordered TS of the rate-determining reaction step: typical $\Delta S^\ddagger$ values for olefin epoxidations by peracids range from -18 to -30 cal/mol [38]. These experimental results provide the basis for the calculations in which one to four molecules of HFIP are added to the transition state of the reaction.

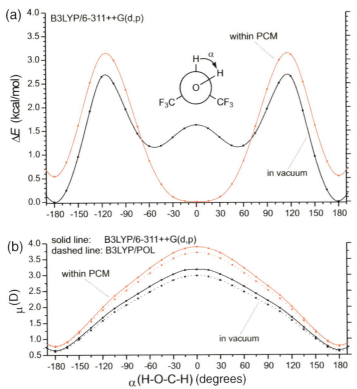

Figure 3.9 Potential energy (a) and dipole moment (b) of HFIP versus <(HOCH) dihedral angle in vacuum (black) and within a PCM (red).

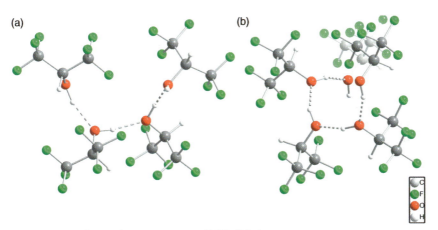

Figure 3.10 Single-crystal X-ray structures of HFIP ((a) view perpendicular to the helix axis; (b) view along the helix axis).

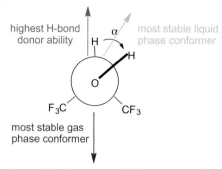

Figure 3.11 Dependence of the properties of monomeric HFIP on the conformation along the CO bond.

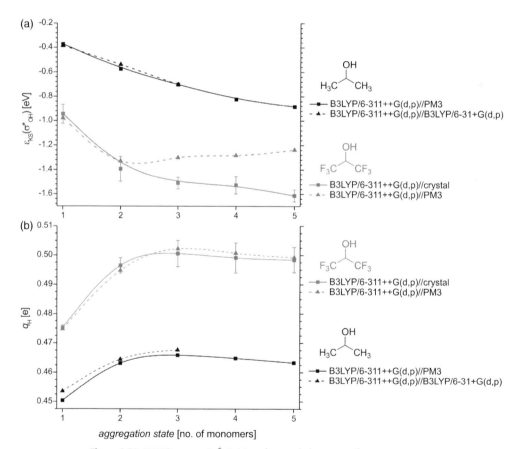

Figure 3.12 LUMO energy (σ^*_{OH}) (a) and natural charge q_H of the hydroxyl proton (b) versus aggregation state of HFIP.

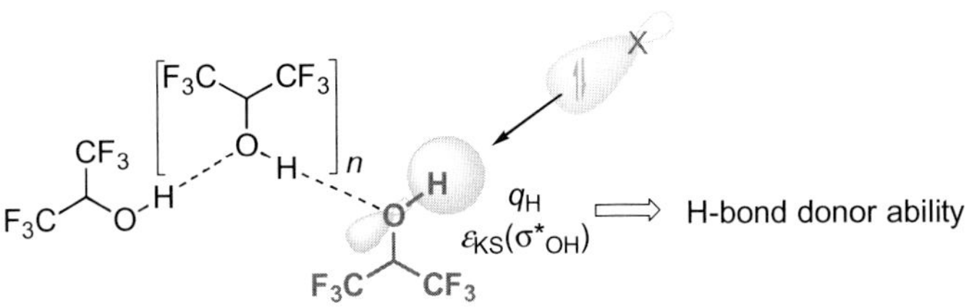

Figure 3.13 Aggregation-induced hydrogen bonding enhancement of HFIP.

The first quantum-chemical investigation of the mechanism of olefin epoxidation in fluoroalcohols was carried out by Shaik et al. [39]. In the absence of kinetic data, a *monomolecular* mode of activation by the fluorinated alcohols for all reaction pathways was assumed [39].

In this chapter, we compare the transition state which does not involve HFIP-participation [**TS(e,0)**] with single-HFIP involvement [**TS(e,1)** and **TS(e,19′)**] (Figure 3.14). Particular emphasis is then put on the twofold HFIP-activated complex (Figure 3.15) for a detailed inspection of the hydrogen bond assisted epoxidation. All relevant characteristics of higher order activation (as shown e.g., in Figure 3.16) are already present in the transition states **TS(e,2)** and **TS(e,1)′** (Figure 3.15).

In the 2:1 precomplex **C2** composed of HFIP and H_2O_2, hydrogen peroxide is coordinated by the two alcoholic hydroxyl groups in a cyclic fashion, one HFIP acting as a hydrogen bond donor toward the leaving OH (hydrogen bond length of 1.767 Å), and the other one as a hydrogen bond acceptor (hydrogen bond length of 1.906 Å), deprotonating the hydroxyl group which is transferred to be the epoxide oxygen atom (**C2**, Figure 3.15). The "internal" hydrogen bond between the two fluorinated alcohols (hydrogen bond length of 1.823 Å) cooperatively increases the hydrogen bond donor ability of the alcohol molecule which activates the leaving OH-group. By this hydrogen bond pattern, a polarization of the O—O bond is achieved (the donated and accepted hydrogen bonds are not equal in length and angle), and an electron deficient oxygen is generated, ready for electrophilic attack on the olefinic double bond. In the corresponding transition state (**TS(e,2)**, Figure 3.15) the shorter hydrogen bond (in which HFIP acts as H-bond donor) is extremely contracted to 1.409 Å (−0.358 Å), whereas the longer hydrogen bond (in which HFIP acts as acceptor) is slightly decreased in length to 1.864 Å (−0.042 Å). The acidity of the donor HFIP molecule is cooperatively increased by shortening of the HFIP internal hydrogen bond from 1.823 to 1.692 Å (−0.131 Å).

A second potential reaction path (**C(2)′**, **TS(e,2)′**, Figure 3.15) for twofold HFIP activation was calculated, which differs from **TS(e,2)** with regard to the hydrogen bond from H_2O_2 to HFIP. Here, a fluorine atom of the trifluormethyl group serves as hydrogen bond acceptor, and not a second hydroxy function. Both transition states are similar in energy but the corresponding precomplex **C(2)′**, consisting of H_2O_2 and two HFIP molecules, lies 18.4 kJ/mol above **C(2)**.

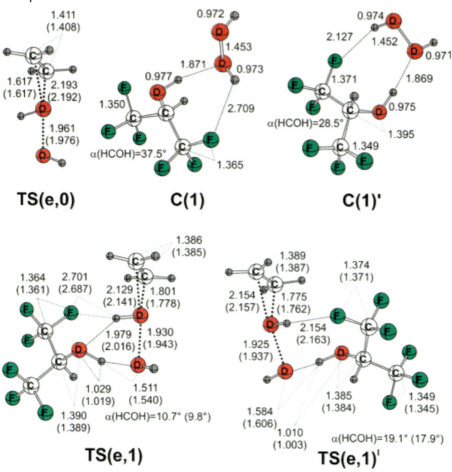

Figure 3.14 Stationary-point structures for the epoxidation of ethene with hydrogen peroxide in the absence and in the presence of one molecule of HFIP, optimized at RB3LYP/6-31+G(d,p) (selected bond lengths in Å; RB3LYP/6-311++G(d,p) results in parentheses).

An analysis of the hydrogen bonding parameters shows that, in all cases where HFIP donates a hydrogen bond to the oxidant, this hydrogen bond is significantly contracted in the transition state, usually by more than 0.3 Å. The result of this significant contraction is the formation of a *low-barrier hydrogen bond* [18], characterized by an increase in covalency which effectively exerts the pronounced stabilization of the highly polar transition states through charge transfer. Hydrogen bonds between two HFIP molecules show the same trend, being regularly shortened by ca. 0.1 Å. This effect clearly indicates a cooperative enhancement of hydrogen bonding. Additionally, we find a reduction of the <(HCOH) dihedral

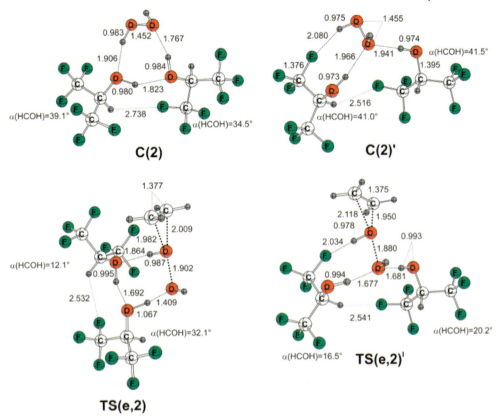

Figure 3.15 Stationary-point structures for the epoxidation of ethene with hydrogen peroxide in the presence of two molecules of HFIP, optimized at RB3LYP/6-31+G(d,p).

angles in 14 of the 16 HFIP molecules within the seven calculated reaction pathways. This result is in agreement with the analysis of the hydrogen bonding properties of HFIP, as the hydrogen bond donor ability is maximized toward the sp conformation of the alcohol.

Proceeding from the transition states to the resulting products, IRC analysis demonstrates that along this reaction path, a subsequent and *barrier-free, cascade-like proton transfer* toward the formation of the epoxide and water takes place. Figure 3.17 shows the overall dependence of the activation parameters on the number of HFIP molecules involved. Interestingly, the activation enthalpy of the epoxidation decreases steadily from zero to fourth order in HFIP. As expected, the activation entropy $-T\Delta S^\ddagger$ shows a continuous increase with increasing numbers of specifically coordinated HFIP molecules. Due to the increasing entropic contribution, the value of $\Delta G^\ddagger$ approaches saturation when three or four HFIP molecules are involved. For methanol, however, no influence of explicit coordination of the

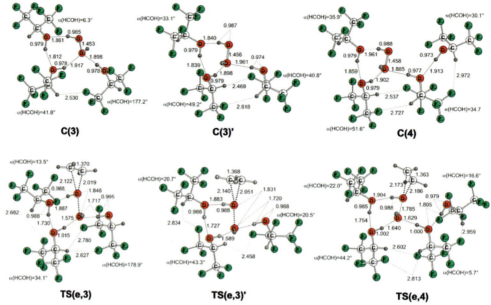

Figure 3.16 Stationary-point structures for the epoxidation of ethene with hydrogen peroxide in the presence of three to four molecule of HFIP, optimized at RB3LYP/6-31+G(d,p).

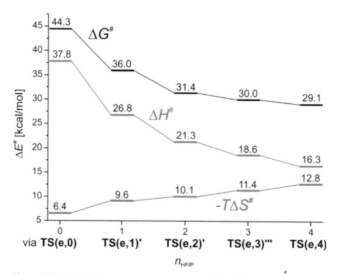

Figure 3.17 Activation parameters versus number of HFIP molecules for the epoxidation of ethane within a solution model at 298 K (RB3LYP/6-311++G(d,p)// RB3LYP/6-31+G(d,p)).

solvent on the activation parameters of oxygen transfer could be found, so it seems solely to act as a polar reaction medium. In line with this result, no significant epoxidation catalysis results from the use of methanol as solvent.

3.5 TADDOL-Promoted Enantioselective Hetero-Diels–Alder Reaction of Danishefsky's Diene with Benzaldehyde – Another Example for Catalysis by Cooperative Hydrogen Bonding

The enantioselective hetero-Diels–Alder reaction of carbonyl compounds with 1,3-dienes represents an elegant access to optically active six-membered oxo-heterocycles. Since the pioneering work of Rawal *et al.* in 2003 [40], the enantioselective HDA reaction catalyzed by diols (such as TADDOLs) has become a flourishing field of research [41].

In the catalytic system shown in Scheme 3.9, a hydrogen bond between one hydroxy function of the diol catalyst and the carbonyl group of the substrate is regarded as the driving force of catalysis. Here, the spatial orientation of the bulky α-1-naphtyl substituents of the TADDOL (α,α,α′,α′-tetraaryl-1,3-dioxolan-4,5-dimethanol) scaffold generates the chiral environment controlling the enantioselectivity of the reaction.

A similar Diels–Alder reaction was investigated at DFT level by Houk and co-workers [42]. Instead of using TADDOL, they selected one methanol molecule, two methanol molecules, and 1,4-butanediol in cooperative and bifurcated coordination as catalysts. It was found that cooperative catalysis is generally the favored route.

Ding, Wu *et al.* [43] used a model system consisting of benzaldehyde and a modified Danishefsky's diene, in which the trimethylsilyl group was replaced by a methyl group. They varied the aryl substituents of the TADDOL catalyst, and the results for the most enantioselective 1-naphtyl substituted catalyst were presented in detail. The central feature of this computational analysis is the use of the ONIOM method (Figure 3.18) [44, 45] for geometry optimization: the substrates and the core of TADDOL were treated with B3LYP/6-31G(d), while the substituents of the catalyst were modeled using semiempirical PM3 [46, 47] level. The energies of the optimized structures were determined by single point calculations with B3LYP/6-31G(d). X-ray crystal structures [21] of the TADDOLs investigated were taken as starting geometries.

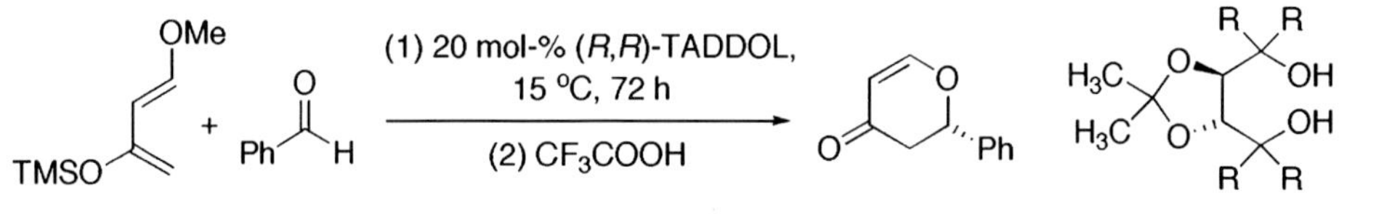

Scheme 3.9 Enantioselective hetero-Diels–Alder (HDA) reaction of Danishefsky's diene with benzaldehyde.

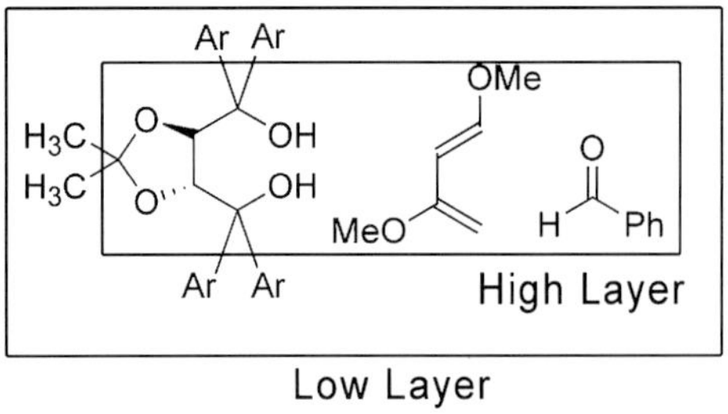

Figure 3.18 Application of the ONIOM method to the model system of the reaction.

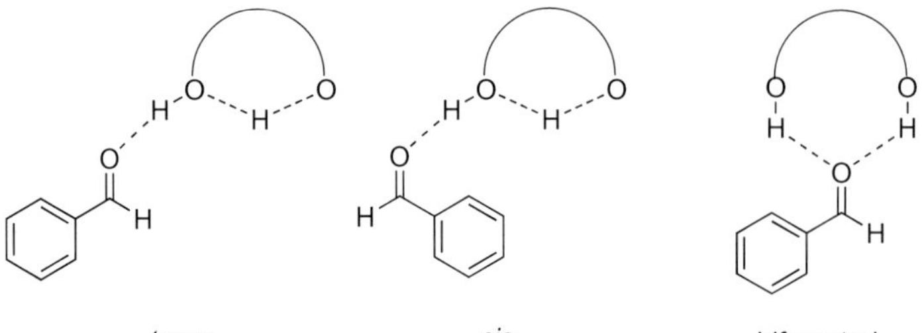

Scheme 3.10 Possible intermolecular hydrogen bonding patterns between benzaldehyde and TADDOL.

In the starting complex, there are three principal possibilities of hydrogen bonding between TADDOL and benzaldehyde (Scheme 3.10). A bifurcated hydrogen bond between the two hydroxy functions and the carbonyl group can be excluded, as TADDOLs are well known to have an *intramolecular* hydrogen bond [21, 48]. Ding, Wu et al. found that structure optimizations resulted, without exception, in the *trans* configuration, even if the initial structure was *cis*. This observation is consistent with the crystal structures available for TADDOL adducts with carbonyl compounds [21, 48].

Upon formation of the *trans* adduct of benzaldehyde with TADDOL, the *intra*molecular hydrogen bond is shortened by 0.128 Å, and the acidity of the substrate binding hydroxy function is increased. The length of the *inter*molecular hydrogen bond to the carbonyl group is 1.825 Å.

To rationalize the enantioselectivity of the TADDOL-catalyzed HDA reaction between Danishefsky's diene and benzaldehyde, eight possible diastereomeric transition states of different regio- and stereochemistry should, in principle, be considered for comprehensive analysis. The cycloaddition between the model diene and benzaldehyde can take place along two regio-isomeric "meta" (C1–O6, C4–C5 bond formation) and "ortho" (C1–C5, C4–O6 bond formation) reaction channels. For both of these pathways, an exo- and an endo-approach can be formulated (Scheme 3.11).

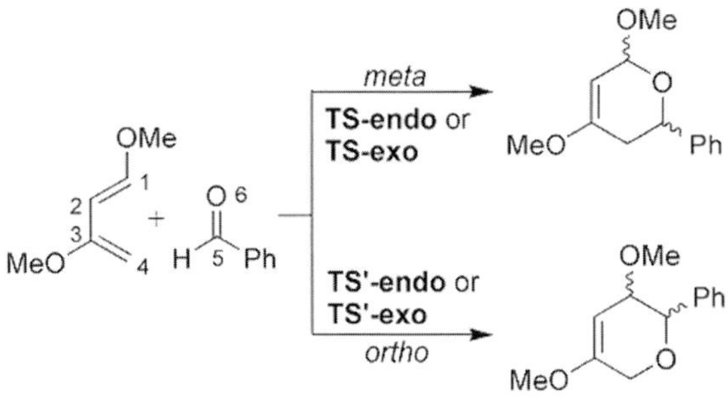

3.5 TADDOL-Promoted Enantioselective Hetero-Diels–Alder Reaction of Danishefsky's Diene

Scheme 3.11 Regio- and stereoselectivity issues of the model hetero-Diels–Alder cycloaddition.

The energy of the localized transition state for the "ortho" route (uncatalyzed reaction) is 14 kcal/mol higher than that of the "meta" channel. Therefore, the "ortho" channel can be excluded. Unlike the uncatalyzed transformation, the TADDOL-catalyzed HDA reaction exhibited a clear energetic preference for the endo- over the exo-approach. Thus, only endo transition states were considered. The number of possible reaction paths/transition states is thus reduced from eight to two, namely endo-approach with *re*- or *si*-face attack of the model diene to the activated benzaldehyde.

As can be seen from Figure 3.19, the activation energy of the reaction in the presence of the 1-naphtyl substituted TADDOL catalyst was reduced by 10.2 kcal/mol, in comparison with the uncatalyzed reaction (20.2 kcal/mol). The reaction proceeds via a concerted but asynchronous pathway, and no zwitterionic intermediate or transition state corresponding to a stepwise Mukaiyama-aldol-type pathway could be located.

The partial charges of the aldehyde carbonyl group are stabilized by an intermolecular hydrogen bond to the TADDOL catalyst. In the organocatalytic reaction, the C–C bond formation has progressed further (due to the more positively polarized carbonyl carbon atom of the benzaldehyde) whereas the C–C bond formation lags behind in the uncatalyzed reaction. The NBO [49] charges indicate that there is a considerable charge transfer of 0.49 e⁻ from the donor diene to the activated aldehyde acceptor in the transition state (**TS-(Si)-4b**), but in the uncatalyzed case the transferred charge does not exceed 0.27 e⁻ (**endo TS**). This is due to the fact that both the cooperative intramolecular hydrogen bond (shortened by about 0.08 Å) and the intermolecular hydrogen bond to the substrate (shortened by about 0.2 Å) reinforce each other cooperatively in the transition state, and stabilize the developing negative charge at the carbonyl oxygen atom during nucleophilic attack. How can the energetic preference of **TS-(si)-4b** over **TS-(re)-4b** be rationalized? Obviously, nonbonding interactions of the 1-naphtyl groups of the TADDOL catalyst with the phenyl ring of benzaldehyde in the *re*-transition state effect the observed enantioselectivity.

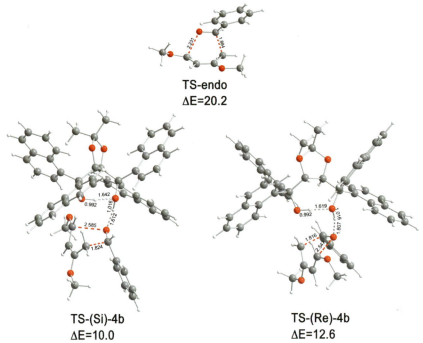

Figure 3.19 Transition states involved in the cycloaddition (endo-mode) of the model diene with benzaldehyde, both in the absence (TS-endo) and in the presence (TS-(Si)-4b, TS-(Si)-4b) of the TADDOL catalyst; corresponding activation energies (kcal/mol) (B3LYP/6-31G(d)//B3LYP/6-31G(d):PM3).

3.6
Epilog

Multiple and specific hydrogen bonding has been recognized as a highly efficient motif not only in enzymatic catalysis, but nowadays also in organocatalysis. Much of our current mechanistic understanding is based on computational analyses of such processes, ideally in combination with kinetic and spectroscopic methods. It appears that naturally evolved catalytic motifs, such as the oxyanion hole, can be "side-tracked" to accelerate a number of "anthropogenic" reactions involving intermediates/transition states with a negatively charged oxygen atom. It is tempting to speculate which other types of enzymatic rate accelerations by hydrogen bonding might be suitable for adaptation to organocatalysis. Along the same lines, one may wonder whether organocatalysts might eventually substitute for enzymes – potentially in a living cell.

References

1. (a) Berkessel, A. and Gröger, H. (2005) *Asymmetric Organocatalysis*, Wiley-VCH Verlag GmbH, Weinheim.
(b) Dalko, P.I. (ed.) (2007) *Enantioselectice Organocatalysis*, Wiley-VCH Verlag GmbH, Weinheim.
(c) List, B. (guest editor) (2007) *Chem. Rev.*, **107**, 5413–5883, issue on Organocatalysis.
2. Mukherjee, S., Yang, J.W., Hoffmann, S. and List, B. (2007) *Chem. Rev.*, **107**, 5471–5569.
3. Erkkilä, A., Majander, I. and Pihko, P.M. (2007) *Chem. Rev.*, **107**, 5416–5470.
4. (a) Eder, U., Sauer, G. and Wiechert, R. (1971) *Angew. Chem. Int. Ed. Engl.*, **10**, 496–497.
(b) Hajos, Z.G. and Parrish, D.R. (1974) *J. Org. Chem.*, **39**, 1615–1621.
5. (a) Taylor, M.S. and Jacobsen, E.N. (2006) *Angew. Chem. Int. Ed.*, **45**, 1520–1543.
(b) Doyle, A.G. and Jacobsen, E.N. (2007) *Chem. Rev.*, **107**, 5713–5743.
(c) Akiyama, T. (2007) *Chem. Rev.*, **107**, 5744–5758.
6. (a) Schreiner, P.R. and Wittkopp, A. (2002) *Org. Lett.*, **4**, 217–220.
(b) Schreiner, P.R. (2003) *Chem. Soc. Rev.*, **32**, 289–296.
(c) Kleiner, C.M. and Schreiner, P.R. (2006) *Chem. Commun.*, 4315–4317.
(d) Kotke, M. and Schreiner P.R. (2007) *Synthesis*, 779–790.
7. (a) Jencks, W.P. (1980) *Acc. Chem. Res.*, **13**, 161–169.
(b) Jencks, W.P. (1976) *Acc. Chem. Res.*, **9**, 425–432.
(c) Hine, J. (1972) *J. Am. Chem. Soc.*, **94**, 5766–5771.
8. (a) Becke, A.D. (1988) *Phys. Rev. A*, **38**, 3098–3100.
(b) Becke, A.D. (1993) *J. Chem. Phys.*, **98**, 5648–5652.
(c) Lee, C., Yang, W. and Parr, R.G. (1988) *Phys. Rev. B*, **37**, 785–789.
9. Levine, I.N. (2000) *Quantum Chemistry*, 5th edn, Prentice-Hall, Upper Saddle River, NJ.
10. Bahmanyar, S. and Houk, K.N. (2001) *J. Am. Chem. Soc.*, **123**, 11273–11283.
11. Bahmanyar, S. and Houk, K.N. (2001) *J. Am. Chem. Soc.*, **123**, 12911–12912.
12. Bahmanyar, S., Houk, K.N. and List, B. (2003) *J. Am. Chem. Soc.*, **125**, 2475–2479.
13. Hoang, L., Bahmanyar, S., Houk, K.N. and List, B. (2003) *J. Am. Chem. Soc.*, **125**, 16–17.
14. Rankin, K.N., Gauld, J.W. and Boyd, R.J. (2002) *J. Phys. Chem. A*, **106**, 5155–5159.
15. Arnó, M. and Domingo, L.R. (2002) *Theor. Chem. Acc.*, **108**, 232–239.
16. Allemann, C., Gordillo, R., Clemente, F.R., Cheong, P.H.-Y. and Houk, K.N. (2004) *Acc. Chem. Res.*, **37**, 558–569.
17. Clemente, F.R. and Houk, K.N. (2004) *Angew. Chem. Int. Ed.*, **43**. 5766–5768.
18. Cleland, W.W., Frey, P.A. and Gerlt, J.A. (1998) *J. Biol. Chem.*, **273**, 25529–25532.
19. Schutz, C.N. and Warshel, A. (2004) *Proteins Struct. Funct. Bioinform.*, **55**, 711–723.
20. Momiyama, N. and Yamamoto, H. (2005) *J. Am. Chem. Soc.*, **127**, 1080–1081.
21. Seebach, D., Beck, A.K. and Heckel, A. (2001) *Angew. Chem. Int. Ed.*, **40**, 92–138.
22. Brunel, J.M. (2005) *Chem. Rev.*, **105**, 857–897.
23. (a) Berkessel, A., Cleemann, F., Mukherjee, S., Müller,T.N. and Lex, J. (2005) *Angew. Chem. Int. Ed.*, **44**, 807–811.
(b) Berkessel, A., Mukherjee, S., Cleemann, F., Müller, T.N. and Lex, J. (2005) *Chem. Commun.*, 1898–1900.
(c) Berkessel, A., Mukherjee, S., Müller, T.N., Cleemann, F., Roland, K., Brandenburg, M. and Neudörfl, J.-M. (**2006**) *Org. Biomol. Chem.*, **4**, 4319–4330.
24. For the related organocatalytic ring-opening of oxazinones, see: Berkessel, A., Cleemann, F. and Mukherjee, S. (2005) *Angew. Chem. Int. Ed.*, **44**, 7466–7469.
25. Reed, A.E., Curtiss, L.A. and Weinold, F. (1988) *Chem. Rev.*, **88**, 899–926.
26. Selected reference with emphasis on the LBHB character of serine proteases' oxy anion hole: Whiting, A.K. and Peticolas, W.L. (1994) *Biochemistry*, **33**, 552–561.
27. Steiner, T. (2002) *Angew. Chem. Int. Ed.*, **41**, 48–76.
28. Berkessel, A., Mukherjee, S., Müller, T.N., Cleemann, F., Roland, K., Brandenburg,

M., Neudörfl, J.-M. and Lex, J. (2006) *Org. Biomol. Chem.*, **4**, 4319–4330.
29 Okino, T., Hoashi, Y. and Takemoto, Y. (2003) *J. Am. Chem. Soc.*, **125**, 12672–12673.
30 Okino, T., Hoashi, Y., Furukawa, T., Xu, X. and Takemoto, Y. (2005) *J. Am. Chem. Soc.*, **127**, 119–125.
31 Takemoto, Y. (2005) *Org. Biomol. Chem.*, **3**, 4299–4306.
32 Hamza, A., Schubert, G., Soós, T. and Pápai, I. (2006) *J. Am. Chem. Soc.*, **128**, 13151–13160.
33 Zhu, R., Zhang, D., Wu, J. and Liu, C. (2006) *Tetrahedron Asymm.*, **17**, 1611–1616.
34 Hammar, P., Marcelli, T., Hiemstra, H. and Himo, F. (2007) *Adv. Synth. Catal.*, **349**, 2537–2548.
35 Berkessel, A., Adrio, J.A., Hüttenhain, D. and Neudörfl, J.M. (2006) *J. Am. Chem. Soc.*, **128**, 8421–8426.
36 Maiti, N.C., Zhu, Y., Carmichael, I., Serianni, A.S. and Anderson, V.E. (2006) *J. Org. Chem.*, **71**, 2878–2880.
37 Berkessel, A. and Adrio, J.A. (2006) *J. Am. Chem. Soc.*, **128**, 13412–13420.
38 Dryuk, V.G. (1976) *Tetrahedron*, **32**, 2855–2866.
39 de Visser, S.P., Kaneti, J., Neumann, R. and Shaik, S. (2003) *J. Org. Chem.*, **68**, 2903–2912.
40 Huang, Y., Unni, A.K., Thadani, A.N., Stankovic, A.R. and Rawal, V.H. (2003) *Nature*, **424**, 146.
41 Unni, A.K., Takenaka, N., Yamamoto, H. and Rawal, V.H. (2005) *J. Am. Chem. Soc.*, **127**, 1336–1337.
42 Gordillo, R., Dudding, T., Anderson, C.D. and Houk, K.N. (2007) *Org. Lett.*, **9**, 501–503.
43 Zhang, X., Du, H., Wang, Z., Wu, Y.-D. and Ding, K. (2006) *J. Org. Chem.*, **71**, 2862–2869. Please note that the "ortho/meta" terminology is used in a different way by the authors of ref. 43. The assignment used in here is based on the original formalism by Diels and Alder.
44 Maseras, F. and Morokuma, K.J. (1995) *J. Comput. Chem.*, **16**, 1170.
45 Svensson, M., Humbel, S., Froese, R.D.J., Matsubara, T., Sieber, S. and Morokuma, K.J. (1996) *J. Phys. Chem.*, **100**, 19357.
46 Stewart, J.J.P. (1989) *J. Comput. Chem.*, **10**, 209.
47 Stewart, J.J.P. (1989) *J. Comput. Chem.*, **10**, 221.
48 Du, H., Zhao, D. and Ding, K. (2004) *Chem. Eur. J.*, **10**, 5964.
49 Reed, A.E., Weinstock, R.B. and Weinold, F. (1985) *J. Chem. Phys.*, **83**, 735.

4
Oxyanion Holes and Their Mimics

Petri Pihko, Sanna Rapakko, and Rik K. Wierenga

4.1
Introduction

One of the most important rate-accelerating effects in enzymatic catalysis arises from electrostatic effects, especially in the form of a preorganized hydrogen bonding geometry in the active site [1]. A wide variety of reactions, ranging from hydrolysis of amides or esters to aldol- and Claisen-type reactions, are catalyzed by enzymes capable of using hydrogen bonding interactions in the active site. Given the central importance of hydrogen bonding in enzymatic catalysis, we believe that a short tour of the essential features of hydrogen bonding interactions in enzyme active sites should also yield important insights into future design of small-molecule catalysts in organic synthesis.

Hydrogen Bonding in Organic Synthesis. Edited by Petri M. Pihko
Copyright © 2009 WILEY-VCH Verlag GmbH & Co. KGaA, Weinheim
ISBN: 978-3-527-31895-7

In this chapter, we review the key properties of a central structural feature in hydrogen bonding powered enzymatic catalysis – *the oxyanion hole*.

4.1.1
What are Oxyanion Holes?

In short, oxyanion holes at active sites of enzymes stabilize negatively charged oxygens in transition states and/or high-energy intermediates [2]. In oxyanion holes, this stabilization takes place via two, or sometimes even three, hydrogen bonds directed at the substrate oxygen atom, when bound at the active site. Generally, these oxyanion holes are the preformed feature of the active site, meaning that its structure already exists in the apo, unliganded enzyme. Hydrogen bonds to negatively charged oxygen anions can be very strong. In the gas phase, hydrogen bonds to charged species are typically three times stronger than the hydrogen bonds between the corresponding neutral species (Table 4.1). Negatively charged oxygen atoms can form relatively strong hydrogen bonds, due to the concentration of the charge in a relatively small volume.

4.1.2
Contributions of Oxyanion Holes to Catalysis

Two important notes should be made to understand the nature of oxyanion holes. First, significant contributions to catalysis can only be expected if *the negative charge develops at the oxygen atom during the course of the reaction*. Otherwise both the starting material (substrate) and the transition state will benefit from equally strong hydrogen bonding with the catalyst. Typical situations where the negative charge develops at the oxygen are (i) addition reactions to carbonyls where tetrahedral intermediates are generated and (ii) reactions involving enolizations or formation of enolate intermediates by abstracting a proton from a carbon atom adjacent to a carbonyl group.

Second, the typical hydrogen bond donors in the oxyanion holes at enzyme active sites are neutral, albeit highly polarized NH groups or OH groups. Although they significantly stabilize the anionic intermediates by hydrogen bonds, the oxyanion does not become fully protonated. The high-energy oxyanionic intermediates, such as tetrahedral intermediates and enolates, are stabilized by these solvating hydrogen bond donors by the enzyme structure. The contribution of the active site oxyanion holes to catalysis arises largely from the preorganization of these hydrogen bond donors (Scheme 4.1).

In aqueous solution, hydrogen bonds between water molecules can compete effectively for hydrogen bonds to carbonyl oxygen. During enolization, a negative charge develops at the oxygen atom, generating a strong attractive force for the polar water molecules. The water molecules will have to reorient themselves toward the developing enolate, and this considerably raises the activation entropy of the reaction. In contrast, in an oxyanion hole, the hydrogen bond donors are already preorganized to stabilize the developing negative charge at oxygen. Little

Table 4.1 Calculated and experimental hydrogen bond strengths.

Bond type	Calculated strength in the gas phase (kJ/mol)	Experimental strength in the gas phase (kJ/mol)	Calculated distance $d_{H \cdots B}$ (Å)	Calculation method/notes
H-O-H⋯O(H)H	−20.6 [3]	−22.7 + −2.9 [4]	1.86	MP2
H-O-H⋯O=C(CH₃)₂	−23.5 [5]		1.96	MP2/6−31++G*
H-O-H⋯N(pyridine)	−30.66 [3]		1.89	MP2
H-O-H⋯⁻O-H	−108.4 [6]	−111.3 + −4.2 [7]	1.30	BLAP3 Sadlej
[H₂O⋯H⋯OH₂]⁺	−104.3 [8]	−132.3 [9]	1.20	C₂ symmetric (Zundel cation)
H₂OH⁺⋯3OH₂ (first solvation shell for H₃O⁺)	−290.22 [10]	−287.7 [11]	N/A	Eigen cation MP2
H⁺(H₂O)ₙ	−1150.1 [12]			See also a commentary on values [13]
H-O-H⋯⁻O-CH=O	−71.4 [14]		1.67	MP2/6−31++G** −76.0 kJ/mol for bidentate binding
CH₃NH₃⁺⋯OH₂	−71.0 [15]	−70.6 [16]	1.72	B3LYP/6−31+G(d)

(a) Enolization in water

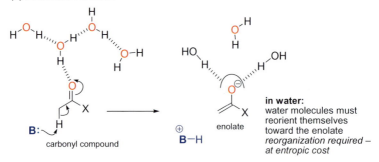

in water:
water molecules must reorient themselves toward the enolate
reorganization required – at entropic cost

(b) Enolization in an oxyanion hole

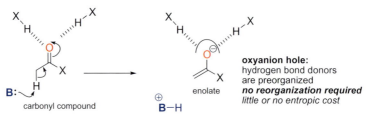

oxyanion hole:
hydrogen bond donors are preorganized
no reorganization required
little or no entropic cost

Scheme 4.1 Key reaction mechanistic differences between enolization in water and enolization in the preorganised active site.

or no reorganization is required, and this considerably lowers the entropic cost that must be paid. In both cases, the enthalpic advantage from stronger hydrogen bonds to the negatively charged oxygen atom is obtained, but only in the case of the preorganized oxyanion hole the simultaneous loss of entropy is prevented.

According to the seminal mutagenesis studies by Fersht [17, 18] on the binding properties of tyrosyl-tRNA synthetase, deletion of each charged/uncharged hydrogen bond of a donor/acceptor interaction weakens binding by approximately 15–19 kJ/mol, whereas deletion of a neutral hydrogen bond donor/acceptor interaction weakens the binding energy by approximately 2–6 kJ/mol. If we assume that the initially neutral substrate becomes a fully charged transition state (or even a high-energy intermediate) during the course of the reaction at the oxyanion hole, we should also expect similar levels of stabilization by the hydrogen bond donors at the oxyanion hole.

The contributions of hydrogen bond donors to catalysis can be estimated by site-directed mutagenesis studies in cases where the hydrogen bond donor is located in the amino acid side chain. Deletion of the main chain NH is only possible by substituting the amino acid with a proline. In all cases, the effects of the substitution to key enzyme kinetic parameters, k_{cat} and K_m, should be checked. Typically, the oxyanion hole residues contribute only little to the binding of substrate [19–21]. This is reflected in the K_m values, which typically remain very similar

Table 4.2 Contributions of different oxyanion hole residues to catalysis. $\Delta\Delta G^{\ddagger}$, calculated from the kinetic data, is the increase in the free energy of the transition state barrier, due to the mutation.

Enzyme	Mutated or deleted hydrogen bond donor site	$\Delta\Delta G^{\ddagger}$ (kJ/mol)
Oxyanion holes with tetrahedral intermediates		
Subtilisin	Asn155 → Thr	20 [20]
	Asn155 → Asp	15 [20]
	Asn155 → Leu	14 [22]
	Asn155 → Ala	16 [23]
Papain	Gln19 → Ser	16 [24]
	Gln19 → Ala	10 [24]
Lipase	Thr40 → Ala	19 [25]
	Thr40 → Val	18 [25]
Cutinase	Ser42 → Ala	14 [19]
Oxyanion holes with enolate intermediates		
Citrate synthase	His274 → Gly	16 [26]
Enoyl-CoA hydratase	Gly141 → Pro	34 [21]

for the variant and wild-type enzyme. The k_{cat} and K_m values can be used to estimate the contribution of a particular side chain to catalysis as follows:

$$\Delta\Delta G^{\ddagger} = -RT \ln\left[\frac{(k_{cat}/K_m)\text{variant}}{(k_{cat}/K_m)\text{wild type}}\right].$$

The conversion of the kinetic data into $\Delta\Delta G^{\ddagger}$-values (Table 4.2) assumes that the rate-limiting step is the same in wild type and variant. It also assumes that the mutation does not cause structural rearrangements. Only in very few cases have the kinetic studies on the transition state stabilization by the oxyanion hole contributions been complemented by protein crystallographic studies of the liganded wild-type and mutated variant. One such example, discussed in more detail below, concerns the studies on the Ser42Ala variant of cutinase, in which case it was found that the structural changes are minimal [19].

4.1.3
Properties of Hydrogen Bonds of Oxyanion Holes

The strength of the hydrogen bond between the protein hydrogen bond donor and the oxyanion will determine how much stabilization is achieved. Hydrogen bond strengths are highly dependent on the precise geometry and environment [29]. In this context, an intense current discussion topic is the possibility of formation of low-barrier hydrogen bonds (LBHBs) in enzyme transition states [27] (Scheme 4.2). LBHBs are very strong hydrogen bonds, which can only form when the pK_a's of the donor and the acceptor match very closely, and have been characterized when studying the interactions between small molecules. LBHB hydrogen bonds

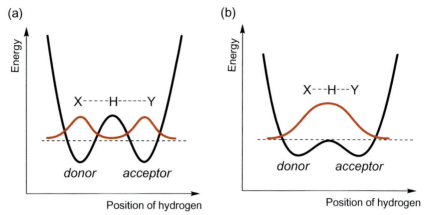

Scheme 4.2 Bond energy as a function of hydrogen position (black solid line), assuming identical pK_a values for the donor and acceptor, relative to the lowest vibrational energy level of the hydrogen atom (highlighted by a dotted line). (a) A standard, symmetric hydrogen bond; (b) the corresponding low-barrier hydrogen bond (LBHB). The red line represents the probability density function [27, 28].

are typically indicated between two oxygen atoms or between O and N atoms when the distances are smaller than 2.55 and 2.65 Å, respectively [2]. Interestingly, atomic resolution crystallography has not found evidence for such short hydrogen bonds in oxyanion holes that bind the tetrahedral intermediate [28]. In addition, the 0.90 Å resolution crystal structure of α-lytic protease, complexed with a transition state analog, the hydrogen atoms of the two oxyanion hole hydrogen bond donors can be located close to the main chain nitrogen atoms [28]. These LBHBs may exist only in the enzyme transition state complex and are therefore difficult to observe experimentally when studying the enzyme substrate interactions [2]. However, in crystallographic studies of complexes of $\Delta^{5,3}$-ketosteroid isomerase (KSI) with transition state analogs, such very short hydrogen bonds have been identified between a tyrosine side chain and a transition state analog [30]. Similarly, there is some evidence for a LBHB between the key residues of the catalytic triad (the Asp-His-Ser triad) in serine proteases [31, 32], although the interpretation of this evidence has been subject to debate [33, 28].

A second important issue related to the existence of LBHBs is the discussion on proton transfer from the hydrogen bond donor to the oxyanion. For enolizing enzymes, this issue has been proposed to be important by Gerlt and Gassman [34]. NMR experiments have found evidence for proton transfer from the Tyr14 phenolic hydroxyl group to the oxygen of the transition state analog dihydroequilenin as bound to $\Delta^{5,3}$-ketosteroid isomerise [35]. In computational studies on enzymes with oxyanions, the existence of discrete anionic intermediates is usually observed [36, 33], at least with enolate intermediates [37, 38]. In the following discussion, the existence of anionic, albeit strongly hydrogen bonded tetrahedral intermediates and enolates is assumed.

4.2
A More Detailed Description of the Two Classes of Oxyanion Holes in Enzymes

4.2.1
A Historical Perspective

The importance of oxyanion holes in enzymes was first discovered in chymotrypsin by David Blow and co-workers and in subtilisin by Jo Kraut and co-workers [39]. In these enzymes, a tetrahedral intermediate is generated after nucleophilic attack of a deprotonated serine side chain on the peptide carbonyl group. It was recognized from the beginning that the geometry of the active site complexes was possibly better complementary to the tetrahedral intermediate than to the planar peptide substrate [40].

Later, oxyanion holes were also discovered in other proteases, such as the cysteine protease papain, and in esterases and lipases, enzymes capable of esterification or ester hydrolysis. Interestingly, in these esterases, sometimes up to three hydrogen bond donors can be located within 3 Å of the carbonyl oxygen atom, whereas such triple hydrogen bonding motifs have not yet been found in the proteases.

In the classical oxyanion hole of the serine proteases of the chymotrypsin family, the hydrogen bond donors are two main chain peptide NH groups protruding out of the same turn. Additionally, for example in papain and subtilisin, amide NH groups of the side chains of glutamine and asparagines, respectively, have also been identified as hydrogen bond donors. It should be noted that aligning the NH groups toward the oxyanion hole induces electrostatic strain that is only relieved when the oxyanion binds. This feature is of assistance in catalysis.

Enzymes with oxyanion holes are now known to catalyze a wide range of reactions with substrates that have a carbonyl moiety. The examples discussed in this chapter include thioesters, oxygen esters, peptides, and ketones (Figure 4.1). Two classes of high-energy intermediates with oxyanions are generated in these reactions (Table 4.3), a tetrahedral intermediate and an enolate. These reactions are

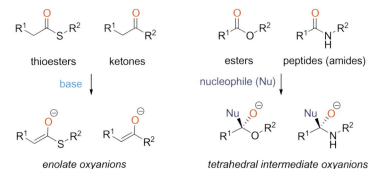

Figure 4.1 High-energy oxyanion intermediates are formed by, for example, enolization of thioesters and ketones/aldehydes, or by addition of nucleophiles to esters and amides.

Table 4.3 The oxyanion hole geometry for selected tetrahedral intermediates and enolate intermediates.

	Nucleophile/base	Hydrogen bond donors				Water
		Main chain group			Side chain group	
		1st NH	2nd NH	3rd NH		
Oxyanion holes for tetrahedral intermediates						
Serine proteases, hydrolyzing a peptide bond						
Trypsin (1PPE) [41, 42]	Nucleophile	Ser195	Gly193		NH	OH
Chymotrypsin (1GGD) [43, 44]	Ser195	Ser195	Gly193			
Proteinase A (2SGA) [45, 46]	Ser195	Ser195	Gly193			
Subtilisin (1SBT) [47, 48]	Ser221	Ser221			Asn155 ND2	
Carboxypeptidase II (3SC2) [49, 50]	Ser146	Ser146	Gly53			
Cysteine proteases, hydrolyzing a peptide bond						
Papain (9PAP) [51, 52]	Cys25	Cys25			Gln19 NE2	
Esterases, hydrolyzing an ester bond						
Streptomyces scabies esterase (1ESC) [53]	Ser14	Ser14	Gly66		Asn106 ND2	
Acetylcholinesterase (2ACE) [54, 55]	Ser200	Ala201	Gly118	Gly119		
Candida rugosa lipase (1CRL) [56]	Ser209	Ala210	Gly124	Gly123		
Human pancreatic lipase	Ser152	Leu153	Phe77			
Pseudomonas glumae lipase (1TAH) [57, 58]	Ser87	Gln88	Leu17			
Mucor miehei lipase (1TGL) [59]	Ser144	Leu145	Ser82			Ser82 OG
Humicola lanuginosa lipase (1TIB) [60]	Ser146	Leu147	Ser83			Ser83 OG
Rhizopus delemar lipase (1TIC) [60, 61]	Ser145	Leu146	Thr83			Thr83 OG1
Candida antarctica B lipase (1TCB) [62]	Ser105	Gln106	Thr40			Thr40 OG1
Cutinase (2CUT) [63]	Ser120	Gln121	Ser42			Ser42 OG

Oxyanion holes for enolate intermediates	Enolizing base	Main chain group			Side chain group		Water
		1st NH	2nd NH	3rd NH	NH	OH	
Enolization of thioesters							
CoA/ACP dependent enzymes							
Citrate synthase (3CSC) [64]	Asp375				His274 ND1		1 water
Dehydrogenase (3MDE) [65]	Glu376	Glu376				Ribityl 2-OH of FAD	
Isomerase, enoyl-CoA isomerase (1SG4) [66]	Glu136	Leu66	Gly111				
Hydratase (1DUB) [67]	Glu164	Ala98	Gly141				
Thiolase (1DM3) [68]	Cys378				His348 NE2		1 water
Enolization of ketones							
Triosephosphate isomerase (1NEY, 1TPH) [69, 70]	Glu167				Lys13 NZ	His95 NE2	
Ketosteroid isomerase (1OGX) [30]	Asp40					Asp103 OD1+Tyr16 OH	
Phosphoglucoisomerase (1KOJ) [71]	Glu357						3 waters
Ribose-5-phosphate isomerase (2BES) [72]	Glu75	Gly70	Ser71		Ser71 OG		2 waters

discussed in more detail in the following sections. In all these discussed examples the enzymes are metal ion independent (i.e., organocatalytic), and no other cofactors are required either.

4.2.2
Oxyanion Holes with Tetrahedral Intermediates

A very well-studied example is the oxyanion hole in serine proteases. In chymotrypsin the oxyanion hole is formed by main chain NH groups of two peptide units, Gly193 and Ser195 [36]. More generally, the classical oxyanion hole stabilizes the negatively charged oxyanion of a tetrahedral intermediate and is formed by two main chain peptide NH groups (cf. Scheme 4.3).

The tetrahedral intermediate is generated by a nucleophilic attack on the carbonyl carbon atom of the activated nucleophile, which in the case of chymotrypsin and trypsin is the catalytic Ser195 (Table 4.3). It is important to note that a small structural rearrangement occurs when the planar carbonyl moiety is converted into the tetrahedral intermediate (Scheme 4.4).

In subtilisin the oxyanion hole is formed by a main chain N(Ser221) and side chain NH group of ND2(Asn155) [74]. Another classic example is the protease papain, in which enzyme the oxyanion hole is also formed by a main chain peptide NH and a side chain amide of a glutamine. In enzyme active sites, which were discovered later, the oxyanion holes were generated by three hydrogen bonding donors, such as acetylcholine esterase [75], in which case three main chain NH groups (Gly118, Gly119, Ala201) generate the oxyanion hole. In the above three examples the substrate carbonyl carbon atom is part of a peptide unit (chymotrypsin, subtilisin, papain) or an oxygen-ester unit (acetylcholine esterase). Other examples concern carbonyl carbon atoms of a thioester unit (thioesterase) [76, 77] and the 3-keto moiety of an acetoacyl unit (thiolase) [78]. The nucleophile in these examples is very often a serine (chymotrypsin, subtilisin, acetyl choline esterase, thioesterase) and less frequently a cysteine (papain, thiolase), or an aspartate/glutamate (thioesterase) [77].

Experimental kinetic data on the importance of the tetrahedral intermediate stabilizing oxyanion hole are available for variants of subtilisin, cutinase, lipase, and papain (Table 4.2). In these studies side chain hydrogen bond oxyanion hole donors have been mutated. The subtilisin studies have been done by Wells and co-workers, who have mutated the Asn155. In the Asn155Thr and Asn155Gln variants the K_m values stay nearly unchanged, but the k_{cat} value is reduced by a factor of 10^3 [20]. In other words, mutations in the hydrogen bond donors do not appear to affect binding, but have a large effect on catalytic efficiency.

In cutinase, the hydrogen bond donors consist of two main chain NH groups and a serine side chain. In analogy with substilisin, removal of the serine hydroxyl group by site-directed mutagenesis reduces the rate by a factor of approximately 10^3 while K_m remains almost unchanged. Also, comparison of the crystal structures of variant and wild-type enzymes shows that the mutation of serine to alanine causes only marginal structural changes [19]. These cutinase studies are one of

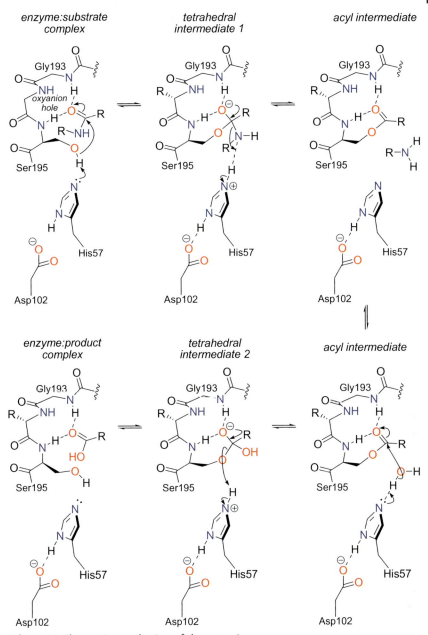

Scheme 4.3 The reaction mechanism of chymotrypsin.

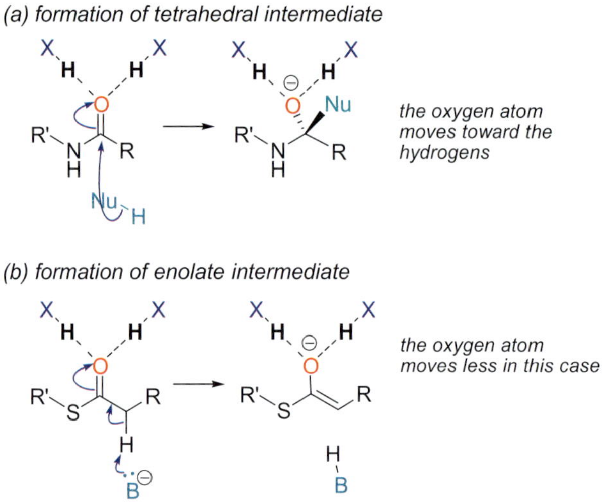

Scheme 4.4 (a) The formation of the tetrahedral intermediate requires substantial movement of the oxygen atom. The oxyanion hole acts as a Circean trap, attracting the oxygen atom toward the hydrogen bond donors; (b) In contrast, in the formation of the enolate intermediate, the position of the oxygen atom is affected to a smaller extent. The alkyl chain (R) has to move toward the plane of the enolate [73, 83].

the few examples in which the enzyme kinetic studies have been complemented by crystallographic studies. Based on the kinetic data, the transition state stabilization of the hydrogen bond donor of the oxyanion hole is approximately 19 kJ/mol in cutinase, which agrees with the 15–19 kJ/mol range reported by Fersht for a hydrogen bond between a neutral donor and a charged acceptor [17].

The oxyanion hole geometry of three complexes is visualized in Figures 4.2–4.4. Figure 4.2 displays the active site of trypsin complexed with a peptide inhibitor [41]. In Figure 4.3, the active site of chymotrypsin complexed with a neutral aldehyde adduct is displayed [43], and in Figure 4.4, cutinase (a lipase) with a covalently bound phosphate, a transition state analog is depicted [63].

Trypsin and chymotrypsin are the classical serine proteases. These enzymes hydrolytically cleave a peptide bond in a two step reaction. In the first, rate limiting, step the catalytic serine is acylated. In the visualized trypsin-peptide inhibitor complex (Figure 4.2) the carbonyl oxygen of the scissile peptide bond points into the oxyanion hole formed by two main chain atoms, but the peptide bond remains uncleaved. This peptide moiety is almost planar and the carbonyl carbon atom is in van der Waals contact with the catalytic serine side chain oxygen. The very high affinity of the inhibitor for the enzyme (K_d is approximately 3×10^{-12} M) likely renders the complex very rigid and prevents the reaction from proceeding further. The visualized complex of chymotrypsin (Figure 4.3) concerns a structure in which an aldehyde substrate analog, N-acetyl-L-leucyl-L-phenylalaninal [43], is used to

4.2 A More Detailed Description of the Two Classes of Oxyanion Holes in Enzymes | 55

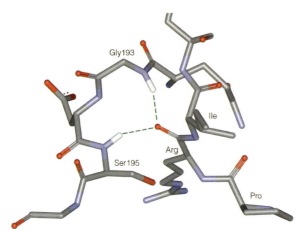

Figure 4.2 Structure of the oxyanion hole of the active site of trypsin, complexed with a peptide inhibitor (PDB: 1PPE). The hydrogen atoms (in white) are only included when relevant for the hydrogen bonding geometry of the oxyanion hole. The dotted lines highlight the key interactions in the oxyanion hole.

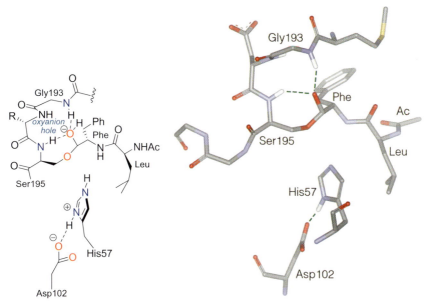

Figure 4.3 Structure of the oxyanion hole of the active site of chymotrypsin complexed with a peptide inhibitor (PDB: 1GGD) (the same view as used in Figure 4.2). As discussed in the text, it is believed [43] that the oxygen atom in the oxyanion hole is protonated and therefore neutral.

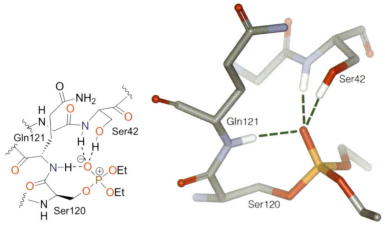

Figure 4.4 Structure of the oxyanion hole of cutinase, with a covalently bound transition state analog, diethyl phosphate (PDB: 2CUT).

capture the transition state. In this complex, the aldehyde moiety has formed a covalent bond with the catalytic serine, forming a hemiacetal, and the oxygen atom of the tetrahedral product is hydrogen bonded to two main chain NH hydrogen bond donors. From additional experiments it is concluded that this oxygen is protonated and therefore neutral. The careful refinement of this structure, at 1.5 Å resolution, also revealed that this oxygen adopts a double conformation, being partly bound in the oxyanion hole (the (S)-hemiacetal adduct) and partly pointing in the opposite direction (the (R)-hemiacetal adduct).

The third example concerns cutinase [63] a serine lipase, which catalyzes the hydrolytic cleavage of an ester bond of a triacylglycerol molecule (Scheme 4.5). The complex of cutinase with a transition state analog visualizes the structure of the negatively charged tetrahedral intermediate, which is mimicked in this case by a phosphate moiety [63]. This complex has been formed by reacting cutinase with diethyl-*p*-nitrophenyl phosphate, in which a covalent adduct (Figure 4.4) is formed between the catalytic serine and the diethylphosphate moiety, whereas the *p*-nitrophenyl group is released. The tetrahedral phosphate moiety mimics the tetrahedral intermediate. In this complex the oxyanion hole is formed by three hydrogen bond donors, two being main chain NH proton donors, as well as a serine side chain hydroxyl group.

4.2.3
Oxyanion Holes with Enolate Intermediates

Typically, the hydrogen bond donor for the tetrahedral intermediate oxyanion is the amide NH group. In contrast, in enolate oxyanion holes, a much more variable range of hydrogen bond donors is observed, such as water, a tyrosine hydroxyl,

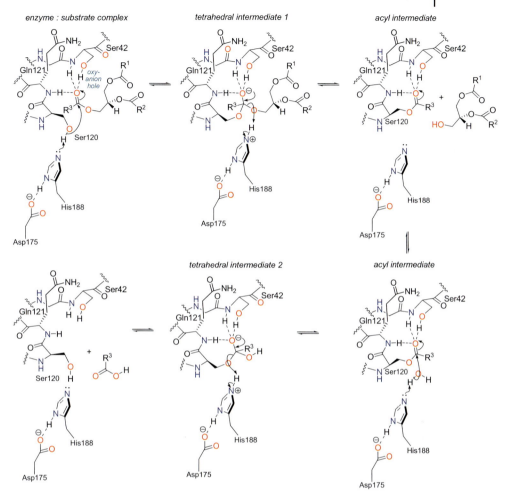

Scheme 4.5 The reaction mechanism of cutinase, illustrating the preferred cleavage site. R^1, R^2, and R^3 are linear alkyl chain moieties.

histidine, or even the carboxylic acid group of aspartic acid! These hydrogen bond donors also span a wider pK_a range. The pK_a's of typical OH and peptide NH groups are in the range of 15–18, whereas that of tyrosine OH is about 10, and aspartic acid is about 5 [79].

In addition, many oxyanion holes have been discovered in which water is one of the hydrogen bonding donors. These waters are clearly defined in the respective electron density maps. In some enzymes these waters are part of a trail of water molecules [80, 81]. When water is used as the hydrogen bond donor, very often the intermediate that is stabilized is not the tetrahedral intermediate but an enolate, which is derived by base-catalyzed deprotonation of the Cα-atom of the carbonyl function (Figure 4.1). The classical oxyanion holes of this sort can be found in the active site of citrate synthase, where the contributing H-bond donors are water and ND1(His274) and in KSI where the hydrogen bond donors are the side chains of a tyrosine and an aspartate.

The larger hydrogen bond donor variability of the enolate oxyanion hole is also seen from the active site geometry of the enzyme triosephosphate isomerase, in which case the side chains of the positively charged NZ-atom of a lysine (Lys13) and the NE2 atom of a histidine (His95) atom are critically important for catalysis [82], whereas in another isomerase (phosphoglucoisomerase) only waters (three waters within 3.5 Å) are seen as hydrogen bond donors [71] (Table 4.3). This variability is further illustrated by the active sites of the enzymes of the large crotonase superfamily. These enzymes catalyze a wide range of reactions. In the active sites of these enzymes, the oxyanion hole is formed by two main chain NH groups, emerging out of two different loops, which are conserved in the superfamily [73]. Finally, it is also interesting to note that in some enzymes, such as enzymes of the thiolase superfamily, both types of oxyanion holes are formed in the same active site. For thiolase these are referred to as oxyanion hole 1 (for the enolate, formed by a water and a histidine side chain NE2-atom) and oxyanion hole 2 (for the tetrahedral intermediate, formed by two main chain NH groups) [78].

There are only few experimental studies where the importance of the oxyanion hole for stabilizing the enolate oxyanion has been quantified. Tonge and collaborators [21] have studied the properties of the Gly141Pro variant of hydratase. In this variant, one of the two peptide NH groups of the oxyanion hole (N(Gly141)) has been disabled because of the Gly141Pro mutation. Compared to the wild-type enzyme, k_{cat} is reduced 10^6-fold, whereas K_m is not affected. In addition, Raman studies suggest that the active site of the variant remains intact.

Detailed binding with substrate and transition state analogs has also been reported on KSI [83, 84] using a wide range of techniques, highlighting the subtle interplay of the electrostatic and geometric properties of the enolate stabilizing active site.

4.2.3.1 Examples of Enolate Oxyanion Holes

Two examples of enolizing oxyanion holes are given in Figures 4.5 and 4.6. The enolizing enzymes enoyl-CoA isomerase [66] (Figure 4.5) and citrate synthase [64] (Figure 4.6) are both CoA-dependent enzymes, catalyzing thioester-dependent reactions. Enoyl-CoA isomerase is a $\Delta^{3,2}$-enoyl-CoA isomerase converting $\Delta^{3,4}$-enoyl-CoA into $\Delta^{2,3}$-enoyl-CoA (Scheme 4.6). It is an important enzyme in lipid degrading pathways, enabling the processing of unsaturated fatty acid side chains, and it functions in an auxiliary pathway of the β-oxidation cycle.

The active site of enoyl-CoA isomerase is a good example of an active site built on the framework of the crotonase fold. It is now well established that this crotonase fold provides an active site framework that has been used by Nature for a wide range of different chemical reactions, as reviewed recently [73, 85]. The reaction of this enoyl-CoA isomerase is initiated by a catalytic base, Glu136, abstracting a proton from the Cα-carbon, generating the negatively charged enolate

4.2 A More Detailed Description of the Two Classes of Oxyanion Holes in Enzymes

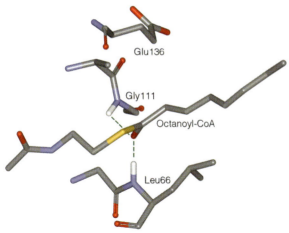

Figure 4.5 The active site of enoyl-CoA isomerase, complexed with octanoyl-CoA (PDB: 1SG4). In this structure the oxygen atom of Glu136 which acts as the catalytic base points to the C2 carbon of the ligand and is therefore well positioned for proton abstraction. This C—H bond, which corresponds to the cleaved C—H bond of the real substrate, is approximately orthogonal to the C=O moiety of activated thioester bond (Corey-Sneen angle) [64b].

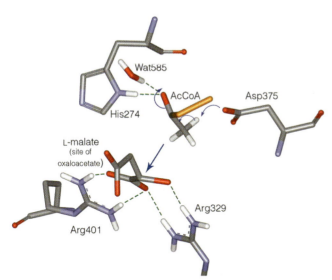

Figure 4.6 The active site of citrate synthase, complexed with acetyl-CoA and the substrate analog L-malate (PDB: 3CSC). Arg329, together with NE2(His320) and ND1(His236), contributes to a second oxyanion hole for the tetrahedral oxyanion that is generated by the nucleophilic attack of enolized acetyl-CoA on the carbonyl carbon atom of oxaloacetate. The initial movement of electrons in the enolization step is visualized by curved arrows, and the aldol-bond forming event that would take place with the actual oxaloacetate substrate is visualized by a straight arrow.

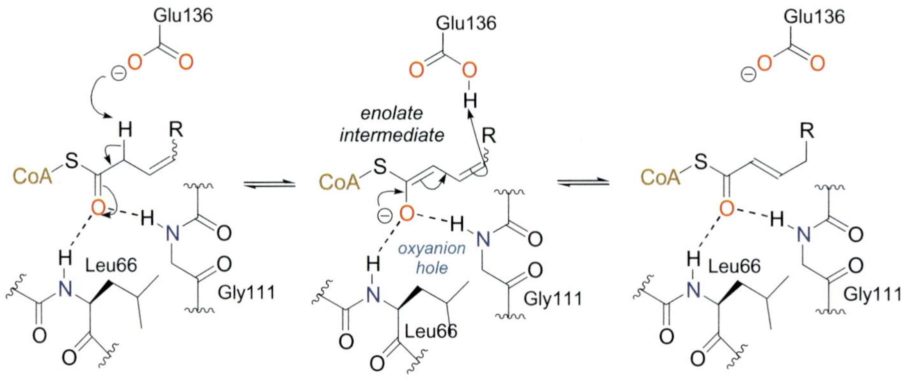

Scheme 4.6 The reaction mechanism and mode of catalysis by enoyl-CoA isomerase.

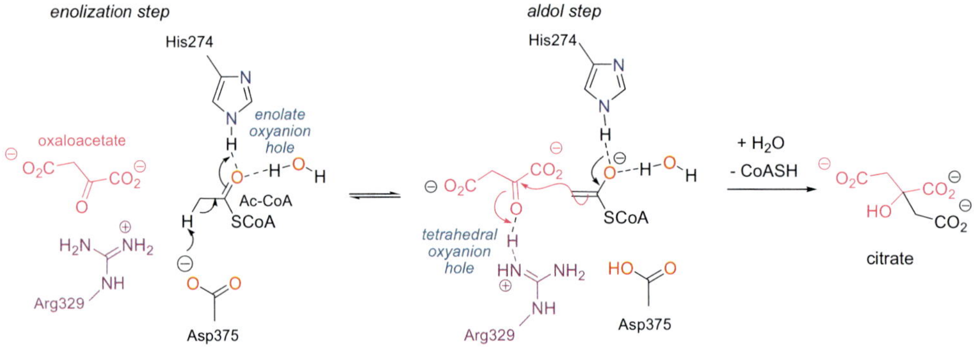

Scheme 4.7 Abbreviated reaction mechanism of citrate synthase.

intermediate. The structure of the complex of enoyl-CoA isomerase with a saturated octanoyl-CoA molecule captures the position of the two main chain NH hydrogen bond donors with respect to the thioester oxygen moiety (Figure 4.5).

Citrate synthase catalyzes the aldol reaction between acetyl-CoA and oxaloacetate. The acyl-CoA link is then cleaved by hydrolysis, releasing a molecule of citrate and free CoA.

In many enolizing enzymes, one of the hydrogen bond donors is a water molecule. This is also the case in the citrate synthase active site (Scheme 4.7). The second hydrogen bond donor is the ND1 side chain atom of His274. In this enzyme, the reaction is initiated by attack at the Cα-proton by Asp375, generating the enolate intermediate. The enolization step is the rate-limiting step of the reaction. In the catalytic cycle, this enolate intermediate then proceeds to form a carbon–carbon bond in the aldol reaction with oxaloacetate, generating citryl-CoA.

Figure 4.6 displays the structure of the complexed active site of citrate synthase complexed with L-malate and acetyl-CoA. This active site is rather solvent exposed and the oxyanion hole water molecule is facing the bulk solvent region. L-malate is an inactive analog of the true substrate, and oxaloacetate, and therefore the aldol reaction does not occur. Nevertheless, this active site geometry visualizes a competent active site, as it has been shown that hydrogen exchange with solvent water of the hydrogen at the Cα-position of acetyl-CoA is observed when incubating citrate synthase with acetyl-CoA and L-malate, indicating that the enzyme active site can abstract the α-proton of acetyl-CoA and exchange it with bulk solvent, although the subsequent aldol step cannot occur.

4.3 Oxyanion Hole Mimics

Several research groups have studied the function of oxyanion holes by generating small-molecule mimics of the key components. In the following discussion, selected examples of oxyanion hole mimics are discussed.

4.3.1 Mimics of Enzymatic Oxyanion Holes and Similar Systems

In 1989, Rebek and co-workers reported a simple system based on Kemp's triacid that served as a mimic of an enolizing enzyme [86]. This early mimic, however, had the enolizing substrate covalently attached to the triacid skeleton. In addition, the mimic did not possess any oxyanion hole functionalities. However, 2 years later the Rebek group reported a true enolizing catalyst that hosted a carboxylic acid as the oxyanion hole component (Scheme 4.8) [87]. The rate of enolization of the quinuclidone substrate was enhanced by a factor of 10^3 in the presence of 2.5 mM of the receptor (R = n-Pr).

At around the same time, Breslow and co-workers described bifunctional cyclodextrin-based catalysts that were capable of hydrolysis of a bound phosphate ester [88]. In later studies, an AD isomer (Scheme 4.9) of a β-cyclodextrin bisimidazole catalyst turned out to be the fastest catalyst for enolization of p-tert-butylacetophenone (Scheme 4.9) [89]. Here, the extra binding is provided by the β-cyclodextrin

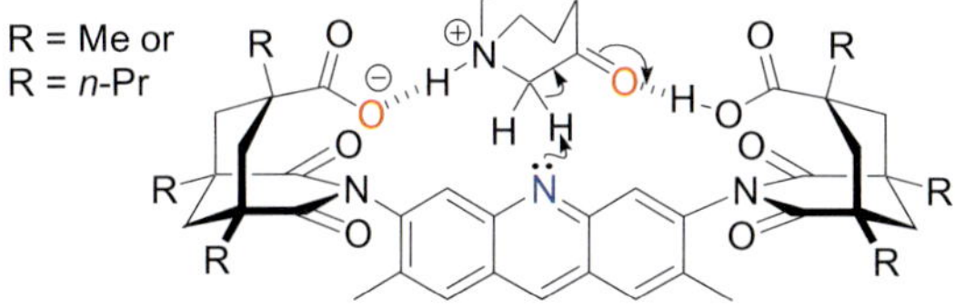

Scheme 4.8 Rebek's trifunctional enolization catalyst including a general base (acridine nitrogen), hydrogen bond donor (carboxylic acid), and a binding group (carboxylate).

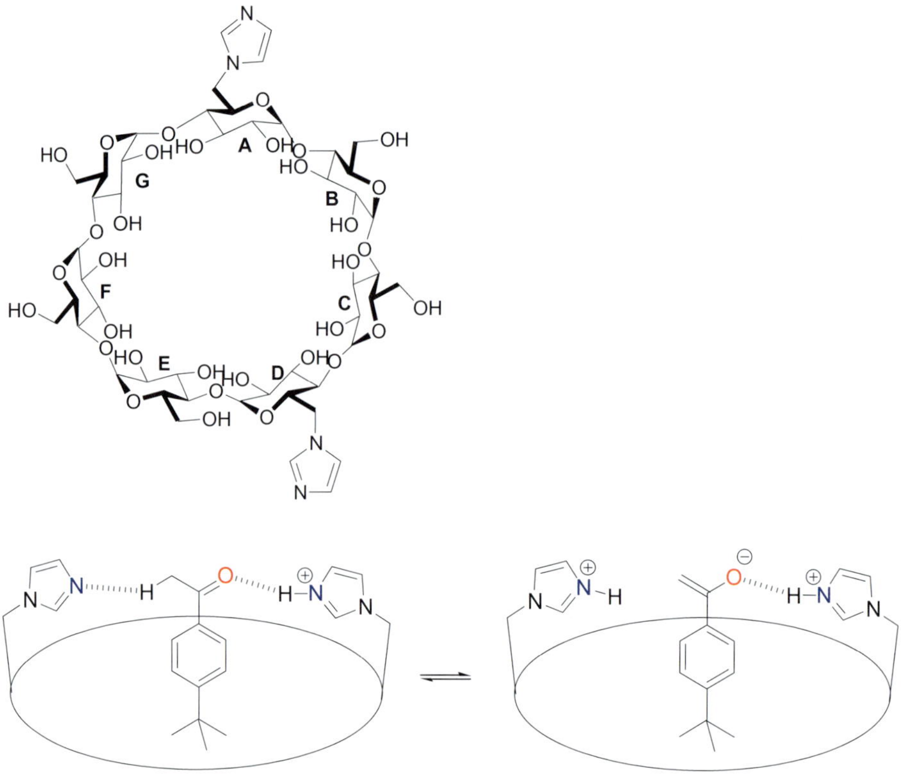

Scheme 4.9 Breslow's bifunctional β-cyclodextrin catalyst readily catalyzes the enolization of the guest, *t*-butylacetophenone. The enolate form of acetophenone is shown; it is possible that the enol form is generated as described in the original reference [89].

cavity, and the two imidazole units function as the general base/hydrogen bond donor pair required for enolization.

The pioneering work by the Rebek and Breslow groups did not fully establish the optimal geometry of the enolization catalysts. Later, in a series of studies, Anslyn and co-workers have demonstrated that enolization catalysts where the hydrogen bond donors are directed toward the lone pairs of the carbonyl oxygen are not particularly effective [90]. More effective catalysts might result from directing the hydrogen bond donors to the π plane of the carbonyl. Thus, two carefully designed oxyanion hole mimics, bearing hydrogen bond donors that operate in the carbonyl plane or in the π plane, respectively, had remarkably different effects on the pK_a of the guest diketones (Scheme 4.10) [90, 91]. In addition, simple ketones bearing a proximal phenolic hydroxyl group, optimally positioned for hydrogen bonding to the carbonyl lone pairs, gave rate enhancements for enolization only of the order 10- to 100-fold (Scheme 4.11) [92].

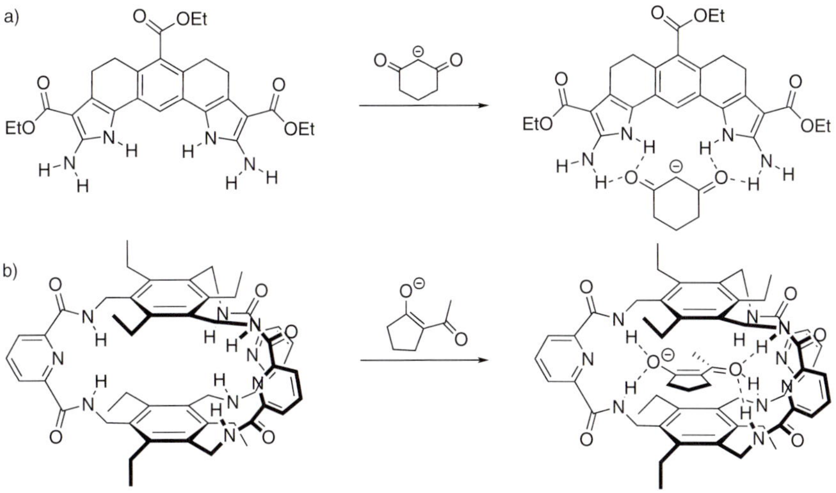

4.3 Oxyanion Hole Mimics

Scheme 4.10 Oxyanion hole receptors developed by the Anslyn group. These receptors operate either (a) in the plane of the carbonyl/enolate or (b) in the pi plane of the carbonyl/enolate. The receptor 2 induces a pK_a shift of 2.9 for the diketone shown, whereas receptor 1 induces a pK_a shift of only 1 pK_a unit. Note that different diketones have to be used to ensure maximum complementarity with the receptor [90].

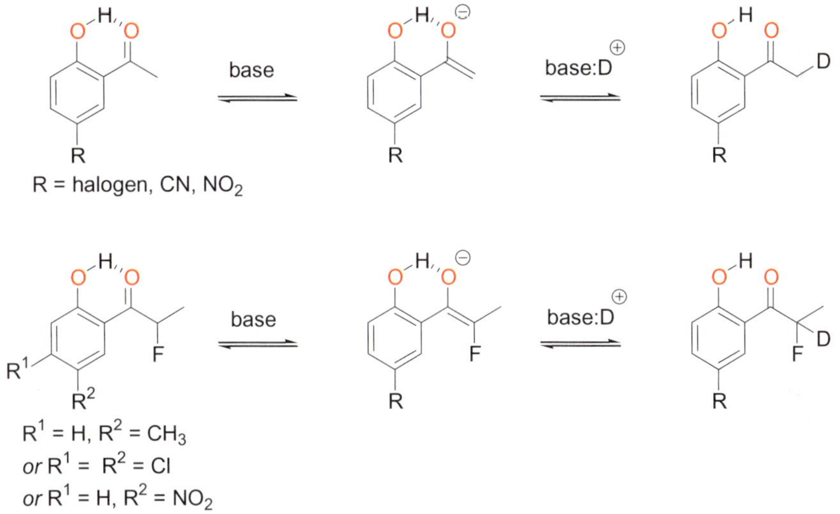

Scheme 4.11 A series of ketones with intramolecular hydrogen bonds in the plane of the carbonyl display remarkably low levels of rate enhancement [92].

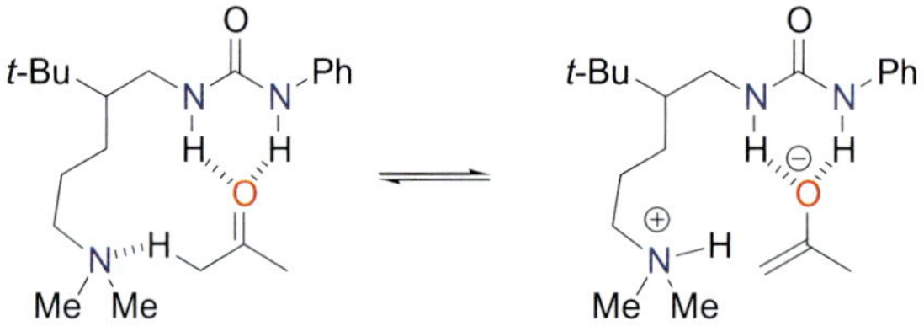

Scheme 4.12 Oxyanion hole mimics developed by Zhu and Drueckhammer [94].

Zhu and Drueckhammer have described a rational approach to the design of oxyanion hole mimics for enolate formation. They used the program CAVEAT developed by Paul Bartlett and co-workers [93] for identifying potential subunits that could be used to link the oxyanion hole hydrogen bond donors and the general base required for enolization. The resulting catalyst bears a thiourea group as the hydrogen bond donor and a basic dimethylamino group (Scheme 4.12) [94]. Disappointingly, the presence of the thiourea moiety exhibited only 5-fold rate enhancement compared to the rate in the presence of an external hydrogen bond donor. Presumably, the thiourea moiety, in spite of its success in other areas of hydrogen bond catalysis (see Chapter 6 in this volume), does not provide adequate hydrogen bonding environment for enolization.

A series of diaminoxanthones were studied by Simón and co-workers as potential oxyanion hole mimics (Scheme 4.13) [95]. In contrast to the previous approaches toward oxyanion hole mimics, the diaminoxanthones bear an additional hydrogen bond acceptor group that could be utilized to enhance binding to the receptor/catalyst. Indeed, these receptors are already quite effective for the catalysis of conjugate addition of pyrrolidine to α,β-unsaturated valerolactone, with k_{cat}/k_{uncat} values in the range of 10^3–10^4 [96].

Kim, Chin, and co-workers have described a highly interesting oxyanion hole mimic that transforms L-amino acids to D-amino acids [97]. The mechanism involves stabilization of the enolate intermediate by an internal hydrogen bond array generated by urea group (Scheme 4.14). In the presence of an external base, such as triethylamine, the receptors readily promote the epimerization of α-amino acids, favoring the D-amino acids due to unfavorable steric interactions in the receptor–L-amino acid complex. These receptors can also be viewed as chiral mimics of pyridoxal phosphate [98].

4.3.2
Utilization of Oxyanion Holes in Enzymes for Other Reactions

The ability of enzymes to catalyze other reactions than those for which they have been evolved is called *enzymatic promiscuity* [99–102].

The groups of Hult, Berglund, and Brinck have collaborated to utilize lipases, such as *Candida antarctica* lipase B (CALB) for a number of promiscuous reactions, all of which utilize the good oxyanion hole present in the enzyme active site [103–106]. The enzymes are also used in organic solvents [107, 108]. A selection of investigated promiscuous reactions is summarized in Scheme 4.15. Although

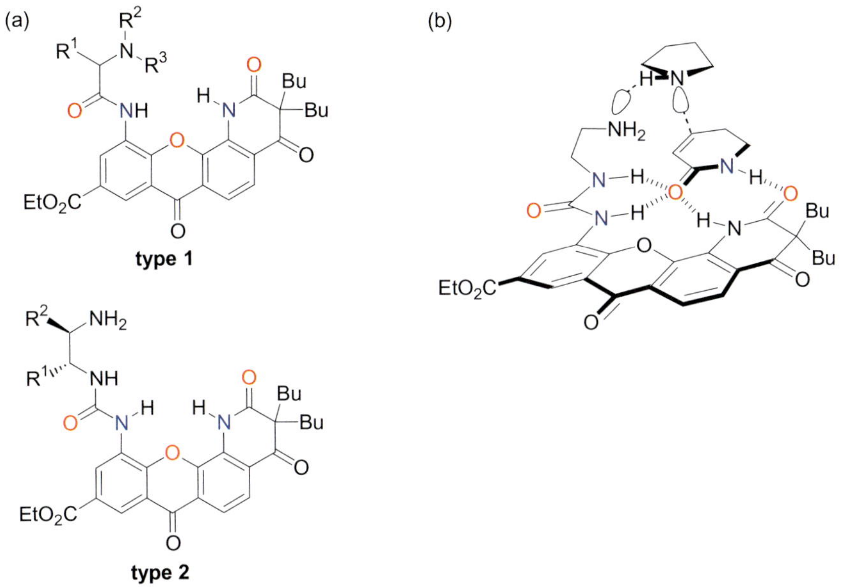

Scheme 4.13 (a) Two different types of xanthone-based oxyanion hole receptors developed by Simón and co-workers and (b) possible mode of catalysis of a conjugate addition reaction between pyrrolidine and α,β-unsaturated valerolactam by one of the receptors.

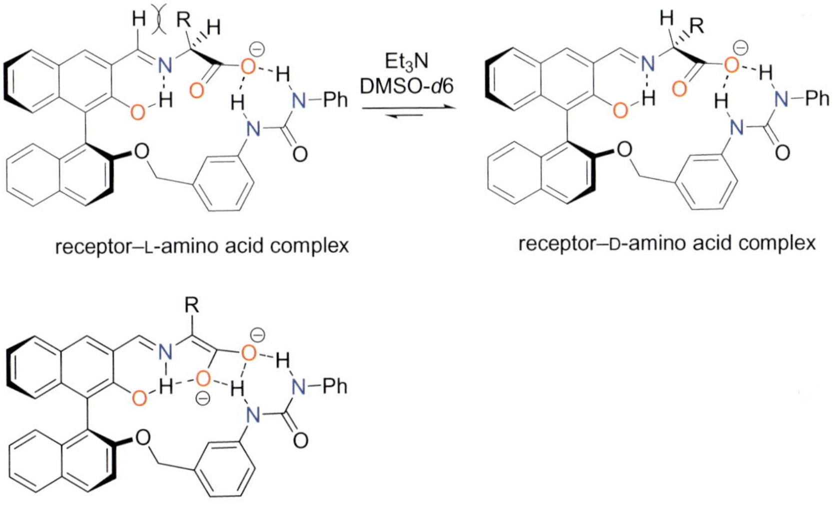

possible enolate-like intermediate

Scheme 4.14 Amino acid receptors developed by Kim, Chin, and co-workers. D-Amino acids are favored as a result of steric strain present in the L-amino acid-derived complex [97].

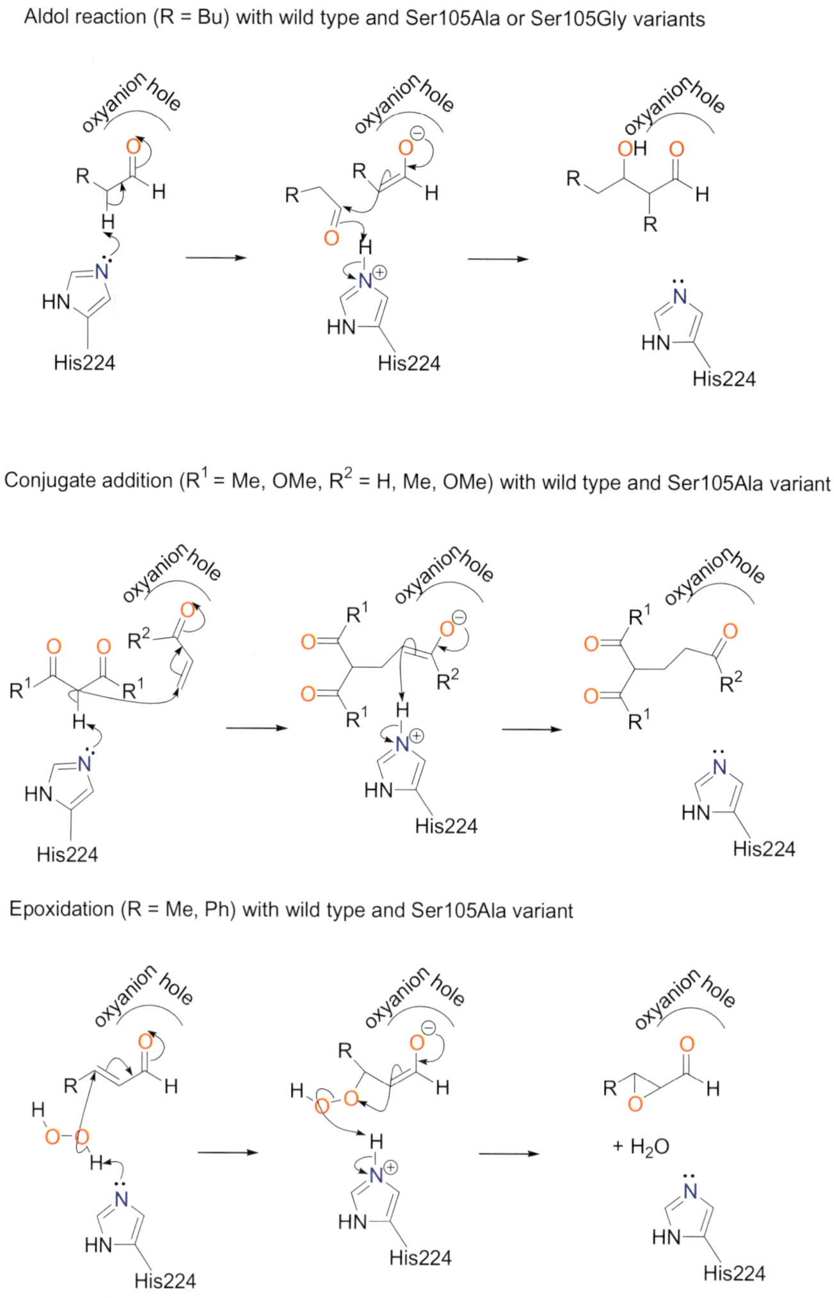
Aldol reaction (R = Bu) with wild type and Ser105Ala or Ser105Gly variants
oxyanion hole
His224
Conjugate addition (R1 = Me, OMe, R2 = H, Me, OMe) with wild type and Ser105Ala variant
oxyanion hole
His224
Epoxidation (R = Me, Ph) with wild type and Ser105Ala variant
oxyanion hole
+ H2O
His224

Scheme 4.15 Examples of promiscuous enzymatic reactions conducted with the oxyanion hole of *Candida antarctica* lipase B: (a) the aldol reaction [104]; (b) the conjugate addition reaction (Michael addition) [105]; (c) the epoxidation reaction [106].

from a synthetic point of view these reactions leave much to be desired (e.g., the enantioselectivity of the reactions has not been determined and typically the products have not been isolated), the fact that the oxyanion hole system has been capable of a diverse range of transformations involving enolate intermediates raises hopes that similar reactions could one day be effectively conducted using small-molecule catalysts as well. Remarkably, small organic catalysts for enzymatic epoxidation reaction with similar substrates were discovered as late as 2005 [109], but these reactions employed the iminium mechanism [110].

4.4
Concluding Remarks

In this chapter, we have highlighted the key features of hydrogen bonding catalysis by oxyanion holes in enzymes. These enzymes catalyze many highly valuable transformations, including carbon–carbon bond forming reactions, and for many of these enzymes no synthetic catalysts are available. The active site geometries discussed in this chapter are the result of extensive evolutionary pressure and represent highly optimized solutions to challenging catalytic problems. As an example, although chemists routinely use strong bases and/or strong Lewis acids to promote selective enolizations, in Nature neither of these are required for effective enolization! Understanding the correlation between the geometry and the proficient catalytic properties of these active sites is important for the design of small molecule enzyme mimetics as well as for the exploitation of these principles in synthetic organic chemistry.

Acknowledgments

We thank the Academy of Finland for support of this work (grant 1122921). We also thank Satyan Sharma and Andre Juffer (University of Oulu) as well as Antti Pohjakallio (Helsinki University of Technology) for helpful discussions and for assistance with the figures.

References

1 For recent discussions, see: (a) Warshel, A., Sharma, P.K., Kato, M., Ziang, Y., Liu, H. and Olsson, M.H.M. (2006) *Chem. Rev.*, **106**, 3210–3235. (b) Warshel, A., Sharma, P.K., Chu, Z.T. and Åqvist, J. (2007) *Biochemistry*, **46**, 1466–1476.

2 Frey, P.A. and Hegeman, A.D. (2007) *Enzymatic Reaction Mechanisms*, Oxford University Press, Oxford.

3 Raub, S. and Marian, C.M. (2007) *J. Comput. Chem.*, **28**, 1503–1515.

4 Curtiss, L.A., Frurip, D.J. and Blander, M. (1979) *J. Chem. Phys.*, **71**, 2703.

5 Coutinho, K., Saavedra, N. and Canuto, S. (1999) *J. Mol. Struct. (Theochem)*, **466**, 69–75.

6 Wei, D., Proynov, E.I., Milet, A. and Salahub, D.R. (2000) *J. Phys. Chem. A*, **104**, 2384–2395.

7 Pudzianowsk, A.T. (1995) *J. Chem. Phys.*, **102**, 8029.
8 Xie, Y., Remington, R.B. and Schaefer, H.F. III (1994) *J. Phys. Chem.*, **101**, 4878–4884.
9 Grimsrud, E.P. and Kebarle, P. (1973) *J. Am. Chem. Soc.*, **95**, 7939.
10 Gresh, N., Leboeuf, M. and Salahub, D. (1994) Modeling the Hydrogen Bond, ACS Symposium Series 569, American Chemical Society, Washington, DC, pp. 82–112.
11 Kebarle, P. (1977) *Ann. Rev. Phys. Chem.*, **28**, 455.
12 Tissandier, M.D., Cowen, K.A., Feng, W.Y., Gundlach, E., Cohen, M.H., Earhart, A.D., Coe, J.V. and Tuttle, T.R. Jr. (1998) *J. Phys. Chem. A*, **102**, 7787–7794.
13 Camaioni, D.M. and Schwerdtfeger, C.A. (2005) *J. Phys. Chem. A*, **109**, 10795–10797.
14 Pan, Y. and McAllister, M.A. (1997) *J. Am. Chem. Soc.*, **119**, 7561–7566.
15 Kim, K.Y., Cho, U.-I. and Boo, D.W. (2001) *Bull. Korean Chem. Soc.*, **22**, 597–604.
16 Meot-Ner, M. (Mautner) (1984) *J. Am. Chem. Soc.*, **106**, 1265.
17 Fersht, A.R., Shi, J.-P., Knill-Jones, J., Lowe, D.M., Wilkinson, A.J., Blow, D.M., Brick, P., Carter, P., Waye, M.M.Y. and Winter, G. (1985) *Nature*, **314**, 235–238.
18 Fersht, A.R. (1999) *Structure and Mechanism in Protein Science*, W.H. Freeman and Company, New York.
19 Nicolas, A., Egmond, M., Verrips, C.T., de Vlieg, J., Longhi, S., Cambillau, C. and Martinez, C. (1996) *Biochemistry*, **35**, 398–410.
20 Welsh, J.A., Cunningham, B.C., Graycar, T.P. and Estell, D.A. (1986) *Phil. Trans. R. Soc. London A*, **317**, 415–423.
21 Bell, A.F., Wu, J., Feng, Y. and Tonge, P.J. (2001) *Biochemistry*, **40**, 1725–1733.
22 Bryan, P., Pantoliano, M.W., Quill, S.G., Hsiao, H.Y. and Poulos, T. (1986) *Proc. Natl. Acad. Sci. U SA*, **83**, 3743–3745.
23 Rao, S.N., Singh, U.C., Bash, P.A. and Kollman, P.A. (1987) *Nature*, **328**, 551–554.
24 Menard, R., Carriere, J., Laflamme, P., Plouffe, C., Khouri, H.E., Vernet, T., Tessier, D.C., Thomas, D.Y. and Storer, A.C. (1991) *Biochemistry*, **30**, 8924–8928.
25 Magnusson, A., Hult, A.K. and Holmquist, M. (2001) *J. Am. Chem. Soc.*, **123**, 4354–4355.
26 Evans, C.T., Kurz, L.C., Remington, S.J. and Srere, P.A. (1996) *Biochemistry*, **35**, 10661–10672.
27 Cleland, W.W., Frey, P.A. and Gerlt, J.A. (1998) *J. Biol. Chem.*, **273**, 25529–25532.
28 Fuhrmann, C.N., Daugherty, M.D. and Agard, D.A. (2006) *J. Am. Chem. Soc.*, **128**, 9086–9102.
29 Shan, S.O. and Herschlag, D. (1999) *Meth. Enzymol.*, **308**, 246–276.
30 Kraut, D.A., Sigala, P.A., Pybus, B., Liu, C.W., Ringe, D., Petsko, G.A. and Herschlag, D. (2006) *PLoS Biol.*, **4**, e99.
31 Frey, P.A., Whitt, S.A. and Tobin, J.B. (1994) *Science*, **264**, 1927–1930.
32 Cleland, W.W. and Kreevoy, M.M. (1994) *Science*, **264**, 1887–1890.
33 Topf, M., Várnai, P. and Richards, W.G. (2002) *J. Am. Chem. Soc.*, **124**, 14780–14788.
34 (a) Gerlt, J.A. and Gassman, P.G. (1993) *Biochemistry*, **32**, 11943–11952. (b) Gerlt, J.A. and Gassman, P.G. (1993) *J. Am. Chem. Soc.*, **115**, 11552–11568.
35 Zhao, Q., Abeygunawardana, C., Talalay, P. and Mildvan, A.S. (2002) *Proc. Natl. Acad. Sci. U SA*, **93**, 8220–8224.
36 (a) For reviews, see: Frey, P.A. (2004) *J. Phys. Org. Chem.*, **17**, 511–520. (b) Hedstrom, L. (2002) *Chem. Rev.*, **102**, 4501–4523. (c) Sedolisins (serine-carboxyl peptidases), such as kumamolisin-As, offer an interesting counterexample. In this carboxypeptidase the oxyanion hole includes an aspartate group which appears to protonate the developing oxyanion. For a computational study of this system, see: Guo, H., Wlodawer, A. and Guo, H. (2005) *J. Am. Chem. Soc.*, **127**, 15662–15663.
37 (a) Mulholland, A.J., Lyne, P.D. and Karplus, M. (2000) *J. Am. Chem. Soc.*, **122**, 534–535.

(b) van der Kamp, M.W., Perruccio, F. and Mulholland, A.J. (2007) *J. Mol. Graph. Model*, **26**, 676–690.
38 Feierberg, I. and Åqvist, J. (2002) *Theor. Chem. Accounts*, **108**, 71–84.
39 Blow, D. (2000) *Structure*, **8**, 77–81.
40 Henderson, R. (1970) *J. Mol. Biol.*, **54**, 341–354.
41 Bode, W., Greyling, H.J., Huber, R., Otlewski, J. and Wilusz, T. (1989) *FEBS Lett.*, **242**, 285–292.
42 Stroud, R.M., Kay, L.M. and Dickerson, R.E. (1974) *J. Mol. Biol.*, **83**, 185–208.
43 Neidhart, D., Wei, Y., Cassidy, C., Lin, J., Cleland, W.W. and Frey, P.A. (2001) *Biochemistry*, **40**, 2439–2447.
44 Blow, D.M. (1976) *Acc. Chem. Res.*, **9**, 145–152.
45 Moult, J., Sussman, F. and James, M.N. (1985) *J. Mol. Biol.*, **182**, 555–566.
46 Blanchard, H. and James, M.N. (1994) *J. Mol. Biol.*, **241**, 574–587.
47 Alden, R.A., Biktoft, J.J., Kraut, J., Robertus, J.D. and Wright, C.S. (1971) *Biochem. Biophys. Res. Commun.*, **45**, 337–344.
48 Bryan, P., Pantaliano, M.W., Quill, S.G., Hsiao, H. and Poulos, T. (1986) *Proc. Natl. Acad. Sci. USA*, **83**, 3743–3745.
49 Liao, D.I. and Remington, S.J. (1990) *J. Biol. Chem.*, **265**, 6528–6531.
50 Liao, D.I., Breddan, K., Sweet, R.M., Bullock, T. and Remington, S.J. (1992) *Biochemistry*, **31**, 9796–9812.
51 Kamphuis, I.G., Kalk, K.H., Swarte, M.B. and Drenth, J. (1984) *J. Mol. Biol.*, **179**, 233–256.
52 Menard, R., Carriere, J., Plouffe, P., Laflamme, C., Khouri, H.E., Vernet, T., Tessier, D.C., Thomas, D.Y. and Storer, A.C. (1991) *Biochemistry*, **30**, 8924–8928.
53 Wei, Y., Schottel, J.L., Derewenda, U., Swenson, L., Patkar, S. and Derewenda, Z.S. (1995) *Nat. Struct. Biol.*, **2**, 218–223.
54 Sussman, J.L., Harel, M., Frolow, F., Oefner, C., Goldman, A., Toker, L. and Sisman, I. (1991) *Science*, **253**, 872–879.
55 Raves, M.L., Harel, M., Pang, Y.-P., Sisman, I., Kozikowski, A.P. and Sussman, J.L. (1997) *Nat. Struct. Biol.*, **4**, 57–63.
56 Grochulski, P., Li, Y., Schrag, J.D., Bouthillier, F., Smith, P., Harrison, D., Rubin, B. and Cygler, M. (1993) *J. Biol. Chem.*, **268**, 12843–12847.
57 Noble, M.E., Cleasby, A., Johnson, L.N., Egmond, M.R. and Frenken, L.G. (1993) *FEBS Lett.*, **331**, 123–128.
58 Noble, M.E., Cleasby, A., Johnson, L.N., Egmond, M.R. and Frenken, L.G. (1994) *Protein Eng.*, **7**, 559–562.
59 Brady, L., Brzozowski, A.M., Derewenda, Z.S., Dodson, E., Tolley, S. and Turkenburg, J.P. (1990) *Nature*, **343**, 767–770.
60 Derewenda, U., Swenson, L., Wei, Y., Green, R., Kobos, P.M., Joerger, R., Haas, M.J. and Derewenda, Z.S. (1994) *J. Lipid Res.*, **35**, 524–534.
61 Joerger, R.D. and Haas, M.J. (1994) *Lipids*, **29**, 377–384.
62 Uppenberg, J., Hansen, M.T., Patkar, S. and Jones, T.A. (1994) *Structure*, **2**, 293–308.
63 Martinez, C., Nicolas, A., van Tilbeurgh, H., Egloff, M.-P., Cudrey, C., Verger, R. and Cambillau, C. (1994) *Biochemistry*, **33**, 83–89.
64 (a) Karpusas, M., Holland, D. and Remington, S.J. (1991) *Biochemistry*, **30**, 6024–6031.
(b) Corey, E.J., Sneen, R.A. (1956) *J. Am. Chem. Soc.*, **78**, 6269–6278.
65 Kim, J.J., Wang, M. and Paschke, R. (1993) *Proc. Natl. Acad. Sci. USA*, **90**, 7523–7527.
66 Partanen, S.T., Novikov, D.K., Popov, A.N., Mursula, A.M., Hiltunen, J.K. and Wierenga, R.K. (2004) *J. Mol. Biol.*, **342**, 1197–1208.
67 Engel, C.K., Mathieu, M., Zeelen, J.P., Hiltunen, J.K. and Wierenga, R.K. (1996) *EMBO J.*, **15**, 5135–5145.
68 Modis, Y. and Wierenga, R.K. (2000) *J. Mol. Biol.*, **297**, 1171–1182.
69 Jogl, G., Rozovsky, S., McDermott, A.E. and Tong, L. (2003) *Proc. Natl. Acad. Sci. USA*, **100**, 50–55.
70 Zhang, Z., Sugio, S., Komives, E.A., Liu, K.D., Knowles, J.R., Petsko, G.A. and Ringe, D. (1994) *Biochemistry*, **33**, 2830–2837.
71 Arsenieva, D., Hardre, R., Salmon, L. and Jeffery, C.J. (2002) *Proc. Natl. Acad. Sci. USA*, **99**, 5872–5877.

72 Roos, A.K., Burgos, E., Ericsson, D.J., Salmon, L. and Mowbray, S.L. (2005) *J. Biol. Chem.*, **280**, 6416–6422.

73 Hamed, R.B., Batchelar, E.T., Clifton, I.J. and Schofield, C.J. (2008) *Cell Mol. Life Sci.*, **65**, 2507–2527.

74 Wells, J.A. and Estell, D.A. (1988) *Trends Biochem. Sci.*, **13**, 291–297.

75 Harel, M., Quinn, D.M., Nair, H.K., Sisman, I. and Sussman, J.L. (1996) *J. Am. Chem. Soc.*, **118**, 2340–2346.

76 Lee, L.C., Lee, Y.L., Leu, R.J. and Shaw, J.F. (2006) *Biochem. J.*, **397**, 69–76.

77 Thoden, J.B., Zhuang, Z., Dunaway-Mariano, D. and Holden, H.M. (2003) *J. Biol. Chem.*, **278**, 43709–43716.

78 Haapalainen, A.M., Meriläinen, G. and Wierenga, R.K. (2006) *Trends Biochem. Sci.*, **31**, 64–71.

79 pK_a Data compiled by Williams, R. http://research.chem.psu.edu/brpgroup/pKa_compilation.pdf (accessed on July 14, 2009).

80 Kursula, P., Ojala, J., Lambeir, A.M. and Wierenga, R.K. (2002) *Biochemistry*, **41**, 15543–15556.

81 Kursula, P., Sikkilä, H., Fukao, T., Kondo, N. and Wierenga, R.K. (2005) *J. Mol. Biol.*, **347**, 189–201.

82 Knowles, J.R. (1991) *Nature*, **350**, 121–124.

83 Sigala, P.A., Kraut, D.A., Caaveiro, J.M., Pybus, B., Ruben, E.A., Ringe, D., Petsko, G.A. and Herschlag, D. (2008) *J. Am. Chem. Soc.*, **130**, 13696–13708.

84 Sigala, P.A., Fafarman, A.T., Bogard, P.E., Boxer, S.G. and Herschlag, D. (2007) *J. Am. Chem. Soc.*, **129**, 12104–12105.

85 Holden, H.M., Benning, M.M., Haller, T. and Gerlt, J.A. (2001) *Acc. Chem. Res.*, **34**, 145–157.

86 (a) Tadayoni, B.M., Parris, K. and Rebek, J. Jr. (1989) *J. Am. Chem. Soc.*, **111**, 4530–4505.
(b) Tadayoni, B.M., Huff, J. and Rebek, J. Jr. (1991) *J. Am. Chem. Soc.*, **113**, 2247–2253.

87 Wolfe, J., Muehldorf, A. and Rebek, J. Jr. (1991) *J. Am. Chem. Soc.*, **113**, 1453–1454.

88 Anslyn, E. and Breslow, R. (1989) *J. Am. Chem. Soc.*, **111**, 5972–5973.

89 Breslow, R. and Graff, A. (1993) *J. Am. Chem. Soc.*, **115**, 10988–10989.

90 Snowden, T.S., Bisson, A.P. and Anslyn, E.V. (1999) *J. Am. Chem. Soc.*, **121**, 6324–6325.

91 Snowden, T.S., Bisson, A.P. and Anslyn, E.V. (2001) *Bioorg. Med. Chem.*, **9**, 2467–2478.

92 Zhong, Z., Snowden, T.S., Best, M.D. and Anslyn, E.V. (2004) *J. Am. Chem. Soc.*, **126**, 3488–3495.

93 Lauri, G. and Bartlett, P. (1994) *J. Comput.-Aided Mol. Des.*, **8**, 51–66.

94 Zhu, Y. and Drueckhammer, D.G. (2005) *J. Org. Chem.*, **70**, 7755–7760.

95 Simón, L., Muñiz, F.M., Sáez, S., Raposo, C., Sanz, F. and Morán, J.R. (2005) *Helv. Chim. Acta*, **88**, 1682–1701.

96 Simón, L., Muñiz, F.M., Sáez, S., Raposo, C. and Morán, J.R. (2007) *Eur. J. Org. Chem.*, 4821–4830.

97 Park, H., Kim, K.M., Lee, A., Ham, S., Nam, W. and Chin, J. (2007) *J. Am. Chem. Soc.*, **129**, 1518–1519.

98 Shaw, J.P., Petsko, G.A. and Ringe, D. (1997) *Biochemistry*, **36**, 1329–1342.

99 Khersonsky, O., Roodvelds, C. and Tawfik, D.S. (2006) *Curr. Opin. Chem. Biol.*, **10**, 498–508.

100 James, L.C. and Tawfik, D.S. (2003) *Trends Biochem. Sci.*, **28**, 361–368.

101 Hult, K. and Berglund, P. (2007) *Trends Biotechnol.*, **25**, 231–238.

102 Bornscheuer, U.T. and Kazlauskas, R.J. (2004) *Angew. Chem. Int. Ed.*, **43**, 6032–6040.

103 Magnusson, A., Hult, K. and Holmquist, M. (2001) *J. Am. Chem. Soc.*, **123**, 4354–4355.

104 Branneby, C., Carlqvist, P., Magnusson, A., Hult, K., Brinck, T. and Berglund, P. (2003) *J. Am. Chem. Soc.*, **125**, 874–875.

105 Svedendahl, M., Hult, K. and Berglund, P. (2005) *J. Am. Chem. Soc.*, **127**, 17988–17989.

106 Svedendahl, M., Carlqvist, P., Branneby, C., Allnér, O., Frise, A., Hult, K., Berglund, P. and Brinck, T. (2008) *Chem. Biochem.*, **9**, 2443–2451.

107 Carrea, G. and Riva, S. (2000) *Angew. Chem. Int. Ed.*, **39**, 2226–2254.
108 Hudson, E.P., Eppler, R.K. and Clark, D.S. (2005) *Curr. Opin. Biotechnol.*, **16**, 637–643.
109 Marigo, M., Franzén, J., Poulsen, T.B., Zhuang, W. and Jørgensen, K.A. (2005) *J. Am. Chem. Soc.*, **127**, 6964–6965.
110 Erkkilä, A., Majander, I. and Pihko, P.M. (2007) *Chem. Rev.* **107**, 5416–5470.

5
Brønsted Acids, H-Bond Donors, and Combined Acid Systems in Asymmetric Catalysis

Hisashi Yamamoto and Joshua N. Payette

5.1
Introduction

This chapter examines recent advances in chiral Brønsted acid (phosphoric acids) and hydrogen bonding catalysis including combined acid systems [1]. Further, only noncovalent catalytic systems are covered which appropriately excludes covalent, enamine-based systems where hydrogen bonding is also important. While chiral Brønsted acids may not be considered strict hydrogen bond catalysts, limitations in our current understanding of the extent of proton transfer in the transition state have allowed these systems to elude formal classification, and therefore their discussion in this volume is warranted. Further, strong hydrogen bonding interactions within the catalyst–substrate ion pairs are often invoked to rationalize high reactivities and selectivities.

In contrast to some related reviews, which use reaction class or electrophiles as organizational elements, this chapter is divided into three main sections according to catalyst class: (i) Brønsted acid catalysis by phosphoric acid and phosphoramide derivatives, (ii) N–H hydrogen bond catalysis by organic base and ammonium systems, and (iii) combined acid catalysis including Brønsted-acid-assisted Brønsted acid, Lewis-acid-assisted Brønsted acid, and Lewis-acid-assisted Brønsted acid systems (Figure 5.1).

The reader is encouraged to contemplate the mechanistic similarities and differences between each type of catalyst system as this will lead to a coherent and logical understanding of this seemingly disjointed field. Thiourea catalysis is not discussed as it is covered in the following chapter.

I. Brønsted Acid

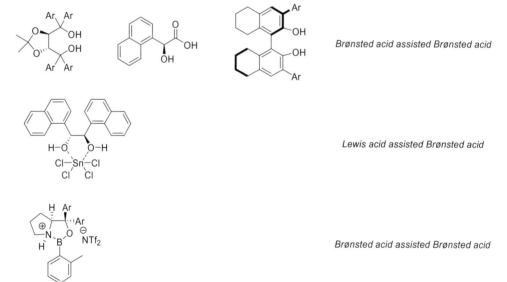

Figure 5.1 Overview of catalyst systems covered in this chapter.

5.2
Brønsted Acid (Phosphoric Acid and Derivatives)

5.2.1
Binapthylphosphoric Acids

Simple (S)-(+) and (R)-(−)-binapthylphosphoric acids (**1a**) (BNP) have long been employed as chiral resolving agents for biologically and therapeutically active compounds (Scheme 5.1) [2]. In 1991, Wilen disclosed the X-ray structure of the salt of BNP and Troger's base which indicated a hydrogen bonding interaction between the phosphate anion and the protonated amine [3]. However, it was not until 2004 that the groups of Akiyama and Terada independently reported the design of phosphoric acid diesters (**1a–e**) derived from (R)-BINOL bearing bulky 3,3′ substituents as a novel class of chiral Brønsted acid catalysts [4, 5]. Both authors demonstrated that these chiral acids effectively catalyzed Mannich reactions of imines with various nucleophiles to afford products in excellent yields and enantioselectivities. In the several years following these initial reports, many groups have expanded the scope and application of these catalysts to a diverse set of organic transformations. Low catalyst loadings and mild, metal-free reaction conditions have made these catalysts attractive alternatives to established Lewis-acid-based methodologies. Characteristics unique to phosphoric acid (**1**) have permitted discovery of new chemical methodologies hitherto unknown. Thus, the structural and electronic properties accounting for the reactivity and stereo-induction of **1** include [6]: (i) a strongly Brønsted acidic site [7] proximal to a Lewis basic site. These adjacent acid/base sites allow for simultaneous activation/orientation of both nucelophile and electrophile through hydrogen bonding. (ii) A well-defined chiral pocket produced by the binaphthyl skeleton and the appended bulky 3,3′ substituents. (iii) A ring structure attached to the phosphoric acid moiety to prevent free rotation at the α-position of the phosphorus center. This feature is not found in other Brønsted acids such as carboxylic and sulfonic acids (Figure 5.2).

5.2.1.1 Mannich Reaction
Akiyama first reported that 30 mol% of **1d** catalyzed the addition of various monosubsitutued silyl ketene acetals to aldimines derived from o-hydroxyaniline to afford the corresponding β-amino esters with high *syn* selectivities in excellent yields and ee's (Scheme 5.1) [3]. It was found that the choice of substituents at the 3,3′ positions is crucial in obtaining high levels of asymmetric induction. Notably, when *N*-benzylideneaniline, lacking the *ortho* hydroxyl group, was used as electrophile, the ee of the product was drastically decreased to 39% ee. DFT studies were undertaken to elucidate the mode of electrophilic activation as well as the observed *re*-facial selectivity. Calculations revealed that protonation of the aldimine generates a dicoordinated cyclic 9-membered ring zwitterionic TS [8]. Nucleophilic attack followed by silyl transfer to the o-hydroxy moiety and

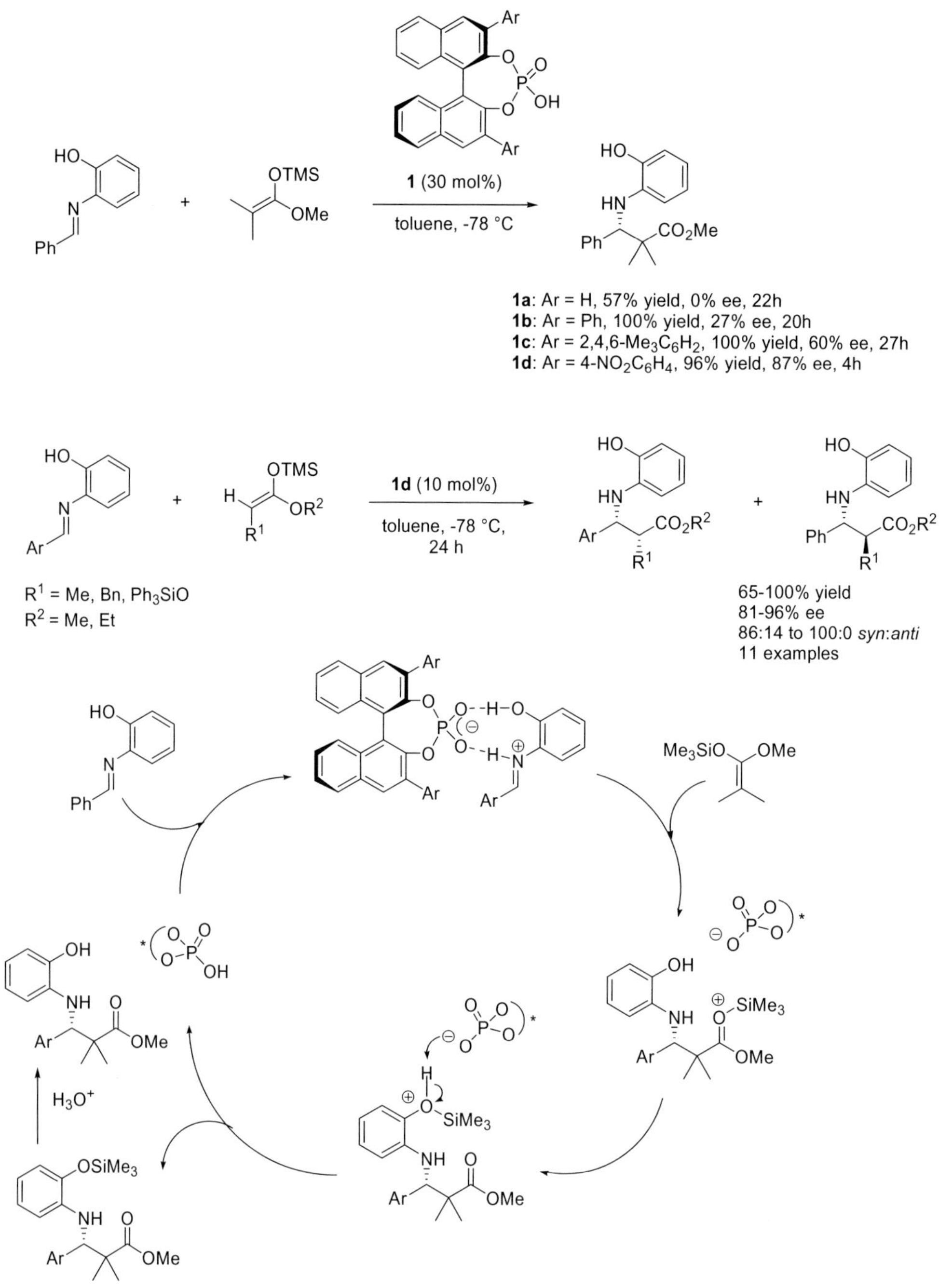
1 (30 mol%)
toluene, -78 °C
1a: Ar = H, 57% yield, 0% ee, 22h
1b: Ar = Ph, 100% yield, 27% ee, 20h
1c: Ar = 2,4,6-Me3C6H2, 100% yield, 60% ee, 27h
1d: Ar = 4-NO2C6H4, 96% yield, 87% ee, 4h
1d (10 mol%)
toluene, -78 °C,
24 h
R1 = Me, Bn, Ph3SiO
R2 = Me, Et
65-100% yield
81-96% ee
86:14 to 100:0 syn:anti
11 examples
H3O+

Scheme 5.1

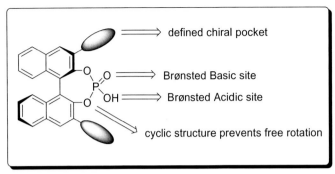

Figure 5.2 Important structural and electronic properties of BINOL-derived chiral phosphoric acids.

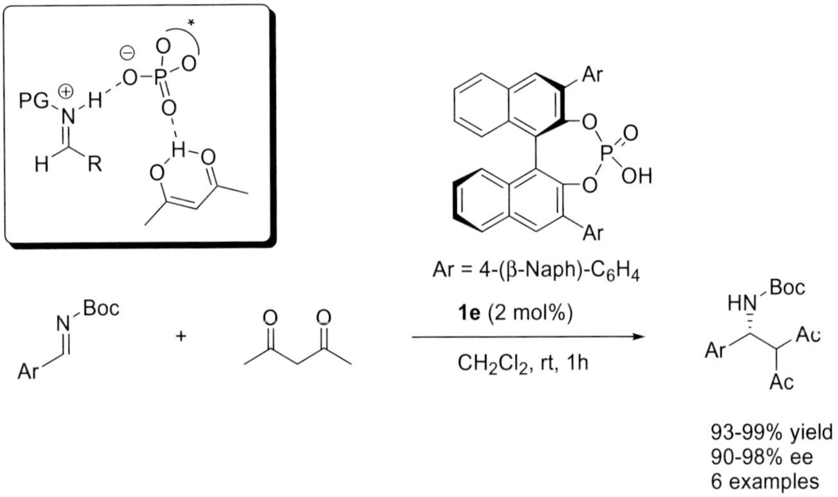

Scheme 5.2

comcomitant reprotonation of the catalyst continues the catalytic cycle. Further, hydrogen bonding as well as π stacking interactions between the N-aryl moiety and the 3,3′ substituents of the catalyst are significant in fixing the geometry of the aldimine in the transition state. Additionally, *si*-facial attack in the TS is disfavored due to strong steric repulsion of the 3,3′ substituents.

Shortly thereafter, Terada demonstrated that the Mannich reaction between several N-Boc aryl imines and acetoacetone was effectively catalyzed by only 2 mol% of **1e** (Scheme 5.2) [4]. In view of Akiyama's work, this study is particularly significant because it suggested that **1e** may act as a bifunctional catalyst [9] not only to form a chiral ion pair with the electrophile but also to activate the nucleophile through hydrogen bonding of the α-proton with Lewis basic phosphoryl oxygen.

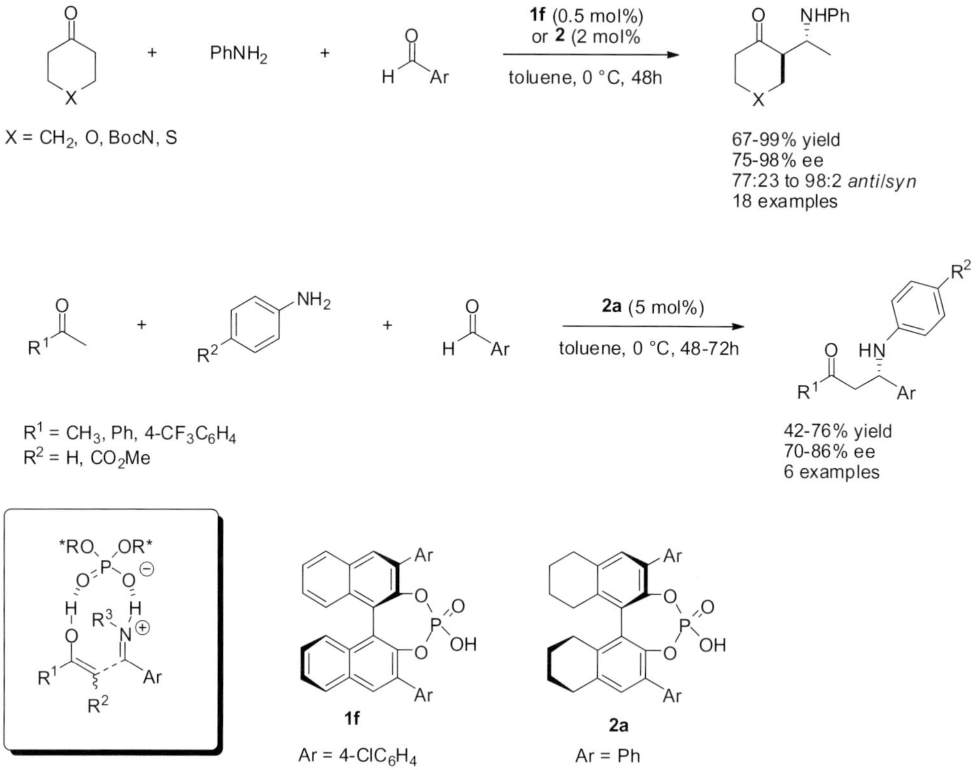

Scheme 5.3

Later in 2007, Gong utilized **1f** and saturated derivative **2** in a direct Mannich reaction between *in situ* generated N-aryl imines and cyclic ketones as well aromatic ketones (Scheme 5.3) [10]. It was found that electron poor anilines as coupling partners gave the highest enantioselectivities. The authors postulate that acid promoted enolization of the ketone forms the reactive enol which adds to the protonated aldimine.

5.2.1.2 Hydrophosphonylation

Akiyama applied **1g** to the hydrophosphonylation of N-PMP aryl imines with diisopropyl phosphate to generate optically active α-amino phosphate esters in good to high yields and ee's (Scheme 5.4) [11]. Interestingly, it was found that the use of trialkyl phosphates as nucelophiles resulted in a deterioration of both reactivity and enantioselectivity. Hence, it is believed that the OH moiety of the reactive phosphite tautomer derived from diisopropyl phosphate is essential in realizing high asymmetric induction. Further, when the N-substituent on the aldimine was changed to an o-hydroxy group, enantioselectivities significantly decreased. Based on these observations the authors suggest **1g** functions as a bifunctional catalyst to activate as well as orient both nucleophile and electrophile.

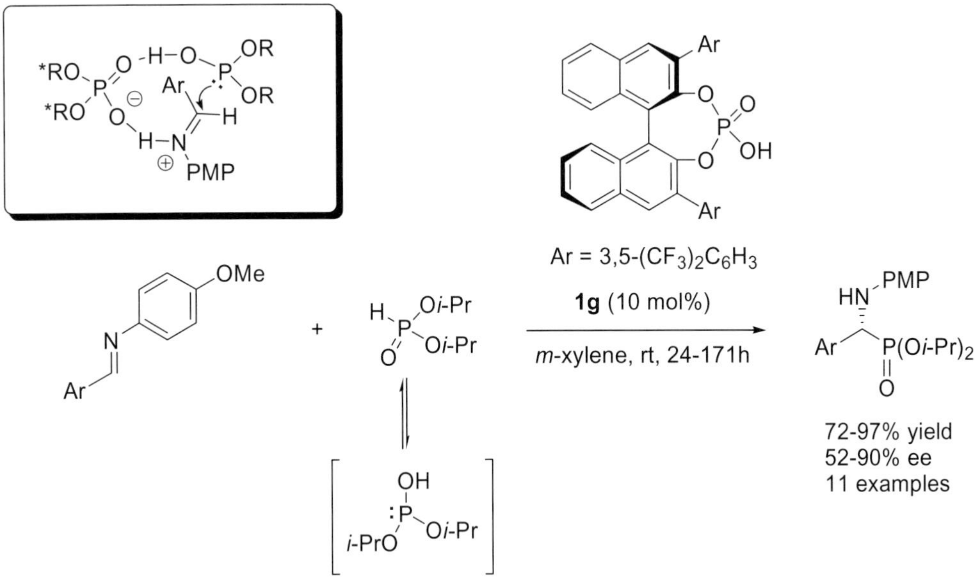

Scheme 5.4

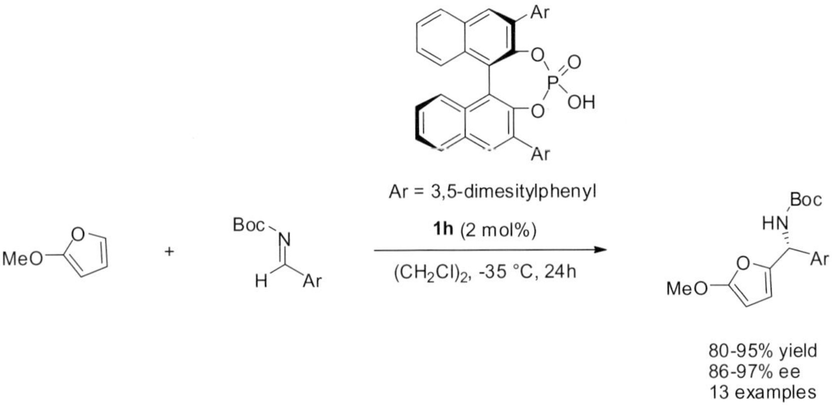

Scheme 5.5

5.2.1.3 Friedel–Crafts

The first organocatalytic 1,2-aza-F-C reaction of aldimines was reported by Terada using catalyst **1h** (Scheme 5.5) [12]. Remarkably, N-Boc aryl imines containing electron-rich or electron-poor aryl groups substituted at either the *ortho*, *meta*, or *para* position were well tolerated yielding F-C products in consistently high yields and excellent ee's. It is noteworthy that the absolute configuration of these products was opposite to that previously found in the same group's work on the Mannich reaction.

Later, the same group reported the Friedel–Crafts addition of unprotected indoles to enecarbamates containing aliphatic substituents (Scheme 5.6) [13]. Use

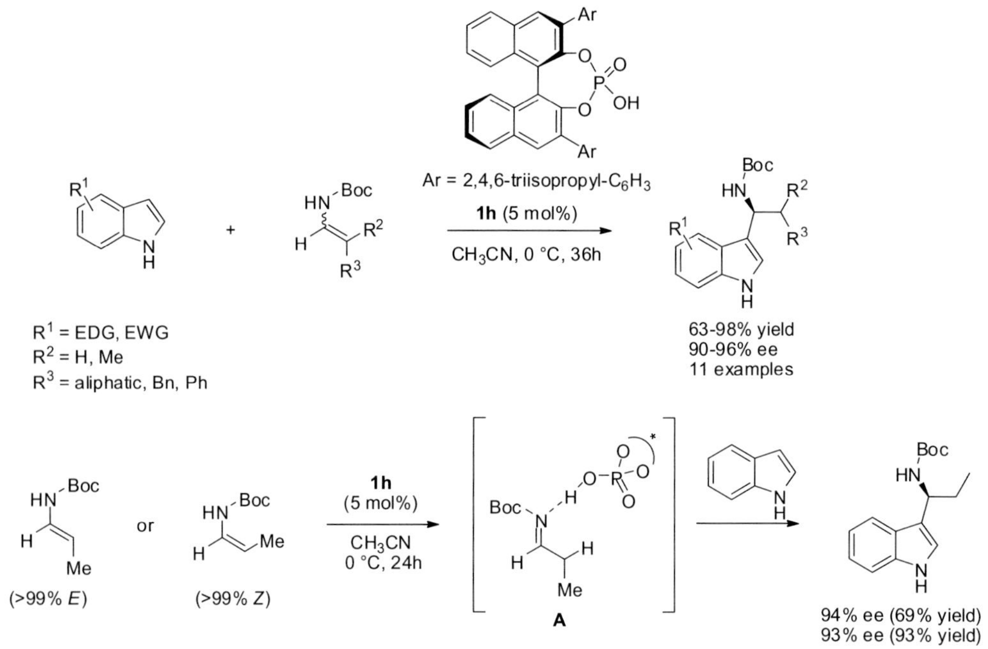

Scheme 5.6

of either isomericzally pure *E* or *Z* enecarbamates gave identical products in similar ee's, possessing the same absolute configuration. Thus, the authors believe both reactions proceed through common intermediate A, whereby protonation of the enecarbamate generates the corresponding aliphatic imine.

Thereafter, the Zhou group reported a similar reaction between indole and various α-aryl enamides catalyzed by catalyst **1h** to afford F–C products possessing a nitrogen-containing quaternary carbon center (Scheme 5.7) [14, 15]. In agreement with Terada's observations, the authors found that use of an *N*-methylated enamide was unreactive, providing further evidence that the corresponding ketimine is the true reactive intermediate. To assess the role of the N–H moiety of indole, *N*-methylated indole was applied under the standard reaction conditions. Remarkably, even at elevated temperatures no addition product was obtained. Similar to Akiyama's hydrophosphonylation reaction, the authors postulate that activation of the indole nucleophile via hydrogen bonding by the phosphoryl oxygen of **1j** is essential in obtaining high levels of reactivity and enantioselectivity.

Additionally, You reported a F-C reaction of indole with electronically diverse aryl aldimines (Scheme 5.8) [16]. In general, products could be obtained in excellent yields and enantioselectivities in short reaction times.

Subsequently, the research groups of Terada and Antilla reported highly enantioselective Friedel–Crafts reactions between *N*-substituted indoles and *N*-acyl aryl imines (Scheme 5.9) [17]. Interestingly, Antilla states that under his reaction condi-

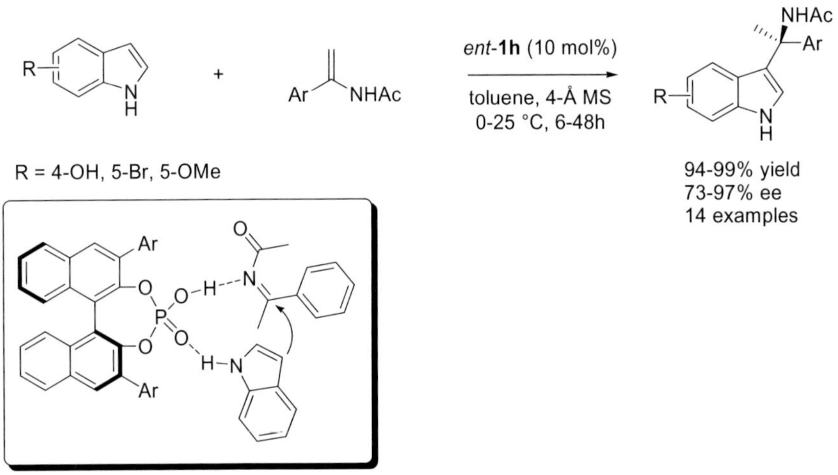

5.2 Brønsted Acid (Phosphoric Acid and Derivatives) | 81

Scheme 5.7

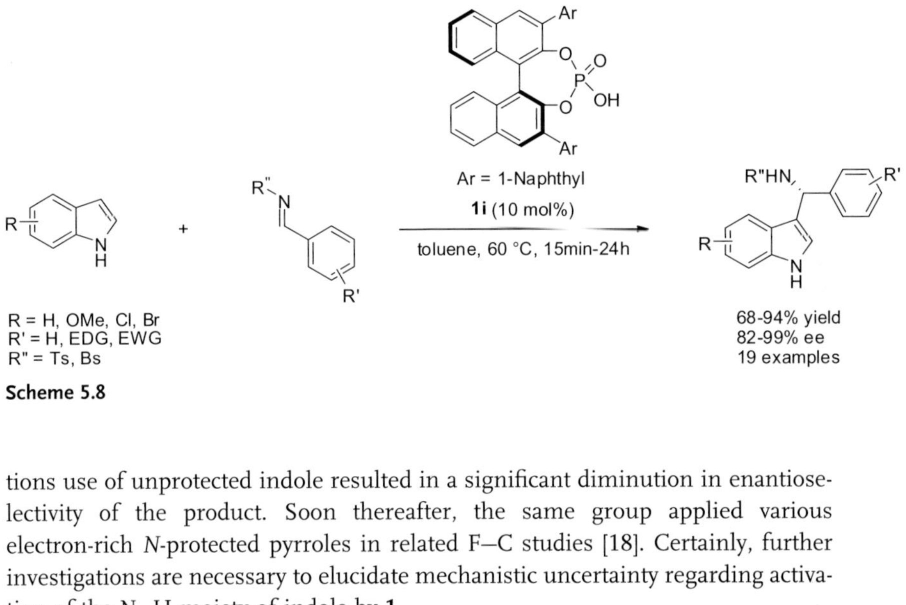

R = 4-OH, 5-Br, 5-OMe

ent-**1h** (10 mol%), toluene, 4-Å MS, 0-25 °C, 6-48h

94-99% yield
73-97% ee
14 examples

Scheme 5.8

R = H, OMe, Cl, Br
R' = H, EDG, EWG
R" = Ts, Bs

Ar = 1-Naphthyl
1i (10 mol%), toluene, 60 °C, 15min-24h

68-94% yield
82-99% ee
19 examples

tions use of unprotected indole resulted in a significant diminution in enantioselectivity of the product. Soon thereafter, the same group applied various electron-rich N-protected pyrroles in related F–C studies [18]. Certainly, further investigations are necessary to elucidate mechanistic uncertainty regarding activation of the N–H moiety of indole by **1**.

List developed a direct Pictet–Spengler reaction catalyzed by ent-**1h** which tolerates a host of aliphatic, branched, and unbranched aldehydes as well as electron-poor aromatic aldehydes to afford the resulting chiral tetra-β-carbolines in good to excellent yields and ee's (Scheme 5.10) [19]. Preliminary studies revealed that treatment of unsubstituted trpytamine **3** with TFA gave only the aldol condensation product **4** [20]. However, geminal disubsitution of tryptamine with an electron withdrawing group favored the cyclization product presumably due to electronic reasons and the Thorpe–Ingold effect.

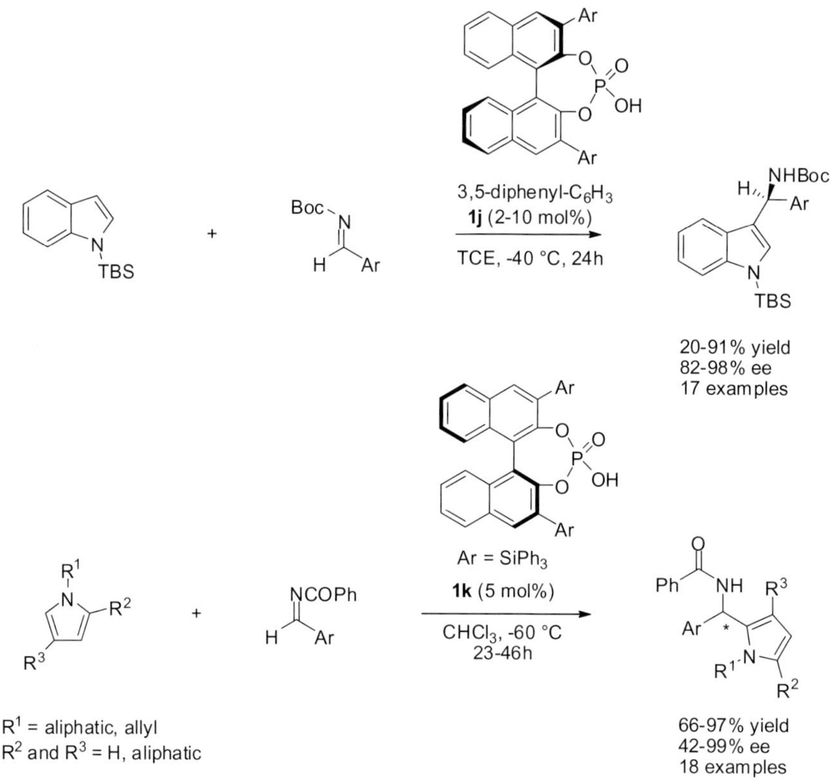
Ar
O
P
O
OH
O
Ar
TBS
Boc
N
H
Ar
3,5-diphenyl-C_6H_3
1j (2-10 mol%)
TCE, -40 °C, 24h
NHBoc
H
Ar
N
TBS
20-91% yield
82-98% ee
17 examples
Ar = $SiPh_3$
1k (5 mol%)
$CHCl_3$, -60 °C
23-46h
NCOPh
H
Ar
Ph
NH
R^3
Ar
R^1
N
R^2
66-97% yield
42-99% ee
18 examples
R^1 = aliphatic, allyl
R^2 and R^3 = H, aliphatic

Scheme 5.9

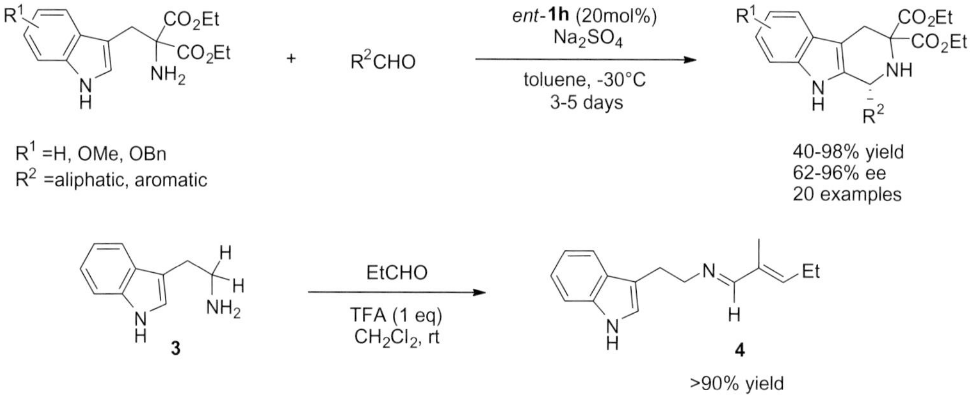
R^1
CO_2Et
CO_2Et
NH_2
N
H
R^2CHO
ent-**1h** (20mol%)
Na_2SO_4
toluene, -30°C
3-5 days
NH
R^2
40-98% yield
62-96% ee
20 examples
R^1 =H, OMe, OBn
R^2 =aliphatic, aromatic
H
H
NH_2
3
EtCHO
TFA (1 eq)
CH_2Cl_2, rt
N
H
Et
4
>90% yield

Scheme 5.10

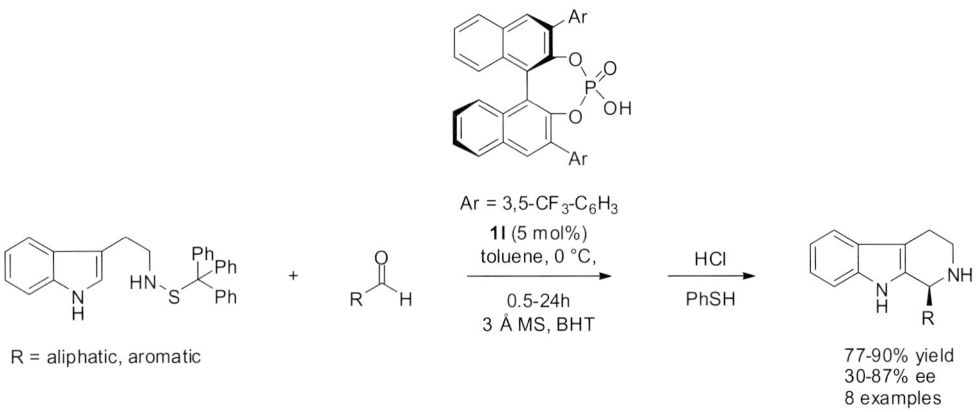

Scheme 5.11

Based on the previous findings by Koomen [21], the Hiemstra group subsequently reported the Pictet–Spengler reaction of N-tritylsulfenyl tryptamines and various aliphatic and aromatic aldehydes by **1l** (Scheme 5.11) [22]. Notably, the authors found that stabilization of the N-sulfenyliminium ion by the sulfenyl substituent facilitated preferential cyclization over enamine formation.

Exploiting the Lewis basic phosphoryl oxygen of **1m**, Terada reported the direct alkylation of α-diazoesters with N-acyl imines to afford β-amino-α-diazoesters in high yields and ee's (Scheme 5.12) [23]. Earlier, Johnston had observed that catalytic TfOH promoted aziridine formation (Aza–Darzens reaction) between diazoacetates and N-benzyl imines [24]. The authors propose that aziridine formation is circumvented through C–H bond cleavage by the phosphoryl oxygen of **1** (Intermediate A). However, as noted by the authors, the low nucleophilicity of N-acyl imines might also be considered as the cause of this selective transformation.

5.2.1.4 Diels–Alder

The groups of Rueping [25] and Gong [26] have developed the aza-hetero-Diels–Alder reaction of aryl imines and cyclohexenone to give isoquinuclidines in good *endo/exo* selectivities and high yields and ee's by **1** and **1a**, respectively (Scheme 5.13). In the presence of acid, cyclohexenone enolizes to afford the dienol which subsequently undergoes a Mannich reaction with the protonated aldimine followed by intramolecular aza-Michael addition to produce the formal Diels–Alder adducts.

In the Diels–Alder reaction between Danishefsky's diene and N-aryl imines derived from o-hydroxyaniline catalyzed by **1h**, Akiyama observed substantial increases in chemical yield and enantioselectivty by the addition of stoichiometric amounts of acetic acid (Scheme 5.14) [27]. The authors concede that the role of the protic acid is unclear.

Later, Akiyama reported that chiral pyrdinium phosphate ion **1n** was an effective catalyst for the aza-Diels–Alder reaction between the highly reactive Brassard's

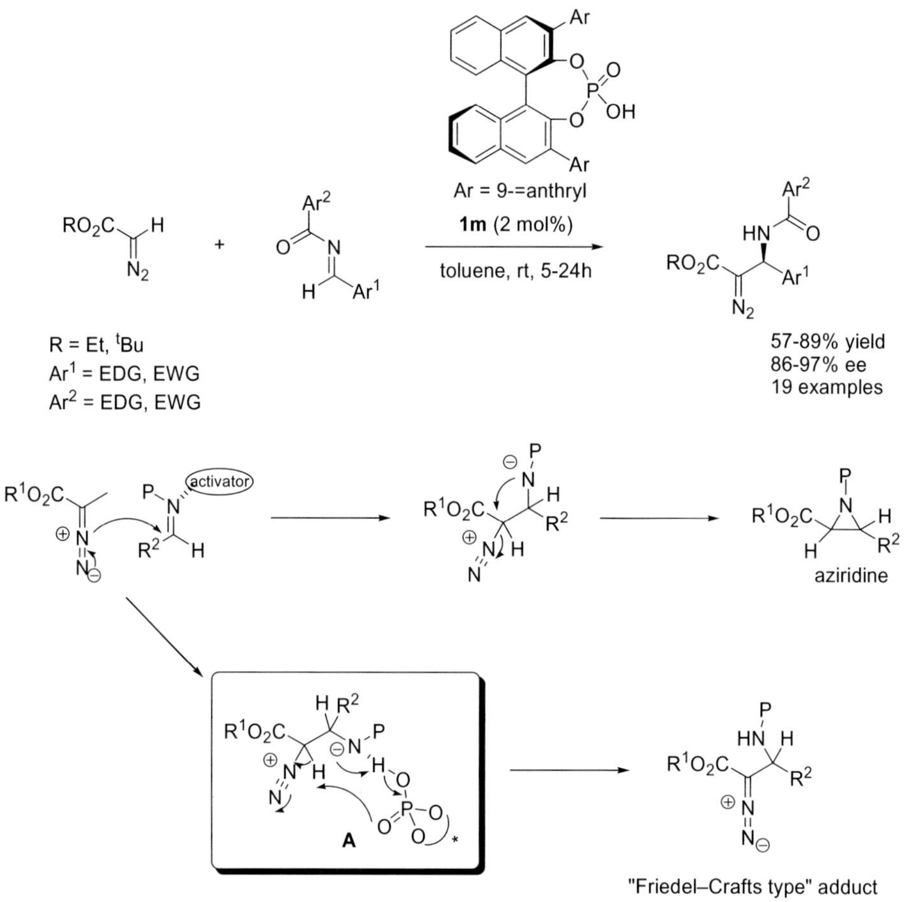

5 Brønsted Acids, H-Bond Donors, and Combined Acid Systems in Asymmetric Catalysis

Ar = 9-=anthryl
1m (2 mol%)
toluene, rt, 5-24h

R = Et, ^{t}Bu
Ar^1 = EDG, EWG
Ar^2 = EDG, EWG

57-89% yield
86-97% ee
19 examples

aziridine

"Friedel–Crafts type" adduct

Scheme 5.12

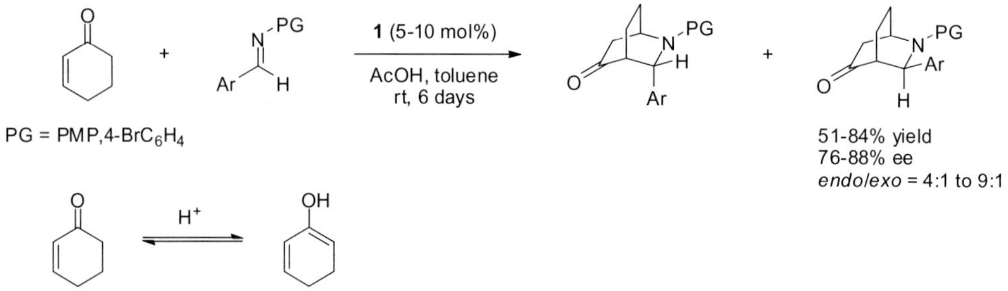

PG = PMP, 4-BrC$_6$H$_4$

1 (5-10 mol%)
AcOH, toluene
rt, 6 days

51-84% yield
76-88% ee
endo/exo = 4:1 to 9:1

Scheme 5.13

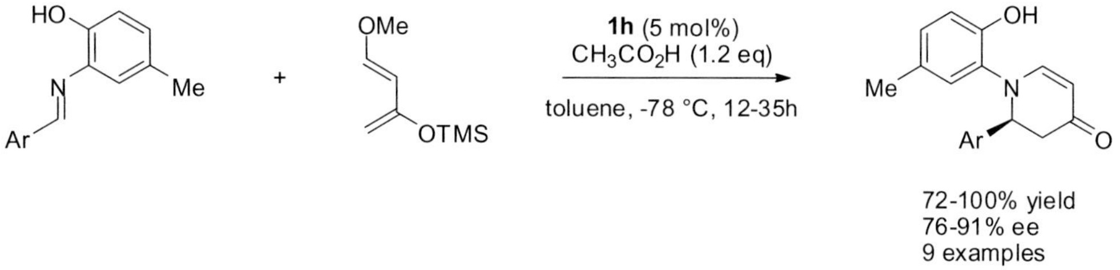

5.2 Brønsted Acid (Phosphoric Acid and Derivatives) | 85

Scheme 5.14

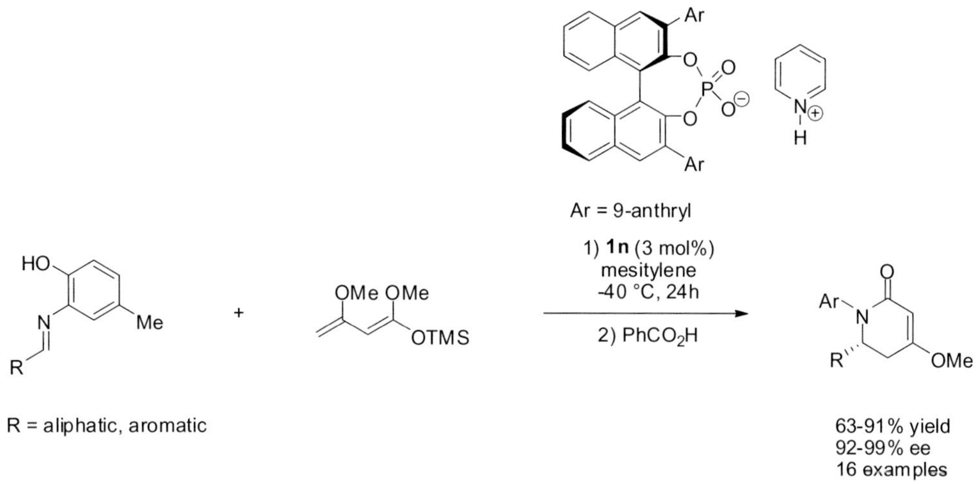

Reagents: HO-C6H3(Me)-N=CH-Ar + CH2=CH-C(OTMS)=CH-OMe → HO-C6H3(Me)-N(Ar-CH)-CH2-C(=O)-CH=CH (dihydropyridone)

Conditions: **1h** (5 mol%), CH₃CO₂H (1.2 eq), toluene, −78 °C, 12–35 h

72–100% yield
76–91% ee
9 examples

Scheme 5.15

Catalyst: BINOL-phosphoric acid with Ar = 9-anthryl, pyridinium counterion

Reagents: HO-C6H3(Me)-N=CH-R + CH2=C(OMe)-CH=CH-OMe with OTMS → dihydropyridone bearing OMe

Conditions: 1) **1n** (3 mol%), mesitylene, −40 °C, 24 h; 2) PhCO₂H

R = aliphatic, aromatic

63–91% yield
92–99% ee
16 examples

diene and aldimines derived from *o*-hydroxyaniline to afford dihydropyridones (Scheme 5.15) [28]. Based on a simple NMR study, it was shown that in the presence of phosphoric acid **1m** only 12% of Brassard's diene remained after 1 h, whereas in the presence of **1n** 75% of the diene could still be detected. Therefore, the authors postulate that the weaker acidity of **1n** renders it more compatible with acid-labile Brassard's diene.

Akiyama applied **1m** in the inverse-demand aza-Diels–Alder reaction of various acyclic and cyclic vinyl ethers with *N*-aryl imines derived from *o*-hydroxyaniline to provide optically active tetrahydroquinoline derivatives (Scheme 5.16) [29]. Since aldimines derived from *p*-methoxyaniline gave no cycloaddition product, a nine-membered cyclic TS (akin to that proposed for the author's Mannich reaction) was invoked to rationalize the high levels of enantio-control.

5.2.1.5 Miscellaneous Reactions

Kobayashi earlier demonstrated the first use of enamides and enecarbamates as nucleophiles in several enantioselective copper-catalyzed reactions [30]. Inspired by this precedent, Terada reported that 0.1 mol% of **1m** effectively

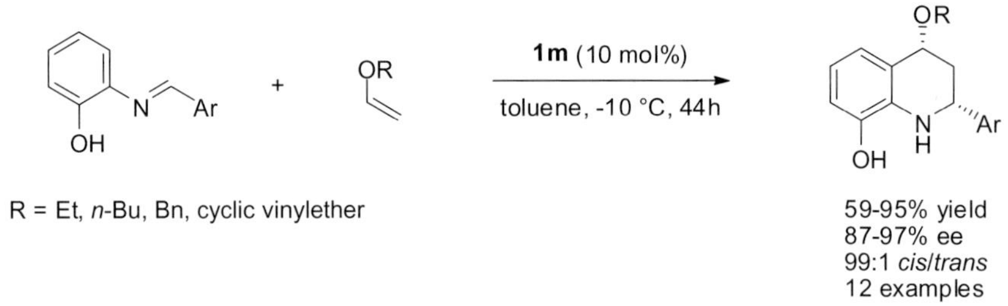

Scheme 5.16

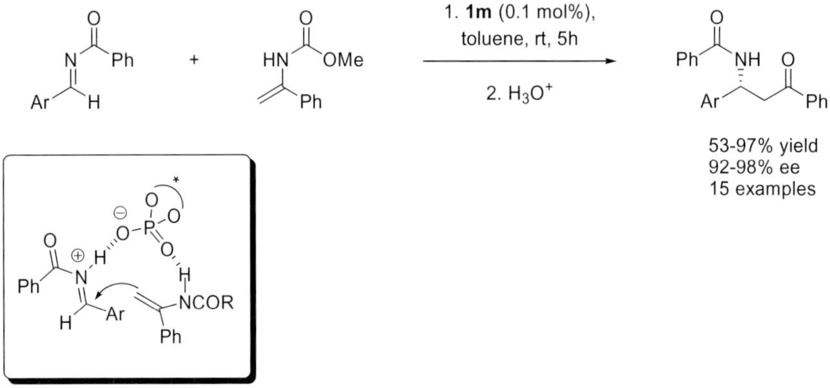

R = Et, n-Bu, Bn, cyclic vinylether

59–95% yield
87–97% ee
99:1 cis/trans
12 examples

Scheme 5.17

53–97% yield
92–98% ee
15 examples

catalyzed the aza-ene-type reaction of disubstituted enecarbamates and N-acyl imines to give synthetically useful β-aminoketimine adducts in high yields and excellent ee's (Scheme 5.17) [31]. The authors propose that the reaction proceeds through the transition state, where **1m** serves as a bifunctional catalyst to activate the imine as well as the enecarbamate through hydrogen bonding interactions.

In related studies, Terada described the synthesis of optically pure piperidines via a tandem aza-ene-type reaction/cyclization sequence (Scheme 5.18) [32]. The reaction of a monosubstituted enecarbamate and an N-acyl aldimine affords aza-ene-type intermediate **5**, which reacts with a second equivalent of enecarbamate to give aldimine **6**. Subsequent intramolecular cyclization terminates the aze-ene-type reaction sequence to furnish trans-piperidine **7** in high enantio- and diasterio-selectivities.

The Gong group utilized **2a** in the multicomponent Biginelli reaction to provide structurally diverse 3,4-dihydropyrimidin-2(1H)-ones in high yields and ee's (Scheme 5.19) [33]. Notably, the authors found that bulky 3,3′ substituents on the catalyst had a deleterious effect with respect to yields and enantioselectivities.

Given the rich chemical diversity accessible from aminohydrazones, Rueping investigated the imino-azaenamine reaction of N-Boc protected imines and

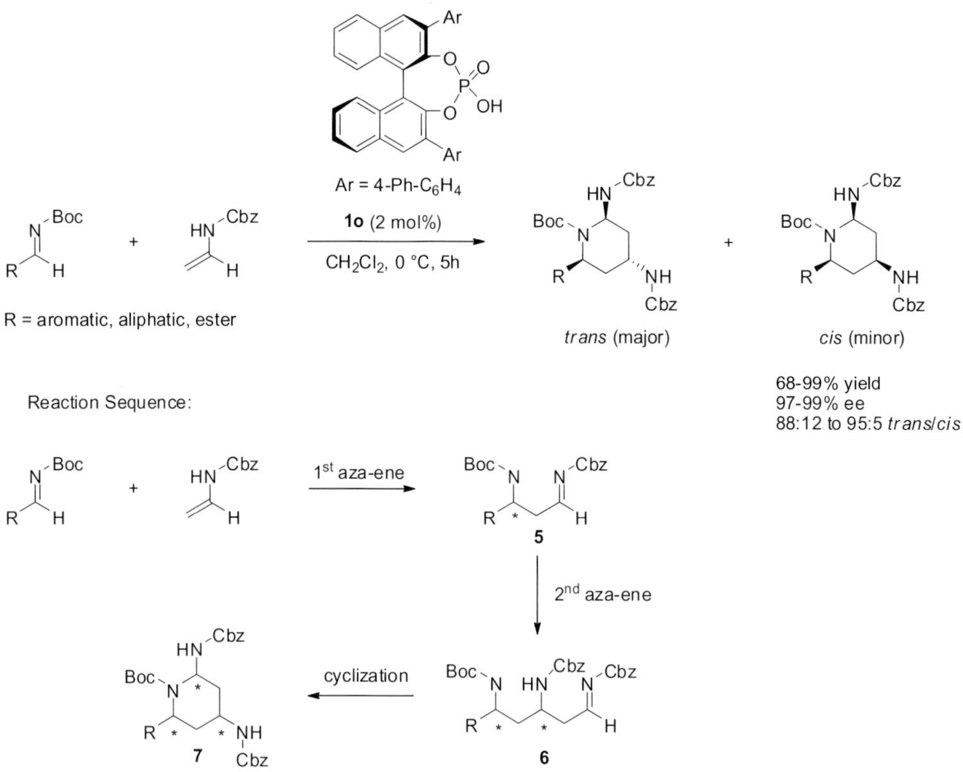

5.2 Brønsted Acid (Phosphoric Acid and Derivatives)

Scheme 5.18

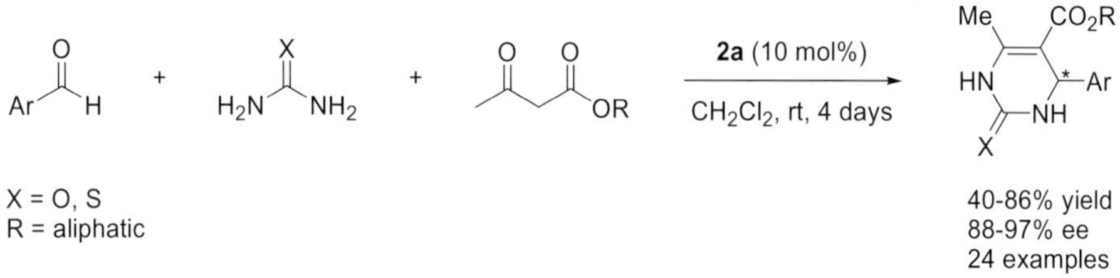

Scheme 5.19

pyrrolidine-derived methlene-hydrazines catalyzed by **2b** (Scheme 5.20) [34]. Interestingly, the polarity of the solvent was found to have a profound effect on enantioselectivities, with chloroform giving substantially better results than toluene.

Rueping has recently reported an interesting alknylation reaction of α-imino esters employing both phosphoric acid **1p** and AgOAc as orthogonal cocatalysts [35]. As seen in the catalytic cycle in Scheme 5.21, generation of chiral iminium ion pair **I** nucleophilic and alkynyl-silver species **II** proceeds simultaneously. Subsequent nucelophilic addition completes both parallel cycles [36].

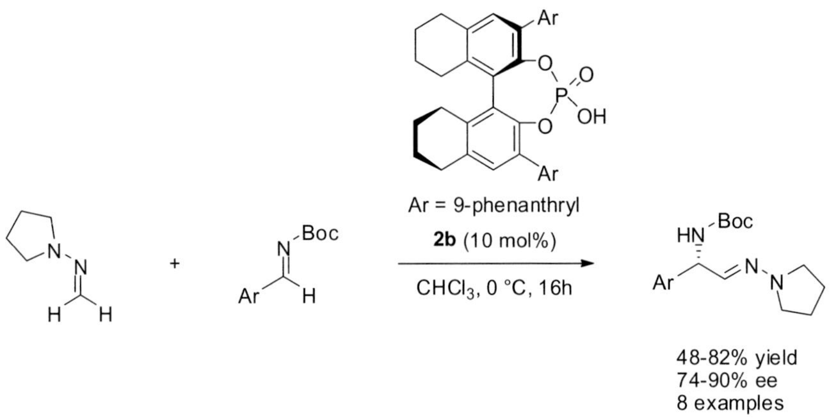

Scheme 5.20

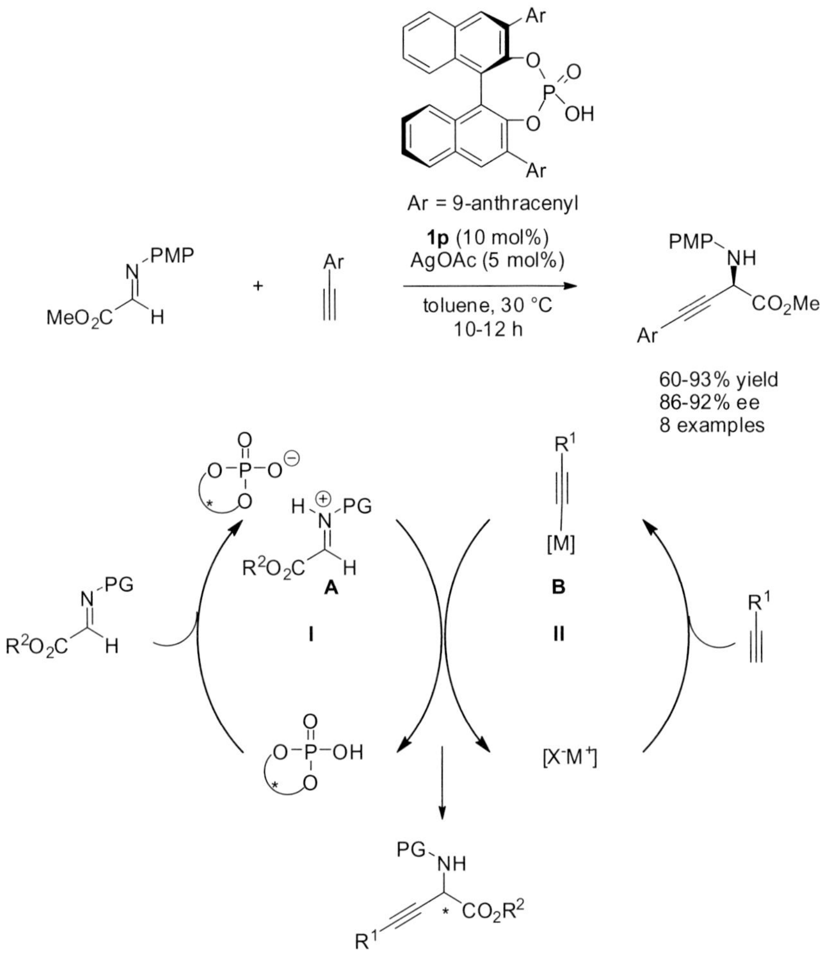

Scheme 5.21

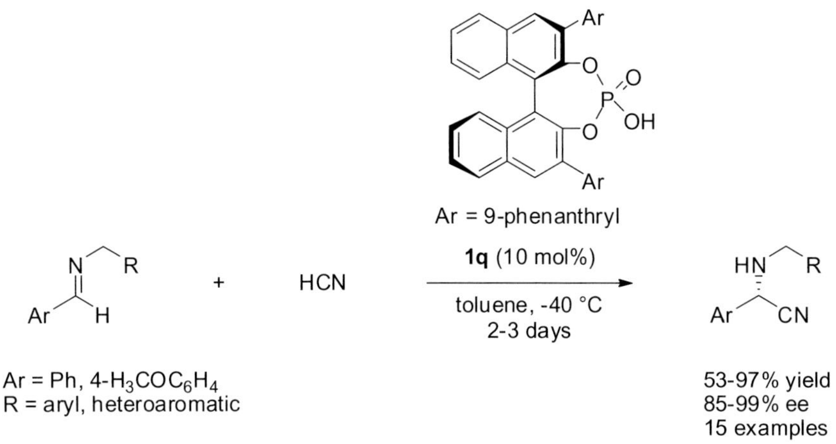

Scheme 5.22

Ar = Ph, 4-H$_3$COC$_6$H$_4$
R = aryl, heteroaromatic

Ar = 9-phenanthryl
1q (10 mol%)
toluene, -40 °C
2-3 days

53-97% yield
85-99% ee
15 examples

In other work, Rueping has described the asymmetric hydrocyanation of a variety of aromatic aldimines mediated by phosphoric acid **1q** (Scheme 5.22) [37].

5.2.1.6 Nonimine Electrophiles

Although phosphoric acids have found broad applicability for a wide range of asymmetric transformations, most reactions are limited to electrophilic activation of imines. Expanding the scope of phosphoric acid catalysis to other classes of electrophiles, Akiyama and Terada subsequently reported activation of nitroalkene [38] and carbonyl [39] electrophiles, respectively (Scheme 5.23).

5.2.1.7 Transfer Hydrogenation

In 2004, List reported that several ammonium salts including dibenzylammonium trifluoroacetate catalyzed the chemoselective 1,4 reduction of α,β-unsaturated aldehydes with Hantszch esters as hydride sources [40]. It is assumed that substrate activation via iminium ion formation results in selective 1,4 addition of hydride. Subsequently, List [41] and MacMillan [42] reported asymmetric versions of this reaction promoted by chiral imidazolidinone salts. In this context, several reports of this metal-free reductive process catalyzed by chiral phosphoric acids have appeared in the recent literature.

The research groups of Rueping [43] and List [44] first reported the phosphoric acid-catalyzed transfer hydrogenation of methyl ketimines with a Hantzsch dihydropyridine as a hydride source. Following these initial studies, Rueping further expanded this protocol to the reduction of a wide range of heterocycles including: benzoxazines, benzothiazines, benzoxazinones [45], pyridines [46], and quinolines (Scheme 5.24) [47]. These methods allow facile access to several families of natural products. Remarkably, catalyst loadings as low as 0.1 mol% can be used without any detrimental effects with respect to yield and ee. As an illustrative example, the cycle of the cascade hydrogenation of quinoline derivatives to generate the

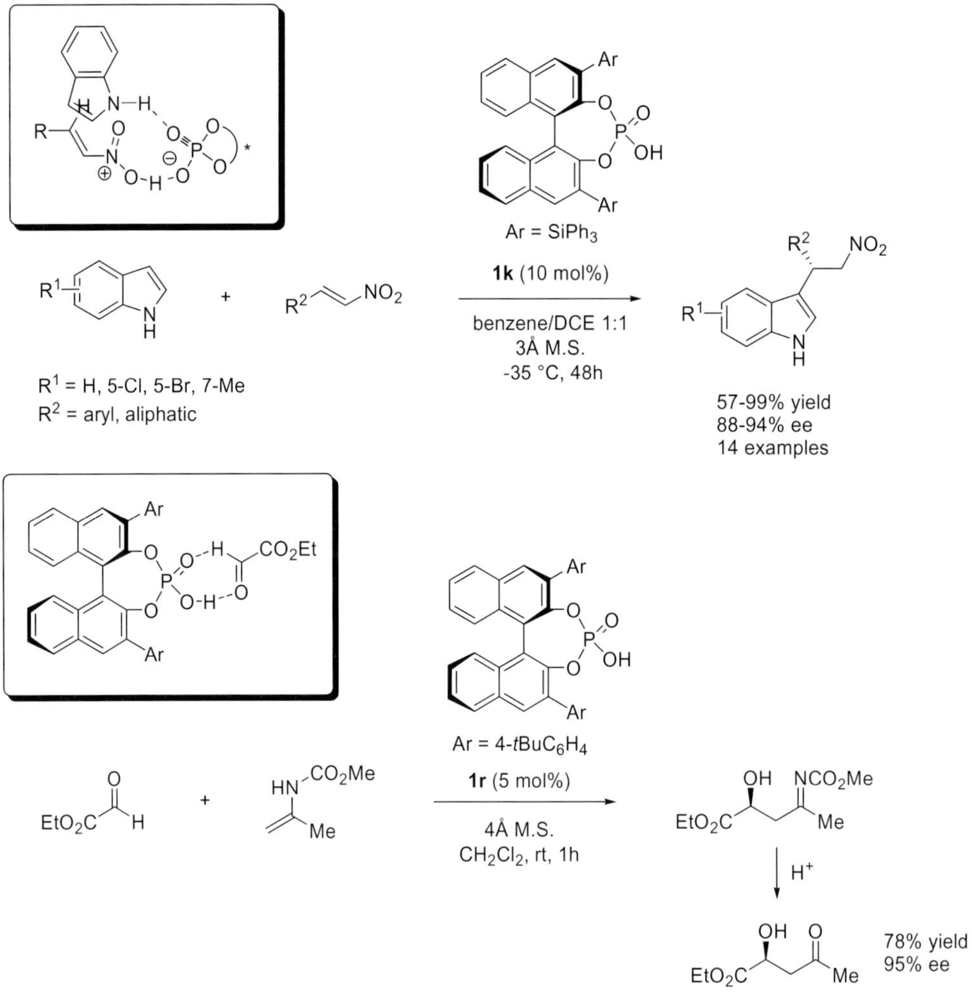

Scheme 5.23

corresponding tetrahydroquinoline is shown in Scheme 5.25. Initial 1,4 hydride addition to chiral ion pair **A** produces enamine (**I**) which upon protonation/isomerization gives chiral iminium ion pair **B**. Subsequent 1,2 hydride transfer by a second equivalent of Hantzsch ester yields the desired saturated product.

The MacMillan laboratory has produced an interesting study on the reductive amination of a broad scope of aromatic and aliphatic methyl ketones catalyzed by ent-**1k**, utilizing Hantzsch ester as a hydride source (Scheme 5.26) [48]. Application of corresponding ethyl ketones gave very low conversions. Computational studies indicated that while catalyst association with methyl ketones exposes the C=N *Si* face to hydride addition, substrates with larger alkyl groups are forced to adopt conformations where both enantiofaces of the iminium π

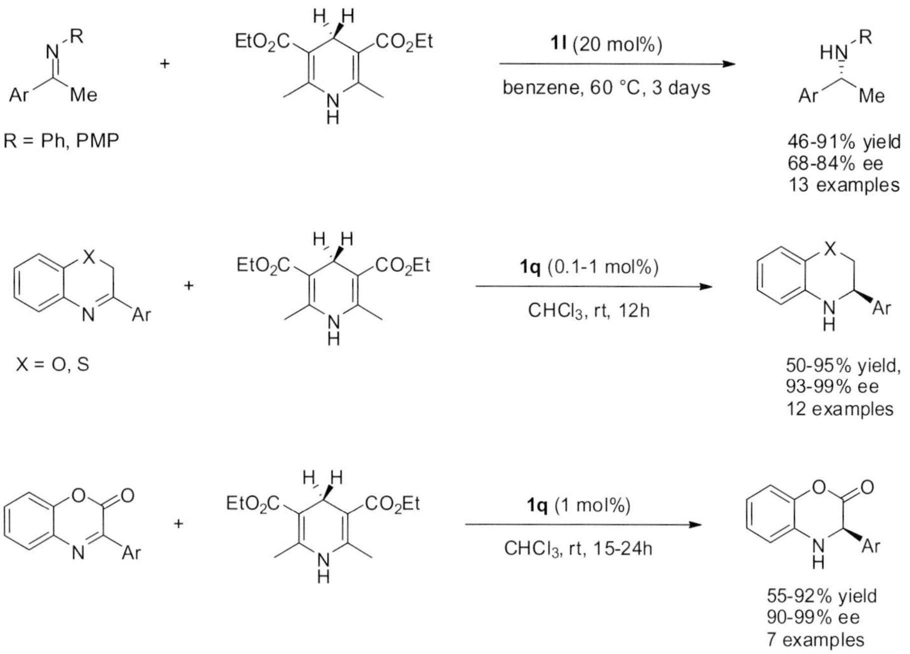

Scheme 5.24

system are shielded. This hypothesis was supported by an experiment involving chemoselective reduction of diketone **9** to yield monoaminated **10** with 18:1 selectivity for coupling at the methyl ketone site. These results also help to explain the high enantiofacial selectivites obtained for methyl alkyl ketone substrates.

List later reported the asymmetric reductive amination of a wide spectrum of aromatic and aliphatic α-branched aldehydes via dynamic kinetic resolution (Scheme 5.27) [49]. The initial imine condensation product is believed to undergo fast racemization in the presence of the acid catalyst **1h** through an imine/enamine tautomerization pathway. Preferential reductive amination of one of the imine enantiomers furnishes the optically pure β-branched amine.

Recently, List has described a cascade reaction promoted by phosphoric acid **1** in combination with stoichiometric amounts of achiral amine, which transforms various 2,6-diketones to the corresponding *cis*-cyclohexylamines (Scheme 5.28) [50]. This three-step process involves initial aldolization via enamine catalysis to give conjugate iminium ion intermediate A. Next, asymmetric conjugate reduction followed by a diastereoselective 1,2 hydride addition completes the catalytic cycle.

5.2.2
Nonbinol-Based Phosphoric Acids

While BINOL-derived chiral phosphoric acids have received great attention, a handful of reports implementing alternative chiral backbones have appeared [51].

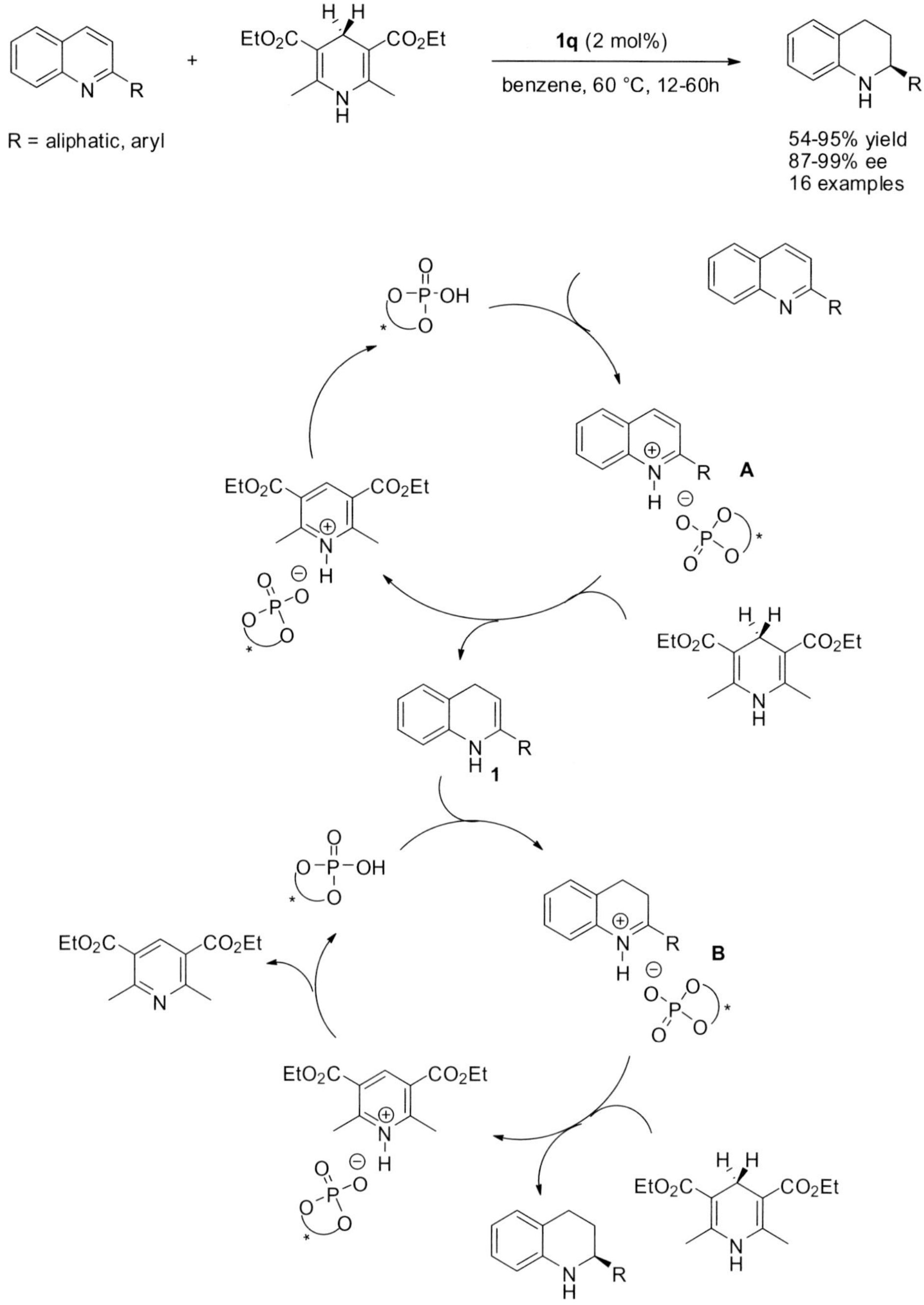

1q (2 mol%)
benzene, 60 °C, 12-60h
R = aliphatic, aryl
54-95% yield
87-99% ee
16 examples
A
1
B

Scheme 5.25

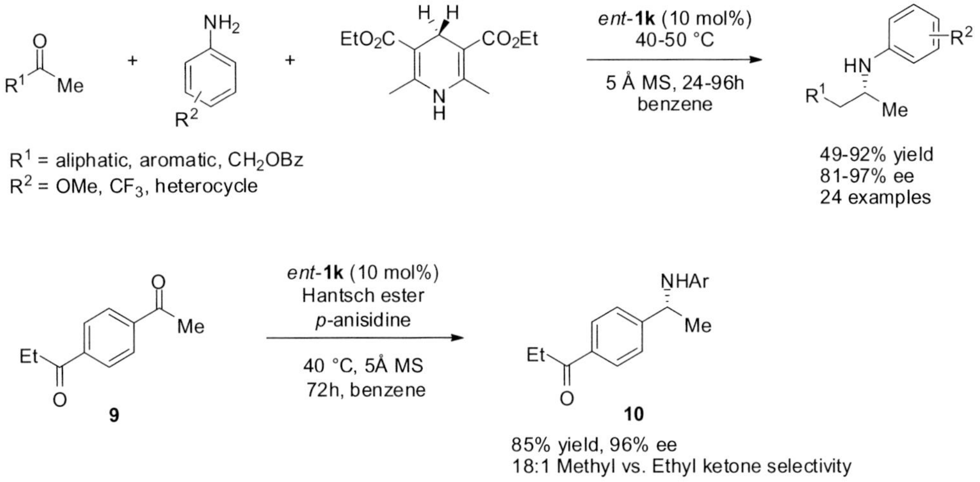

5.2 Brønsted Acid (Phosphoric Acid and Derivatives) | 93

Scheme 5.26

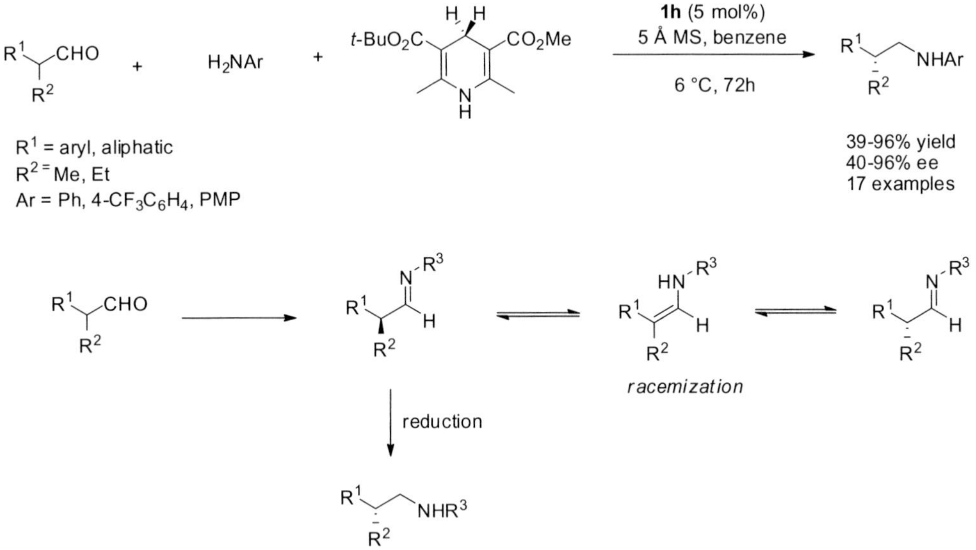

Scheme 5.27

Most notably, the Antilla laboratory has employed VANOL and VAPOL phosphoric acid derivatives in several novel asymmetric transformations. In addition, TADDOL and phosphordiamide phosphoric acid derivatives have been applied in several Mannich-type reactions.

VAPOL-derived phosphoric acid **11** was shown to catalyze the amidation of Boc-protected N-aryl imines with sulfonamide, phthalimide, and maleimide nucleophiles to furnish the corresponding chiral aminals in excellent yields and ee's (Scheme 5.29) [52, 53]. This represents the first general catalytic and asymmetric

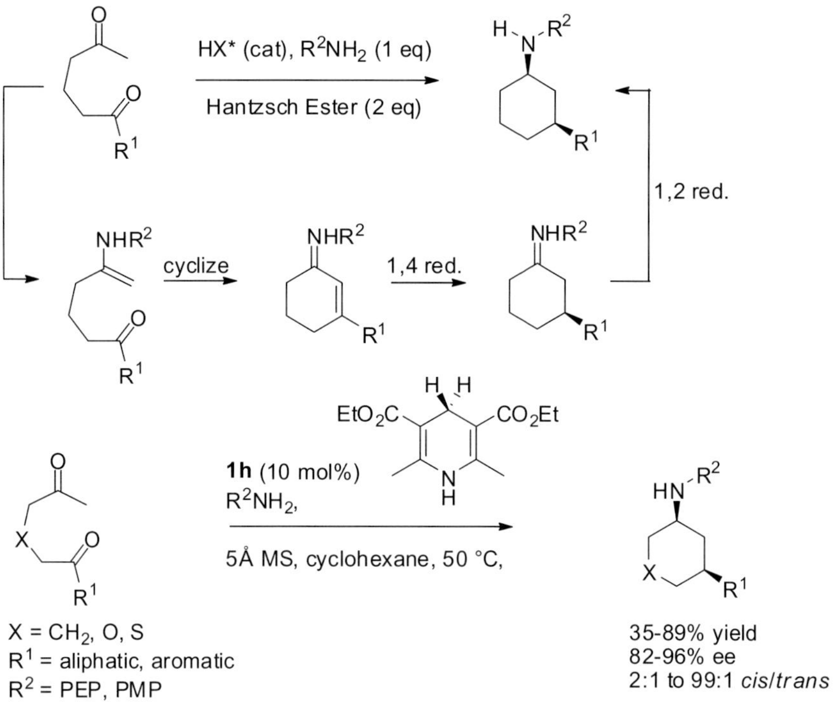
HX* (cat), R2NH2 (1 eq)
Hantzsch Ester (2 eq)
1,2 red.
cyclize
1,4 red.
1h (10 mol%)
R2NH2,
5Å MS, cyclohexane, 50 °C,
X = CH2, O, S
R1 = aliphatic, aromatic
R2 = PEP, PMP
35-89% yield
82-96% ee
2:1 to 99:1 cis/trans

Scheme 5.28

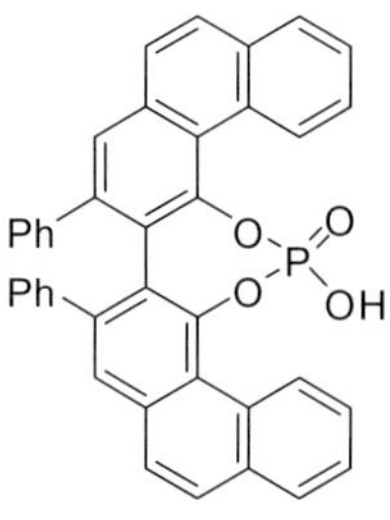

11

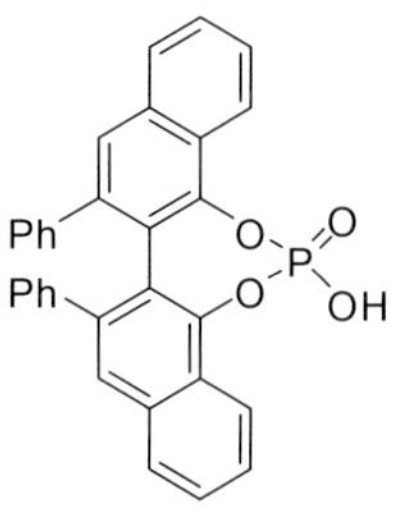

12

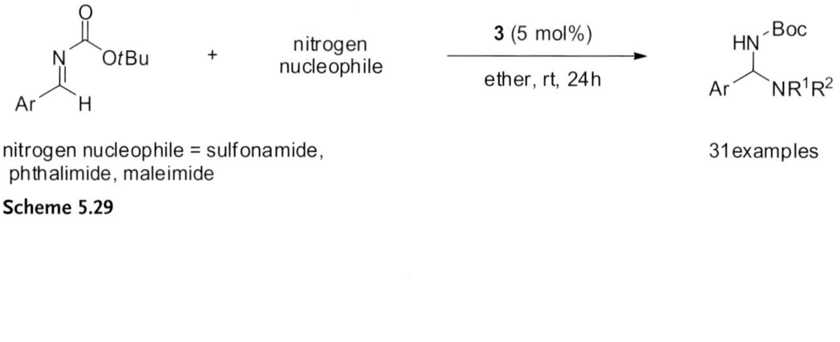
nitrogen nucleophile
3 (5 mol%)
ether, rt, 24h
nitrogen nucleophile = sulfonamide, phthalimide, maleimide
31examples

Scheme 5.29

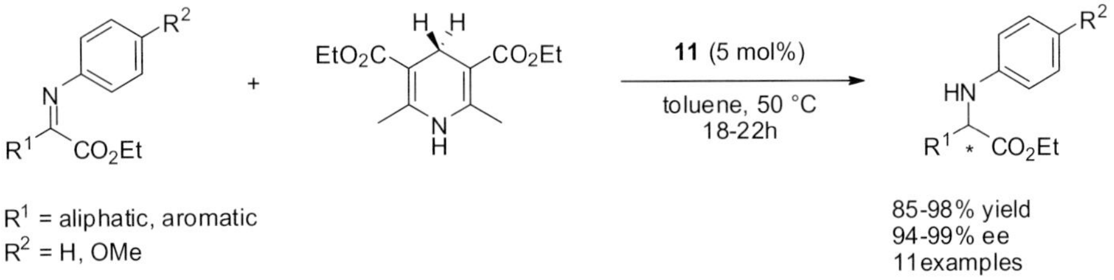

Scheme 5.30

system for the addition of amide nucleophiles to imines. Utilization of several well-known BINOL phosphoric acids provided only moderate levels of enantioselection (60–71% ee versus 94% for VAPOL) [54].

11 was later applied in the highly enantioselective transfer hydrogenation of various acyclic aliphatic and aromatic α-imino esters to provide the corresponding chiral α-amino esters (Scheme 5.30) [55].

Lastly, Antilla has disclosed a novel asymmetric desymmetrization of a wide range of aliphatic, aromatic, and heterocyclic *meso*-aziridines with TMS-N_3 promoted by **11** and related **12** (Scheme 5.31) [56]. Uniquely, this is one of only several reports of electrophilic activation of nonimine substrates by a chiral phosphoric acid. Mechanistic studies suggest that silylation of **11** or **12** by displacement of azide generates the active catalytic species **A**. Consequently, the aziridine is activated through coordination of it carbonyl with chiral silane **A** to produce intermediate **B**. Nucleophilic ring opening by azide furnishes the desymmetrized product and regenerates **11** or **12**.

5.2.3
N-Triflyl Phosphoramide

Prior to Yamamoto's entry into this field, the scope of chiral phosphoric acid catalysis was strictly limited to electrophilic activation of imine substrates. By designing a catalyst with higher acidity it was suspected that activation of less Lewis basic substrates might be possible. To this end, Yamamoto reported incorporation of the strongly electron accepting *N*-triflyl group [57] into a phosphoric acid derivative to yield the highly acidic *N*-triflyl phosphoramide **13** (Scheme 5.32) [58]. This novel Brønsted acid catalyzes the Diels–Alder reaction between ethyl vinyl ketone and various acyclic siloxy dienes to furnish adducts in uniformly high yields and ee's. Further, the corresponding chiral phosphoric acid was unable to catalyze this reaction.

Yamamoto also applied *N*-triflyl phosphoramide **13b** to the 1,3 dipolar addition of various electron-deficient diaryl nitrones and ethyl vinyl ether (Scheme 5.33) [59]. Whereas an AlMe–BINOL catalyst system had earlier been shown to provide adducts in high *exo* selectivities [60], **13b** promoted selective formation of the *endo* product. It is reasoned that the much smaller acidic proton obviates steric repul-

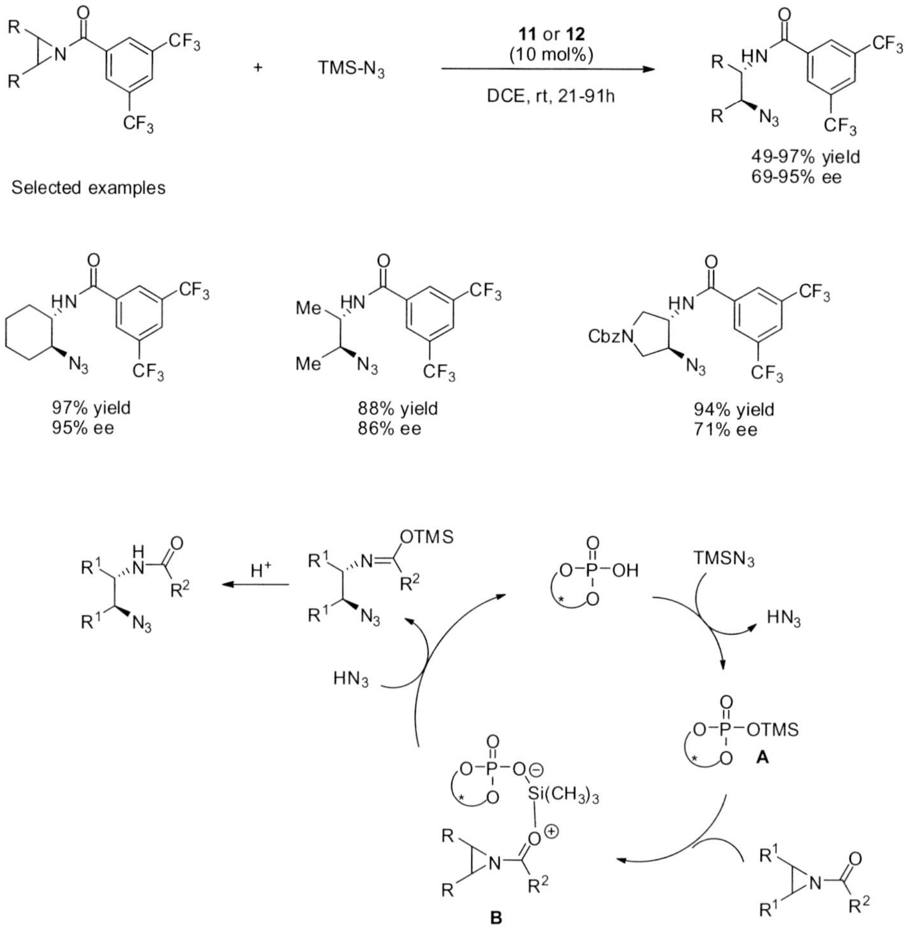

Scheme 5.31

Selected examples

49–97% yield
69–95% ee

97% yield
95% ee

88% yield
86% ee

94% yield
71% ee

sions present in the *endo* TS between a bulky Lewis acid and the alkoxy group of the vinyl ether.

Thereafter, Yamamoto reported the first metal-free Brønsted acid catalyzed asymmetric protonation reactions of silyl enol ethers using chiral Brønsted acid **13c** in the presence of achiral Brønsted acid media (Scheme 5.34) [61]. Importantly, replacement of sulfur and selenium into the *N*-triflyl phosphoramide increases both reactivities and enantioselectivities for the protonation reaction.

Rueping employed *N*-triflyl phosphoramide **13d** in the Nazarov cyclization to afford *cis*-cyclopentenones with moderate diastereselectivities in excellent yields and ee's. This represents the first example of an organocatalytic electrocyclization reaction [62]. Notably, related asymmetric metal-catalyzed Nazarov cyclizations often provide the *trans*-product [63]. Later, Rueping applied *N*-triflyl phosphoramide **13e**

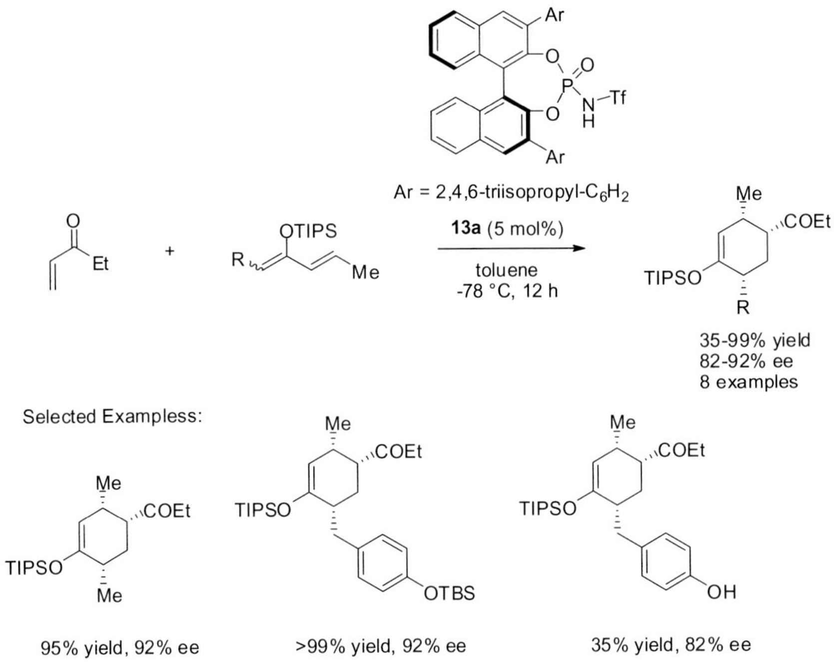

5.2 Brønsted Acid (Phosphoric Acid and Derivatives)

Ar = 2,4,6-triisopropyl-C$_6$H$_2$

13a (5 mol%)

toluene
−78 °C, 12 h

35-99% yield
82-92% ee
8 examples

Selected Exampless:

95% yield, 92% ee

>99% yield, 92% ee

35% yield, 82% ee

Scheme 5.32

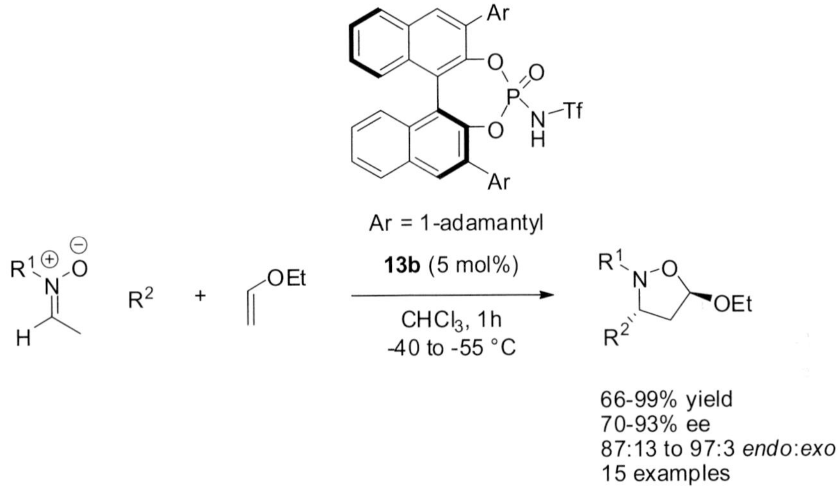

Ar = 1-adamantyl

13b (5 mol%)

CHCl$_3$, 1h
−40 to −55 °C

66-99% yield
70-93% ee
87:13 to 97:3 *endo:exo*
15 examples

Scheme 5.33

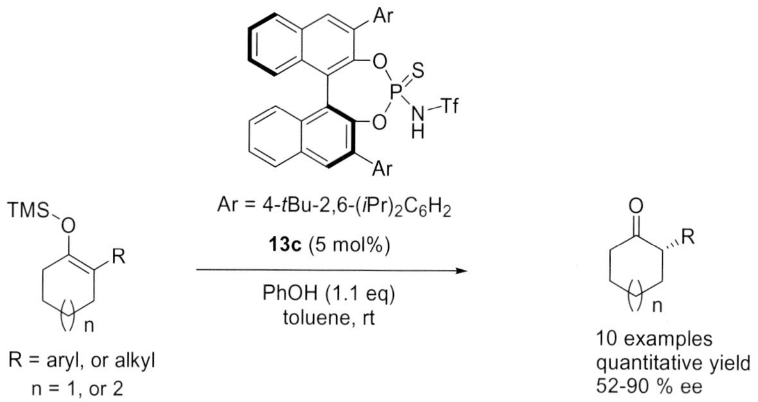

Scheme 5.34

TMS-O-cyclohexenyl-R + catalyst 13c (5 mol%), Ar = 4-*t*Bu-2,6-(*i*Pr)$_2$C$_6$H$_2$, PhOH (1.1 eq), toluene, rt → 2-R-cyclohexanone

R = aryl, or alkyl
n = 1, or 2

10 examples
quantitative yield
52–90 % ee

(Catalyst 13c: BINOL-derived phosphoramide with N–Tf group, Ar substituents)

to the enantioselective carbonyl-ene reaction of α-ketoesters to afford α-hydroxyesters in excellent yields and enantioselecitivities (Scheme 5.35) [64].

5.2.4
Asymmetric Counteranion-Directed Catalysis

In 2006, List introduced the concept of asymmetric counteranion-directed catalysis. Utilizing the morpholine salt of chiral phosphate anion **14** as catalyst, 1,4-transfer hydrogenation of various α,β-unsaturated aldehydes was affected wtih high asymmetric induction (Scheme 5.36) [65]. Thus, in a reaction involving a cationic intermediate, a chiral counteranion could be employed to exert stereochemical control through ion pairing. The utility of this methodology was demonstrated in the selective reduction of citral to provide optically pure cintrolellal, an intermediate in the industrial synthesis of menthol.

List later applied this strategy to the 1,4 reduction of various acyclic and cyclic α,β-unsaturated ketones using the diasteriomeric salt **15** (Scheme 5.37) [66]. Notably, use of the opposite (S) phosphate counterion resulted in a matched/mismatched ion pair combination, forming the same enantiomer of the product but in significantly diminished ee.

Recently, List has reported the asymmetric epoxidation of 1,2-disubstituted and 1,2,2-trisubstituted α,β-unsaturated aldehydes employing chiral dibenzylammonium phosphate **16** (Scheme 5.38) [67]. In previous studies of enal reduction as well as epoxidations involving 1,2-disubstituted enals, the stereogenic center was formed in the initial conjugate addition step. However, in the present case of trisubstiuted enals, the conjugate addition product (**B**) is achiral. It is only in the ensuing cyclization step to give iminium ion **C** that chirality can be established. The authors postulate that asymmetry is induced through catalyst-assisted cyclization of intermediate **B**.

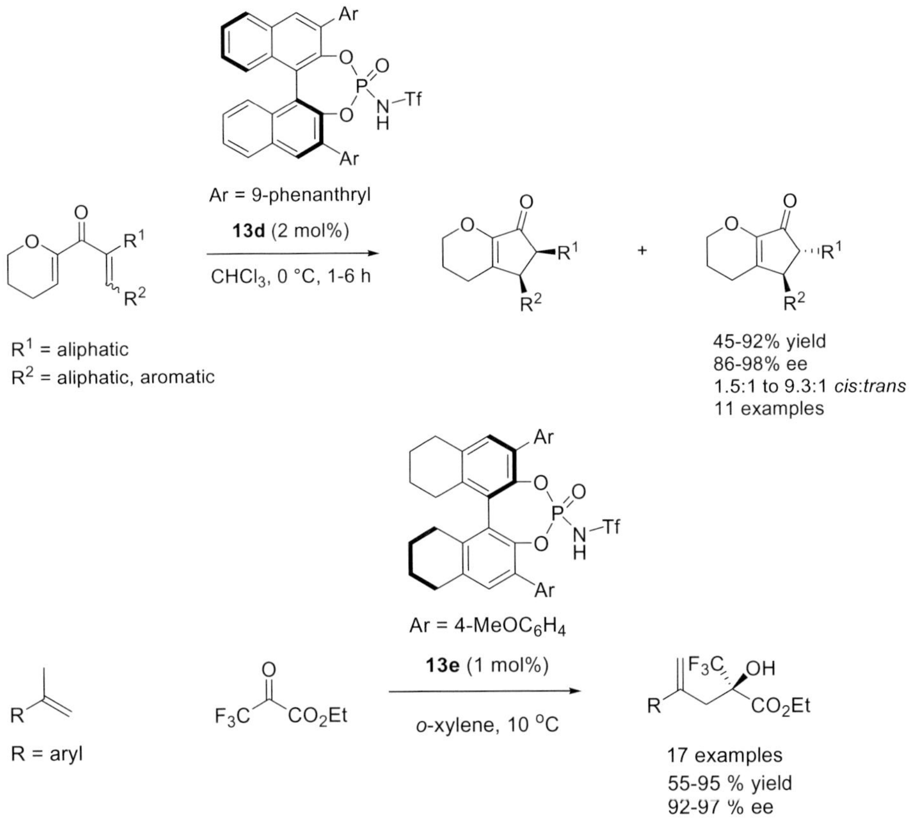

Scheme 5.35

5.3
N–H Hydrogen Bond Catalysts

5.3.1
Guanidine Organic Base

The guanidine functional group is classified as a neutral, organic superbase due to strong resonance stabilization of the conjugate guanidinium ion (pK_a = 13.5) [68]. Compounds containing this moiety, notably 1,1,3,3-tetramethylguanidine (TMG), have been reported to catalyze a wide range of base-mediated organic reactions [69]. In enzymatic systems, the guanidinium group, present in the side chain of arginine, has been shown to function as a recognition site for oxoanionic substrates. Crystallographic data have revealed that substrate binding occurs through parallel double hydrogen bond donation from the guanidinium group to the oxoanion [60]. In light of its strong basicity and double hydrogen bonding

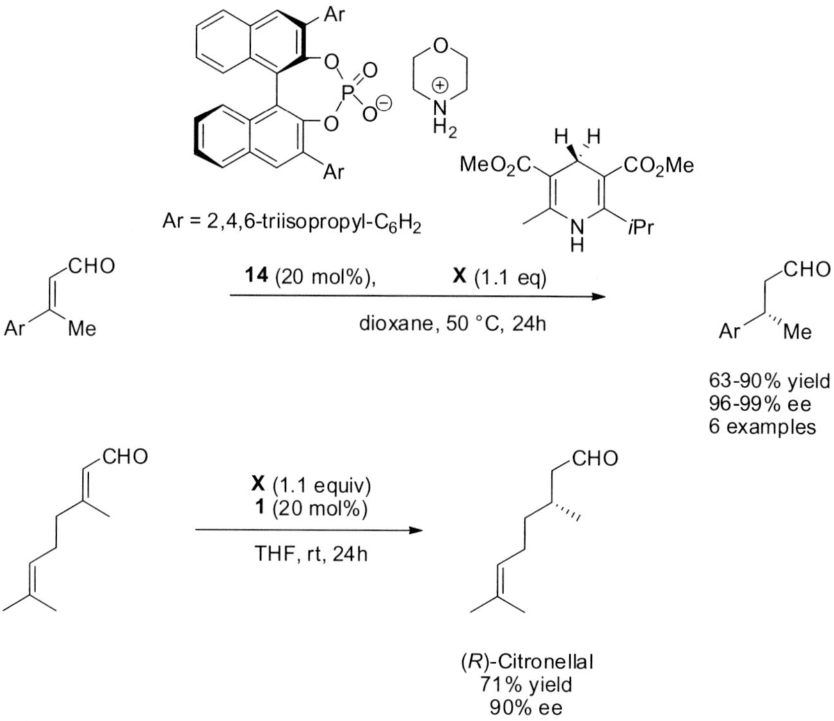

Scheme 5.36

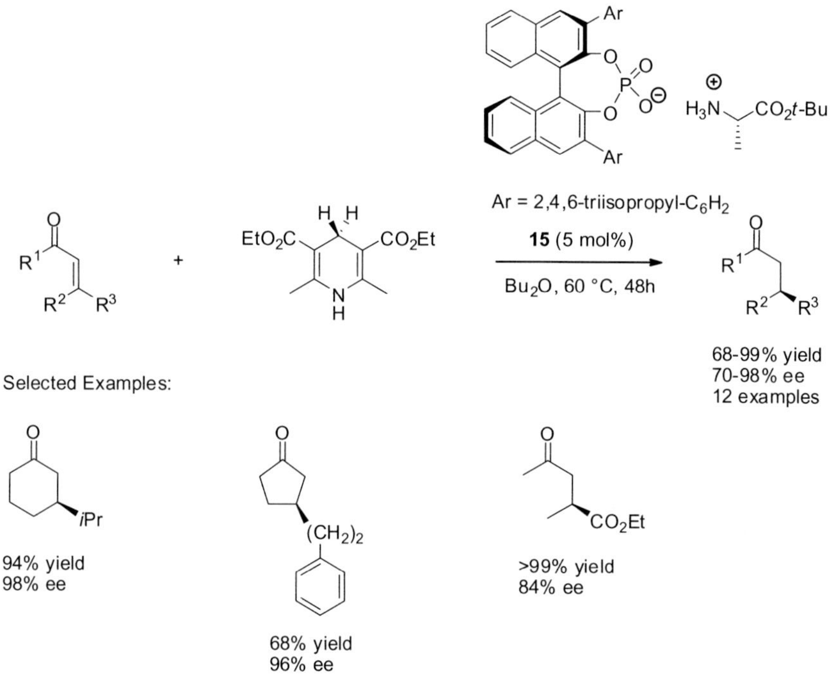

Scheme 5.37

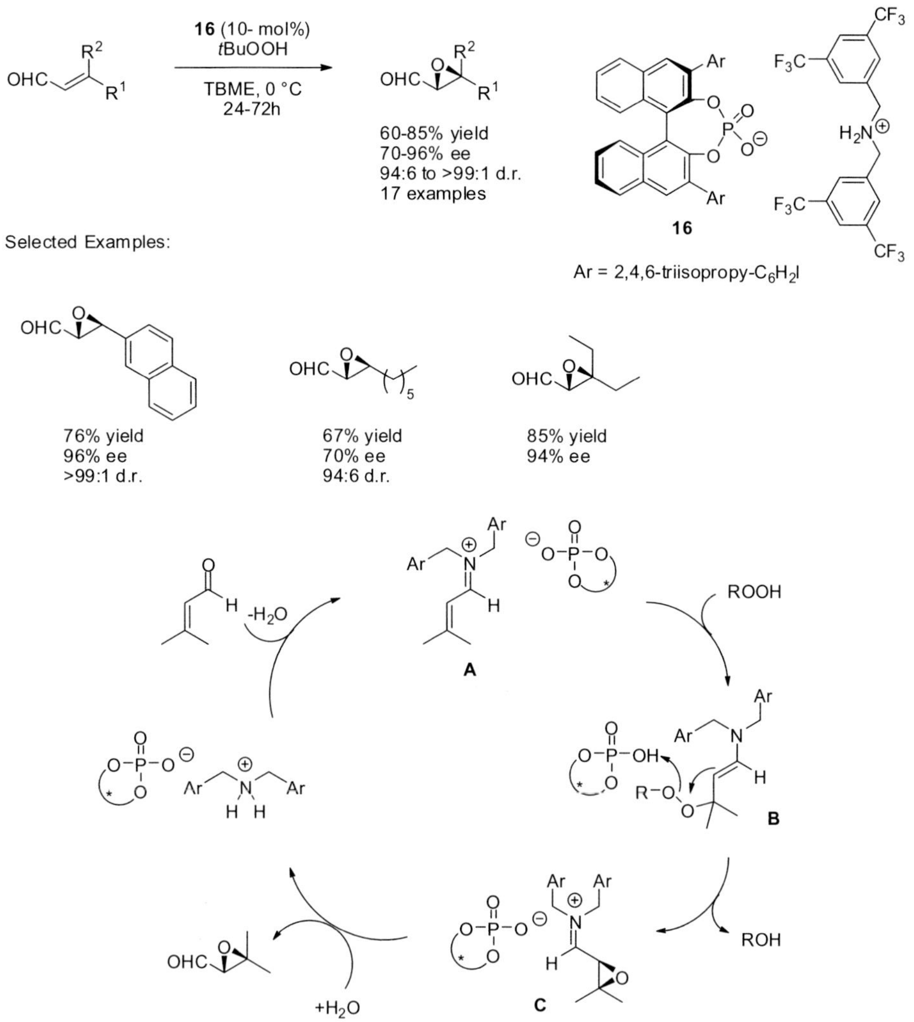

5.3 N–H Hydrogen Bond Catalysts

Scheme 5.38

Conditions: **16** (10 mol%), tBuOOH, TBME, 0 °C, 24–72 h
60–85% yield; 70–96% ee; 94:6 to >99:1 d.r.; 17 examples

Ar = 2,4,6-triisopropy-C$_6$H$_2$I

Selected Examples:
- 76% yield, 96% ee, >99:1 d.r.
- 67% yield, 70% ee, 94:6 d.r.
- 85% yield, 94% ee

capability of the resulting guanidinium ion, the guanidine group has vast potential in asymmetric catalysis. Accordingly, since the initial studies of Lipton in 1996, a small but impressive collection of chiral guanidine base catalyzed reactions have continued to appear up to the present. Whereas chiral phosphoric acids **1** generally serve to first activate the electrophile and then orient the nucleophile through coordination with the phosphoryl oxygen, chiral guanidine base catalysts often function in a similar but reversed manner.

Inoue's cyclic dipeptide **17** had been shown to catalyze the addition of HCN to aldehydes to form optically active cyanohydrins [70]. Lipton found that this same catalyst was unable to catalyze the mechanistically related Strecker reaction

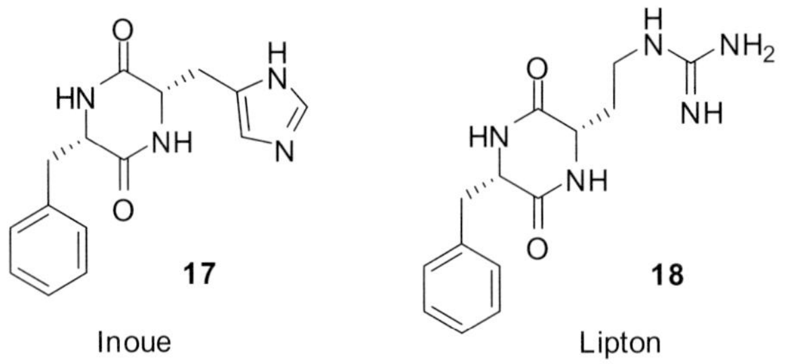

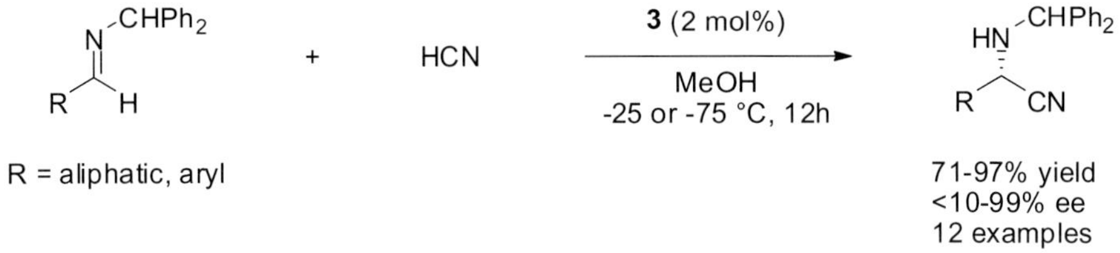

Scheme 5.39

because the low basicity of the imidazole side chain failed to accelerate proton transfer during the addition of HCN to the aldimine. By simply replacing the imidazole moiety with the more basic guanidine group, only 2 mol% of **18** catalyzed the enantioselective addition of HCN to a variety of aromatic and aliphatic N-benzhydryl imines (Scheme 5.39) [71–73]. Further, the N-benzhydryl was easily removed under the same conditions as nitrile hydrolysis.

Several years later, Corey disclosed the C_2 symmetric bicyclic guanidine **19** as an effective bifunctional catalyst for the Strecker reaction (Scheme 5.40) [74]. According to the catalytic cycle, HCN should hydrogen bond to the catalyst to form guanidinium–cyanide complex **A**. A subsequent increase in acidity of the catalyst N—H proton allows donation of a hydrogen bond to the aldimine to form TS assembly **B**. Enantiofacial attack of CN to the bound aldimine gives the Strecker product.

Recently, Tan reported related bicyclic guanidine **20** as a chiral Brønsted base to promote the highly enantioselective Diels–Alder reaction of various anthrones and maleimides (Scheme 5.41) [75]. Interestingly, use of dithranol led to the exclusive formation of the enantio-enriched Michael adducts.

Tan also found that guanidine **21**, acting as a base to activate the σ [3], λ [3] tautomers of diaryl phosphine oxides, catalyzes the asymmetric phospha-Michael reaction of aryl nitroalkenes (Scheme 5.42) [76]. He later employed **21** to realize highly enantioselective Michael additions of dithiomalonate and β-keto thioesters with a range of acceptors, including maleimides, cyclic enones, furanone, and acyclic 1,4-dicarbonylbutenes [77].

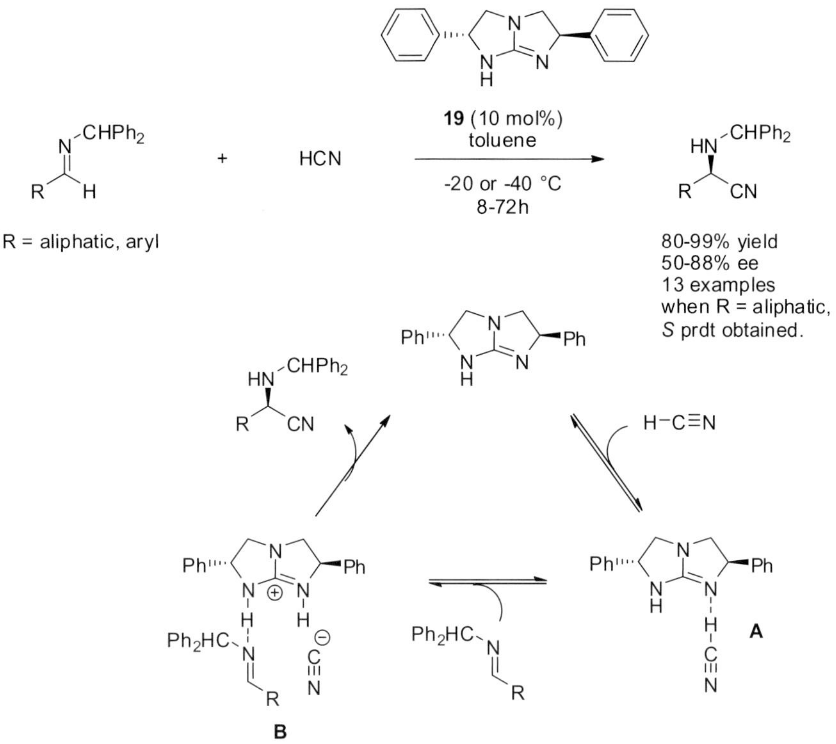

5.3 N–H Hydrogen Bond Catalysts

Scheme 5.40

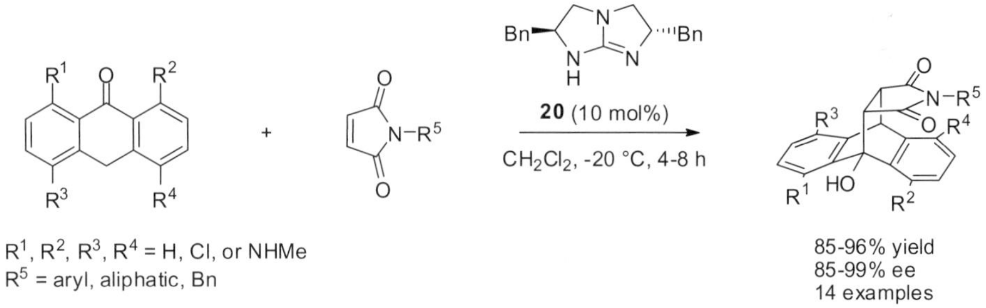

80-99% yield
50-88% ee
13 examples
when R = aliphatic,
S prdt obtained.

Scheme 5.41

R^1, R^2, R^3, R^4 = H, Cl, or NHMe
R^5 = aryl, aliphatic, Bn

85-96% yield
85-99% ee
14 examples

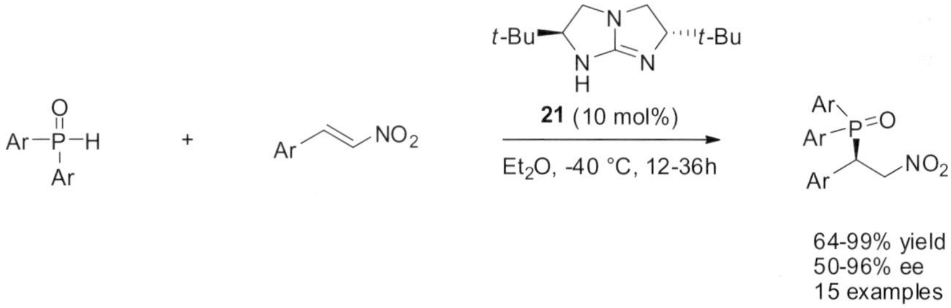

Scheme 5.42

64-99% yield
50-96% ee
15 examples

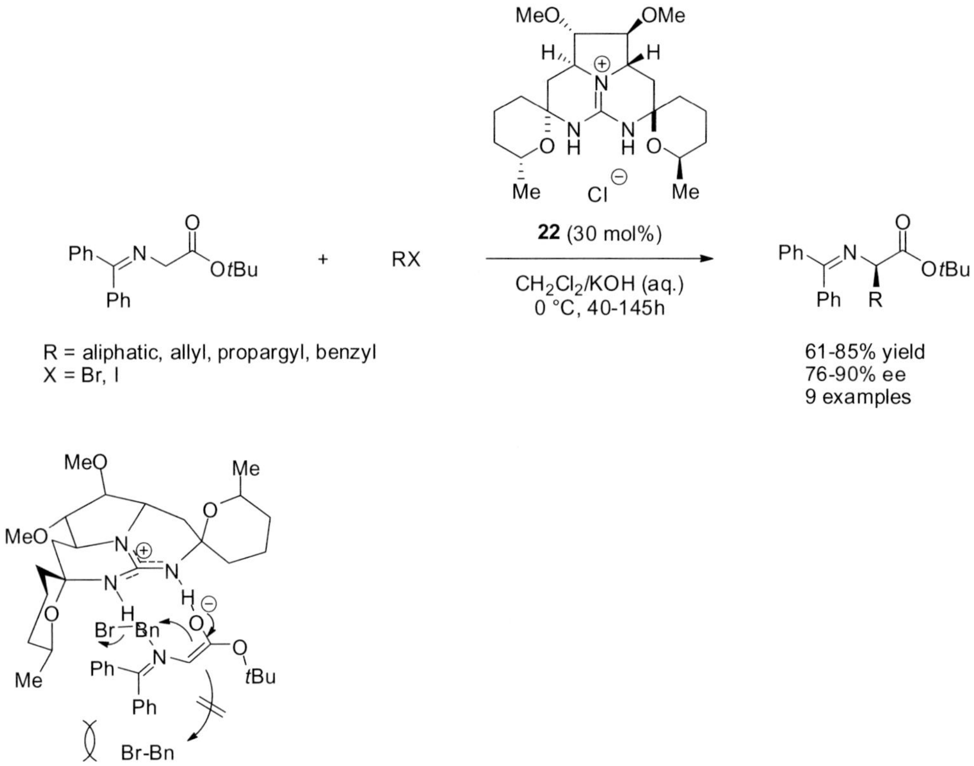

Scheme 5.43

R = aliphatic, allyl, propargyl, benzyl
X = Br, I

61–85% yield
76–90% ee
9 examples

Nagasawa designed a new class of guanidine containing phase transfer catalysts (PTCs) based on the parent skeleton of the marine natural product ptilomycalin A [78, 79]. In the presence of PTC **22** the alkylation of *tert*-butyl glycinate benzophenone Schiff base with a variety of aliphatic, allylic, propargylic, and benzylic halides proceeded to give products in consistently high yields and ee's (Scheme 5.43). The authors postulate that the action of KOH converts the guanidinium salt PTC **22** into its free base form. This free base then forms a complex with the Z enolate of the Schiff base via ionic and hydrogen bonding interactions and subsequently the electrophile adds to the less hindered face.

Terada perceived that an inherent limitation to the development of chiral guanidine catalysis is its highly planar, symmetric structure. He reasoned that introduction of an axially chiral binaphthyl backbone containing bulkyl substituents at the 3,3′ positions to the guanidine skeleton would break the planar symmetry, creating an effective chiral environment for asymmetric transformations. Guanidine base catalyst **23**, containing the bulky 3,5-bis(3,5-di-*tert*-butylphenyl)-phenyl at the 3,3′ positions, was found to induce high levels of asymmetry in the catalytic 1,4 addition of malonic esters, 1,3-diketones, and β-ketoesters to various aromatic- and aliphatic-substituted nitroalkenes (Scheme 5.44) [80].

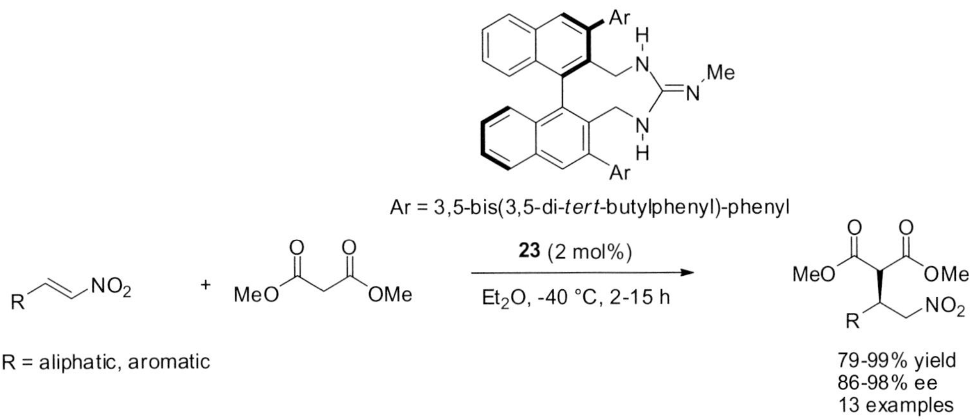

5.3 N–H Hydrogen Bond Catalysts

Ar = 3,5-bis(3,5-di-*tert*-butylphenyl)-phenyl

R⌒NO$_2$ + MeO-C(O)-CH$_2$-C(O)-OMe →[23 (2 mol%)][Et$_2$O, -40 °C, 2-15 h] MeO-C(O)-CH(R-CH$_2$NO$_2$)-C(O)-OMe

R = aliphatic, aromatic

79-99% yield
86-98% ee
13 examples

Scheme 5.44

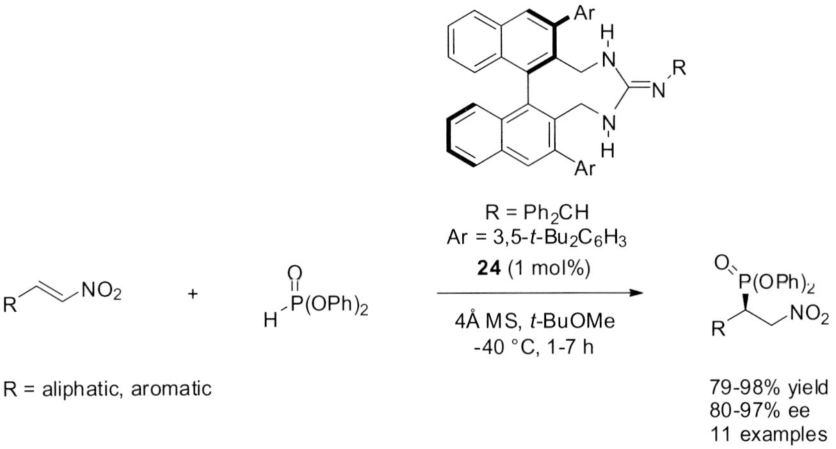

R = Ph$_2$CH
Ar = 3,5-*t*-Bu$_2$C$_6$H$_3$
24 (1 mol%)

R⌒NO$_2$ + H-P(O)(OPh)$_2$ →[4Å MS, *t*-BuOMe][-40 °C, 1-7 h] (PhO)$_2$(O)P-CH(R)-CH$_2$NO$_2$

R = aliphatic, aromatic

79-98% yield
80-97% ee
11 examples

Scheme 5.45

Thereafter, Terada applied related binaphthyl guanidine base catalyst **24** to the asymmetric 1,4 addition of diphenylphosphite to nitroalkenes (Scheme 5.45) [81].

In light of the advances made with cat **24**, enantioselective guanidine base-mediated additions of nonsymmetric 1,3-dicarbonyl compounds pose a formidable challenge. To address this problem, Terada has developed axially chiral guanidine **25** containing a seven-membered ring structure which upon protonation will possess C_2 symmetry [82]. Remarkably, **25** facilitated the addition of various acyclic and cyclic β-ketoesters to di-*tert*-butyl azodicarboxylate with high asymmetric induction (Scheme 5.46). While two nonequal binding modes of the substrate and catalyst are possible, the authors postulate that stereoelectronic differences between unsymmetrical substituents as well as conformational constraints imposed by the bulkyl 3,3′ substituents favor one complexation mode over the other.

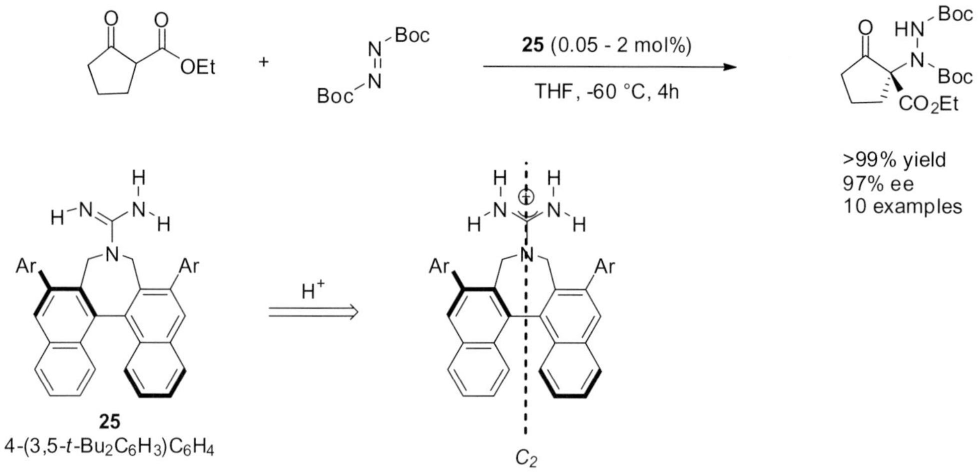

Scheme 5.46

25
4-(3,5-*t*-Bu$_2$C$_6$H$_3$)C$_6$H$_4$

5.3.2
Ammonium Salt Catalysis

The last section illustrated that upon protonation, guanidine base catalysts form the chiral conjugate guandinium ion. In many cases, this form of the catalyst orients the substrates through hydrogen bonding to facilitate formation of products enantioselectively. In contrast, ammonium salt catalysts are typically crystalline salts prepared by protonation of the corresponding amine with a strong acid; however, their mode of catalyst–substrate complexation via hydrogen bonding closely resembles that found with guanidine base catalysts.

Initial studies by Gobel demonstrated that amidinium ions possessing the non-coordinating counteranion tetrakis(3,5-bis(trifluoromethyl)phenyl)borate (TFPB⁻) significantly enhanced the rates of the Diels–Alder reaction [83]. Extending this concept to enantioselective catlaysis, chiral amidinium ion was synthesized and applied in the Diels–Alder reaction of **28** and **29** to give **30**, a key intermediate in the synthesis of (−)-norgestrel, as well as undesired isomer **31** in moderate ee (Scheme 5.47) [84]. The authors invoke a double-hydrogen-bonding mode of dienophilic activation. Subsequently, Gobel synthesized several other chiral amidinium ion catalysts (**27**) and applied them to similar DA reactions [85, 86].

Intrigued by Gobel's findings, Johnston has designed a bisamidine ligand which upon addition of an equimolar amount of triflic acid forms the bench-stable salt **32** (Scheme 5.48) [87]. This chiral amidinium salt efficiently catalyzes the direct aza-Henry reaction of nitroalkanes and electron-deficient N-Boc imines to give the β-aminonitroalkane in high stereo- and enantio-control. It is believed that the potential for bidentate coordination of the proton by **32** produces a well-organized, stereochemically defined catalyst/substrate complex [88].

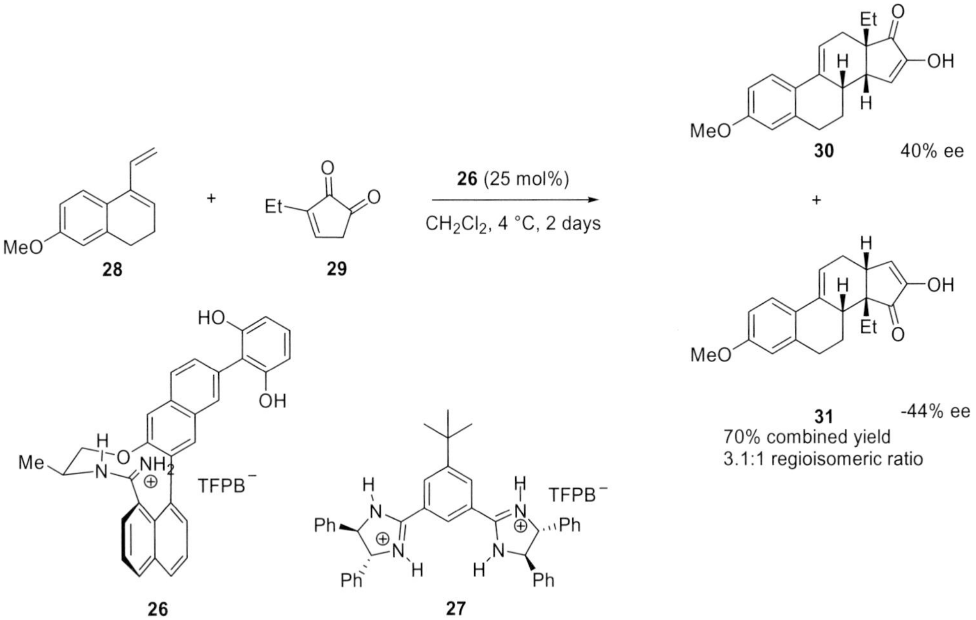

5.3 N–H Hydrogen Bond Catalysts | 107

Scheme 5.47

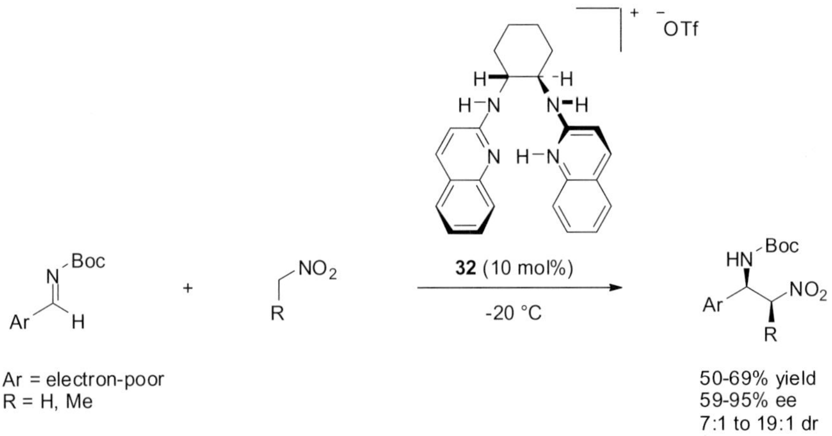

Scheme 5.48

32 was next formed by the addition of nitroacetic acid esters to *N*-Boc imines. However, while favorable product enantioselectivities could be attained, diastereoselectivities suffered due to catalyst-induced epimerization. Catalyst **33** was later developed and upon sodium borohydride/cobalt(II) chloride-mediated reduction was found to furnish the resulting α,β-diamino esters in high levels of enantio- and diastereoselection (Scheme 5.49) [89]. The use of a bifunctional catalyst to

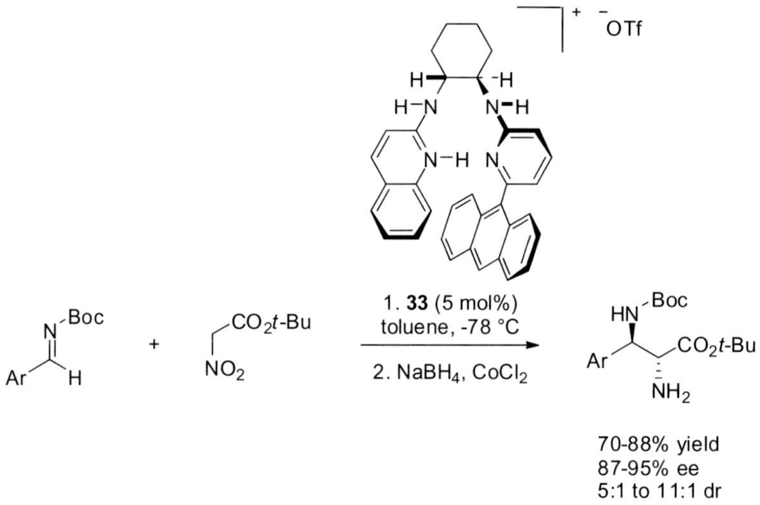

Scheme 5.49

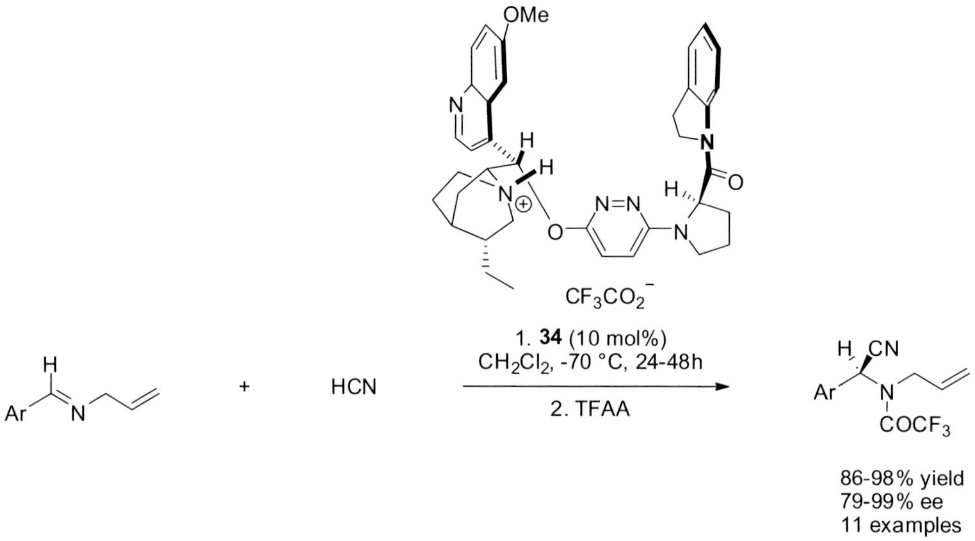

Scheme 5.50

selectively deprotonate substrate over product offers a new direction for more direct syntheses of epimerizable chiral materials.

In previous work, Corey used the free base form of **34** as an effective chiral ligand in the OsO$_4$-promoted dihydroxylation of olefins [90]. He later found that ammonium salt **34** catalyzed the addition of HCN to aromatic N-allyl imines (Scheme 5.50) [91]. The U-shaped pocket of the catalyst is essential in fixing the orientation of the hydrogen-bonded activated aldimine via π–π interactions.

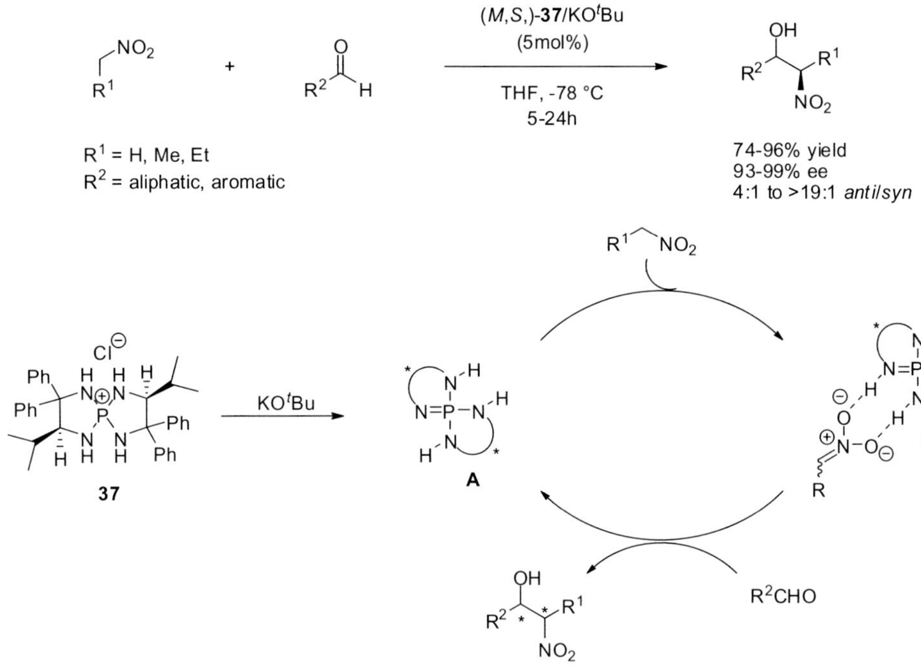

Scheme 5.51

5.3.3
Chiral Tetraaminophosphonium Salt

Ooi has recently reported application of chiral P-spiro tetraaminophosphonium salt **37** as a catalyst for the highly enantio- and diasterioselective direct Henry reaction of a variety of aliphatic and aromatic aldehydes with nitroalkanes (Scheme 5.51) [92]. Addition of the strong base KO^{t}Bu generates *in situ* the corresponding catalytically active triaminoiminophosphorane base **A**. Ensuing formation of a doubly hydrogen-bonded ion pair **B** positions the nitronate for stereoselective addition to the aldehyde. This catalyst system bears many similarities to guanidine base catalysis.

5.4
Combined Acid Catalysis

Implementation of the concept of combined acids in the field of asymmetric catalysis has been known for over 20 years. Several excellent reviews containing historical background and theoretical perspectives on this subject have appeared [93]. Fundamentally, a combined acid system involves the association of an acceptor atom **A** with a donor atom **D** that is chemically bonded to another acceptor atom **A***. Formation of the **A–D-A*** complex will result in a transfer of electron density from the **D-A*** subunit to the **A–D** subunit. In this arrangement, **A*** is a much stronger acceptor relative to its original state (**D-A***). Use of either Lewis or Brønsted acids as acceptors yields four possible manifestations of combined acid systems: Lewis-acid-assisted Lewis acid (LLA), Brønsted-acid-assisted Lewis acid (BLA), Lewis-acid-assisted Brønsted acid (LBA), and Brønsted-acid-assisted Brønsted acid (BBA). While examples of LLA, BLA, and LBA acid systems in asymmetric catalysis have been known for some time, it was not until 2003 that Rawal reported the first example of a chiral BBA acid system. In the sections to follow recent advances in the areas of chiral BBA, LBA, and BLA catalysis will be covered. However, given the focus of this review, LLA systems will be excluded.

5.4.1
Brønsted-Acid-Assisted Brønsted Acid Catalysis

Spectroscopic, crystallographic, and computational studies on the hydrogen bonding networks of various water and carbohydrate systems have led to the concept of σ-cooperativity or nonadditivity. This principle holds that the total hydrogen bond energy of a chain of H-bonds will be greater than the total energies of the individual links [94].

$$E(H\cdots A)_n >_n E(H\cdots A).$$

Essentially, this is a specific example of the general combined acid system **A–D-A*** where **A** and **A*** are now hydrogen atoms. This principle is operative in the addition of HCl to alkenes where kinetic studies by Pocker in 1969 have revealed a second-order dependence on the acid. It is likely that activation of one molecule of HCl through hydrogen bonding with a second molecule results in an increase in acidity, thereby facilitating the rate-determining proton transfer step [95]. Moreover, recently fluorinated alcoholic solvents have been shown to provide significant rate enhancements in oxidation reactions with hydrogen peroxide [96]. Kinetic data suggest that higher order solvent aggregates serve to activate the oxidant through hydrogen bonding [97]. Thus, it may be surmised that incorporation of two proximal HX (X = electronegative atom) groups into an organic catalyst will result in a highly Brønsted acidic single hydrogen bond donor species (BBA).

In order to achieve asymmetric induction through chiral BBA catalysis, the directionality of the acidic proton must be strongly defined. Figure 5.3 lists protic activation of an aldimine (illustrative of a general Lewis basic electrophiles) using three types of chiral acids: a weak acid (type I), a strong acid (type II), and a BBA (also LBA) (type III) [98]. In type I, only moderate levels of enantioselection can be obtained due to rotation of the hydrogen bond about the R*-O axis. In type II, high levels of enantioselectivity may be obtained by proton transfer to form a chiral ion pair. In type III, the R*-O axis is rotationally fixed due to the formation of an additional hydrogen bond. This arrangement produces a conformationally rigid

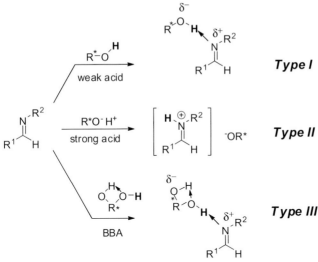

Figure 5.3 General modes of electrophilic activation by a chiral Brønsted acid.

OH···N hydrogen bond, which can lead to high levels of asymmetric induction. Therefore, incorporation of a BBA functionality into the appropriate chiral scaffold has tremendous potential to result in both a highly reactive and enantioselective hydrogen-bonding catalyst.

It should be noted that while formation of a strongly acidic BBA species of type II is conceivable, crystallographic evidence from several BBA/substrate complexes implicates hydrogen bonding activation in agreement with a type III mode of activation.

For organizational reasons, several catalytic systems possessing multiple hydrogen bond donating functionalities have been included in this section which may not be classified as BBAs. For example, the bistriflylamide Mikami catalyst could perhaps be classified as a double-hydrogen bond donor catalyst akin to thiourea catalysis.

5.4.1.1 Diol Activation of Carbonyl Electrophiles

While studying the solvent effect on the kinetics of the hetero-Diels–Alder (HDA) reaction between 1-amino-3-siloxybutadiene and *p*-anisaldehyde, significant rate increases were observed in solvents capable of hydrogen bonding. Further, the measured rate differences bore no correlation with the solvent's dielectric constant. It was hypothesized that hydrogen bonding between a solvent molecule and the aldehyde carbonyl served to lower the carbonyl LUMO, much akin to Lewis acid activation [99]. These findings prompted the Rawal group to extend this new concept to the arena of asymmetric catalysis. Consequently, upon examination of various chiral alcohols, TADDOL (**38a**) was shown to catalyze the HDA reaction of 1-amino-3-siloxybutadiene and various aldehydes to provide, after conversion to the corresponding dihydropyrones, adducts in excellent yields

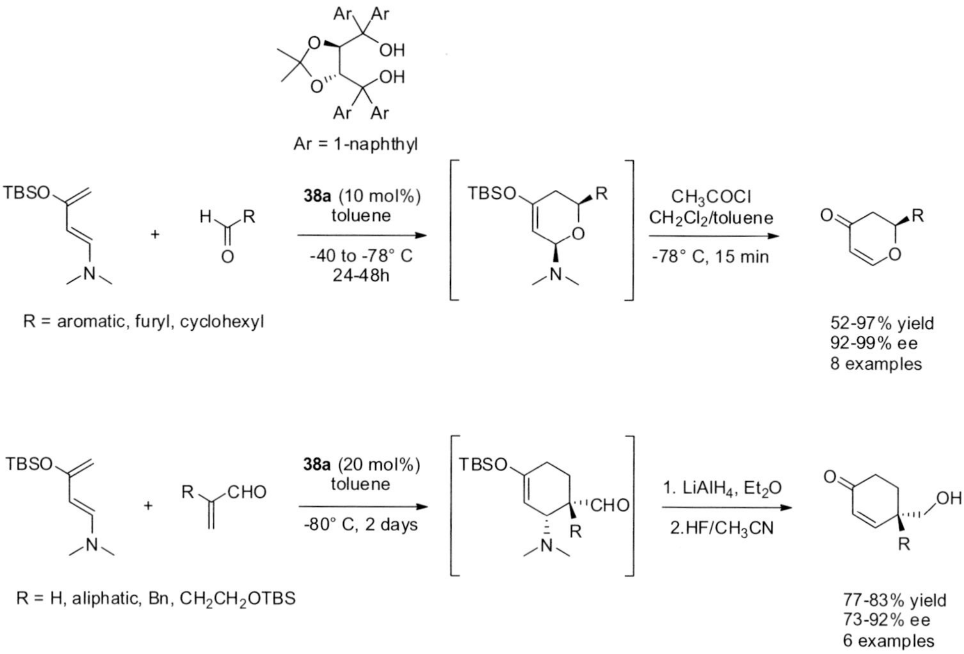

Scheme 5.52

and ee's (Scheme 5.52). Evaluation of the monomethyl and dimethyl derivatives of **38a** gave poor results, illustrating that the hydrogen bonding capability of **38** is essential for catalysis [100, 101]. Soon thereafter, Rawal also reported Diels–Alder reactions of 1-amino-3-siloxybutadiene and various acroleins could also be catalyzed by TADDOL to afford products containing all carbon stereocenters [102].

Collaboration between the Rawal and Yamamoto groups leads to the development of axially chiral diols (BAMOL) **39a** and **39b** sharing in common the bis(diarylhydroxymethyl) functionality with TADDOL, which permits easy modification of the electronic and steric properties of the catalyst [103]. Like TADDOL, this catalyst was also an effective promoter of the HDA reaction of 1-amino-3-siloxybutadiene and a diverse host of aliphatic, alkynyl, and aromatic aldehydes (Scheme 5.53). Importantly, an X-ray structure of a 1 : 1 complex of a BAMOL derivative and benzaldeyde was obtained. The solid-state structure reveals the presence of an *intramolecular* hydrogen bond between the two hydroxyls as well as a single *intermolecular* hydrogen bond between one hydroxyl and the aldehyde carbonyl. As was earlier postulated for TADDOL, these data suggest that carbonyl activation occurs through a single-point hydrogen bond.

Ding extended the scope of TADDOL catalyzed HDA reactions to the use of Brassard's diene with a variety of aromatic aldehydes (Scheme 5.54) [104]. Starting

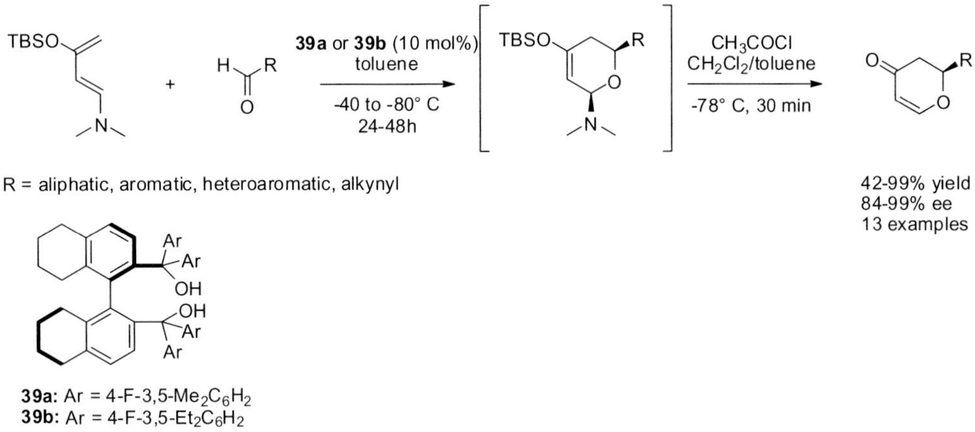
TBSO
N
H
R
O
39a or 39b (10 mol%)
toluene
-40 to -80° C
24-48h
TBSO
R
O
N
CH3COCl
CH2Cl2/toluene
-78° C, 30 min
O
R
O
R = aliphatic, aromatic, heteroaromatic, alkynyl
42-99% yield
84-99% ee
13 examples
Ar
Ar
OH
OH
Ar
Ar
39a: Ar = 4-F-3,5-Me2C6H2
39b: Ar = 4-F-3,5-Et2C6H2

Scheme 5.53

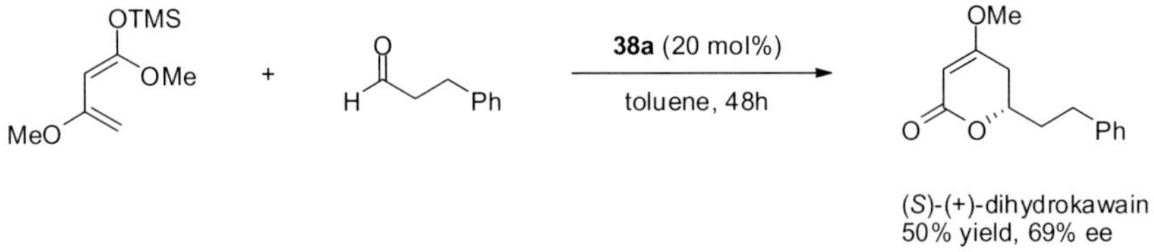
OTMS
OMe
MeO
H
Ph
38a (20 mol%)
toluene, 48h
OMe
O
O
Ph
(S)-(+)-dihydrokawain
50% yield, 69% ee

Scheme 5.54

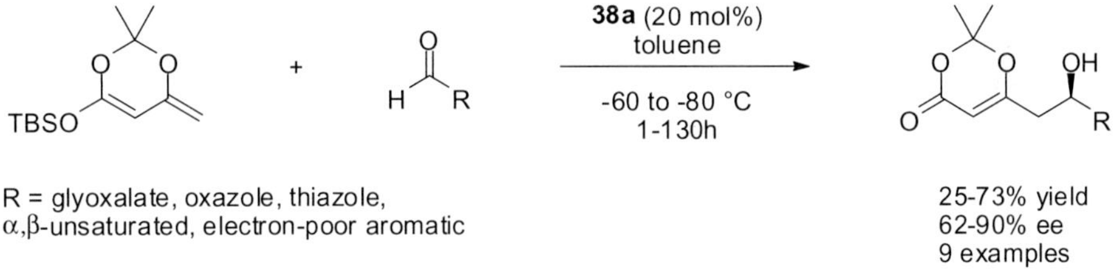
O
O
TBSO
O
H
R
38a (20 mol%)
toluene
-60 to -80 °C
1-130h
O
O
OH
O
R
R = glyoxalate, oxazole, thiazole,
α,β-unsaturated, electron-poor aromatic
25-73% yield
62-90% ee
9 examples

Scheme 5.55

from 3-phenylpropionaldeyde, the natural product (*S*)-(+)-dihydrokawain could be obtained in 69% ee in a single step.

The Rawal group next applied diol catalysis to the enantioselective vinylogous Mukaiyama aldol (VMA) reaction of electron-deficient aldehydes [105]. Screening of various known chiral diol derivatives, including VANOL, VAPOL, BINOL, BAMOL, and TADDOL, revealed that **38a** was the only catalyst capable of providing products in acceptable levels of enantioselection (Scheme 5.55). Subsequent to this work, Scettri reported a similar study of TADDOL-promoted VMA reactions with Chan's diene [106].

The success of their initial VMA studies led the Rawal group to further probe hydrogen bonding catalysis of the Mukaiyama aldol reaction between the highly

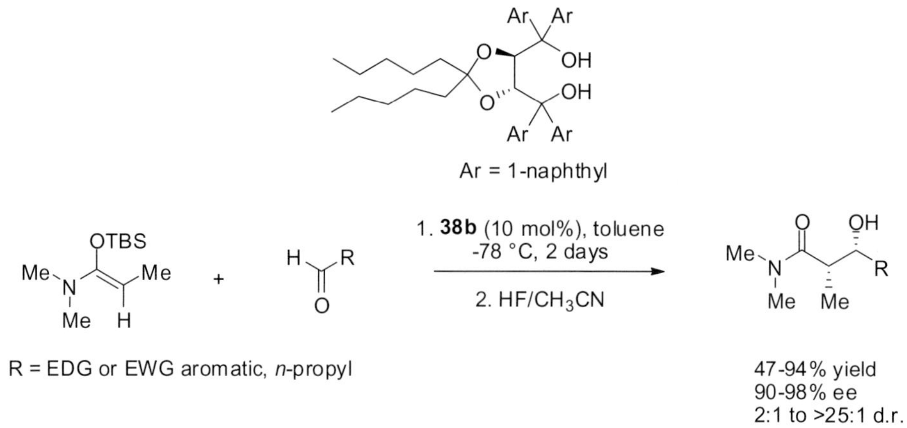

Scheme 5.56

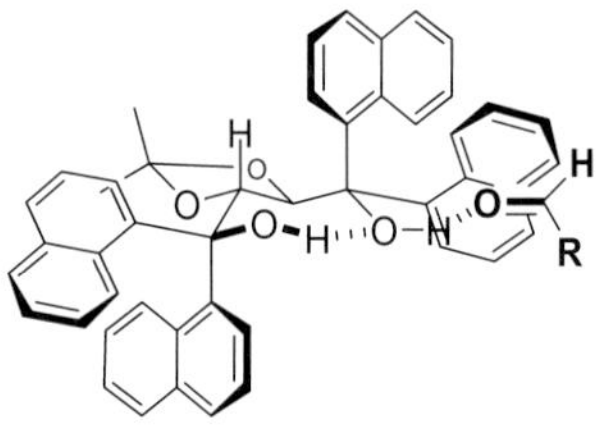

Ar = 1-naphthyl

R = EDG or EWG aromatic, *n*-propyl

1. **38b** (10 mol%), toluene
−78 °C, 2 days
2. HF/CH$_3$CN

47–94% yield
90–98% ee
2:1 to >25:1 d.r.

Figure 5.4 Representation of the X-ray structure of a 1:1 complex between TADDOL and *p*-anisaldehyde.

nucelophilic *O*-silyl-*N*,*O*-ketene acetals and various aldehydes. With slightly modified TADDOL catalyst **38b**, aldol products were obtained in excellent yields and enantioselectivies with high *syn* selectivities (Scheme 5.56) [107]. Application of Schwartz's reagents converted the amide aldol products into the corresponding chiral aldehydes with little or no epimerization of the α-stereocenter. Additionally, the first crystal structure of a 1:1 complex of a TADDOL derivative and an aldehyde was obtained. Similar to the BAMOL complex, an intramolecular hydrogen bond between TADDOL's two hydroxyl groups as well as an intermolecular single point hydrogen bond between the remaining hydroxyl group and the aldehyde carbonyl was observed (Figure 5.4). The function of the intramolecular hydrogen bond is twofold. A natural consequence of this type of hydrogen bonding network is an increase in the acidity of the free hydroxyl group. Consequently, this proton is able to activate the aldehyde carbonyl through stronger hydrogen bonding. Secondly, a well-defined chiral environment is created through rotational constraint of the acidic hydroxyl induced by the intramolecular hydrogen bond. Rawal postulated that stabilization of the TADDOL:aldehyde complex is additionally realized through a π–π* donor–acceptor interaction between the electron-rich proximal equatorial 1-naphthyl ring and the electron-deficient aldehyde carbonyl, which would also selectively shield one face of the dienophile. In contrast, recent computational studies of Houk suggest a stabilizing CH–π interaction between the aldehyde CH and pseudoequatorial naphthalene ring to account for the observed levels of enantioselection [108].

As early as 2000, Ikegami showed that Baylis–Hillman reactions could be promoted by cooperative catalysis with tributylphosphine and *rac*-BINOL in high yield [109]. ^{1}H NMR studies suggested that the observed rate acceleration might begin to be understood in terms of BINOL's ability to act as a Brønsted acid. Inspired by these initial findings, the Schaus group developed the highly enantioselective Baylis–Hillman reaction of cyclohexenone and a diverse set of aldehydes catalyzed by saturated BINOL derivatives **40a** and **40b** (Scheme 5.57) [110]. Monomethylation of one of the hydroxyl groups of a related derivative of **40** resulted in an erosion of the yield and enantioselectivity of the product, thereby implicating both hydroxyl groups in catalytic function.

In related work, Sasai developed several bifunctional BINOL-derived catalysts for the aza-Morita–Baylis–Hillman (aza-MBH) reaction [111]. In early studies, careful optimization of the catalyst structure regarding the location of the Lewis base unit revealed **41** as an optimal catalyst for the aza-MBH reaction between acyclic α,β-unsaturated ketones and N-tosyl imines. Systematic protection or modification of each basic and acidic moiety of **41** revealed that all four heterofunctionalities were necessary to maintain both chemical and optical yields. As seen in Scheme 5.58, MO calculations suggest that one hydroxyl groups forms a

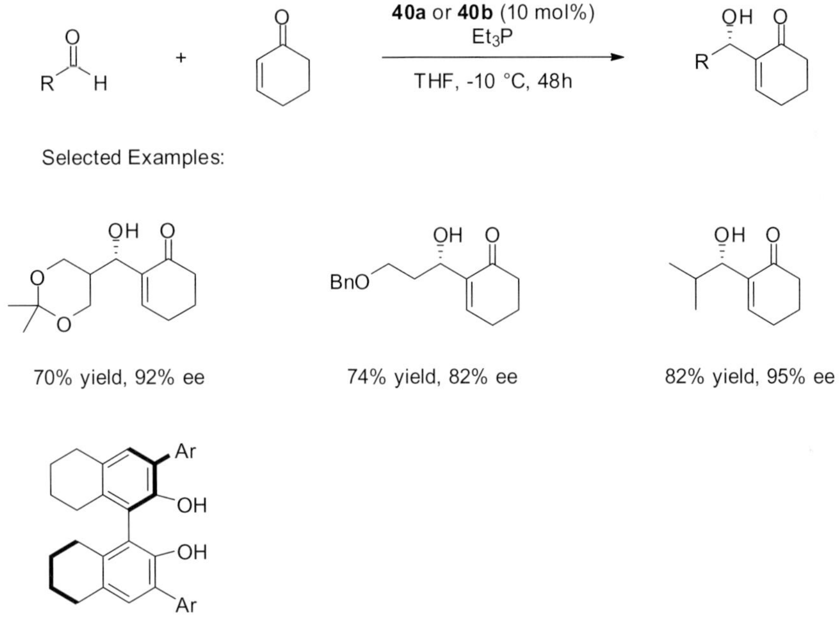

40a: Ar = 3,5-MeC$_6$H$_3$
40b: 3,5-CF$_3$C$_6$H$_3$

Scheme 5.57

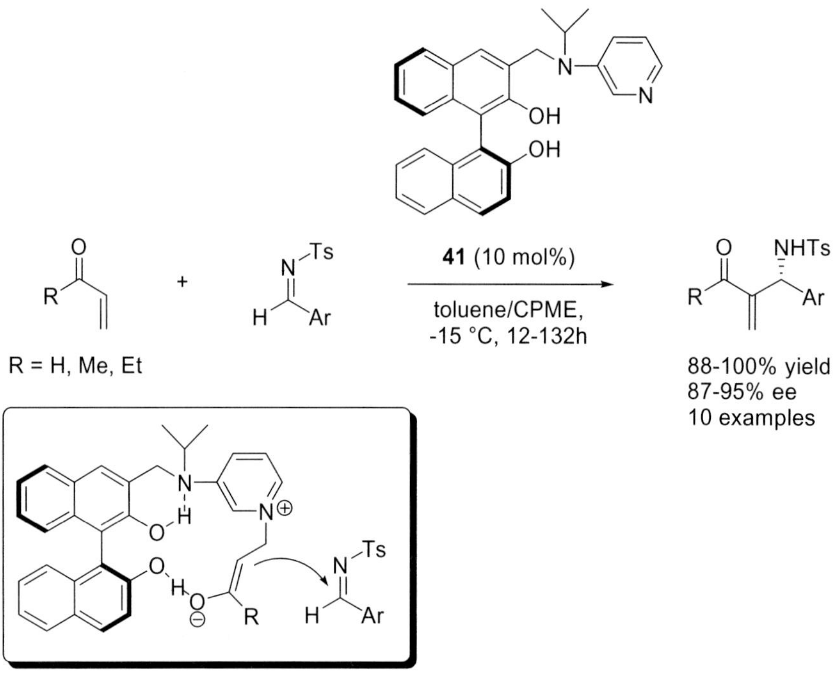

Scheme 5.58

hydrogen bond with the amino-nitrogen to fix the conformation of **41**. The remaining pyridinyl nitrogen and hydroxy group serve to activate the α,β-unsaturated ketone as a nucleophile.

In a subsequent publication, Sasai reported bifunctional BINOL-derived catalyst **42** containing a Lewis basic phosphine unit [112]. Application of this catalyst in the aza-MBH to similar substrates as found in the preceding example furnished products in uniformly high yields and ee's (Scheme 5.59). Further, monomethyl analogs of **42** resulted in diminished catalytic activity. The authors postulate that the catalyst activates the nucleophile through P-Michael addition and concomitant stabilization of the enolate via double hydrogen bonding (as the authors suggest) or possibly an intramolecular hydrogen bonding network as shown in Scheme 5.59.

5.4.1.2 Diol Activation of Other Electrophiles

Nitroso-Aldol Reaction In the course of the Yamamoto group's studies on the nitroso-aldol (NA) reaction of enamines substantial rate increases were observed upon addition of stoichiometric amounts of achiral Brønsted acid. Furthermore, exclusive regioselective formation of the N- versus O-adducts could be controlled by the choice of MeOH or AcOH, respectively, as Brønsted acids (Scheme 5.60) [113]. Subsequently, enantioselective versions of both N- and O-nitroso aldol

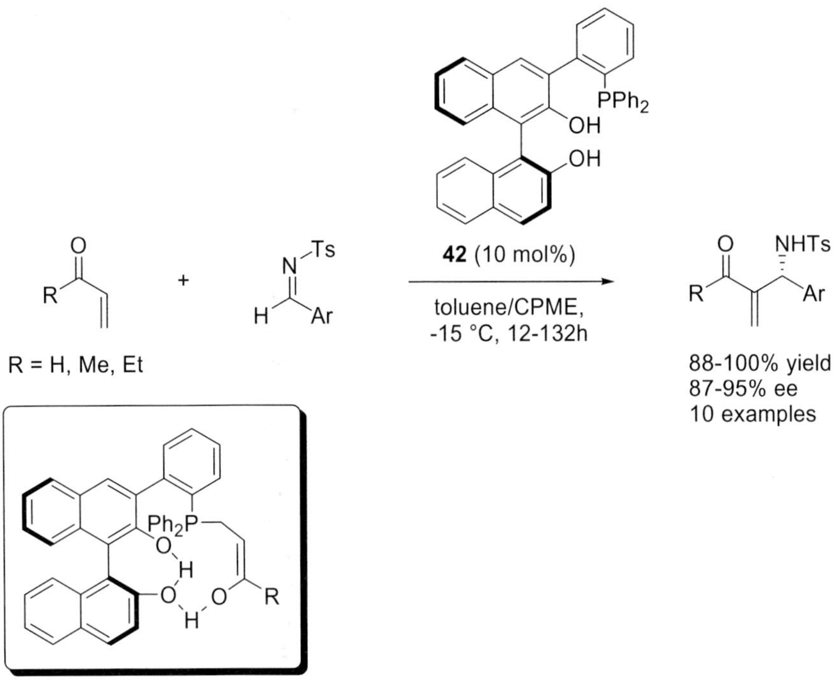

Scheme 5.59

syntheses were developed. Regarding the *N*-NA reaction, screening of various chiral alcohols revealed that TADDOL **38a** provided *N*-NA adducts derived from several piperidine cyclohexene enamines and nitrosobenzene in high yield and ee with complete regioselective control. Further, chiral glycolic acid **43** was found to catalyze exclusive formation of the *O*-NA adduct in excellent yields and ee's [103]. To rationalize the factors governing the regiochemical outcome of these systems, preliminary computational and isotope effect studies have been undertaken [114]. These data indicate that Lewis basic sites of both alcohol and carboxylic acid catalysts are able to coordinate to the *o*-hydrogen of nitrosobenzene. In the case of the *N*-NA reaction, two molecules of alcohol (or one diol) participate to form a nine-membered ring hydrogen bond network between the *o*-hydrogen and the *anti* lone pair of the nitroso group. As for the *O*-NA reaction, the carboxylic acid forms an eight-membered chelate through Lewis basic coordination with the *o*-hydrogen and hydrogen bond activation of the nitrogen.

Soon thereafter, the Yamamoto group reported an extension of this work to the highly diastereo- and enantioselective synthesis of nitroso Diels–Alder-type bicycloketones using dienamines in the presence of the BINOL derivative **44** (Scheme 5.61) [115]. This reaction was thought to proceed through a sequential *N*-NA/hetero-Michael reaction mechanism. Support for this mechanism was provided from an experiment employing bulkyl 4,4-diphenyl dienamine where the *N*-NA

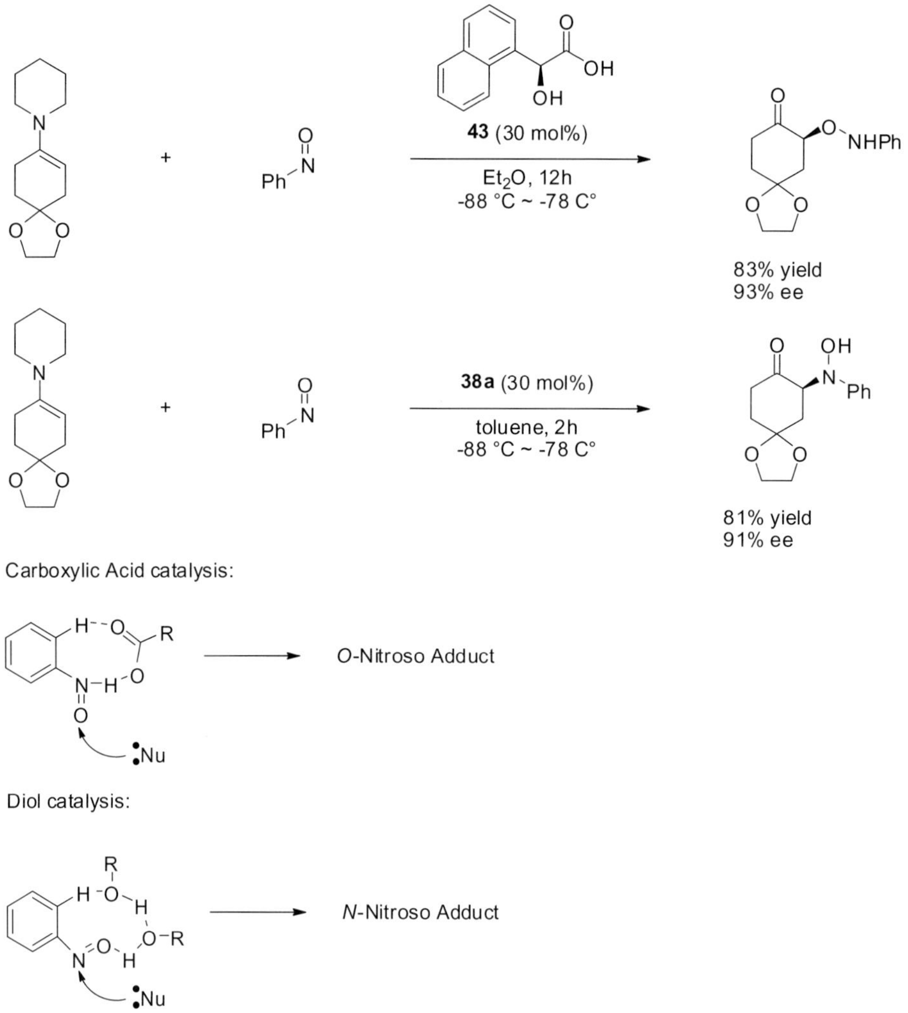

Scheme 5.60

Carboxylic Acid catalysis: → O-Nitroso Adduct

Diol catalysis: → N-Nitroso Adduct

adduct was obtained as the sole product in 27% yield and 61% ee. Presumably, subsequent hetero-Michael addition was suppressed for steric reasons.

Dixon reported that saturated BINOL **45** sufficiently activates various N-Boc aryl imines toward Mannich reaction with acetophenone-derived enamines to yield β-amino aryl ketones in good yields and enantioselectivities (Scheme 5.62) [116]. The same group applied a BINOL-derived tetraol catalyst to the addition of methyleneaminopyrrolidine to N-Boc aryl imines. Interestingly, appendage of two extra diarymethanol groups to the BINOL scaffold resulted in a marked increase in enantiomeric excess [117].

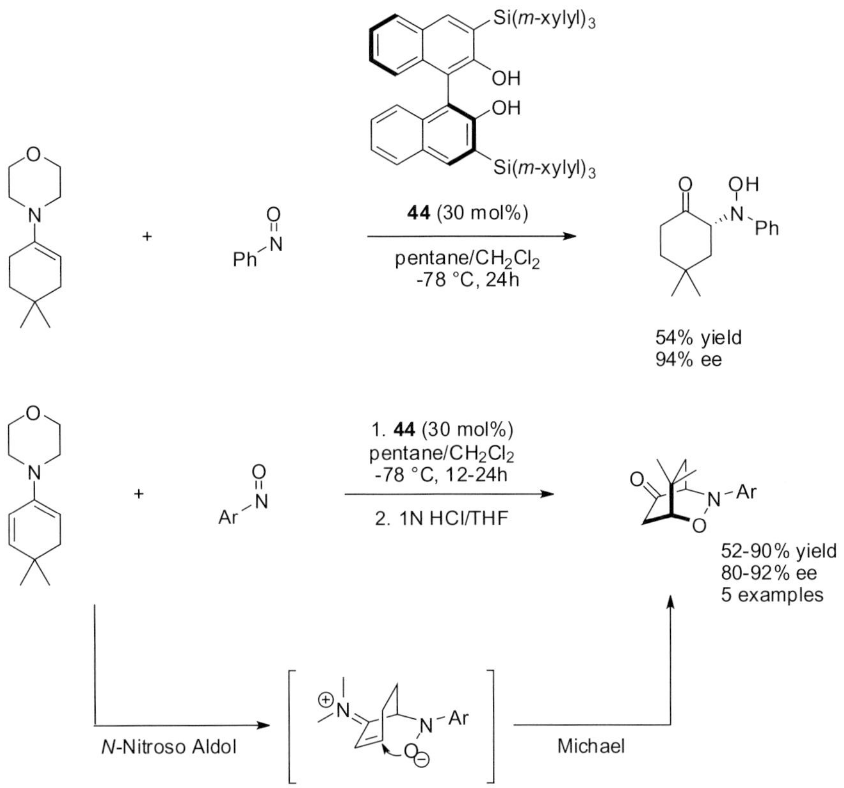

5.4 Combined Acid Catalysis

Scheme 5.61

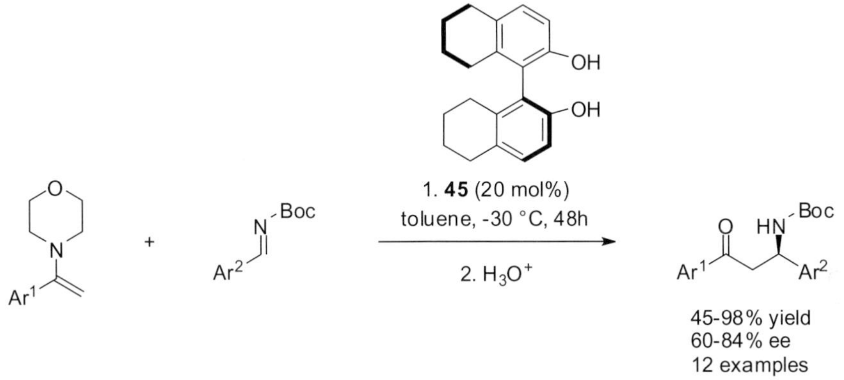

Scheme 5.62

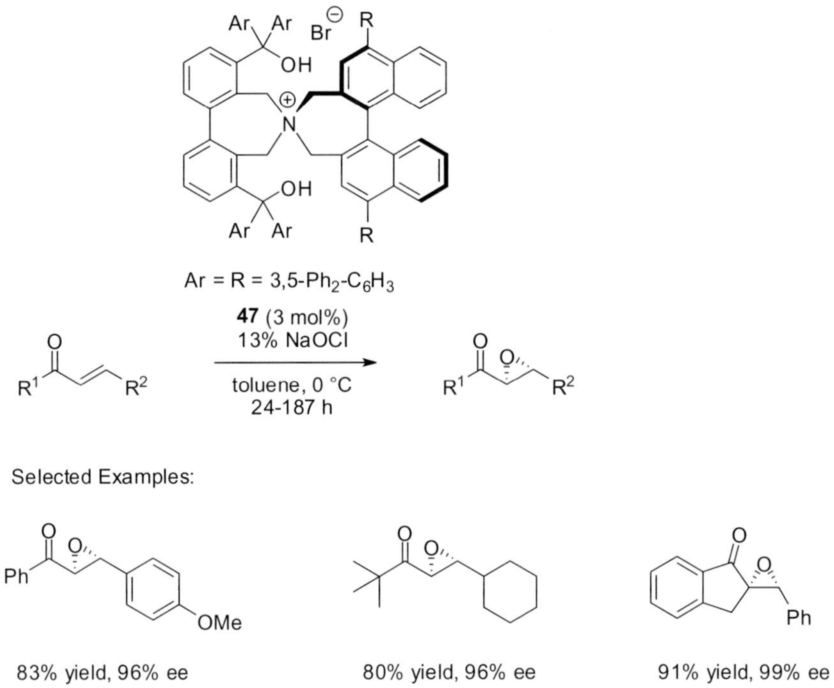

Ar = R = 3,5-Ph$_2$-C$_6$H$_3$

47 (3 mol%)
13% NaOCl
toluene, 0 °C
24-187 h

Selected Examples:

83% yield, 96% ee

80% yield, 96% ee

91% yield, 99% ee

Scheme 5.63

Maruoka introduced a new heterochiral quaternary ammonium bromide PTC **47** consisting of a core *N*-spirocycle flanked by biphenyl, containing diphenylmethanol subsitutuents at the 3,3′ positions, and binapthyl subunits. This PTC has been applied in the highly asymmetric epoxidation of a diverse set of α,β-unsaturated ketones [118] as well as in the Michael addition of diethyl malonate to chalcone derivatives (Scheme 5.63) [119].

5.4.1.3 Miscellaneous BBA and Related Systems

Ishihara and Yamamoto reported the design of new BINOL-derived Brønsted acid **48** where replacement of one of the hydroxyl groups with the strongly withdrawing bis(trifluoromethanesulfonyl)methyl group was expected to result in an increase in catalytic activity. Application of this catalyst in the enantioselective Mannich reaction of *N*-phenyl imines and ketene silyl acetals in the presence of stoichimetric amounts of achiral proton source as a silicon trapping agent afforded the β-aminoester in high yields and good ee's (Scheme 5.64). Synthesis and evaluation of the corresponding 2′-methoxy and 2′-methyl analogs of **48** suggest that the catalytic activity of **48** is attributable to intramolecular hydrogen bonding. As seen in Scheme 5.64, while two BBA forms are possible, the authors believe that **B** is the preferred conformation because the bis(trifyl)-methyl proton is more acidic than the hydroxyl proton and the hydroxyl oxygen more Lewis basic than the trifyl oxygen.

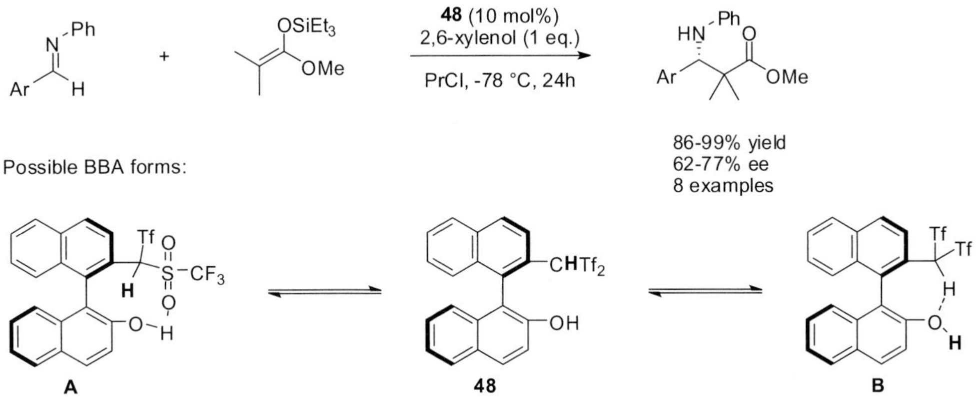

Scheme 5.64

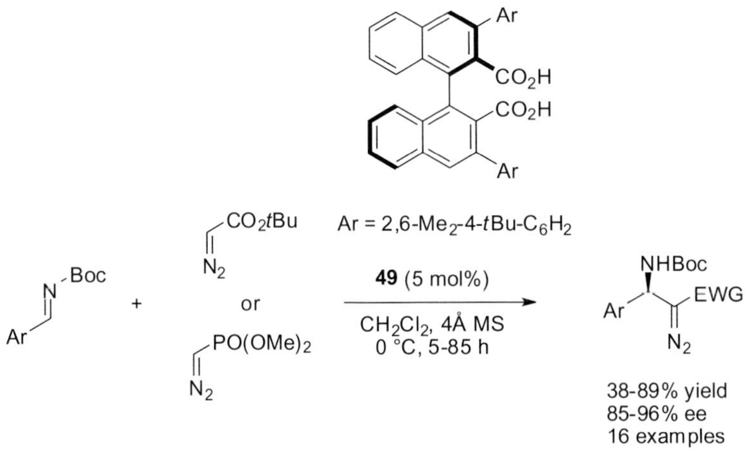

Possible BBA forms: **A**, **48**, **B**

Scheme 5.65

Ar = 2,6-Me$_2$-4-tBu-C$_6$H$_2$

38–89% yield
85–96% ee
16 examples

In light of the recent developments in thiourea, diol, and phosphoric-acid-mediated catalysis, far fewer studies have focused on the use of chiral carboxylic acids as suitable hydrogen bond donors. To this end, Maruoka synthesized binaphthyl-derived dicarboxylic acid **49** which catalyzes the asymmetric Mannich reaction of N-Boc aryl imines and tert-diazoacetate (Scheme 5.65) [120]. The authors postulate that catalytic activity is enhanced by the presence of an additional carboxylic acid moiety given that use of 2-napthoic acid as catalyst provided only trace amounts of product.

In 1995, Mikami reported that chiral lanthanide complex **35** catalyzed the enantioselective hetero-Diels–Alder reaction of Danishefsky's diene and various glyoxylates [121]. Interestingly, the addition of water to the reaction mixture resulted in increases in both chemical yield and enantioselectivity. It was later reasoned that

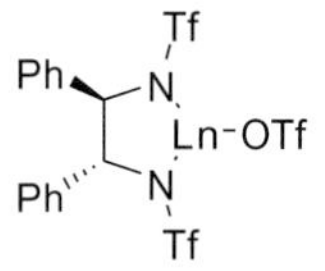

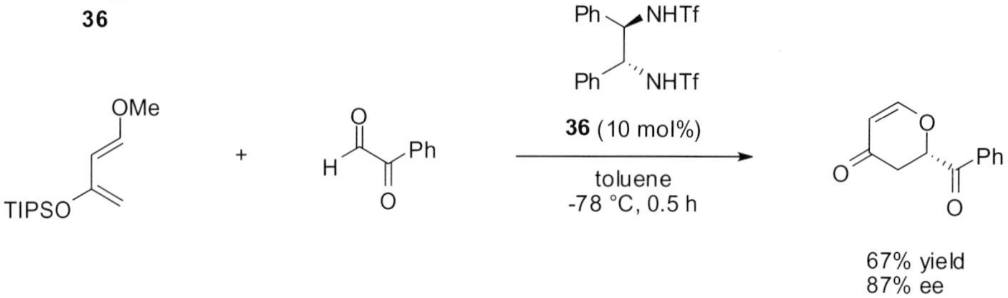

Scheme 5.66

the water served as a proton source to generate the Brønsted acidic bis-trifylamide **36** as the catalytically active species. Mikami later reported that **36** promotes the hetero-Diels–Alder reaction of Danishefsky's diene and aryl glyoxalates to provide highly functionalized adducts in good yields and ee's (Scheme 5.66) [122]. While the reactivity of **36** could conceivably arise from an intramolecular hydrogen bonding interaction akin to that proposed for TADDOL, a series of NMR titration studies eliminated this possibility. Instead, the authors postulate that carbonyl activation by **36** occurs through a double hydrogen bonding interaction. Support for this hypothesis was later provided by Jørgensen who observed an interesting correlation between the reactivity and N–C–C–N dihedral angle of several bis-tiflylamides [123].

5.4.2
Lewis-Acid-Assisted Brønsted Acid Catalysis [124]

In 1994, Yamamoto reported that a chiral LBA complex **50**, formed *in situ* by mixing equimolar amounts of BINOL and SnCl$_4$, was effective as a stoichiometric reagent for the enantioselective protonation of silyl enol ethers [125]. Analogous to type-III catalysis, coordination of a Lewis acid with a Brønsted acid allows enantiofacial recognition of one prochiral face of an olefin by constraining the rotation and directionality of the catalyst–proton bond. Later in 2003, Yamamoto described modified LBA **51a** and **51b** which was shown to be superior to LBA **50** for the enantioselective protonation of silyl enol ethers and ketene disilyl acetals (Scheme 5.67) [126]. Incorporation of a monoalkyl ether moiety in LBA **51a** and **b** is believed to result in tighter complexation of the Lewis acid relative to LBA **50**. Based on an X-ray crystal structure of a related LBA complex of **51**, the authors propose TS to explain the absolute stereopreference of the reaction. As indicated by the X-ray, complexation of the ligand occurs at an equatorial–equatorial site on tin to form

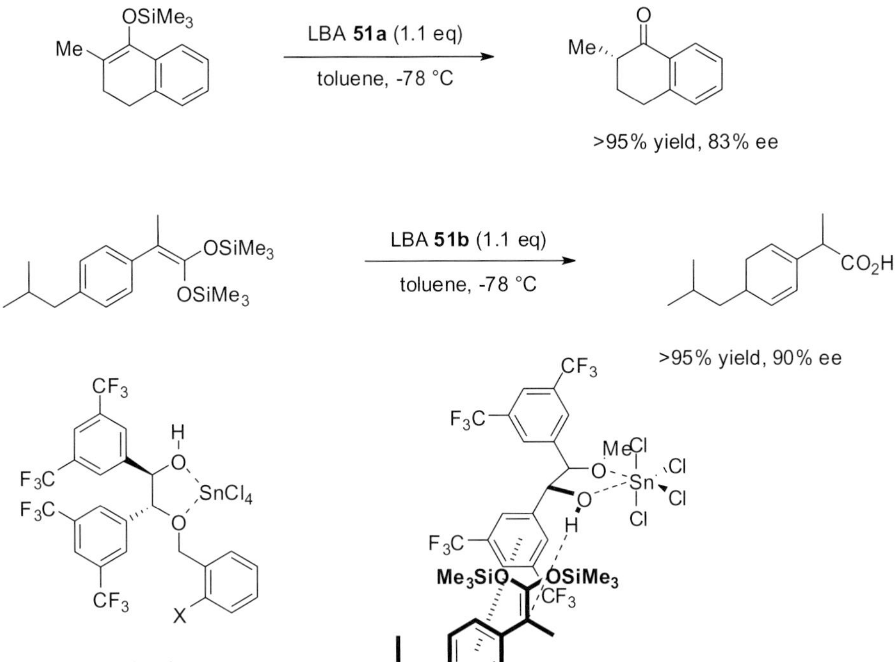

Scheme 5.67

a five-membered chelate. Additionally, the stereochemical course of the protonation step should be controlled by a linear OH/π interaction between the catalyst and the olefin. A similar chiral LBA system using $TiCl_4$ has also been reported by Yamamoto [127].

In 1999, Yamamoto reported the first example of an enantioselective biomimetic polyene cyclization using chiral LBAs as artificial cyclases. The LBA cyclase is believed to participate in the initial enantioselective protonation of the terminal isoprenyl group which induces concomitant diastereoselective cyclization [128]. Subsequent work by the Yamamoto group led to the development of LBA 52 as an artificial cyclase for hydroxypolyprenoids (Scheme 5.68) [129]. LBA 52 mediated cyclization of the the appropriate achiral hydroxypolypreniods permitting the short total syntheses of (−)-Chromazonarol, (+)-8-*epi*-puupehedione, and (−)-11′-deoxytaondiol (not shown).

(R)- and (S)-LBA 52 were also applied to the diastereoselective cyclization of optically active hydroxypolyprenoid 53 to afford after several steps the natural bicyclic sesquiterpene ethers (−)-caparrapi oxide and (+)-8-epicaparrapi oxide, respectively (Scheme 5.69) [130]. Notably, LBAs 52 and *ent*-52 were able to

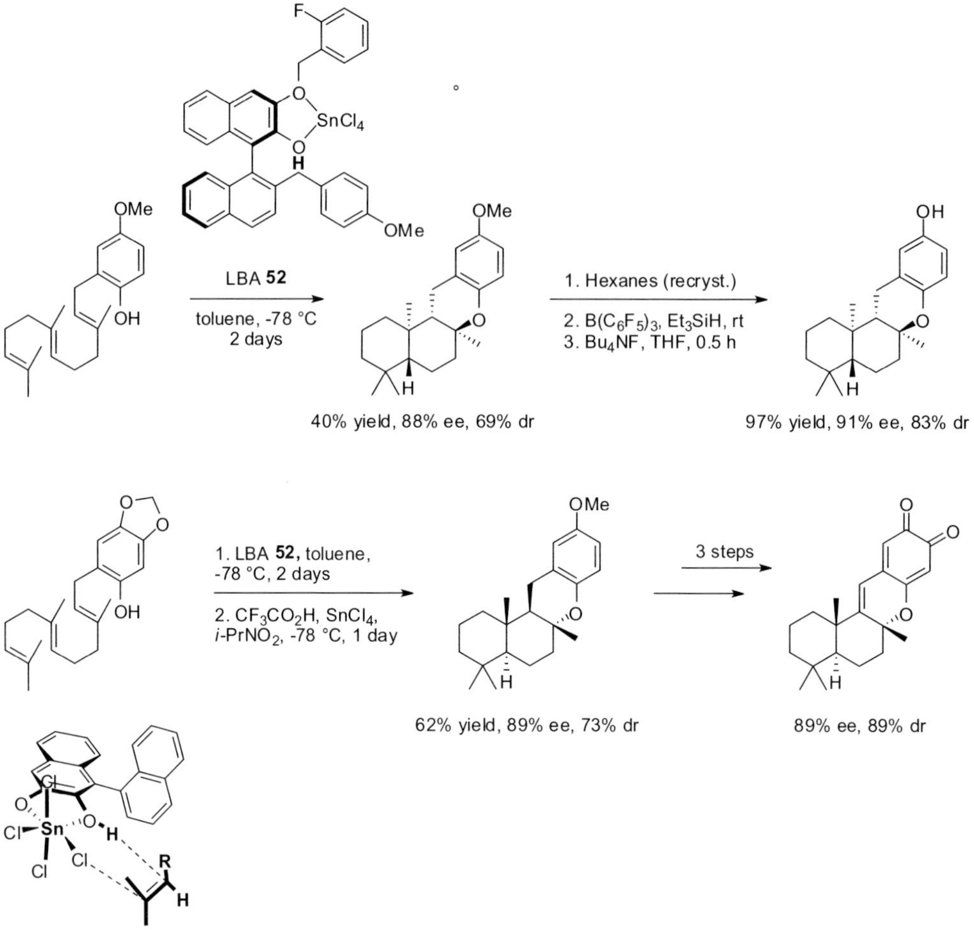

Scheme 5.68

overcome substrate control to provide the desired diasteromers with high selectivities.

In related work, the Yamamoto group developed LBA **54** containing a pyrogallol skeleton for the enantioselective polyene cyclizations of 4-(homogeranyl)toluene and 2-geranylphenol derivatives (Scheme 5.70) [131].

In earlier work, the Hall group found that catalytic amounts of triflic acid promoted the addition of allylboronates to aldehydes [132]. While allylboration reactions catalyzed by chiral Lewis acids in general led to only low levels of enantioselection [133], Hall found that chiral LBA **1** catalyzes the asymmetric addition of allyl- and crotylboronates to various aldehydes to provide products in excellent yields and moderate to high ee's (Scheme 5.71) [134]. Further, double diastereoselective crotylboration could also be achieved with high selectivities using the

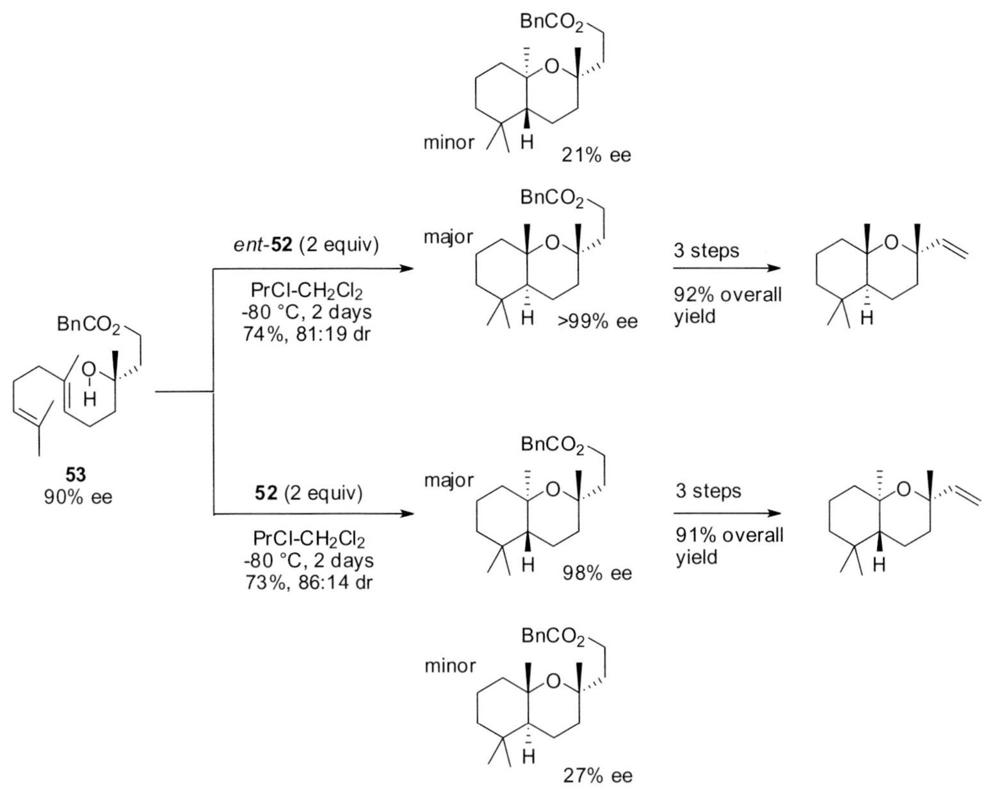

Scheme 5.69

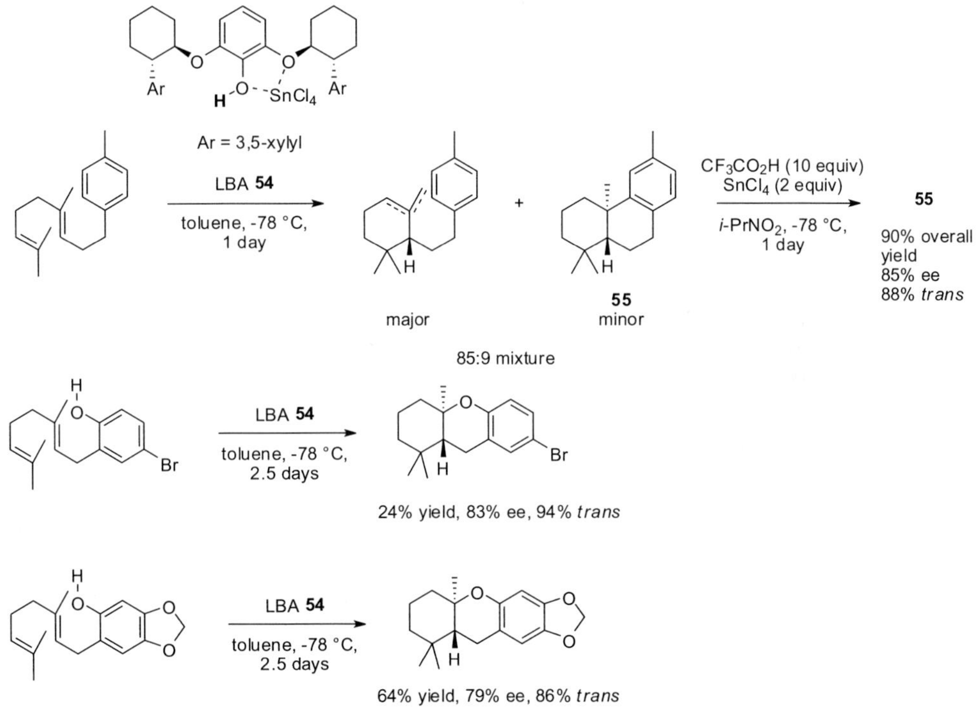

Scheme 5.70

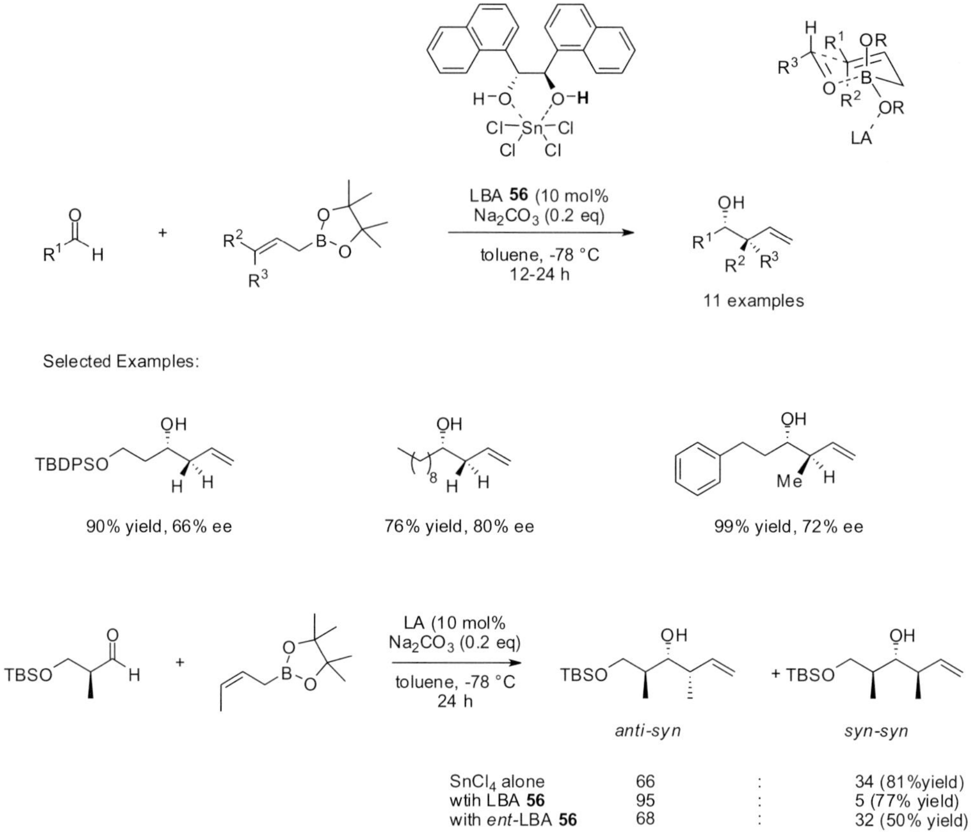

Scheme 5.71

matched LBA **56**. Mechanistic studies indicate that the reaction proceeds through a chair-like transition state where electrophilic activation of boron is achieved through coordination of the LBA with one of the boronate esters [135]. The authors postulate that the hindered nature of the pinacolboronate precludes effective coordination with a chiral Lewis acid complex whereas use of a smaller activator such as a proton is sterically allowed.

5.4.3
Brønsted-Acid-Assisted Lewis Acid Catalysis (Cationic Oxazaborolidine) [136]

5.4.3.1 Diels–Alder Reactions

The CBS-reduction [137] of prochiral ketones is a well-known process which employs a chiral oxazaborolidine as catalyst and $BH_3 \cdot THF$ or catecholborane as stoichiometric reductants. It is believed that the active catalytic species is a LLA, resulting from coordination of the oxazaborolidine nitrogen with the boron reagent to render the oxazaborolidine boron atom highly Lewis acidic [87]. Similarly, Corey

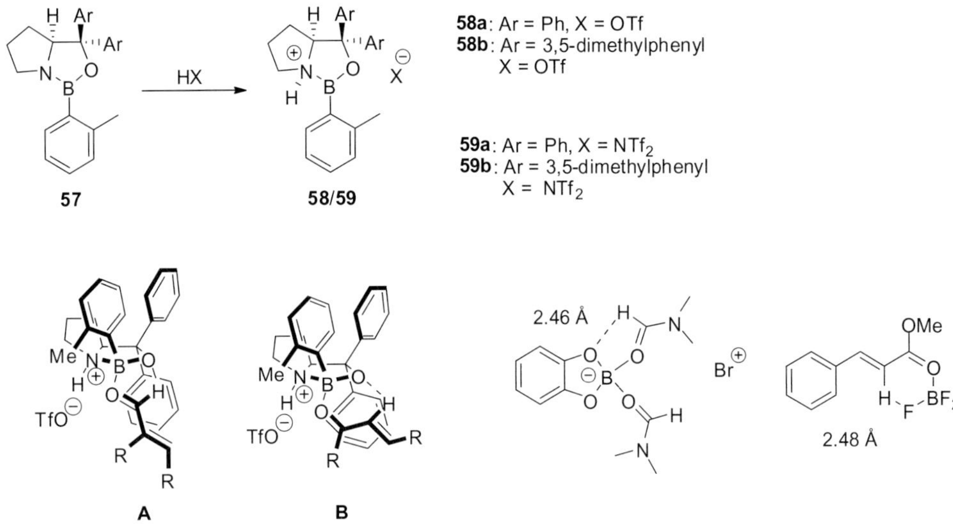

Figure 5.5 General considerations in cationic oxazaborolidine catalysis.

reported that addition of the strong protic acids, TfOH or Tf$_2$NH, to chiral oxazaboroldine **57** generates N-protonated cationic Lewis acid **58/59** (Figure 5.5). A series of reports have shown these chiral Lewis acids effectively catalyze a broad spectrum of Diels–Alder reactions [138]. This system activates various dienophiles, including α-subsituted α,β-unsaturated enals, α,β-unsaturated esters, carboxylic acids, ketones (cyclic and acyclic), and lactones as well as various quinones toward 4 + 2 cycloaddition with a wide range of cyclic and acyclic dienes (Scheme 5.72). Corey later reported AlBr$_3$ activation of **57** to form a LLA combined acid system. However, this catalyst system is beyond the scope of this chapter [139].

Interestingly, the facial selectivities observed in the Diels–Alder reaction of 2-substituted acroleins were opposite to those found for all other nonaldehydic dienophiles. To rationalize these differing absolute stereochemical outcomes, Corey proposed pretransition-state assemblies **A** and **B**. Discrimination of one of the two lone pairs of a dienophile carbonyl by BLA **58/59** is postulated to arise from a hydrogen bonding interaction between the formyl or α-olefinic hydrogen of the dienophile and the oxazaborolidinyl oxygen. In this scenario, it can be expected that acrolein which contains both formyl and α-olefinic hydrogens would provide adducts in low optical purities due to the existence of two competitive binding modes. Accordingly, the cycloaddition product of cyclopentadiene and acrolein could be obtained in only 69% ee. This hypothesis is further supported by the X-ray structures of complexes of BF$_3$ with methacrolein, methyl cinnamate, and benzylidene acetone indicating an attractive H–F interaction between the formyl hydrogen and α-olefinic hydrogens, respectively [123b, 140].

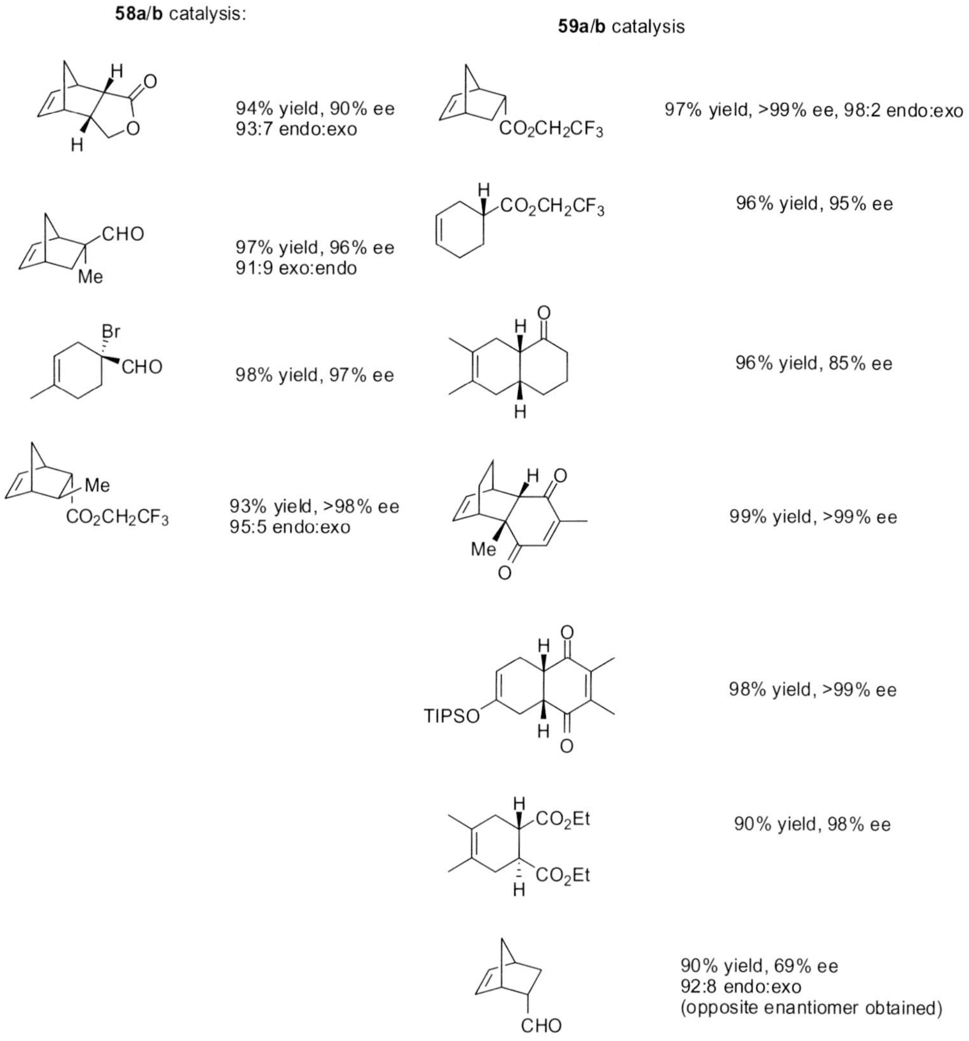

5 Brønsted Acids, H-Bond Donors, and Combined Acid Systems in Asymmetric Catalysis

58a/b catalysis:

59a/b catalysis

94% yield, 90% ee
93:7 endo:exo

97% yield, >99% ee, 98:2 endo:exo

97% yield, 96% ee
91:9 exo:endo

96% yield, 95% ee

98% yield, 97% ee

96% yield, 85% ee

93% yield, >98% ee
95:5 endo:exo

99% yield, >99% ee

98% yield, >99% ee

90% yield, 98% ee

90% yield, 69% ee
92:8 endo:exo
(opposite enantiomer obtained)

Scheme 5.72

In subsequent work, Corey examined the Diels–Alder reaction of various unsymmetrical quinones with 2-triisopropylsilyloxy-1,3-butadiene as test diene (Scheme 5.73) [141]. Achieving high enantioselectivities and diasterioselectivities with this class of substrates is particularly challenging due to the presence of four nonequal lone pairs as well as two reactive HC CH subunits. Remarkably, in all cases BLA **59a** selectively afforded Diels–Alder adducts as single regioisomers in excellent yields and ee's. Based on these results, Corey has formulated a set of selection rules to explain the preferred course of catalyst/dienophile complexation [142]. Briefly stated: (i) catalyst coordination at the more basic of the two 1,4 quinone oxygens will predominate so long as an α-olefinic hydrogen is present

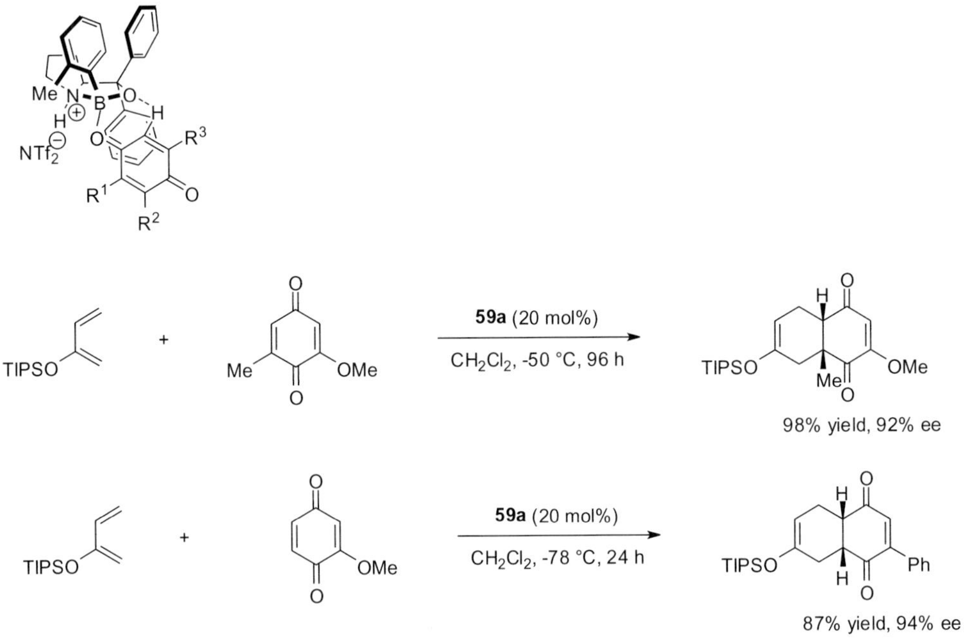

Scheme 5.73

and (ii) the double bond of the quinone containing the least basic substituents will be more reactive.

Corey later reported that BLA **59a** catalyzes the intramolecular Diels–Alder reaction of several triene aldehydes and esters with high asymmetric induction to yield the corresponding 6/5-*trans*-fused bicyclic structures (Scheme 5.74) [143].

In recent work, Jacobsen has applied modified **59c** to the transannular Diels–Alder (TADA) reaction [144]. Various-sized macrocycles containing α,β-unsaturated lactone or ketone dienophilic moieties could be cyclized to furnish tricyclic *endo* products containing medium-to large-sized rings in good to excellent diastereo- and enantioselectivities. Interestingly, BLA **59c**, containing a 2-FC$_6$H$_4$ boron substituent, provided TADA products in significantly higher yields and enantioselectivities than the corresponding *o*-tolyl-substituted BLA **59a** (49% versus 90% ee) (Scheme 5.75). Additionally, chiral macrocyclic substrates could undergo cycloaddition to give products with significantly higher diastereomeric ratios than those obtained under thermal or conventional Lewis acid catalysis.

Yamamoto had earlier reported that Lewis acid activation of valine-derived oxazaborolidine **60** yielded a highly reactive and moisture-tolerant LLA catalyst **61** for the Diels–Alder reaction (Scheme 5.76) [145]. In later studies, activation of **60** with the super Brønsted acid, C$_6$F$_5$CHTf$_2$, was found to produce the even more reactive catalytic species BLA **62**. During studies toward an enantioselective route to Platensimycin [146], BLA **62** was found to catalyze the Diels–Alder reaction between various monosubstituted dienes and ethyl acrylate to afford adducts

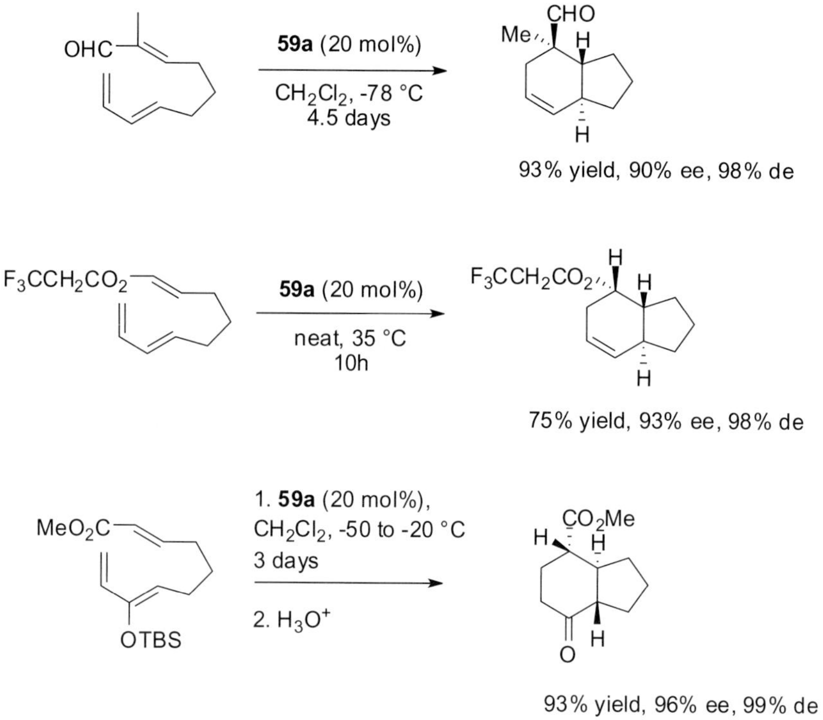

Scheme 5.74

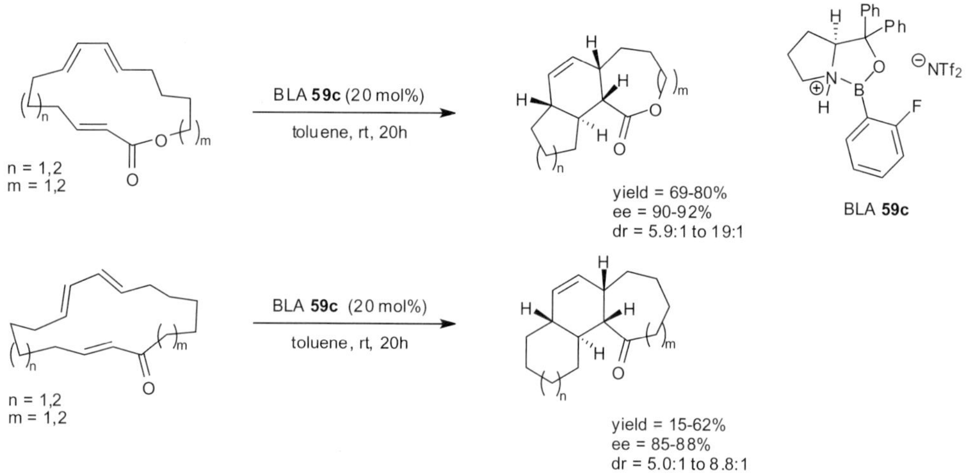

Scheme 5.75

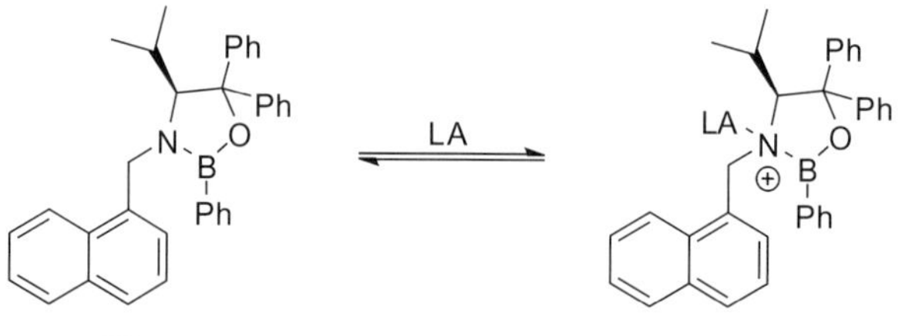

60
Scheme 5.76

LLA **61**

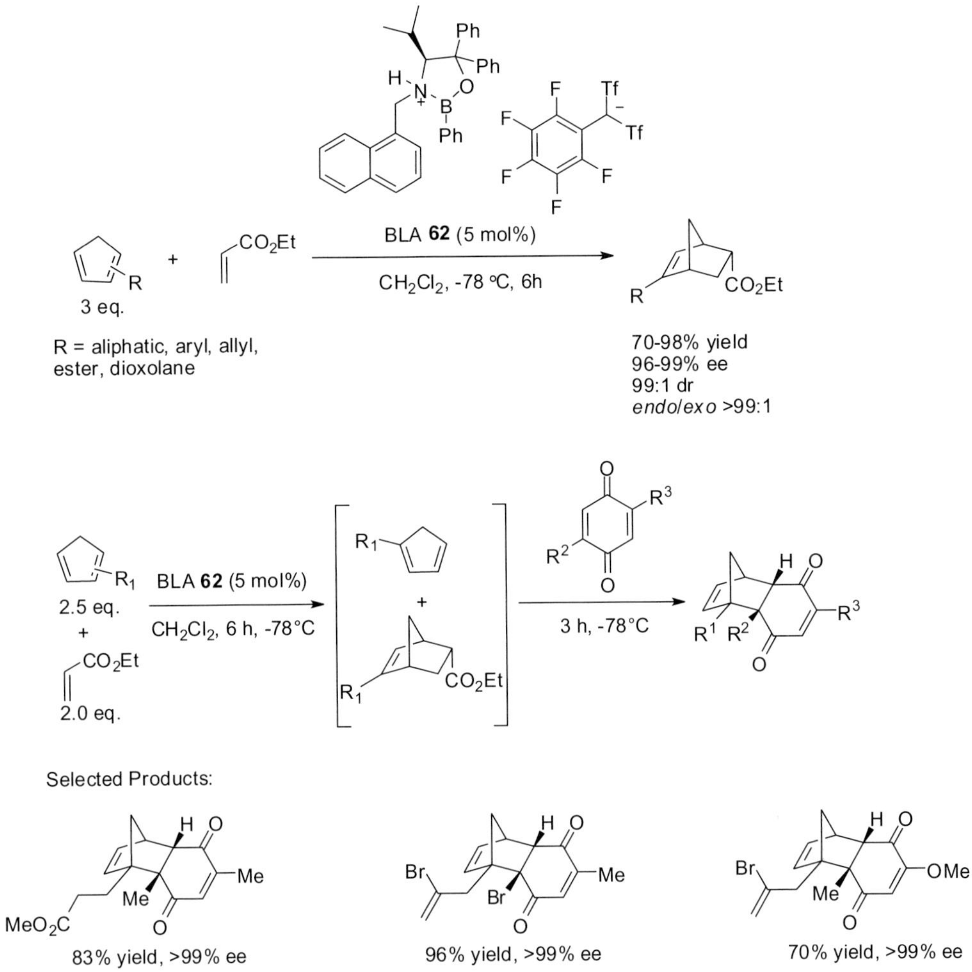

5.4 Combined Acid Catalysis | 131

Scheme 5.77

derived from 2-substituted cyclopentadiene as single isomers in excellent yields and ee's (Scheme 5.77) [147]. To account for the observed regio-discrimination between 1- and 2-substituted cyclopentadienes, the hypothetical transition state **A** was proposed (Figure 5.6). Thus, in **TS1**, the substituents of 1-substituted cyclopentadienes will experience significant steric interactions with the phenyl group appended to boron. However, 2-substituted cyclopentadienes should be oriented with their substituents pointed away from the catalyst–dienophile complex, resulting in a much lower energy transition state as indicated in **TS2**.

Diels–Alder adducts of 1-substituted cyclopentadienes could also be obtained through a one-pot procedure whereby ethyl acrylate is initially employed to consume all 2-substituted cyclopentadienes. Subsequently, various 2,5-disubstituted benzoquinones are added to react with remaining 1-substituted cyclopenta-

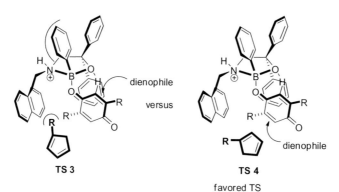

Figure 5.6 Hypothetical transition states.

dienes. Remarkably, a reaction occurs selectively at the double bond coordinated *anti* to BLA **62** to provide adducts containing adjacent all-carbon quaternary stereocenters in high yields and excellent enantioselectivities. To rationalize the observed positional discrimination, the hypothetical transition state **B** was proposed. In accordance with **TSA**, **TS3** should be disfavored due to a strong steric interaction with the substituents of 1-substituted cyclopentadienes and the phenyl ring of the catalyst. However, in **TS4**, the *anti*-coordinated benzoquinone double bond is placed far enough from the steric sphere of the catalyst to allow the approach of 1-substituted cyclopentadienes to be unimpeded.

5.4.3.2 Miscellaneous Reactions

In subsequent work, Corey applied BLA-type cationic oxazaborolidines to several other reactions, including [3 + 2] cycloaddition, cyanosilylation, Michael addition, and β-lactone formation.

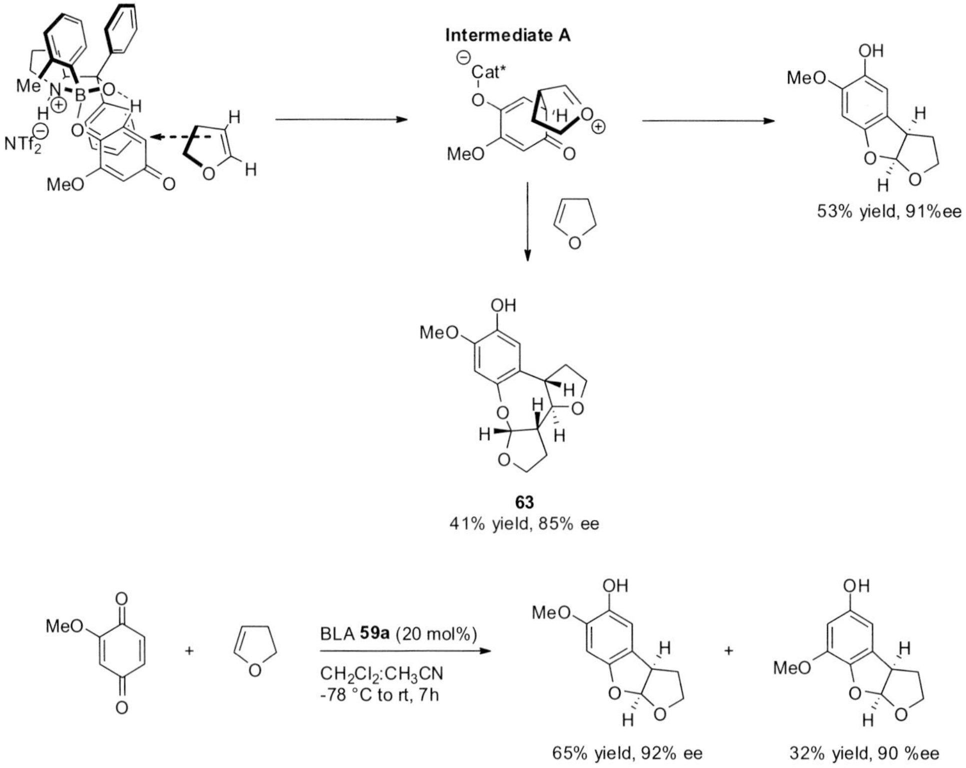

Scheme 5.78

BLA **59a** was shown to catalyze the [3 + 2] cycloaddition of several 1,4-benzoquinones with 2,3-dihydrofuran to afford chiral phenolic tricycles (Scheme 5.78) [148]. This new methodology was showcased in a short, enantioselective syntheis of Aflatoxin B_2. A mechanistic understanding of this reaction was acquired through a simple trapping experiment employing a large excess of 2,3-dihydrofuran. Two major products were isolated from this experiment including the expected cycloaddition adduct (53% yield, 91% ee) as well as tricyclic **63** (41% yield, 85% ee) which incorporates two dihydrofuran units. Formation of **63** is believed to result from trapping of putative dipolar reaction intermediate **A**.

Highly enantioselective cyanosilylation of aldehydes [149] and methyl ketones [150] has also been achieved utilizing cationic oxazaborolidines **59b** and **58b**, respectively (Scheme 5.79). While the use of TMSCN led to low levels of enantioselection, the addition of catalytic amounts of various phosphine oxides to generate *in situ* the more reactive $R_3P(OTMS)(N{=}C)$: as cyanide donor led to substantial increases in asymmetric induction. Similar to the postulated coordination modes of formyl or α-olefinic hydrogen-containing substrates, a hydrogen bonding interaction between the ketone α-hydrogen and the oxazaborolidinyl oxygen is proposed to function as an organizing element.

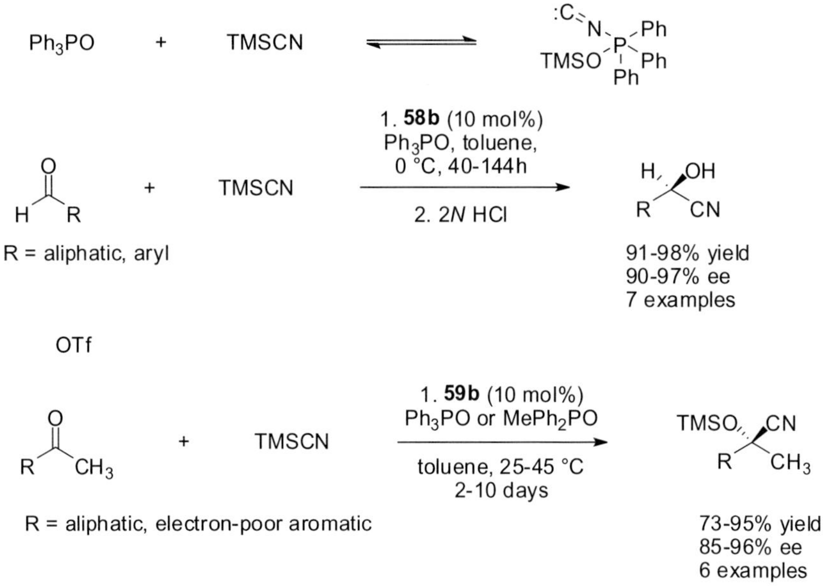

Ph₃PO + TMSCN ⇌ [phosphine oxide–TMSCN adduct]

Scheme 5.79

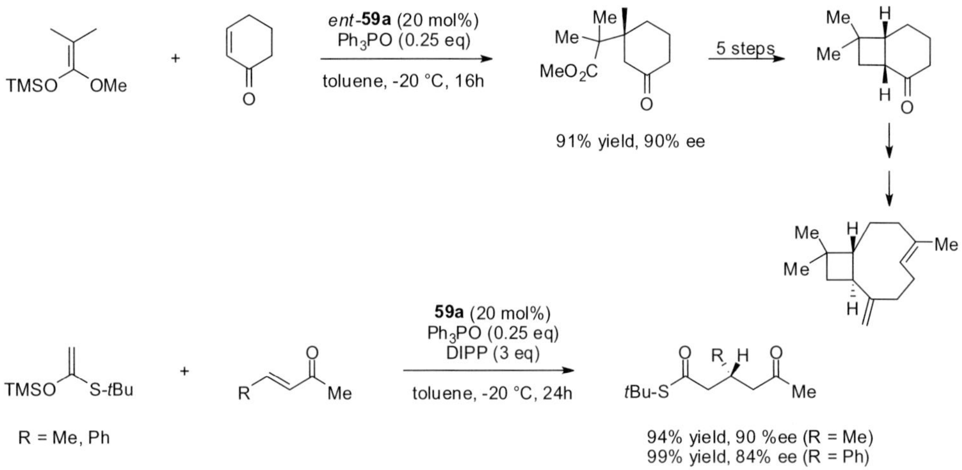

RCHO + TMSCN →
1. **58b** (10 mol%), Ph₃PO, toluene, 0 °C, 40–144 h
2. 2N HCl
→ R-CH(OH)-CN

R = aliphatic, aryl

91–98% yield
90–97% ee
7 examples

R-C(=O)-CH₃ (with OTf) + TMSCN →
1. **59b** (10 mol%), Ph₃PO or MePh₂PO, toluene, 25–45 °C, 2–10 days
→ R-C(OTMS)(CN)-CH₃

R = aliphatic, electron-poor aromatic

73–95% yield
85–96% ee
6 examples

Scheme 5.80

TMSO-C(=CMe₂)-OMe + cyclohexenone →
ent-**59a** (20 mol%), Ph₃PO (0.25 eq), toluene, −20 °C, 16 h
→ Michael adduct (Me, MeO₂C substituted cyclohexanone)
91% yield, 90% ee
→ 5 steps → bicyclic intermediate → caryophyllene skeleton

TMSO-C(=CR)-S-tBu (R = Me, Ph) + R-CH=CH-C(=O)-Me →
59a (20 mol%), Ph₃PO (0.25 eq), DIPP (3 eq), toluene, −20 °C, 24 h
→ tBu-S-C(=O)-CH(R)-CH₂-CH(H)-C(=O)-Me type adduct

94% yield, 90% ee (R = Me)
99% yield, 84% ee (R = Ph)

In contrast, phosphine oxides were later applied in Michael additions of silyl ketene acetals to cyclic and acyclic α,β-unsaturated ketones promoted by **59a** as scavengers of any catalytically active Me₃Si species formed during the reaction (Scheme 5.80) [151]. Further, this methodology now allows the realization of an enantioselective synthesis of caryophyllene.

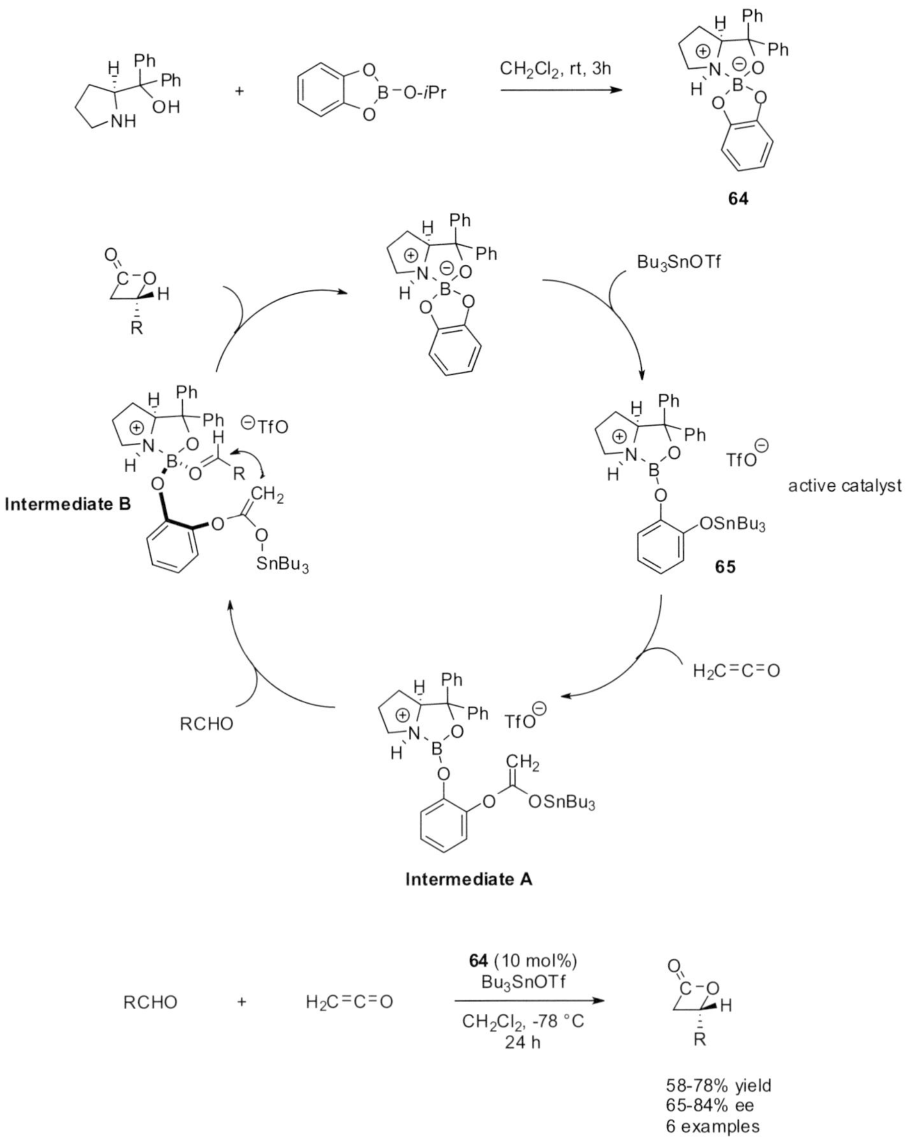

Scheme 5.81

Lastly, Corey has developed BLA species **65**, derived from zwitterionic oxazaborolidine **64** and tri-*n*-butyltin triflalte, as a novel catalytic system for the enantioselective synthesis of β-lactones from ketene and aldehydes (Scheme 5.81) [152]. The reaction of BLA **65** with ketene generates intermediate **A**. The subsequent addition of the ketene acetal unit to the coordinated aldehyde (intermediate **B**) followed by extrusion of the β-lactone completes the catalytic cycle.

References

1. (a) For general reviews, see: Schreiner, P.R. (2003) *Chem. Soc. Rev.*, **32**, 289–296.
(b) Pihko, P.M. (2004) *Angew. Chem. Int. Ed.*, **43**, 2062–2064.
(c) Bolm, C., Rantanenm, T., Schiffers, I. and Zani, L. (2005) *Angew. Chem. Int. Ed.*, **44**, 1758–1763.
(d) Pihko, P.M. (2005) *Lett. Org. Chem.*, **2**, 398–403.
(e) Taylor, M.S. and Jacobsen, E.N. (2006) *Angew. Chem. Int. Ed.*, **45**, 1520–1543.
(f) Doyle, A.G. and Jacobsen, E.N. (2007) *Chem. Rev.*, **107**, 5713–5743.
(g) see Reference 6.
(h) Steiner, T. (2002) *Angew. Chem. Int. Ed.*, **41**, 48–76.
(i) see References 93 and 118. Yu, X. and Wang, W. (2008) *Chem. Asian J.*, **3**, 516–532.
2. Jacques, J. and Fouquey, C. (1989) *Org. Synth.*, **67**, 1–12.
3. Wilen, S.H. and Qi, J.Z. (1991) *J. Org. Chem.*, **56**, 485–487.
4. Akiyama, T., Itoh, J., Yokota, K. and Fuchibe, K. (2004) *Angew. Chem. Int. Ed.*, **43**, 1566–1568.
5. Uraguchi, D. and Terada, M. (2004) *J. Am. Chem. Soc.*, **45**, 5356–5357.
6. For reviews, see: (a) Akiyama, T. (2007) *Chem. Rev.*, **107**, 5744–5758.
(b) Connon, S. (2006) *Angew. Chem. Int. Ed.*, **45**, 3909–3912.
(c) Akiyama, T., Itoh, J. and Fuchibe, K. (2006) *Adv. Synth. Catal.*, **348**, 999–1010. (d) Terada, M. (2008) *Chem. Comm.*, 4097–4112.
7. The pK_a of $(EtO)_2P(O)OH$ is 1.3, see: Quin, L.D. (2000) *A Guide to Organophosphorous Chemistry*, John Wiley & Sons, Inc., New York, Chapter 5, p. 133.
8. Yamanaka, M., Itoh, J., Fuchibe, K. and Akiyama, T. (2007) *J. Am. Chem. Soc.*, **129**, 6756–6764.
9. For a review on asymmetric bifunctional catalysis, see: Shibasaki, M., Kanai, M. and Funabashi, K. (2002) *Chem. Commun.*, 1989–1999.
10. Guo, Q.-X., Liu, H., Guo, C., Luo, S.-W., Gu, Y. and Gong, L.-Z. (2007) *J. Am. Chem. Soc.*, **129**, 3790–3791.
11. Akiyama, T., Moriota, H., Itoh, J. and Fuchibe, K. (2005) *Org. Lett.*, **7**, 2583–2585.
12. Uraguchi, D., Sorimachi, K. and Terada, M. (2004) *J. Am. Chem. Soc.*, **126**, 11804–11805.
13. Terada, M. and Sorimachi, K. (2007) *J. Am. Chem. Soc.*, **129**, 292–293.
14. Jia, Y.-X., Zhong, J., Zhu, S.-F., Zhang, C.-M. and Zhou, Q.-L. (2007) *Angew. Chem. Int. Ed.*, **46**, 5565–5567.
15. For an example using cinchona alkaloids, see: Wang, Y.-Q., Song, J., Hong, R., Li, H. and Deng, L. (2006) *J. Am. Chem. Soc.*, **128**, 8156–8157.
16. Kang, Q., Zhao, Z.-A. and You, S.-L. (2007) *J. Am. Chem. Soc.*, **129**, 1484–1485.
17. (a) Terada, M., Yokoyama, S., Sorimachi, J. and Uraguchi, D. (2007) *Adv. Synth. Catal.*, **349**, 1863–1867.
(b) Rowland, G.B., Rowland, E.B., Liang, Y., Perman, J.A. and Antilla, J.C. (2007) *Org. Lett.*, **9**, 2609–2611.
18. Li, G., Rowland, G.B., Rowland, E.B. and Antilla, J.C. (2007) *Org. Lett.*, **9**, 4065–4068.
19. Seayad, J., Seayad, A.M. and List, B. (2006) *J. Am. Chem. Soc.*, **128**, 1086–1087.
20. Taylor, M.S. and Jacobsen, E.N. (2004) *J. Am. Chem. Soc.*, **126**, 10558–10559.
21. Gremmen, C., Wilolemse, B., Wanner, M.J. and Koomen, G.-J. (2000) *Org. Lett.*, **2**, 1955–1958.
22. Wanner, M.J., van der Haas, R.N.S., de Cuba, K.R., van Maarseveen, J.H. and Hiemstra, H. (2007) *Angew. Chem. Int. Ed.*, **46**, 7485–7487.
23. Uraguchi, D., Sorimachi, K. and Terada, M. (2005) *J. Am. Chem. Soc.*, **127**, 9360–9361.
24. Williams, A.L. and Johnston, J.N. (2004) *J. Am. Chem. Soc.*, **126**, 1612–1613.
25. Rueping, M. and Azap, C. (2006) *Angew. Chem. Int. Ed.*, **45**, 7832–7835.

26 Liu, H., Cun, L.-F., Mi, A.-Q., Jiang, Y.-Z. and Gong, L.-Z. (2006) *Org. Lett.*, **8**, 6023–6026.

27 Akiyama, T., Tamura, Y., Itoh, J., Morita, H. and Fuchibe, K. (2006) *Synlett*, 141–143.

28 Itoh, J., Fuchibe, K. and Akiyama, T. (2006) *Angew. Chem. Int. Ed.*, **45**, 4796–4798.

29 Akiyama, T., Morita, H. and Fuchibe, K. (2006) *J. Am. Chem. Soc.*, **128**, 13070–13071.

30 (a) Matsubara, R., Nakamura, Y. and Kobayashi, S. (2004) *Angew. Chem. Int. Ed.*, **43**, 1679–1681.
(b) Matsubara, R., Nakamura, Y. and Kobayashi, S. (2004) *Angew. Chem. Int. Ed.*, **43**, 3258–3260.
(c) Matsubara, R., Vital, P., Nakamura, Y., Kiyohara, H. and Kobayashi, S. (2004) *Tetrahedron*, **60**, 9769–9784.

31 Terada, M., Machioka, K. and Sorimachi, K. (2006) *Angew. Chem. Int. Ed.*, **45**, 2254–2257.

32 Terada, M., Machioka, K. and Sorimachi, K. (2007) *J. Am. Chem. Soc.*, **129**, 10336–10337.

33 Chen, X.-H., Xu, X.-Y., Liu, H., Cun, L.-F. and Gong, L.-Z. (2006) *J. Am. Chem. Soc.*, **128**, 14802–14803.

34 Rueping, M., Sugiono, E., Theissmann, T., Kuenkel, A., Kockritz, A., Pews-Davtyan, A., Nemati, N. and Beller, M. (2007) *Org. Lett.*, **9**, 1065–1068.

35 Rueping, M., Antonchick, A.P. and Brinkmann, C. (2007) *Angew. Chem. Int. Ed.*, **46**, 6903–6906.

36 For metal calysis with chiral phosphate anions, see: Hamilton, G.L., Kang, E.J., Mba, M. and Toste, F.D. (2007) *Science*, **317**, 496–499.

37 Rueping, M., Sugiono, E. and Azap, C. (2006) *Angew. Chem. Int. Ed.*, **45**, 2617–2619.

38 Itoh, J., Fuchibe, K. and Akiyama, T. (2008) *Angew. Chem. Int. Ed.*, **47**, 4016–4018.

39 Terada, M., Soga, K. and Momiyama, N. (2008) *Angew. Chem. Int. Ed.*, **47**, 4122–4125.

40 Yang, J.W., Hechavarria Fonseca, M.T. and List, B. (2004) *Angew. Chem. Int. Ed.*, **43**, 6660–6662.

41 Yang, J.W., Hechavarria Fonseca, M.T., Vignola, N. and List, B. (2005) *Angew. Chem. Int. Ed.*, **43**, 108–110.

42 Ouellet, S.G., Tuttle, J.B. and MacMillan, D.W.C. (2005) *J. Am. Chem. Soc.*, **127**, 32–33.

43 Rueping, M., Sugiono, E., Azap, C., Theissmann, T. and Bolte, M. (2005) *Org. Lett.*, **7**, 3781–3783.

44 Hoffman, S., Seayad, A.M. and List, B. (2005) *Angew. Chem. Int. Ed.*, **44**, 7424–7427.

45 Rueping, M., Antonchick, A.P. and Theissmann, T. (2006) *Angew. Chem. Int. Ed.*, **45**, 6751–6755.

46 Rueping, M. and Antonchick, A.P. (2007) *Angew. Chem. Int. Ed.*, **46**, 4562–2565.

47 Rueping, M., Antonchick, A.P. and Theissmann, T. (2006) *Angew. Chem. Int. Ed.*, **45**, 3683–3686.

48 Storer, R.I., Carrera, D.E., Ni, Y. and MacMillan, D.W.C. (2006) *J. Am. Chem. Soc.*, **128**, 84–86.

49 Hoffman, S., Nicoletti, M. and List, B. (2006) *J. Am. Chem. Soc.*, **128**, 13074–13075.

50 Zhou, J. and List, B. (2007) *J. Am. Chem. Soc.*, **129**, 7498–7499.

51 (a) For a TADDOL-based phosphoric acid-diester, see: Akiyama, T., Saitoh, Y., Morita, H. and Fuchibe, K. (2005) *Adv. Synth. Catal.*, **347**, 1523–1526.
(b) For a phosphorodiamidic acid derivative, see: Terada, M., Sorimachi, K. and Uraguchi D. (2006) *Synlett*, 133–136.

52 Rowland, G.B., Zhang, H., Rowland, E.B., Chennamadhavuni, S., Wang, Y. and Antilla, J.C. (2005) *J. Am. Chem. Soc.*, **127**, 15696–15697.

53 Liang, Y., Rowland, E.B., Rowland, G.B., Perman, J.A. and Antilla, J.C. (2007) *Chem. Commun.*, 4477–4479.

54 For a related highly enantioselective reduction of α-imino esters with a BINOL-derived phosphoric acid diester. see: Kang, Q., Zhao, Z.-A. and You, S.-L. (2007) *Adv. Synth. Catal.*, **349**, 1657–1660.

55 Li, G., Liang, Y. and Antilla, J.C. (2007) *J. Am. Chem. Soc.*, **129**, 5830–5831.
56 Rowland, E.B., Rowland, G.B., Rivera-Otero, E. and Antilla, J.C. (2007) *J. Am. Chem. Soc.*, **129**, 12084–12085.
57 Yagupolski, L.M., Petrik, V.N., Kondratenko, N.V., Soovali, L., Kaljurand, I., Leito, I. and Koppel, I.A. (2002) *J. Chem. Soc., Perkin Trans. 2*, 1950–1955 and references cited therein.
58 Nakashima, D. and Yamamoto, H. (2006) *J. Am. Chem. Soc.*, **128**, 9626–9627.
59 Jiao, P. and Yamamoto, H. (2008) *Angew. Chem. Int. Ed.*, **47**, 2411–2413.
60 Simonsen, K.B., Bayon, P., Hazell, R.G., Gothelf, K.V. and Jørgensen, K.A. (1999) *J. Am. Chem. Soc.*, **121**, 3845–3853.
61 Cheon, C.H. and Yamamoto, H. (2008) *J. Am. Chem. Soc.*, **130**, 9246–9247.
62 Rueping, M., Ieawsuwan, W., Antonchick, A.P. and Nachtsheim, B.J. (2007) *Angew. Chem. Int. Ed.*, **46**, 2097–2100.
63 Aggarwal, V.K. and Belfield, A.J. (2003) *Org. Lett.*, **5**, 5075–5078.
64 Rueping, M., Thiessmann, T., Kuenkel, A. and Keonigs, R.M. (2008) *Angew. Chem. Int. Ed.*, **47**, 6798–6801.
65 Mayer, S. and List, B. (2006) *Angew. Chem. Int. Ed.*, **45**, 4193–4195.
66 Martin, N.J.A. and List, B. (2006) *J. Am. Chem. Soc.*, **128**, 13368–13369.
67 Wang, X. and List, B. (2008) *Angew. Chem. Int. Ed.*, **47**, 1119–1122.
68 Schmidtchen, F.P. and Berger, M. (1997) *Chem. Rev.*, **97**, 1609–1646.
69 For reviews on guanidines in organic synthesis, see: (a) Ishikawa, T. and Isobe, T. (2002) *Chem. Eur. J.*, **8**, 552–557.
(b) Ishikawa, T. and Kumamoto, T. (2006) *Synthesis*, **5**, 737–752.
70 (a) Oku, J. and Inoue, S. (1981) *J. Chem. Soc., Chem. Commun.*, 229–230. (b) Tanaka, K., Mori, A. and Inoue, S. (1990) *J. Org. Chem.*, **55**, 181–185. (c) Danda, H. (1991) *Synlett*, 263–264. (d) Danda, H., Nishikawa, H. and Otaka, K. (1991) *J. Org. Chem.*, **56**, 6740–6741.
71 Iyer, M.S., Gigstad, K.M., Namdev, N.D. and Lipton, M. (1996) *J. Am. Chem. Soc.*, **118**, 4910–4911.
72 For a theoretical study, see: Li, J., Jiang, W.-Y., Han, K.-L., He, G.-Z. and Li, C. (2003) *J. Org. Chem.*, **68**, 8786–8789.
73 Recently, a study challenging Lipton's findings has appeared: Becker, C., Hoben, C., Schollmeyer, D., Scherr, G. and Kunz, H. (2005) *Eur. J. Org. Chem.*, 1497–1499.
74 Corey, E.J. and Grogan, M.J. (1999) *Org. Lett.*, **1**, 157–160.
75 Shen, J., Nguyen, T.T., Goh, Y.-P., Ye, W., Fu, X., Xu, J. and Tan, C.-H. (2006) *J. Am. Chem. Soc.*, **128**, 13692–13693.
76 Fu, X., Jang, Z. and Tan, C.-H. (2007) *Chem. Commun.*, 5058–5060.
77 Ye, W., Jiang, Z., Zhao, Y., Goh, S.L.M., Leow, D., Soh, Y.-T. and Tan, C.-H., (2007) *Adv. Synth. Catal.*, **349**, 2454–2458.
78 Kashman, Y., Hirsh, S., McConnell, O.J., Ohtani, I., Kusumi, T. and Kakisawa, H. (1989) *J. Am. Chem. Soc.*, **111**, 8925–8926.
79 Kita, T., Georgieva, A., Hashimoto, Y., Nakata, T. and Nagasawa, K. (2002) *Angew. Chem. Int. Ed.*, **41**, 2832–2834.
80 Terada, M., Ube, H. and Yaguchi, Y. (2006) *J. Am. Chem. Soc.*, **128**, 1454–1455.
81 Terada, M., Ikehara, T. and Ube, H. (2007) *J. Am. Chem. Soc.*, **129**, 14112–14113.
82 Terada, M., Nakano, M. and Ube, H. (2006) *J. Am. Chem. Soc.*, **128**, 16044–16045.
83 Schuster, T., Kurz, M. and Gobel, M.W. (2000) *J. Org. Chem.*, **65**, 1697–1701.
84 Schuster, T., Bauch, M., Durner, G. and Gobel, M.W. (2000) *Org. Lett.*, **2**, 179–181.
85 (a) Tsogoeva, S.B., Durner, G., Bolte, M. and Gobel, M.W. (2003) *Eur. J. Org. Chem.*, **9**, 1661–1664.
(b) Akalay, D., Durner, G., Bats, J.W., Bolte, M. and Gobel, M.W. (2007) *J. Org. Chem.*, **72**, 5618–5624.
86 For related reviews, see: (a) Pihko, P.M. (2004) *Angew. Chem. Int. Ed.*, **43**, 2062–2064. (b) Bolm, C., Rantanen, T., Schiffers, I. and Zani, L. (2005) *Angew. Chem. Int. Ed.*, **44**, 1758–1763.

87 Nugent, B.M., Yoder, R.A. and Johnston, J.N. (2004) *J. Am. Chem. Soc.*, **126**, 3418–3419.

88 The pK_a of **32** was determined to be 5.78 in DMSO; Hess, A.S., Yoder, R.A. and Johnston, J.N. (2006) *Synlett*, 147–149.

89 Singh, A., Yoder, R.A., Shen, B. and Johnston, J.N. (2007) *J. Am. Chem. Soc.*, **129**, 3466–3467.

90 Huang, J. and Corey, E.J. (2003) *Org. Lett.*, **5**, 3455–3458.

91 Huang, J. and Corey, E.J. (2004) *Org. Lett.*, **6**, 5027–5029.

92 Uraguchi, D., Sakaki, S. and Ooi, T. (2007) *J. Am. Chem. Soc.*, **129**, 12392–12393.

93 For reviews, see: (a) Yamamoto, H. and Futatsugi, K. (2005) *Angew. Chem. Int. Ed.*, **44**, 1924–1942.
(b) Yamamoto, H. (ed.) (2000) *Lewis Acids in Organic Synthesis*, Wiley-VCH Verlag GmbH, Weinheim. (c) Ref. 118.

94 (a) Steiner, T. (2002) *Angew. Chem. Int. Ed.*, **41**, 48–76.
(b) Jeffrey, G.A. (2003) *Crystallogr. Rev.*, **11**, 135–176.
(c) Scheiner, S. (1997) *Hydrogen Bonding. A Theoretical Perspective*, Oxford University Press, Oxford.

95 (a) Pocker, Y., Stevens, K.D. and Champoux, J.J. (1969). *J. Am. Chem. Soc.*, **91**, 4199–4205.
(b) Pocker, Y., Stevens, K.D., and Champoux, J.J. (1969) *J. Am. Chem. Soc.*, **91**, 4205–4210.

96 Neimann, K. and Neumann, R. (2000) *Org. Lett.*, **2**, 2861–2863.

97 Berkessel, A. and Adrio, J.A. (2006) *J. Am. Chem. Soc.*, **128**, 13412–13420.

98 Hasegawa, A., Naganawa, Y., Fushimi, M., Ishihara, K. and Yamamoto, H. (2006) *Org. Lett.*, **8**, 3175–3178.

99 Huang, Y. and Rawal, V.H. (2002) *J. Am. Chem. Soc.*, **124**, 9662–9663.

100 Huang, Y., Unni, A.K., Thadani, A.N. and Rawal, V.H. (2003) *Nature*, **424**, 146.

101 Prior to this work, solid-state structures of TADDOL derivatives indicated the existence of an intramolecular hydrogen bond, implicating a BBA-type catalytic mechanism. For a review, see: Seebach, D., Beck, A. and Heckel, K. (2001) *Angew. Chem. Int. Ed.*, **40**, 92–138.

102 Thadani, A.N., Stankovic, A.R. and Rawal, V.H. (2004) *Proc. Natl. Acad. Sci. U S A*, **101**, 5846–5850.

103 Unni, A.K., Takenaka, N., Yamamoto, H. and Rawal, V.H. (2005) *J. Am. Chem. Soc.*, **127**, 1336–1337.

104 Du, H., Zhao, D. and Ding, K. (2004) *Chem. Eur. J.*, **10**, 5964–5970.

105 Gondi, V.B., Gravel, M. and Rawal, V.H. (2005) *Org. Lett.*, **7**, 5657–5660.

106 Villano, R., Acocella, M.R., Massa, A., Palombi, L. and Scettri, A. (2007) *Tetrahedron Lett.*, **48**, 891–895.

107 McGilvra, J.D., Unni, A.K., Modi, K. and Rawal, V.H. (2006) *Angew. Chem. Int. Ed.*, **45**, 6130–6133.

108 Anderson, C.D., Dudding, T., Gordillo, R. and Houk, K.N. (2008) *Org. Lett.*, **10**, 2749–2752.

109 (a) Yamada, Y.M.A. and Ikegami, S. (2000) *Tetrahedron Lett.*, **41**, 2165–2169.
(b) For preliminary studies implicating the importance of hydrogen bonding, see: Drewes, S.E., Freese, S.D., Emslie, N.D. and Roos, G.H. (1988) *Synth. Commun.*, **18**, 1565–1572.

110 (a) McDougal, N.T. and Schaus, S.E. (2003) *J. Am. Chem. Soc.*, **125**, 12094–12095.
(b) McDougal, N.T., Trevellini, W.L., Rogden, S.A., Kliman, L.T. and Schaus, S.E. (2004) *Adv. Synth. Catal.*, **346**, 1231–1240.

111 Matsui, K., Takizawa, S. and Sasai, H. (2005) *J. Am. Chem. Soc.*, **127**, 3680–3681.

112 Matsui, K., Takizawa, S. and Sasai, H. (2006) *Synlett*, 761–765.

113 Momiyama, N. and Yamamoto, H. (2005) *J. Am. Chem. Soc.*, **127**, 1080–1081.

114 (a) Yamamoto, H. and Kawasaki, M. (2007) *Bull. Chem. Soc. Jpn.*, **80**, 595–607.
(b) For related studies, see: Cheong, P.H.-Y. and Houk, K.N. (2004). *J. Am. Chem. Soc.*, **126**, 13912–13913.

115 Momiyama, N., Yamamoto, Y. and Yamamoto, H. (2007) *J. Am. Chem. Soc.*, **129**, 1190–1195.

116 Tillman, A.L. and Dixon, D.J. (2007) *Org. Biomol. Chem.*, **5**, 606–609.

117 Dixon, D.J. and Tillman, A.L. (2005) *Synlett*, 2635–2638.

118 Ooi, T., Ohara, D., Tamura, M. and Maruoka, K. (2004) *J. Am. Chem. Soc.*, **126**, 6844–6845.
119 Ooi, T., Ohara, D., Fukumoto, K. and Maruoka, K. (2005) *Org. Lett.*, **7**, 3195–3197.
120 Hasimoto, T. and Maruoka, K. (2007) *J. Am. Chem. Soc.*, **129**, 10054–10055.
121 Mikami, K., Kotera, O., Motoyama, Y. and Sakaguchi, H. (1995) *Synlett*, 975–977.
122 Tonoi, T. and Mikami, K. (2005) *Tetrahedron Lett.*, **46**, 6355–6358.
123 (a) Zhuang, W., Hazell, R.G. and Jørgensen, K.A. (2005) *Org. Biomol. Chem.*, **3**, 2566–2571.
(b) Zhuang, W., Poulsen, T.B. and Jørgensen, K.A. (2005) *Org. Biomol. Chem.*, **3**, 3284–3289.
124 For excellent reviews, see: (a) Ishibashi, H., Ishihara, K. and Yamamoto, H. (2002) *Chem. Rec.*, **2**, 177–188.
(b) Brunoldi, E., Luparia, M., Porta, A., Zanoni, G. and Vidari, G. (2006) *Curr. Org. Chem.*, **10**, 2259–2282.
125 (a) Ishihara, K., Kaneeda, M. and Yamamoto, H. (1994) *J. Am. Chem. Soc.*, **116**, 11179–11180.
126 Ishihara, K., Nakashima, D., Hiraiwa, Y. and Yamamoto, H. (2003) *J. Am. Chem. Soc.*, **125**, 24–25.
127 Nakashima, D. and Yamamoto, H. (2006) *Synlett*, 150–152.
128 Ishihara, K., Nakamura, S. and Yamamoto, H. (1999) *J. Am. Chem. Soc.*, **121**, 4906–4907.
129 Ishiibashi, H., Ishihara, K. and Yamamoto, H. (2004) *J. Am. Chem. Soc.*, **126**, 11122–11123.
130 (a) Uyanik, M., Iswhibashi, H., Ishihara, K. and Yamamoto, H. (2005) *Org. Lett.*, **7**, 1601–1604.
(b) Uyanik, M., Ishihara, K. and Yamamoto, H. (2005) *Bioorg. Med. Chem.*, **13**, 5055–5065.
131 Kumazawa, K., Ishihara, K. and Yamamoto, H. (2004) *Org. Lett.*, **6**, 2551–2554.
132 Yu, S.H. and Hall, D.G. (2005) *J. Am. Chem. Soc.*, **127**, 12808–12809.
133 Ishiyama, T., Ahiko, T. and Miyaura, N. (2002) *J. Am. Chem. Soc.*, **124**, 12414–12415.
134 (a) Rauniyar, V. and Hall, D.G. (2006) *Angew. Chem. Int. Ed.*, **45**, 2426–2428.
(b) Rauniyar, V. and Hall, D.G. (2007) *Synlett*, 3421–3426.
135 Hall, D.G. (2007) *Synlett*, 1644–1655.
136 For a review, see: Corey, E.J. (2002) *Angew. Chem. Int. Ed.*, **41**, 1650–1667.
137 For a review, see: Cho, B.T. (2006) *Tetrahedron*, **62**, 7621–7643.
138 (a) Corey, E.J., Shibata, T. and Lee, T.W. (2002) *J. Am. Chem. Soc.*, **124**, 3808–3809.
(b) Ryu, D.H., Lee, T.W. and Corey, E.J. (2002) *J. Am. Chem. Soc.*, **124**, 9992–9993.
(c) Ryu, D.H. and Corey, E.J. (2003) *J. Am. Chem. Soc.*, **125**, 6388–6390.
(d) Ryu, D.H., Kim, K.H., Sim, J.Y. and Corey, E.J. (2007) *Tetrahedron Lett.*, **48**, 5735–5737.
139 Liu, D., Canales, E. and Corey, E.J. (2007) *J. Am. Chem. Soc.*, **129**, 1498–1499.
140 Corey, E.J. and Lee, T.W. (2001) *Chem. Commun.*, 1321–1329.
141 Ryu, D.H., Zhou, G. and Corey, E.J. (2004) *J. Am. Chem. Soc.*, **126**, 4800–4802.
142 For a related study on dienophile complexation with cationic oxazaborolidines, see: Ryu, D.H., Zhou, G. and Corey, E.J. (2005) *Org. Lett.*, **7**, 1633–1636.
143 Zhou, G., Hu, Q.-Y. and Corey, E.J. (2003) *Org. Lett.*, **5**, 3979–3982.
144 Balskus, E.P. and Jacobsen, E.N. (2007) *Science*, **317**, 1736–1740.
145 Futatsugi, K. and Yamamoto, H. (2005) *Angew. Chem. Int. Ed.*, **44**, 1484–1487.
146 Li, P., Payette, J.N. and Yamamoto, H. (2007) *J. Am. Chem. Soc.*, **129**, 9534–9535.
147 Payette, J.N. and Yamamoto, H. (2007) *J. Am. Chem. Soc.*, **129**, 9536–9537.
148 Zhou, G. and Corey, E.J. (2005) *J. Am. Chem. Soc.*, **127**, 11958–11959.
149 Ryu, D.H. and Corey, E.J. (2004) *J. Am. Chem. Soc.*, **126**, 8106–8107.
150 Ryu, D.H. and Corey, E.J. (2005) *J. Am. Chem. Soc.*, **127**, 5384–5387.
151 Liu, D., Hong, S. and Corey, E.J. (2006) *J. Am. Chem. Soc.*, **128**, 8160–8161.
152 Gnanadesikan, V. and Corey, E.J. (2006) *Org. Lett.*, **8**, 4943–4945.

6
(Thio)urea Organocatalysts[1]

Mike Kotke and Peter R. Schreiner

6.1
Introduction and Background

The concept of noncovalent organocatalysis [1–8] utilizing explicit hydrogen-bonding (thio)urea derivatives originates from natural catalytic systems such as ribonucleases, antibodies, and enzymes that can be considered as archetypes of organocatalysts [9]. Crystal structure analyses, spectroscopic investigations [10, 11], and computational studies [12, 13] revealed that enzymes, for instance, typically do not contain strong Lewis acids for the binding (recognition) of the substrate resulting in catalytic activity and stereoselectivity due to weak enthalpic binding and small Gibbs energy changes [14]. In metal-free enzymes the recognition process is dominated by hydrogen bonding ("partial protonation") [15–20] and hydrophobic interactions [21–23] in an often complex interplay with other weak noncovalent interactions such as aromatic π-stacking [24], van der Waals [25, 26], and dipole–dipole interactions [27, 28]; making the details of enzyme catalysis difficult to rationalize [14, 29–31]. These interactions are provided by the enzyme's active site, that is, an ensemble of various functionalities, e.g., hydrogen-bond donors properly arranged in the enzyme-binding pocket capable to coordinate selectively and to activate the embedded substrate(s) for highly specific biochemical transformations via host–guest complexes. Scheme 6.1 exemplarily visualizes the active sites of enzymes using double hydrogen-bonding interactions to the hydrogen bond accepting basic site of various substrates, which leads to substrate activation and acceleration of the respective reactions. Haloalcohol dehalogenase (**1**) activates an epoxide for reversible epoxide ring opening by chloride [32] similar to enzymatic epoxide hydrolyis catalyzed by epoxide hydrolase (Scheme 6.8) [33–35], formate dehydrogenase (**2**) promotes the oxidation of formate [29], and serine protease (**3**) accelerates amide hydrolysis by double hydrogen-bonding, multiple noncovalent catalyst–substrate interactions, and bifunctional catalysis (Scheme 6.1) [36].

1) *This chapter is dedicated to Eva Kotke.*

Hydrogen Bonding in Organic Synthesis. Edited by Petri M. Pihko
Copyright © 2009 WILEY-VCH Verlag GmbH & Co. KGaA, Weinheim
ISBN: 978-3-527-31895-7

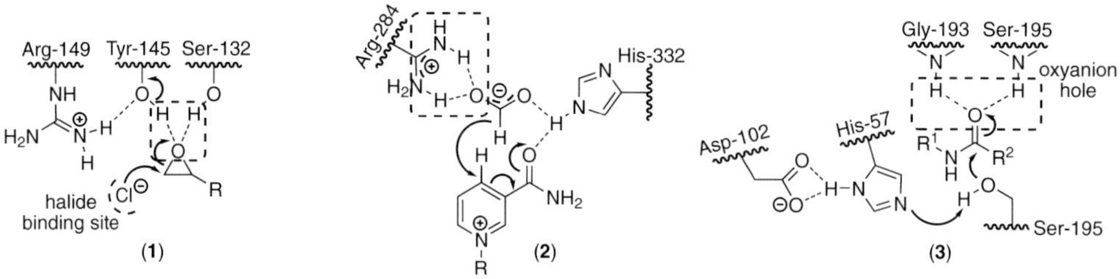

Scheme 6.1 Active sites of enzymes employing a double hydrogen-bonding motif for substrate coordination and activation in various biochemical transformations: Haloalcohol dehalogenase (**1**), formate dehydrogenase (**2**), and serine protease (**3**).

Metal-free catalysis as performed by hydrogen-bonding enzymes shows no apparent product inhibition [37] resulting in high TOF values while proceeding in the natural medium water under aerobic conditions [38, 39]. These characteristics offer attractive perspectives for the development of artificial metal-free catalytic systems as supplementation and/or alternatives to traditional metal(-ion)-containing, often toxic, water-incompatible, and air-sensitive catalysts. Artificial enzymes or enzyme mimetics [40–42] designed according to the Pauling paradigm [43, 44] or related paradigms [45, 46] aim to imitate the entire enzyme architecture including the active sites(s) and the structurally complex protein backbone [47]. This design strategy leads to "synzymes" presumably employing mechanisms similar to their natural counterparts that follow general acid catalysis [48, 49] and Michaelis–Menten kinetics [50] to provide high catalytic efficiencies comparable to natural systems [29, 51]. The structural design of noncovalent and covalent organocatalysts [52–83], however, limits itself to the mimicry of the suggested active site of the natural catalyst, thus resulting in readily accessible, small, purely organic compounds operating in an enzyme-like fashion aiming at high catalytic activities and stereoselectivities at minimal structural complexity, such as *explicit double hydrogen-bonding (thio)urea organocatalysts*, as discussed in details in this chapter.

The success story of explicit double hydrogen-bonding (thio)urea organocatalysts started with seminal studies performed by Hine and co-workers [84], when identifying the clamp-like double hydrogen-bonding motif in cocrystals [85] between conformationally rigid 1,8-biphenylenediol **1** and Lewis basic substrates such as hexamethyl phosphoramide 1,2,6-trimethyl-4-pyridone, and 2,6-dimethyl-γ-pyrone. X-ray crystal structure analyses showed 1,8-biphenylenediol **1** capable of providing two identical strong hydrogen bonds simultaneously to the same oxygen atom of the respective substrate [84]. In 1985, the same group reported diol **1** (15–70 mol% loading) to efficiently catalyze the aminolysis of the epoxide phenyl glycidyl ether with diethyl amine in butanone at 30 °C (**(1)**; Figure 6.1). A Brønsted plot based on the catalysis of this model reaction by a range of substituted phenols suggested that diol catalyst **1** promoted this epoxide opening with an efficiency per OH group that is expected from a phenol 600-fold as acidic, and that both

6.1 Introduction and Background | 143

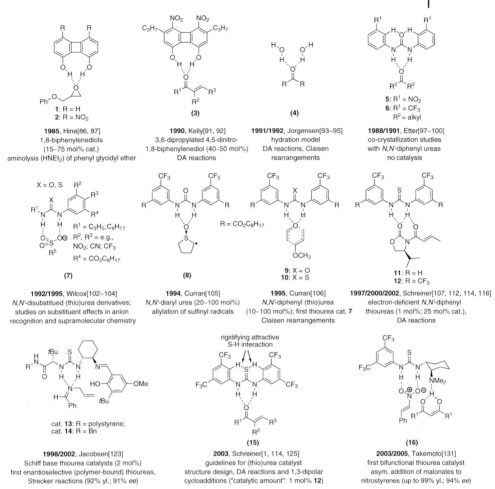

Figure 6.1 Chronological order of milestone achievements toward catalytically active (thio)urea organocatalysts utilizing explicit double hydrogen-bonding interactions for substrate activation.

hydroxy groups are involved in the epoxide activation as depicted in complex **(1)** (Figure 6.1) [33, 86, 87]. Further investigations on the ionization constants (pK_a values) and the double hydrogen-bonding ability of various 1,8-biphenylenediols having electron-rich or electron-deficient substituents, respectively [87, 88], indicated a close correlation between acidity and improved hydrogen-bonding properties in the formation of cocrystals; acidic 4,5-dinitro-1,8-biphenylenediol **2** was found to be the strongest double hydrogen-bond donor (**(2)**; Figure 6.1) [89, 90]. Based on these results Kelly *et al.*, in 1990, synthesized 3,6-dipropyl-4,5-dinitro-1,8-biphenylenediol **3** more soluble than **2** in the reaction solvent CD_2Cl_2 for catalysis (40–50 mol% loading **3**) of Diels–Alder reactions between various α,β-unsaturated aldehydes as well as ketones and predominantly cyclopentadiene at 55 °C. The observed rate enhancements up to 30-fold determined by ^{1}H NMR were interpreted by the activation of the dienophile through double hydrogen bonding of **3** to the carbonyl group giving substrate–catalyst complex **(3)** depicted in Figure 6.1 [91, 92]. This catalyst–substrate association motif was supported by computational studies on the accelerating solvent effect of water compared to aprotic solvents in Diels–Alder reactions [93, 94] and Claisen rearrangements [95]. The resulting hydration model suggested by Jorgensen *et al.*, in 1991/1992, explained the observed rate enhancements in water not with *cooperative* hydrogen bonding effects, but with a significant aqueous solvent effect described as clamp-like *explicit* hydrogen-bonding interaction of two water molecules to the carbonyl group leading to a preferential stabilization of the transition states (**(4)**; Figure 6.1). The pioneering work by Hine and Kelly demonstrated that structural rigidity of a double hydrogen-bonding organocatalyst such as 1,8-biphenylenediols and general acid catalysis [48, 49] ("partial *de*protonation"), instead of specific acid catalysis ("full *de*protonation") typical for Brønsted-acid catalysis [6, 47], are important aspects for the base of (thio)urea organocatalyst design. From 1988 to 1991, the Etter group published seminal X-ray structural studies on hydrogen-bond-directed cocrystallization of imides and N,N′-diphenyl ureas with Lewis basic compounds such as triphenylphospine oxide [96], nitroaromatics, ethers, ketones, and sulfoxides. For the first time, urea derivatives were also identified to be capable of explicit double hydrogen-bonding interactions affording stable host–guest complexes such as **(5)** formed by N,N′-bis(3-nitrophenyl)urea **5** (Figure 6.1) [97, 98]. Further intensive investigations on various hydrogen-bond patterns in cocrystal structures of organic compounds [99] and on hydrogen-bonding-mediated molecular recognition [100] suggested that urea derivative **5** readily formed stable highly qualitative 1 : 1 cocrystals with cyclohexanone due to the electron-withdrawing substituent at the *meta* position. This substituent pattern allows structure stabilizing intramolecular hydrogen bonds between the carbonyl group and the adjacent hydrogen atoms at the *ortho* position and inhibits self-association through intermolecular hydrogen bonds to the carbonyl group. However replacing the nitro group with electron-donating substituents (CH_3, OCH_3) or at the *para* position led to decreased hydrogen-bonding properties, complexation, and cocrystals of reduced quality, while electron-deficient N,N′-bis [(3-trifluoromethyl)phenyl]urea **6** was found to be a good complexing agent [98]. Hydrogen-bonding interactions between a ligand and a substrate have long been known to be responsible – among other noncovalent interactions – for the formation of supramolecular structures called supra- or supermolecules ("Übermolekül"), a term coined by Wolf *et al.*, in 1937 [101]. In often-overlooked experimental and theoretical investigations reported in 1992 by Wilcox and co-workers, a series of electron-rich and electron-deficient N-allyl [102]- and N-octyl-N′-phenyl (thio)ureas [103] were evaluated in solution ($CHCl_3$) for hydrogen-bond based molecular recognition of sulfonates, phosphates, carboxylates, and zwitterionic 4-tributylammonium-1-butanesulfonate as well as for their potential in supramolecular chemistry in dependence on their substituent pattern

[103, 104]. The formation of the complex and the host–guest association, respectively, were estimated in titration experiments by ^{1}H NMR shift data (downfield shifts of the N–H signal) and the shift of the UV/Vis absorptions relative to the substrate-free (thio)urea solutions [102]. In these studies again electron-deficient *meta-* or *para-*substituted thiourea derivatives turned out to be stronger double hydrogen-bond donors compared to the respective ureas and (thio)ureas, and acidity can be considered as a useful parameter to predict the ability of the hydrogen bond and thus improved molecular recognition.

Curran and Kuo, in 1994, introduced *N,N'*-diphenylurea **8** as the first double hydrogen-bonding urea organocatalyst accelerating the allylation of α-sulfinyl radicals with allytributylstannane ((**8**); Figure 6.1). Urea **8** incorporates *meta-*CF$_3$ groups as electron-withdrawing substituents, which are more compatible with the radical reaction than NO$_2$ groups, and also lipophilic ester substituents to improve the solubility in common organic solvents. In the presence of substoichiometric amounts of **8** (20–100 mol%), small rate accelerations and increased *cis/trans* selectivities were observed [105]. The same group, in 1995, utilized urea catalyst **9** (10–50 mol%) in Claisen rearrangements in C$_6$D$_6$ at 80–100 °C resulting in 1.7- to 5.0-fold rate accelerations; at 100 mol% **9** up to 22-fold relative rate enhancements were reported [106]. A derivative of **9** lacking both NH protons due to dimethylation (k_{rel} = 1.0) and a corresponding benzanilide (k_{rel} = 1.6) capable of providing only a single hydrogen bond proved to be catalytically inactive in the model Claisen rearrangement under identical conditions. This finding is in line with the afore-mentioned results on hydrogen-bonded substrate–catalyst complexes and it emphasizes the importance of the double hydrogen-bonding motif on the catalytic activity of (thio)urea organocatalysts. It is notable that thiourea derivative **10** was also examined in the model Claisen rearrangement of 1-methoxy-3-vinyloxy-propene. Although **10** turned out to decompose slowly under the reaction conditions, a small rate-accelerating effect could be estimated (at <10% decomposition: k_{rel} = 3–4). However, this experiment by Curran and Kuo marks the first application of a hydrogen-bonding thiourea derivative as organocatalyst [106]. Incomprehensibly, these promising findings had not been continued and their potential for catalyst development had remained unrecognized until the Schreiner group, in 1997, started research efforts toward explicit hydrogen-bonding thiourea organocatalysts [107]. Instead of urea derivatives, thiourea derivatives were chosen for proof-of-principle studies since (a) they are more soluble in various organic solvent, (b) they are easy to prepare (liquid thiophosgene is much easier to handle than phosgene), (c) the thiocarbonyl group is a much weaker hydrogen-bond acceptor leading to less catalyst self-association and to a higher concentration of free catalyst, and (d) they show a higher acidity (urea: pK_a = 26.9; thiourea: pK_a = 21.0) [108] possibly leading to more stable hydrogen-bonded catalyst–substrate complexes supporting Hine's, Kelly's, Etter's, and Wilcox's results mentioned above. Furthermore, encouraging indications for the strong hydrogen-bonding ability of thiourea derivatives have been provided by different research fields and applications, e.g., supramolecular chemistry, molecular (anion) recognition [109–111], crystal engineering, herbicides, and inclusion compounds.

In 2000, Schreiner and Wittkopp published their first results on thiourea organocatalysis [112]. The modification of Curran's thiourea derivative **7** by removing the coordinating ester groups led to electron-deficient hydrogen-bonding *N,N′*-bis [(3-trifluoromethyl)phenyl]thiourea **11** that was found to accelerate the Diels–Alder reaction of methyl vinyl ketone with cyclopentadiene employing a substoichiometric "catalytic amount" [113] of 1 mol% in cyclohexane (no cat: 18% conv.; **11**: 30% conv./1 h), chloroform (no cat.: 31% conv.; **11**: 52% conv./1 h), and even in the hydrogen-bonding environment of water (no cat.: 74% conv.; **11**: 85% conv.) [112]. Based on computations for the DA reaction between MVK and Cp, catalyst **11** was found to compete effectively with water and it stabilizes MVK by 6.4 kcal mol^{-1} (two water molecules by only 3.7 kcal mol^{-1}) and the polarized DA transition state through explicit double hydrogen-bonding interactions complementary to hydrophobic interactions through water [112, 114]. Subsequent binding studies on the interaction of **11** with a bidentate *N*-acyloxazolidinone dienophile were performed to verify the working hypothesis that a hydrogen-bonding thiourea organocatalyst such as **11** behaves like a Lewis acid (e.g., Et$_2$AlCl$_2$) [115] so that the 1,3-diketone is activated through bidentate coordination to both carbonyl groups accelerating Diels–Alder reactions [116]. Utilizing a combination of NMR methods, low-temperature IR spectroscopy, and computations on reduced model systems, various double hydrogen-bonded model complexes (1:1 ratio) between **11** and the *N*-acyloxazolidinone were analyzed assuming that the thiourea moiety in catalyst **11** adopts the required *syn* orientation (*trans/trans* rotamer) [117]; this was suggested from the crystal structure data of urea **5** [97, 98] and in particular of structurally related thiourea **12** [1, 118].[2] The analyses of the temperature-dependent NH signal situation (^{1}H NMR) revealed that **11** exhibits a large dimerization entropy ($\Delta S = -35.6$ cal mol K^{-1}) leading to efficient self-association only at low temperature (193 K), thus making free catalyst **11** available at room temperature (291 K) for favored complexation of the 1,3-diketone dienophile ($\Delta S = -9.6$ cal mol K^{-1}). A comparison of computed and measured (at 77 K) IR-active carbonyl absorptions employed as an indicator for intermolecular hydrogen-bonding interactions upon binding implies that **11** interacts with both the ring carbonyl ($\Delta v = +24$ cm^{-1}) and the side-chain carbonyl group ($\Delta v = +7.2$ cm^{-1}) in a mode in the bidentate complex leading to stabilization of the selective, but disfavored 1,3-diketone *syn* conformation. This catalyst–substrate interaction is supported by the observed diastereoselectivity (*dr*) of Diels–Alder reactions between cyclopentadiene and the *N*-acyloxazolidinone in the presence of 25 mol% **11**, or **12**, or AlCl$_3$, respectively. **11** gave 74% conv. (48 h, 23 °C) and (*dr* 77 : 23) and AlCl$_3$ 95% conv. (*dr* 92 : 8 at −78 °C; 1 h). An even increased efficiency at 25 mol% loading in the same model reaction was observed for more acidic *N,N′*-bis [3,5-(trifluoromethyl)phenyl]thiourea **12** (78% conv.; *dr* 81 : 19; 48 h, 23 °C) that, later on, was established by Schreiner and co-workers as highly active nonstereoselective privileged [119] organocatalyst for

---

2) The structure of thiourea catalyst **12** was deposited in the Cambridge Crystallographic Data Center (CCDC 206506) and can be retrieved free of charge from there. Key data: 1,3-bis-(3,5-bis-trifluoromethyl-phenyl)-thiourea; Formula: C$_{17}$ H$_8$ F$_{12}$ N$_2$ S$_1$; unit cell parameters: *a* 15.178(2) *b* 8.2203(8) *c* 17.399(2) β 112.495(14); space group P2$_1$/c.

various organic reactions (Section 6.2.1.1). These results demonstrated that small hydrogen-bond donors and metal-containing Lewis acids lead to isostructural complexes with 1,3-diketones despite large differences in the interaction energies. These results also revealed that double hydrogen-bonding thiourea organocatalysts such as **11** and **12** act like weak Lewis acids by lowering the LUMO of the dienophile through electron-deficient complexation affording the observed rate accelerations [116, 120, 121]. Moreover, hydrogen-bonding (thio)urea organocatalysis for the first time had been specifically identified as a desired metal-free approach for the catalysis of a organic transformation, which was traditionally dominated by metal-containing catalysts. In parallel developments, the Jacobsen group, however, originally involved in the development of organometallic catalytic systems searched for a novel tridentate chiral ligand for the efficient chirality transfer in asymmetric Strecker reactions of *N*-allyl-protected aldimines. In 1998, high-throughput screening (HTS) of polystyrene-bound tridentate Schiff bases provided the first enantioselective (92% yield.; 91% *ee*) thiourea organocatalysts **13** and **14** (Figure 6.1) [122]; the high efficiencies of these Schiff base catalysts[3] were ascribed, in 2002, to activate double hydrogen-bonding interactions between the imine and the thiourea catalyst, which supported Schreiner's findings [107, 112, 116, 123] on thiourea catalyst activity (Section 6.2.2.1) [124]. Schreiner and Wittkopp, in 2003, elucidated the thiourea structure–activity relationship in a series of Diels–Alder reactions between cyclopentadiene and five different dienophiles such as MVK and (aza)chalcones in the presence of various symmetrically *N,N′*-dialkyl- and diphenyl-substituted thiourea derivatives (1 mol% loading) as potential catalysts [114]. Kinetic data determined by ^{1}H NMR revealed thiourea derivatives incorporating rigid electron-withdrawing aromatic substituents such as **11** (k_{rel} = 2.5–5.9) and **12** (k_{rel} = 4.8–8.2) to be the most efficient in accelerating the model DA reactions, while dialkyl- and electron-rich *N*-phenyl thioureas bearing *ortho* substituents proved to be poor catalysts (e.g., *N,N′*-diphenyl thiourea: k_{rel} = 1.1–1.5). Product inhibition turned out to be low since the catalytic activity was still present after 80% conversion. This is in line with the weak enthalpic binding of the thiourea catalyst to carbonyl groups (~7 kcal mol^{-1} at rt) [116] and emphasizes the entropy term in the formation of the catalyst–substrate complex. The more rigid the free thiourea catalyst, the more stable is the complex, minimizing the entropy loss upon complexation. Thiourea catalysts having flexible substituents, such as octyl or the phenyl groups have low rotational barriers (e.g., *N,N′*-diphenyl thiourea: 1.5 kcal mol^{-1}), while thiourea derivative **12** appears to be conformationally stable, forms more stable complexes, and gives increased rate enhancements due to a higher barrier (3.4 kcal mol^{-1}) – an approach already applied

---

3) The Jacobsen group introduced the term "Schiff base catalyst" [122] to demonstrate that the structure of this novel catalyst class originates from Schiff base ligands and incorporates a Schiff base moiety; notably, this term does not indicate that these catalysts operate as bases, but their high catalytic efficiencies result from explicit double hydrogen bonding as shown in Section 6.2.2.1. In the following, the scientifically established term "Schiff base (thio)urea" is used to describe this class of organocatalysts.

by Hine and Kelly [86, 91]. The rigidity of **12** and comparable thioureas bearing electron-deficient *meta*- or *para*-substituted phenyl substituents is suggested to root in intermolecular hydrogen-bonding interactions between the positively polarized *ortho* hydrogen atom and the Lewis-basic thiocarbonyl sulfur. This attractive interaction restrains the rotation of the thiourea moiety and favors complexation entropically, while substituents at the *ortho* position lead to repulsive interactions with the thiocarbonyl group decreasing the rotational barrier. In addition to this structural effect, **12** possesses due to CF_3 substitution acidified NH protons capable of providing strong hydrogen bonds. In 2003, Schreiner highlighted these experimental and theoretical findings in the first review [1] on (thio)urea organocatalysts utilizing explicit double hydrogen-bonding interactions and provided rough guidelines and concepts for (thio)urea catalyst design: (a) if the catalyst is able to interact with the starting material, the transition state (TS), and the products, it is necessary that the relative stabilization of the TS is the largest; (b) the bi- or multidentate mode of catalyst–substrate binding increases catalytic efficiency and restricts degrees of freedom; (c) the catalyst structure should be an adequate compromise between rigidity and flexibility; (d) to avoid self-association, the catalyst should not incorporate strong hydrogen-bond acceptors such as ester groups – the noncoordinating acidifying CF_3 group as in the 3,5-(trifluoromethyl)phenyl moiety of **12** appeared to be ideal; (e) weak interactions reduce product inhibition allowing "catalytic amounts" of the catalyst, and also from an environmental point of view [38, 39, 125, 126]; and (f) the catalyst should be water-compatible or even catalytically active in water [127, 128]. These guidelines are not limited to the development of nonstereoselective (thio)urea organocatalysts (Section 6.2.1), but also or in particular are applicable to stereoselective derivatives of benchmark thiourea catalyst **12** (Section 6.2.2) as first demonstrated by Takemoto and co-workers, in 2003, when introducing the first bifunctional hydrogen-bonding thiourea organocatalyst **16** depicted in Figure 6.1 (Section 6.2.2.1) [129].

Regarding the number of publications per year on (thio)urea organocatalysis illustrated in the bar chart (Figure 6.2) mirrors that not only the resurrection of the term "organocatalysis", in 2000 [130], a translation of the old and neglected concept "organic catalysis" coined by Langenbeck with the first authoritative review on "organic catalysts" [131–136], but also the early publications in 1997 to 2003 provided inspiring and encouraging impulses on the scientific community to re-focus on this research field. The steep increase in the related literature also results from the often easy and inexpensive preparation of tailor-made (thio)urea organocatalysts similarly to a unit construction system employing primary amine-functionalized chiral compounds, e.g., chiral pool compounds, as (privileged) chiral scaffolds [119] and isocyanates or isothiocyanates as building blocks, respectively, to obtain the respective chiral (thio)urea derivatives. Notably, these chiral scaffolds such as *trans*-1,2-diaminocyclohexane and related amines [137], cinchona alkaloids [138, 139], 2.2′-binaphthol derivatives [140], amino alcohols [138], oxazolines [141, 142], and (thio)urea derivatives [143, 144] are incorporated in chiral ligands of highly efficient organometallic catalytic

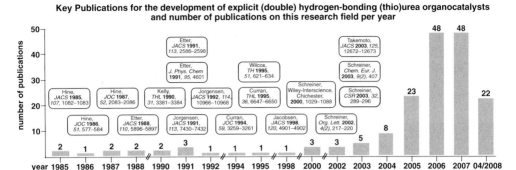

Figure 6.2 Key publications on explicit (double) hydrogen-bonding (thio)urea (organocatalysts) and number of publications per year in this research field.

systems that represent benchmarks for the catalytic performance of (thio)urea organocatalysts. Because of the strong hydrogen-bond donor ability, the 3,5-bis(trifluoromethyl)phenyl thiourea moiety derived from thiourea **12** has gained importance as "substrate anchor" and is incorporated in various highly active and (non)-stereoselective double hydrogen-bonding thiourea derivatives (Figure 6.3). The corresponding achiral, 5-bis(trifluoromethyl)phenyl isothiocyanate is readily accessible by a one-step procedure [145].

In the following book chapter [118] the last 10 years of research on (thio)urea organocatalysis are summarized considering catalyst design concepts, experimental details such as structure optimization studies, screening conditions, reaction conditions, the typical substrate and product spectrum of each procedure as well as proposed mechanistic scenarios for each published methodology (~150 articles).

6.2
Synthetic Applications of Hydrogen-Bonding (Thio)urea Organocatalysts

6.2.1
Nonstereoselective (Thio)urea Organocatalysts

6.2.1.1 Privileged Hydrogen-Bonding N,N'-bis-[3,5-(Trifluoromethyl)phenyl]thiourea

Wittkopp and Schreiner introduced the simple electron-deficient N,N'-bis [3,5-(trifluoromethyl)phenyl]thiourea **9** (Figure 6.3) as an efficient double hydrogen-bonding organocatalyst in a series of Diels–Alder reactions and 1,3-dipolar cycloadditions of

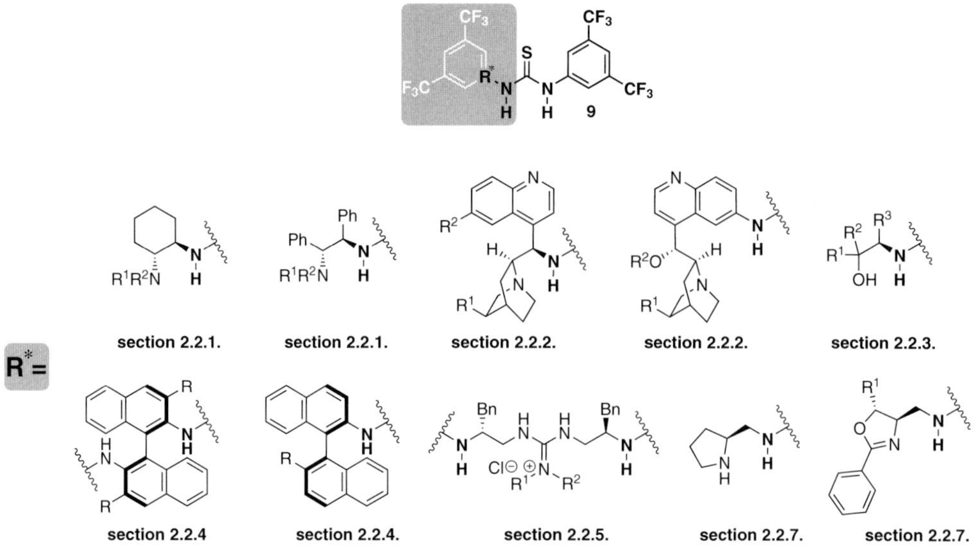

Figure 6.3 Stereoselective, chiral thiourea derivatives of achiral benchmark thiourea organocatalyst N,N'-bis [3,5-(trifluoromethyl)phenyl]thiourea **9**; stereoselective hydrogen-bonding thiourea organocatalysts incorporating the privileged 3,5-bis(trifluoromethylphenyl)thiourea moiety. The (thio)urea catalyst structure is the leitmotif for the chapter organization.

nitrones [1, 114, 116]. These proof-of-principle studies were also accompanied by computational investigations and soon thereafter were taken up by several groups for the development of metal-free, nonstereoselective synthetic applications, which are summarized and presented in the following section. Thiourea catalyst **9** is readily accessible in over 80% yield from inexpensive thiophosgene and 3,5-bis(trifluoromethyl) aniline by a straightforward large-scale (100 mmol) procedure [146].

As demonstrated in a series of kinetic experiments by Wittkopp and Schreiner, nitrone N-benzylideneaniline N-oxide can be activated for 1,3-dipolar cycloadditions through double hydrogen-bonding **9** [1]. Takemoto and co-workers, in 2003, published the nucleophilic addition of TMSCN and ketene silyl acetals to nitrones and aldehydes proceeding in the presence of thiourea organocatalyst **9** (Figure 6.4) [147].

The initial screening reaction of 6-methyl-2,3,4,5-tetrahydropyridine N-oxide and TMSCN (5 equiv.) in the presence of 50 mol% of amide **13**, diphenyl urea **14**, and thiourea derivatives **15**, **8**, and **9**, respectively, in CH_2Cl_2 at −78 °C verified the correlation between N−H acidity and the strength of hydrogen bonds. These studies emphasized the importance of double hydrogen-bonding coordination for effective substrate activation and rate acceleration (Figure 6.5).

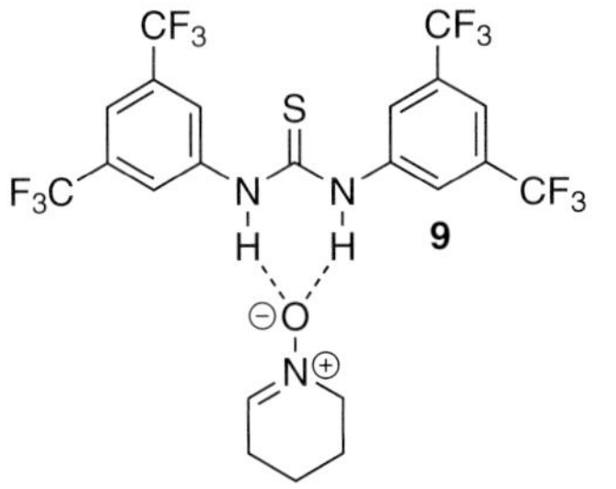

Figure 6.4 Proposed double hydrogen-bonding activation of nitrones through thiourea derivative **9**.

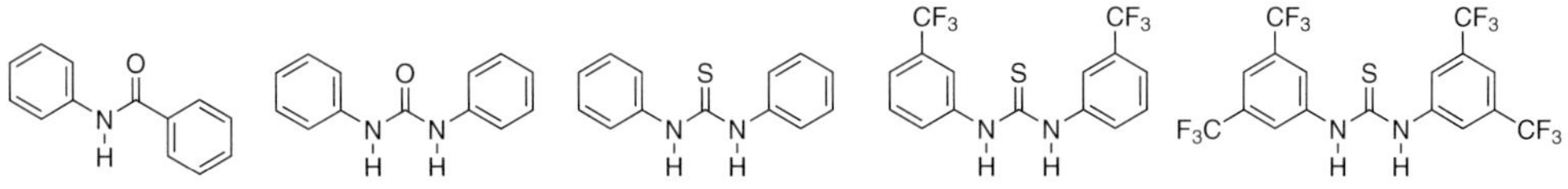

13
83% yield/180 min

14
77% yield/90 min

15
81% yield/45 min

8
75% yield/15 min

9
81% yield/15 min

Figure 6.5 Hydrogen-bond donors (50 mol% loading) screened for the reaction of nitrone 6-methyl-2,3,4,5-tetrahydropyridine *N*-oxide and TMSCN at −78 °C. The yields and reaction times are given for the resulting adducts.

The scope of this cyanation was demonstrated by the reaction of acyclic, cyclic, and chiral nitrones with TMSCN producing the respective desilylated adducts **1–4** in yields ranging from 75 to 96% when utilizing 50 mol% of thiourea catalyst **9** (Scheme 6.2). The uncatalyzed reference experiments gave comparable yields but required longer reaction times (Scheme 6.2). In case of a conjugated nitrone the 1,2-adduct **3** instead of the 1,4-adduct was isolated and only a marginal effect of catalyst **9** on the diastereoselectivity (*syn*/*anti* 42 : 58) was observed in the formation of adduct **4** from a chiral nitrone (Scheme 6.2).

The authors also reported addition reactions of ketene silyl acetals to various **9**-activated nitrones followed by a desilylation and subsequent base-induced cyclization resulting in the products **1–6** in moderate to good yields (52–88%), as shown in Scheme 6.3. Without **9**, for all examples, no product formation occurred under otherwise identical conditions. Performing the reaction at reduced catalyst loading (10 mol%) neither affected reaction time nor product yield (adduct **6**; Scheme 6.3). For the screening reaction, the quantitative recovery of catalyst **9** through column chromatography as well as catalyst reusability without the loss of catalytic activity (1. cycle: 76% yield/20 min; 2. cycle: 86% yield/20 min) were demonstrated.

The **9**-catalyzed Mukaiyama-aldol reaction [74] of benzaldehyde and 1,2-dimethoxy benzaldehyde with a ketene silyl acetal in the presence of 10 mol% thiourea **9** furnished the target product in low yield (36%), while the same reaction

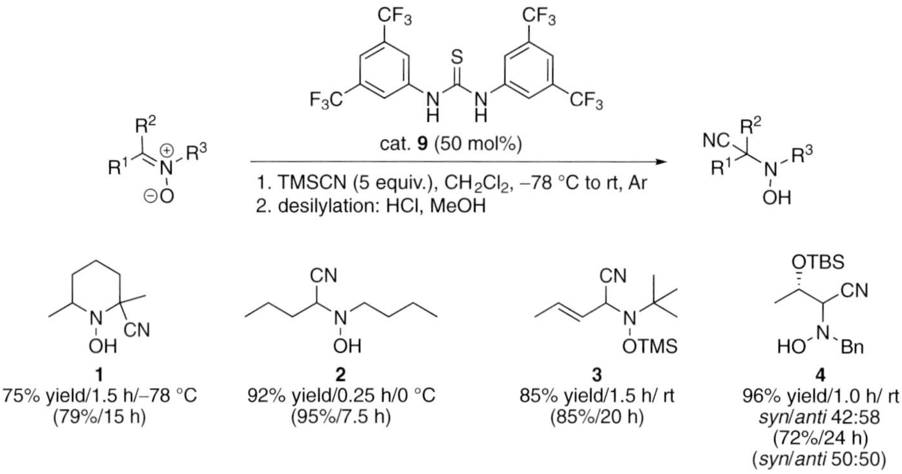

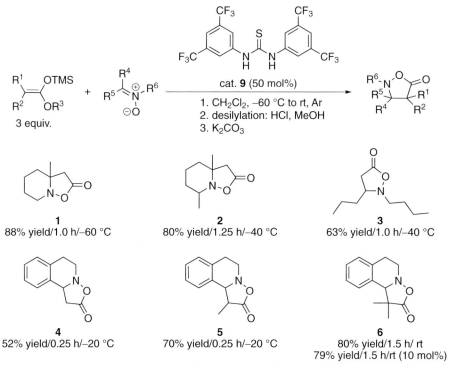

Scheme 6.2 Cyanation of nitrones utilizing TMSCN as cyanide source and catalyst **9**; yields in parentheses refer to uncatalyzed reactions.

Scheme 6.3 Product range of the addition reactions of ketene silyl acetals to various nitrones activated by thiourea catalyst **9**.

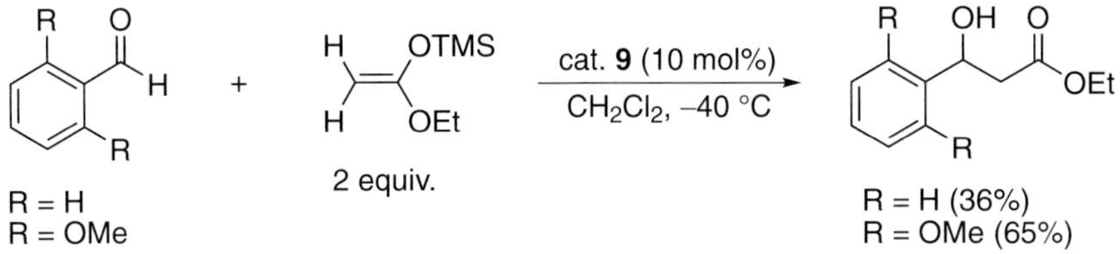

6.2 Synthetic Applications of Hydrogen-Bonding (Thio)urea Organocatalysts

R = H
R = OMe

R = H (36%)
R = OMe (65%)

Scheme 6.4 Mukaiyama-aldol reaction of benzaldehydes with a ketene silyl acetal catalyzed by thiourea **9**.

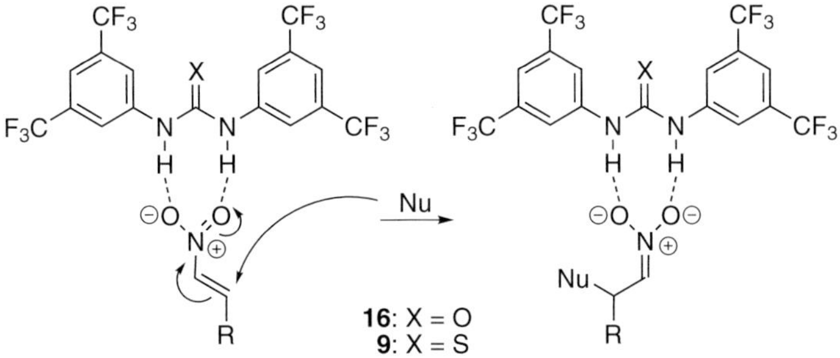

16: X = O
9: X = S

Scheme 6.5 Nitroalkene activation via double hydrogen-bonding enhances electrophilicity at β-position and facilitated Michael-type attack of the (hetero)aromatic nucleophile resulting in Friedel–Crafts adducts.

of an aldehyde bearing the methoxy groups at the *ortho* position of the aromatic ring proceeded smoothly giving 65% yield (Scheme 6.4) [147]. The reason for the higher reactivity of the electron-rich benzaldehyde could be derived from a potential bidentate coordination of **9** at the oxygens of the carbonyl group and the contiguous methoxy group allowing a more activated catalyst–substrate intermediate.

Ricci *et al.* applied thiourea derivative **9** and its oxygen analogue urea **16** in a comparative catalysis study to the syntheses of alkylated aromatic and heteroaromatic N-containing compounds through Friedel–Crafts alkylation [148]. In analogy to the earlier discussed hypothesis (Scheme 6.5), the alkylating nitroalkenes may be activated through double hydrogen-bonding to facilitate the product-forming Michael-type nucleophilic attack on the β-position of β-nitrostyrene and (2-nitroethenyl) cyclohexane.

To demonstrate the catalytic efficiency of thiourea **9** and urea **16** (each 10 mol% loading), Friedel–Crafts alkylation of various aromatic and heteroaromatic substrates was performed at room temperature in toluene as well as under solvent-free conditions. The results for the products **1–7** shown in Scheme 6.6 revealed that in all cases the **9**-catalyzed reactions gave higher yields. In toluene N-methylpyrrole reacted smoothly to give the 2-substituted Friedel–Crafts adduct **1**, while the adducts **3–5** and **7** formed slowly and required longer reaction times (72 h). The

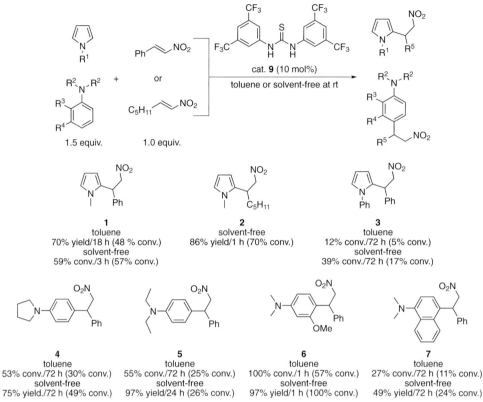

Scheme 6.6 Products resulting from **9**-catalyzed Friedel–Crafts alkylation of aromatic and heteroaromatic N-containing substrates performed in toluene and without solvent.

presence of an electron-releasing function in the aryl moiety, as in the case of m-OMe-N,N'-dimethylaniline increased the reactivity and led to quantitative yields of adduct **6** within 1 h reaction time (Scheme 6.6). Solvent-free conditions remarkably improved the catalytic activity resulting in moderate to excellent yields of the Friedel–Crafts adducts (Scheme 6.6). Notably, catalyst-free control experiments failed in all cases. The same group extended this organocatalytic Friedel–Crafts alkylation to indole and various methyl-indoles serving as nucleophiles toward Michael acceptor β-nitroalkenes [148]. The formation of the products **1–5** proceeded with good conversions (82–93%) in toluene and provided good (83%) to very good yields (94%) under solvent-free conditions (Scheme 6.7). Under these conditions and MW irradiation (100 W) at 10 mol% loading of **9** even the challenging 2-position of 3-methyl indole was alkylated (49% yield/20 min). Since the **9**-promoted Friedel–Crafts alkylation and Michael addition [149–152] worked

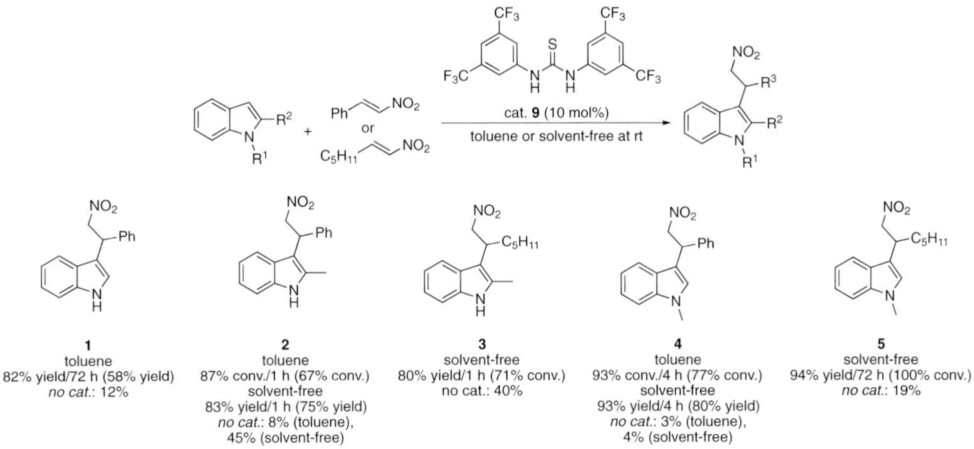

6.2 Synthetic Applications of Hydrogen-Bonding (Thio)urea Organocatalysts | 155

Scheme 6.7 Product range of **9**- and urea **16**-catalyzed Friedel–Crafts alkylations of indole and methyl indoles.

1
toluene
82% yield/72 h (58% yield)
no cat.: 12%

2
toluene
87% conv./1 h (67% conv.)
solvent-free
83% yield/1 h (75% yield)
no cat.: 8% (toluene),
45% (solvent-free)

3
solvent-free
80% yield/1 h (71% conv.)
no cat.: 40%

4
toluene
93% conv./4 h (77% conv.)
solvent-free
93% yield/4 h (80% yield)
no cat.: 3% (toluene),
4% (solvent-free)

5
solvent-free
94% yield/72 h (100% conv.)
no cat.: 19%

under acid-free and mild conditions, respectively, no polymerization side products were detected.

In 2006, the Schreiner group published an environmentally benign protocol toward β-amino alcohol synthesis by utilizing the effects of hydrogen-bonding thiourea **9** and the hydrogen-bonding environment of water for the acceleration of nucleophilic epoxide aminolysis with various primary and secondary amines [153]. Water not only served as the solvent but also promoted the reaction through "hydrophobic hydration" [154, 155] similar to Nature's enzyme catalysis in water. The simple rationale behind the development of this concept is that water avoids mixing with organic solutes, because this will lead to increased structuring and thus a loss of entropy of the water molecules around the solutes. Catalysis in water depends on the ability of the catalysts to tolerate water on the one hand and to remain active on the other. As already shown by Wittkopp and Schreiner in Diels–Alder reactions, hydrogen-bonding thiourea derivatives such as **9** displayed catalytic efficiency at substoichiometric loadings even in water [114], that is, thiourea **9** remained capable of forming explicit double hydrogen-bond patterns to the basic epoxide oxygen and thus facilitating nucleophilic ring-opening (Scheme 6.8). Hence, instead of being mutually exclusive, hydrogen-bonding catalysts and water complement each other.

Running a series of aminolysis reactions of propene oxide (at rt) and cyclohexene oxide (at 40 °C) in a heterogeneous water mixture (emulsion or suspension) and in dichloromethane (solution) under otherwise identical conditions at 10 mol% loading of **9** revealed that the reactions proceeded best in water. The catalytic activity of hydrogen-bonding catalyst **9** was amplified in comparison to dichloromethane to give the corresponding β-amino alcohols **1–10** in good (60%) to excellent (97%) yields within 24 h reaction time as shown in

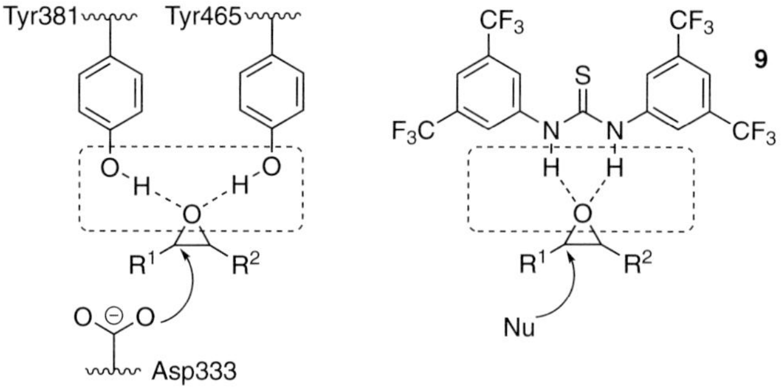

Scheme 6.8 Epoxide recognition for epoxide hydrolase that detoxify living cells by catalyzing alcoholysis to water soluble diols. The working model involves the phenolic H-atoms of two tyrosines activating the epoxide for nucleophilic attack. This principle is realized analogously by double hydrogen-bonding thiourea catalyst **9** in the natural medium water.

Scheme 6.9. These findings of relative accelerations as large as 200-fold supported the assumed catalytic amplification due to hydrophobic interactions, and this eventually led to the term "hydrophobic amplification," a key element in enzyme catalysis, which for the first time was implied in organocatalytic reactions with neutral molecules. The amplifying effect also occurred in the **9**-catalyzed (10 mol%) opening of styrene oxide with thiophenol (DCM: 32%; 76% yield; 24 h; rt) and phenol (DCM: 30%; 74% yield; 24 h; rt). With propene oxide, only the sterically less hindered regioisomer was formed, whereas the reverse was found for styrene oxide. The latter result is likely due to benzyl conjugation that outweighed the steric effect. DFT computations performed on the hydrogen-bonded complexes and transition structures (TS) for the opening of ethylene oxide with NH_3 with and without thiourea in the gas phase, CH_2Cl_2, and water as model clusters identified a polar TS, which was favorably stabilized by water molecules resulting in the lowest overall barrier and an additional rate enhancement due to the TS inclusion into the hydrophobic hydration cavity [153]. Further evidence of "hydrophobic amplification" was provided by the 20–40% decrease in the yields when the aminolysis was carried out, e.g., with morpholine in D_2O (62% yield; 36 h) instead of H_2O (83% yield; 36 h). D_2O has approximately 20% higher viscosity that makes mixing more difficult and reduces the hydrophobic effect.

Kotke and Schreiner applied the principle of double hydrogen-bonding organocatalysis to high-yielding acid-free acetalizations [118, 146]. This approach includes various aromatic, aliphatic, unsaturated, and acid-labile aldehydes and ketones, which could be cleanly acetalized in the presence of ethanol, methanol, 1-propanol, 2-propanol, and 1,2-ethanediol as the alcohol components and as solvents to the respective acetals **1–15** in good to excellent yields (61–95%) at low catalyst loadings of only 0.01–1 mol% at room temperature (Scheme 6.10). The **9**-catalyzed acetalization of aromatic aldehydes and simple aliphatic aldehydes

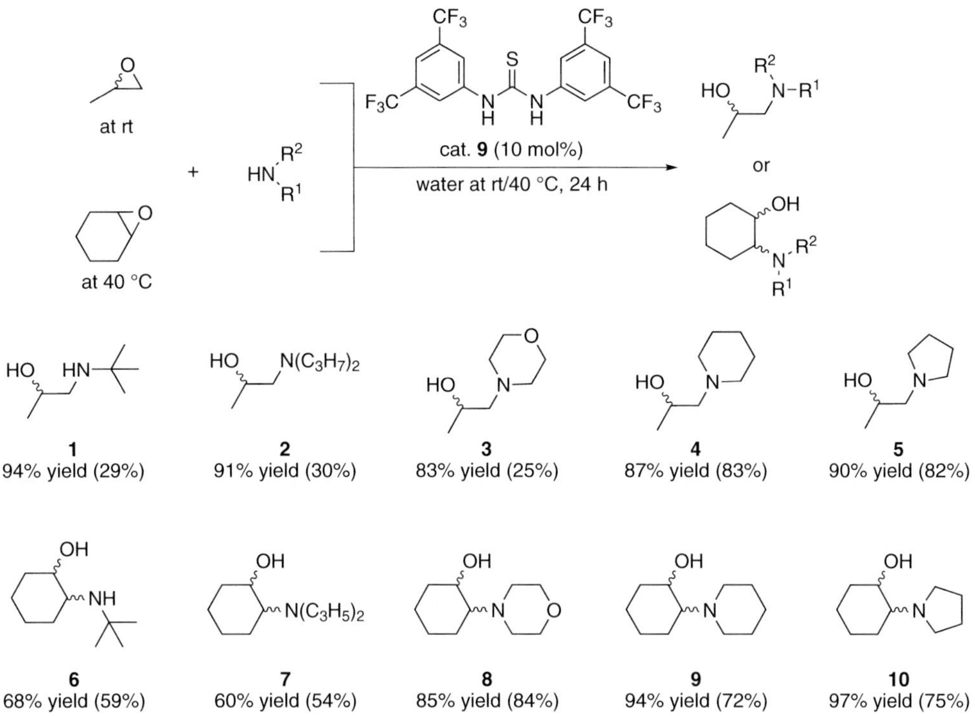

Scheme 6.9 Thiourea **9**-catalyzed aminolysis of propene oxide and cyclohexene oxide conducted in water to utilize "hydrophobic amplification." The yields of uncatalyzed control experiments in water are given in parentheses.

performed equally well resulting in the respective acetals in excellent yields (Scheme 6.10). Considerably less reactive ketones such as cyclohexanone and acetophenone can also be acetalized to the corresponding products (61–65% yield) but as expected, at considerably longer reaction times (92–98 h) (Scheme 6.10). Scale-up in preparative (20 mmol) experiments (Scheme 6.10) also performed well and underlined the synthetic utility of this organocatalytic protocol; the catalyst loading could be reduced routinely to 0.01 mol%, which still gave high yields at marginally extended reaction times (Scheme 6.10). Turnover revealed significant enough to be expressed in terms of turnover number (TON) and, better, turnover frequency (TOF). The authors found for the diethyl acetalization of p-chlorobenzaldehyde (product **3**) and octanal (product **9**) TOFs of 632 h^{-1} (TON = 9800) and 577 h^{-1} (TON = 9700), respectively. A limitation of this protocol concerned the acetalization of electron-rich substrates such as p-tolylbenzaldehyde because these required reaction longer reaction times (250 h) due to their lower electrophilicities. The significant difference in reaction times of aldehydes and ketones translated into the observed chemoselectivity as evident from a competition experiment between

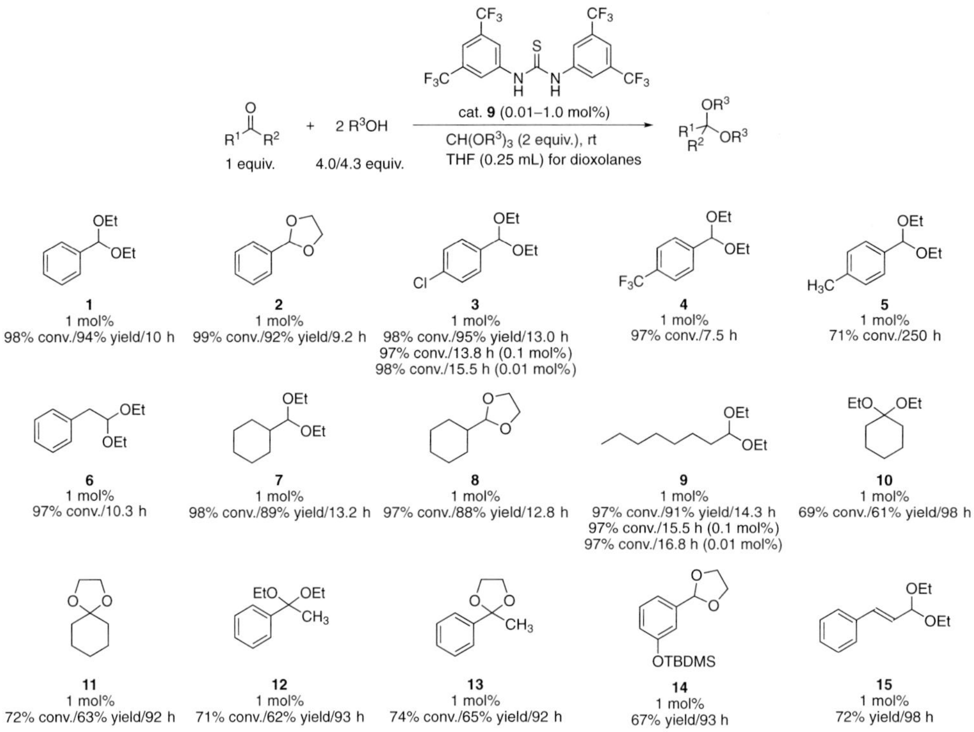

Scheme 6.10 Range of representative acetals prepared from the **9**-catalyzed acid-free acetalization of various aldehydes and ketones. The yields refer to preparative experiments (20 mmol scale).

benzaldehyde and acetophenone to prepare the respective diethyl acetals. After 8 h, an acetal product mixture of 6.1:1 favoring the acetal of diethyl benzaldehyde acetal was detected (^{1}H NMR). The practicality of this acid-free acetalization was further exemplified (Scheme 6.10) with the acetalization of acid-labile TBDMS-protected *m*-hydroxy benzaldehyde to give the desired acetal in 67% yield (93 h), which was reported to react rather sluggishly under Brønsted or Lewis-acid catalysis. The clean conversion of *trans*-cinnamic aldehyde to its diethyl acetal also emphasized the synthetic application of this acid-free conversion (Scheme 6.10). All uncatalyzed reactions run in parallel under otherwise identical conditions produced no product within the time required to complete the catalyzed transformation. Even after one week, the uncatalyzed reactions generally gave <1% of the respective acetals [146].

Mechanistically, the authors favored a thiourea **9**-assisted heterolysis of the orthoester through hydrogen bonding as the entry into the catalysis cycle of the organocatalytic acetalization. The orthoester was suggested to serve as the source of the alcoholate, which rapidly attacks the carbonyl compound to form a

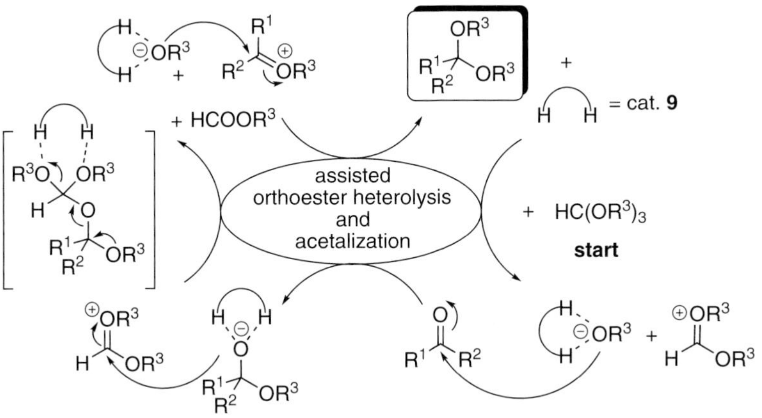

Scheme 6.11 Proposal for the catalytic cycle of the acid-free, organocatalytic acetalization induced by **9**-assisted orthoester heterolysis.

hydrogen-bonded hemiacetal anion as presented in Scheme 6.11. Subsequent nucleophilic attack furnished the acetal and released the catalyst **9** without product inhibition to start a new cycle. Experimental evidence of this mechanistic proposal resulted from the attempts to perform thioacetalization reactions, which were also found to be accelerated in the presence of **9** and in the absence of $HC(OEt)_3$. Adding the orthoester only produced normal diethyl acetal although the thiols are much better nucleophiles [146].

Ricci and co-workers utilized thiourea **9** for the activation of β-nitrostyrene for the conjugate addition of (hetero)aromatic N-containing compounds such as indoles resulting in the Friedel–Crafts alkylated adducts (Schemes 6.6 and 6.7) [148]. The same group used various **9**-activated nitroalkenes as Michael acceptors for the conjugate addition of hydrazones derived from formaldehyde and enolizable aldehydes [156]. In the initial **9**-catalyzed Michael reaction [149–152] (20 mol% catalyst loading), N,N′-dimethyl formaldehyde hydrazone showed the expected reactivity and added via the azomethine carbon to β-nitrostyrene to form the Michael adduct in 90% yield after 18h at room temperature, while the uncatalyzed reaction remained incomplete under otherwise identical conditions. Enolizable hydrazones were identified to react as ene-hydrazine nucleophiles not from the azomethine carbon but from the β-position to afford the γ-nitrohydrazones **1–8** in yields ranging from 58 to 92% (Scheme 6.12). The observed product formation was explained by a mechanistic proposal based on the equilibrium between the enolizable hydrazone and its ene-hydrazine structure, which attacked the **9**-activated nitroalkene on the electrophilic β-position as shown in Scheme 6.13.

Products containing two adjacent stereogenic centers were obtained as mixtures of diastereoisomers that could be separated by column chromatography (*dr* values, Scheme 6.12). The N,N′-dimethyl acetaldehyde hydrazone appeared remarkably

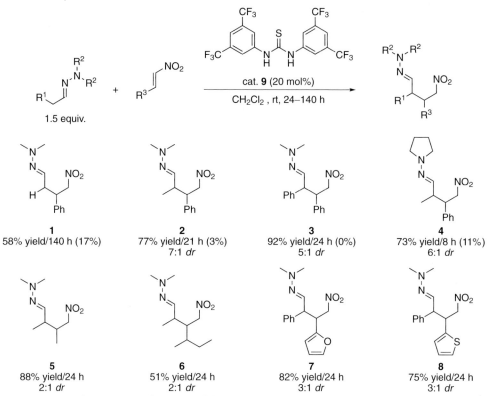

Scheme 6.12 Product range of the **9**-catalyzed Michael addition between nucleophilic hydrazones and various nitroalkenes.

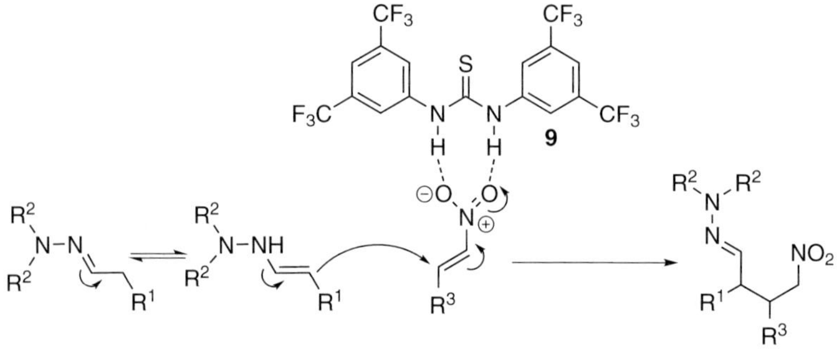

Scheme 6.13 Mechanistic proposal for the catalytic effect of hydrogen-bonding thiourea **9** and the product formation resulting from an equilibrium between the hydrazone and its nucleophilic ene-hydrazine form.

less reactive and provided the resulting adduct **1** in 140 h, while the adducts **2** and **3** were formed smoothly after approximately 24 h (Scheme 6.12). These results could be rationalized by considering different stabilities of the ene-hydrazine moieties of the hydrazones; the less substituted acetaldehyde-derived ene-hydrazine was rather unstable as compared to the more substituted or conjugated ene-hydrazines, which afforded products **2** and **3**. The hydrazone originated from isobutyraldehyde turned out to be unreactive due to steric hindrance of the bulky isopropyl group at the reacting center in the corresponding ene-hydrazine. The Michael reaction of the *N,N'*-dimethyl hydrazone of benzaldehyde or pivalaldehyde with β-nitrostyrene as Michael acceptor failed because no ene-hydrazine was formed. Ionic liquids (BMImBF$_4$ or BMImPF$_6$: 1-butyl-3-methylimidazolium tetra- or hexafluoroborate) accelerated the uncatalyzed formation of product **2**, compared to other solvents (THF, MeOH, DCM), and showed full disappearance of β-nitrostyrene after 24 h; however, they produced poorer yields after work-up (30–50%). The self-evident combination of an ionic liquid with catalyst **9** proved to be unsuccessful. The synthetic utility of this mild organocatalytic approach was emphasized when applying Lewis-acid catalysis (e.g., catalytic amounts of Sc(OTf)$_3$, In(OTf)$_3$, InF$_3$, or Cu(OTf)$_2$) to the formation of adduct **2**; in all cases the starting material underwent acid-induced decomposition resulting in just traces of the desired product [156].

In 2006, List *et al.* presented a thiourea **9**-catalyzed procedure for the synthesis of *N*-acetylated Strecker adducts [158], which are useful intermediates in the synthesis of α-amino acids [159]. This novel approach introduced acetyl cyanide as practical and readily available aldimine cyanation reagent that was first studied by Dornow *et al.* in 1956 [159, 160]. Utilizing thiourea **9** (2–5 mol% loading) and dichloromethane as solvent at 0 °C, a variety of aldimine substrates were transformed to their corresponding *N*-acetyl Strecker adducts **1–8** in yields ranging from 64 to 96% (Scheme 6.14). Both aromatic aldimines with electron-donating or electron-withdrawing substituents (adducts **1–3**), as well as (hetero)aromatic aldimines (adducts **4** and **5**), could be transformed with similar efficiencies. Aliphatic α-branched adducts such as **6** were also accessible (Scheme 6.14). Mechanistically, the authors suggested that the reaction proceeded via an initial reaction of the aldimine with acetyl cyanide to form an acyl iminium–cyanide ion pair, which recombines in a final **9**-catalyzed product-forming step.

An asymmetric version of the aldimine acylcyanation utilizing Jacobsen's Schiff base catalyst **47** (Section 6.2.2.1; Scheme 6.47) [161] and a one-pot, three-component **9**-catalyzed modification of the acylcyanation described above were developed quickly [162]. This three-component acyl-Strecker reaction was performed with various aliphatic and (hetero)aromatic aldehydes, amines, and acetyl cyanide for the *in situ* aldimine formation and hydrogen cyanide generation. The **9**-promoted reaction required the presence of a drying agent to scavenge the water formed during the aldimine synthesis in the absence of an HCN source. The best results in the initial screening reaction of benzaldehyde, benzyl amine, and acetyl cyanide in dichloromethane were observed with molecular sieve (MS) 5 Å (99% conv.; 24 h) instead of MgSO$_4$ (86% conv.; 24 h) and when the aldimine formation

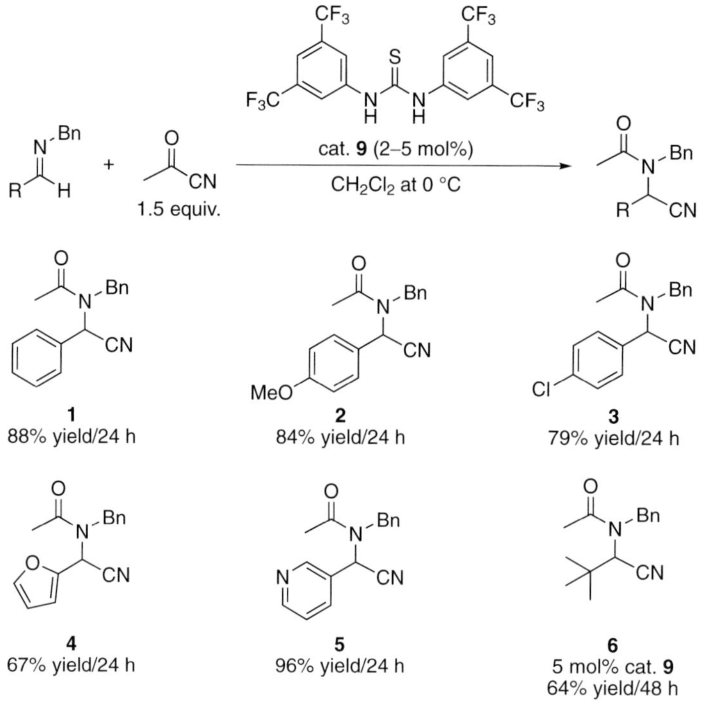

Scheme 6.14 Product range of the **9**-catalyzed acetyl cyanation reaction of aldimines with acetyl cyanide as the cyanide source.

proceeded under stirring of the mixture of aldehyde, amine, MS, and thiourea **9** (5 mol% loading) for 2 h at room temperature before the acetylcyanation step was initiated by adding acetyl cyanide at 0 °C. The uncatalyzed screening reaction showed 42% conversion after 24 h. Under optimized conditions, this method provided the *N*-acetyl-Strecker adducts **1–10** of various aldehydes and amines in moderate to good yields (48–85%) at practical reaction times (36 or 48 h) as presented in Scheme 6.15.

In 2007, another departure from carbonyl-type activation was marked by Kotke and Schreiner in the organocatalytic tetrahydropyran and 2-methoxypropene protection of alcohols, phenols, and other ROH substrates [118, 145]. These derivatives offered a further synthetically useful acid-free contribution to protective group chemistry [146]. The **9**-catalyzed tetrahydropyranylation with 3,4-dihydro-2*H*-pyran (DHP) as reactant and solvent was described to be applicable to a broad spectrum of hydroxy functionalities and furnished the corresponding tetrahydropyranyl-substituted ethers, that is, mixed acetals, at mild conditions and with good to excellent yields. Primary and secondary alcohols can be THP-protected to afford **1–8** at room temperature and at loadings ranging from 0.001 to 1.0 mol% thiourea

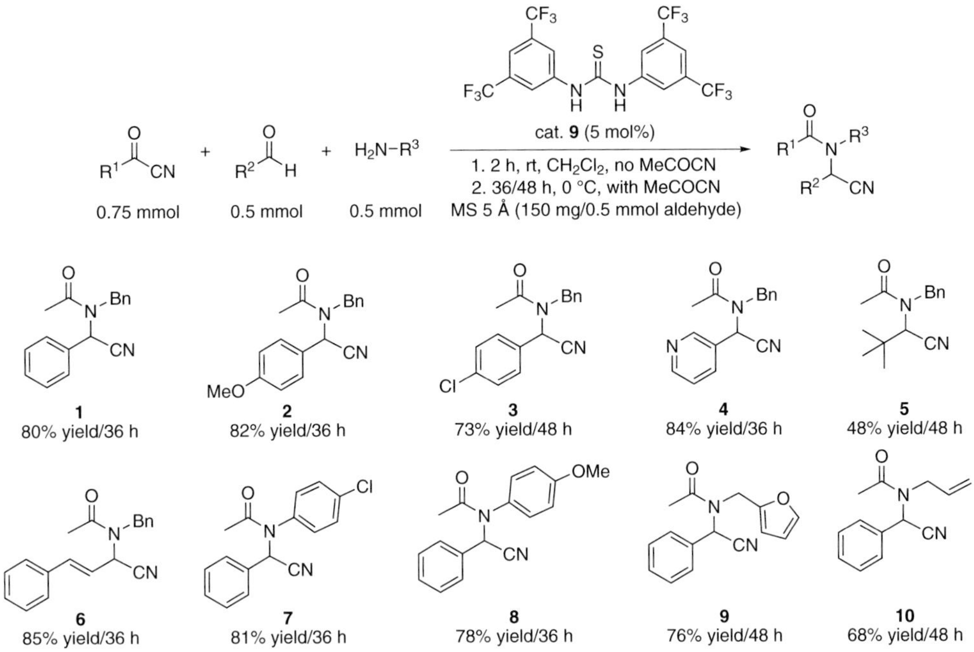

6.2 Synthetic Applications of Hydrogen-Bonding (Thio)urea Organocatalysts

Scheme 6.15 Representative N-acetyl-Strecker products resulting from the **9**-catalyzed three-component acylcyanation reaction.

Products shown:
- **1** 80% yield/36 h
- **2** 82% yield/36 h
- **3** 73% yield/48 h
- **4** 84% yield/36 h
- **5** 48% yield/48 h
- **6** 85% yield/36 h
- **7** 81% yield/36 h
- **8** 78% yield/36 h
- **9** 76% yield/48 h
- **10** 68% yield/48 h

9 (Scheme 6.16). The effective THP protection of benzyl alcohol at very low catalyst loadings down to 0.001 mol% emphasized the catalytic power of **9** (Scheme 6.16) and revealed a calculated maximum TON close to 100 000 and a TOF of around 2000 h^{-1}. Tertiary alcohols, which are normally difficult to protect as THP ethers due to steric hindrance and elimination as a side reaction, could also be THP protected under this conditions supporting the THP ether **1–10** (Scheme 6.17). Particularly remarkable is the tolerance of even the most sterically hindered substrates diamantan-1-ol and triphenylmethanol resulting in the THP ethers **7** and **8**, respectively. Phenol derivatives were also readily converted into their corresponding THP ethers at 50 °C (Scheme 6.17). As shown for phenol, THP protection to its THP ether could be achieved with catalyst loadings down to 0.001 mol%, resulting in a TOF of 5700 h^{-1}, which marks the most efficient organocatalytic reaction to date. Selected scale-up experiments (50 mmol scale), e.g., in the case of phenol, also demonstrated that loadings of only 0.01–0.1 mol% are sufficient and feasible for routine preparative THP protection (Scheme 6.17). Phenol derivatives also provided the important experimental clue that the hydroxy-group acidity was not a factor for the mechanistic interpretation of these reactions because phenols are more acidic than alkanols and electron-deficient phenols such as 4-(trifluoromethyl)phenol (product **3**: 86% yield/45 h) were found

164 | 6 (Thio)urea Organocatalysts

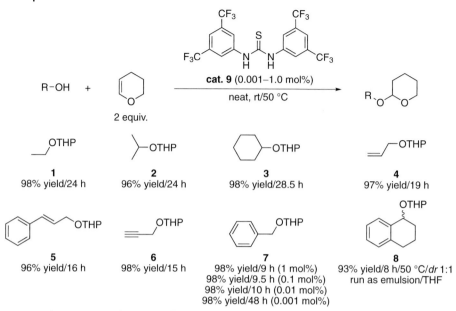

Scheme 6.16 Product range of the **9**-catalyzed tetrahydropyranylation of primary and secondary alcohol substrates.

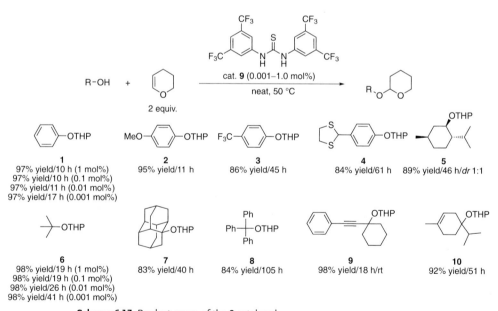

Scheme 6.17 Product range of the **9**-catalyzed tetrahydropyranylation of sterically hindered and phenolic substrates.

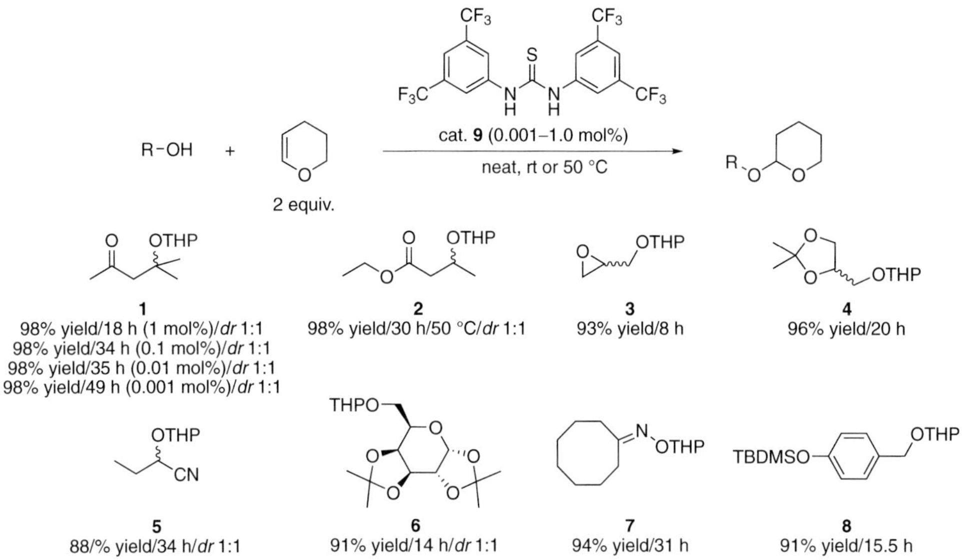

Scheme 6.18 Product range of the tetrahydropyranylation of acid-sensitive substrates catalyzed by thiourea **9**.

to react more slowly than electron-rich 4-methoxyphenol (product **2**: 95%/11 h) (Scheme 6.17).

Due to the mild and acid-free conditions the protocol also tolerated acid-labile hydroxyl-functionalized substrates as typical aldol products (product **1**; TOF = 2000 h^{-1}), β-hydroxy esters, epoxides, acetonides, cyanhydrines, oximes as well as highly acid-sensitive TBDMS-protected benzyl alcohol without detectable side reactions in excellent yields (88–98%) as depicted in Scheme 6.18 [118, 145]. To improve the feasibility of this reaction further, the authors utilized readily prepared 3,5-bis(trifluoromethyl)phenyl isothiocyanate [145] to attach the bis(trifluoromethyl)phenyl thiourea moiety of the catalytic motif in a straightforward protocol to simple amino-terminated polystyrene beads resulting in thiourea derivatives **17** and **18**, respectively (Scheme 6.19). **9**-analog polymer-bound **17** (~10 mol% loading) was found to efficiently catalyzed the THP protection of various hydroxy-functionalized substrates under heterogeneous conditions supporting the desired THP ethers **1–7** in excellent yields (92–98%) (Scheme 6.20); additionally, catalyst **17** was demonstrated to be readily recoverable by simple filtration and reusable (four cycles) for THP protection after washing with dichloromethane without loss of the catalytic activity [61, 62]. Polymer-bound bisthiourea **18** bearing a secondary and tertiary amine group, however, turned out to be catalytically inactive in the examined THP protections. This was consistent with the finding that hydroxy substrates such as amino alcohols incorporating an amine functionality could not be THP protected with the reported protocol.

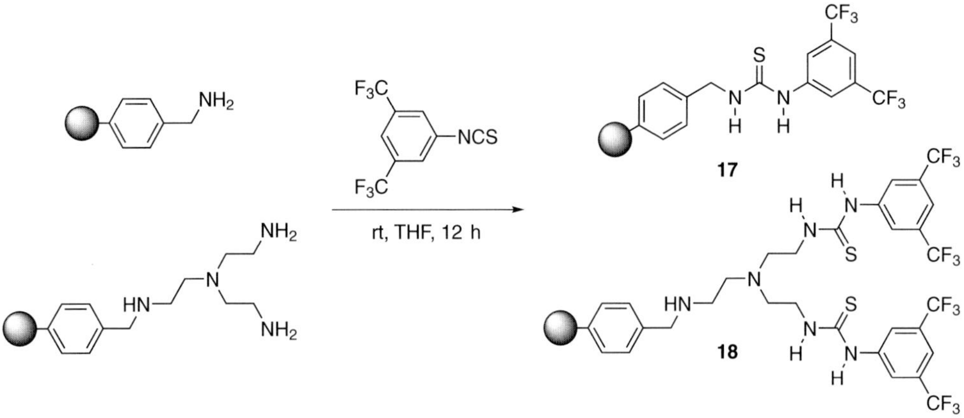

Scheme 6.19 Synthesis of polystyrene-bound thiourea derivatives **17** and **18** screened in the THP protection of hydroxy substrates.

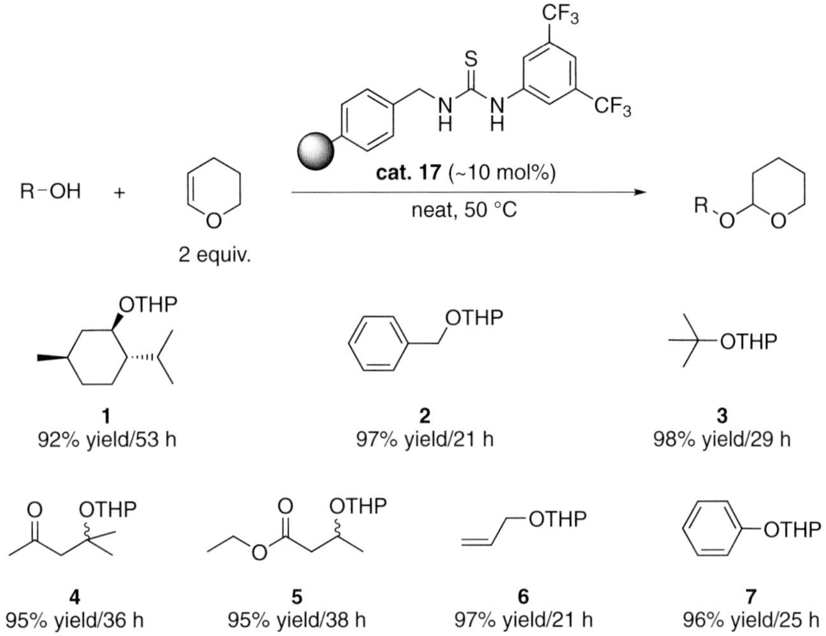

Scheme 6.20 THP ethers obtained from the THP protection of various hydroxy substrates utilizing polymer-bound catalyst **17**.

1 92% yield/53 h

2 97% yield/21 h

3 98% yield/29 h

4 95% yield/36 h

5 95% yield/38 h

6 97% yield/21 h

7 96% yield/25 h

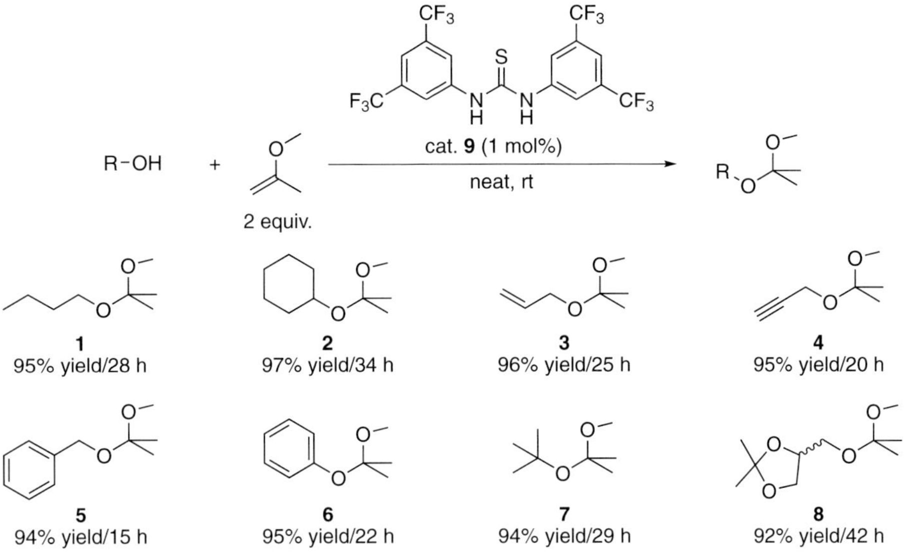

Scheme 6.21 Product range of the **9**-catalyzed MOP protection of hydroxy functionalities.

Products:
- **1** 95% yield/28 h
- **2** 97% yield/34 h
- **3** 96% yield/25 h
- **4** 95% yield/20 h
- **5** 94% yield/15 h
- **6** 95% yield/22 h
- **7** 94% yield/29 h
- **8** 92% yield/42 h

The authors successfully applied their protocol to the alternative enol ether 2-methoxypropene (MOP) to prepare the MOP ether **1–8** from a subset of the various alcohol substrates as depicted in Scheme 6.21. This high-yielding (92–97%) MOP protection occurred smoothly at room temperature; MOP turned out to be so reactive that the uncatalyzed reaction also proceeded albeit at lower rates [118, 145].

A reasonable mechanistic entry into this reaction may start with the complexation of catalyst **9** with the alcohol substrate. This double hydrogen bonding mediated coordination increases the alcohol's acidity as well as polarizability and hence its ability to form a subsequent ternary complex with the enol ether DHP via a pseudoaxial approach and interaction. The catalyst remains attached during the polar addition through a highly polarized transition structure and is finally released from the product complex to initiate a new catalytic cycle (Scheme 6.22). This mechanistic proposal clearly indicates the departure from the often-implied concept of carbonyl (or related functionalities) activation through hydrogen bonding with (thio)urea derivatives and other hydrogen-bonding organocatalysts.[4] Hence, this mechanistic alternative suggested either the hydrogen-bond assisted generation of the free nucleophile (e.g., RO$^-$, CN$^-$) or the stabilization of the active form of the nucleophile through hydrogen-bonding and polar interactions to the

---

[4] Mechanistically, in these reactions the catalytic effect of hydrogen-bonding thiourea **9** results from the stabilization of the oxyanion hole in the TS and not from the activation of the substrate through hydrogen bonding.

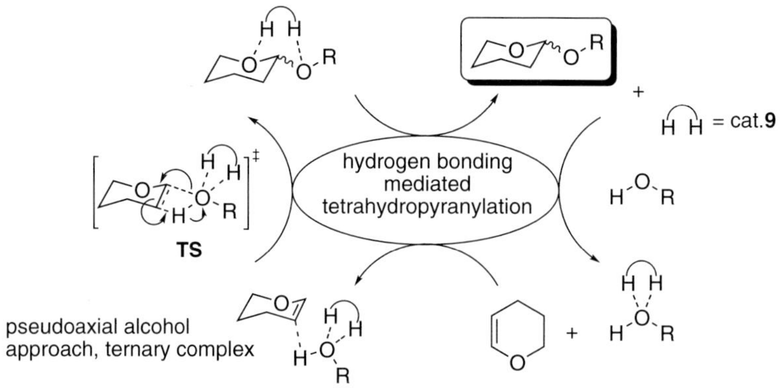

Scheme 6.22 Proposed tetrahydropyranylation cycle catalyzed by hydrogen-bonding thiourea derivative **9**.

respective precursor (e.g., ROH, HC(OR)$_3$, HCN, TMSCN). Density functional theory (DFT) and high-level coupled cluster computations demonstrated that the catalyst preferentially stabilized the developing oxyanion hole [145] in the transition state through double hydrogen bonding without the formation of the charged alkoxide nucleophile. This conclusion was reached on the basis of a comparative computational analysis of the uncatalyzed versus catalyzed model reaction of methanol with DHP [145]. The stabilizing effect of **9** on the key transition structure amounted to ca. 23 kcal mol^{-1}, resulting in a minimized barrier of only 17.7 kcal mol^{-1} (uncatalyzed: 45.2 kcal mol^{-1}), which is in line with the experimentally found efficacy of **9** already at room temperature in contrast to the uncatalyzed THP protection, which showed no product formation under otherwise identical conditions. Closer inspection of the transition structure with **9** revealed that the catalyst helps preorganize the reactants and the overall geometric changes in going from the complexes to the transition structures (least motion principle); **9** is placed sideways and points away from the R group of the substrate making steric hindrance not a critical factor, as found experimentally (Scheme 6.17).

Hiersemann, Strassner, and co-workers, in 2007, reported a combined computational (DFT) and experimental study on the Claisen rearrangement of a 2-alkoxycarbonyl substituted allyl vinyl ether [106] in the presence of thiourea derivative **9** (20 mol% and 100 mol%) as potential hydrogen-bonding organocatalyst (Scheme 6.23) [164]. Proof-of-principle experiments on this Claisen rearrangement performed in a sealed tube at 25 and 45 °C using 1,2-dichloroethane, chloroform, and trifluoroethanol as the solvents revealed that the conversion in the presence of thiourea **9** was nearly identical to the control experiment without **9** under otherwise unchanged conditions (e.g., in trifluoroethanol with 20 mol% **9**: 44% conv.; 45 °C; 6 d; without **9**: 41% conv; 6 d). Employing stoichiometric amounts (100 mol%) loading of thiourea **9** at 45 °C indicated only a poor accelerating effect in the model reaction resulting in 87% conversion (without **9**: 74% conv.) after 7 d reaction time

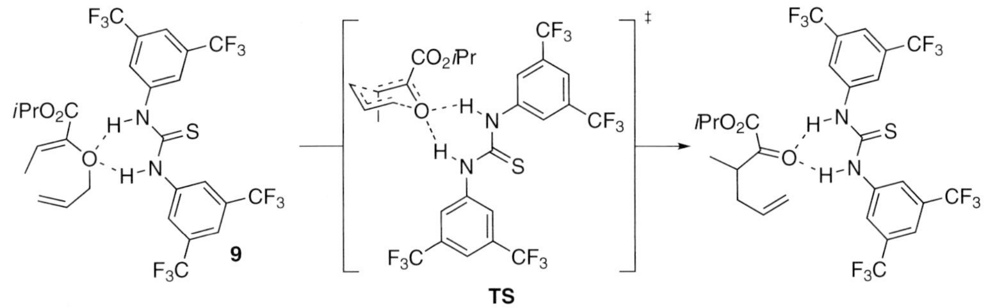

Scheme 6.23 Claisen rearrangement of a 2-alkoxycarbonyl substituted allyl vinyl ether in the presence of thiourea derivative **9**.

in chloroform. These experimental results identified **9** to be catalytically ineffective in the observed Claisen rearrangement and were supported by computations that suggested the transition-state stabilization was not significant enough to overcome the energetic costs of conformational changes and complexation required for the formation of the reactive complex. Since the corresponding TS was found to be characterized by a concerted but asynchronous bond reorganization process that led to TS polarization, the authors suggested that the design of a noncovalent (thio)urea organocatalysts for the Claisen rearrangement must focus on significant polarization of the TS by strong hydrogen bonding to the ether oxygen atom of the 2-alkoxycarbonyl substituted allyl vinyl ether.

In 2007, the Schreiner group published a **9**-catalyzed transfer hydrogenation of aldimines through hydrogen-bonding activation utilizing Hantzsch 1,4-dihydropyridine **19** "Hantzsch ester" [165] as the hydrogen source [166]. While the ketimine derived from acetophenone only gave traces of product, the benzaldehyde imine produced the corresponding amine product **1** even at 0.1 mol% loading of **9**, but at a longer reaction time (Scheme 6.24). Employing dichloromethane as solvent, thiourea **9** (1 mol% loading) as catalyst, and Hantzsch ester as reductant the scope of this acid- and metal-free transfer hydrogenation was explored. In general, a variety of aromatic aldimines underwent this reductive amination, including electron-rich, electron-deficient, as well as *ortho*-, *meta*-, and *para*-substituted aryl aldehydes, and provided the corresponding secondary amines **1–8** in yields ranging from 80 to 93% within 15 h; in addition, aliphatic aldimines could also be reduced to give the respective amines **9** and **10** with good yields (80 and 87%, respectively) as depicted in Scheme 6.24.

A systematic study on enzymatic catalysis has revealed that isolated enzymes, from baker's yeast or old yellow enzyme (OYE) termed nitroalkene reductase, can efficiently catalyze the NADPH-linked reduction of nitroalkenes. For the OYE-catalyzed reduction of nitrocyclohexene, a catalytic mechanism was proposed in which the nitrocyclohexene is activated by nitro-oxygen hydrogen bonds to the enzymes His-191 and Asn-194 [167, 168]. Inspired by this study Schreiner *et al.*

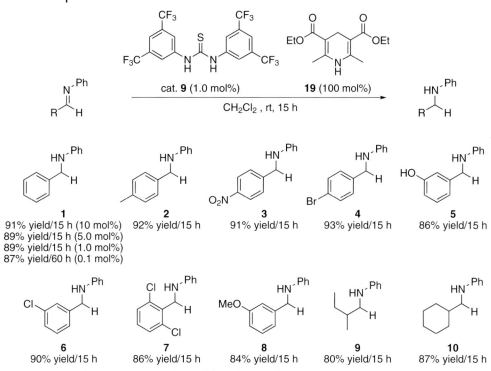

Scheme 6.24 Amines obtained from the transfer hydrogenation of aldimines in the presence of catalyst 9 and Hantzsch ester 19.

mimicked this natural procedure in preparative chemistry with hydrogen-bonding organocatalyst 9 functioning as the "reductase" and Hantzsch ester as NADPH analog to develop a mild, efficient, and selective method for the synthesis of nitroalkanes from α,β-unsaturated nitroalkenes [169]. This biomimetic reduction is not only practical, but may also provide insights into the mechanisms of redox transformations in biological systems. In the initial experiments, trans-β-nitrostyrene as the model substrate was reduced in the presence of 10 mol% catalyst 9 and 1.1 equiv. Hantzsch ester to give the respective nitroalkene in 87% yield after 48 h in toluene (50 °C), while the uncatalyzed reaction running under otherwise identical conditions showed no product formation. Using only 5 mol% catalyst 9 under the same conditions decreased the product yield (75%/48 h). Reductions of nitrostyrene (10 mol% 9; 1.1 equiv. Hantzsch ester 19) in nonpolar media such as benzene (84 yield/24 h) and toluene (78% yield/24), as well as halogenated solvents such as chloroform (76% yield/24 h) and dichloromethane (88%/24 h) proceeded smoothly. Performing the reaction in more polar media led to sluggish reactions and considerably diminished yields. A moderate yield was also obtained in the protic solvent methanol (52%/24 h), but a conversion was also detected

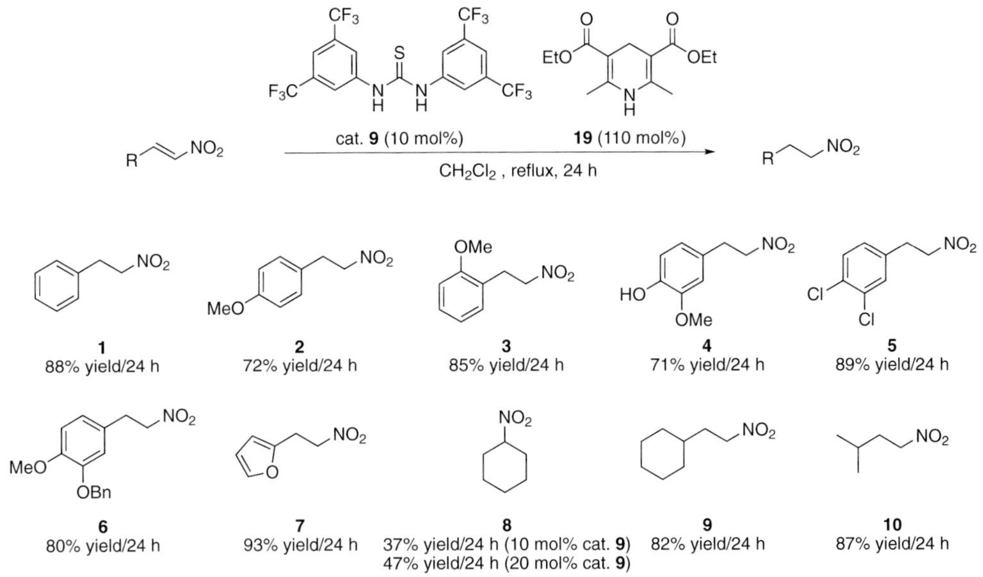

Scheme 6.25 Product range of the biomimetic reduction of α,β-unsaturated nitroalkenes catalyzed by **9** in the presence of **19**.

without catalyst **9** affording the desired nitroalkane **1** (45% yield/24 h) and the corresponding Michael adduct. Under optimized conditions with dichloromethane as solvent, 10 mol% **9**, and 1.1 equiv. Hantzsch ester **19**, various aromatic nitroalkenes were smoothly reduced to the corresponding nitoalkanes **1–10** in yields ranging from 72 to 93% after 24 h (Scheme 6.25). The electronic effect of the substituents and the substituent pattern turned out to have no marked effect on the results of this reaction (Scheme 6.25). The reductions of electron-rich nitrostyrenes took longer and generated the product in lower yield. Aliphatic nitroalkenes were reduced to the corresponding nitroalkenes in good yields (82 and 87%) as shown in Scheme 6.25. Only the reduction of nitrocyclohexene gave the corresponding nitroalkane **8** in poor yield (10 mol% loading: 37%) even at 20 mol% loading of catalyst **9** (47% yield). Reduction of the nitro group and polymerization of the alkenes did not occur in any case and emphasized on the mild conditions of this organocatalytic methodology.

In analogy to the enzymatic mode of substrate binding, activation and reduction [167, 168], the authors proposed enzyme-like hydrogen-bonding interactions between the nitro-group and thiourea **9** to effectively lower the LUMO of the conjugated double bond; this facilitated the reducing hydride transfer from the Hantzsch ester and led to the observed accelerating effect in the nitroalkane formation (Scheme 6.26).

In Nature, epoxide ring-opening is catalyzed by enzymes by employing the (double) hydrogen-bonding motif (Scheme 6.1) for epoxide activation toward

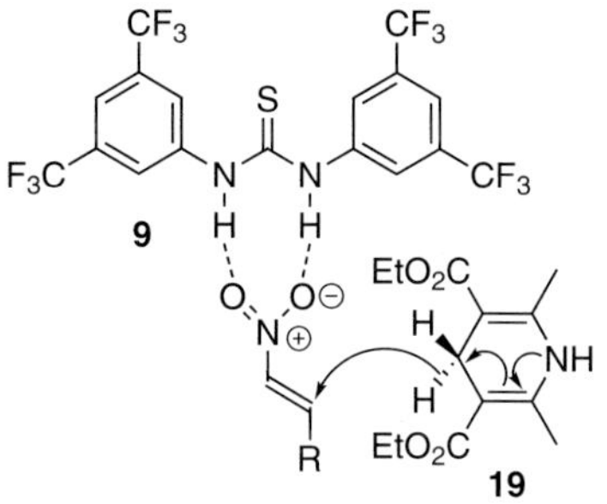

Scheme 6.26 Proposed model of the biomimetic reduction of conjugated nitroalkenes in the presence of thiourea catalyst **9** and **19**.

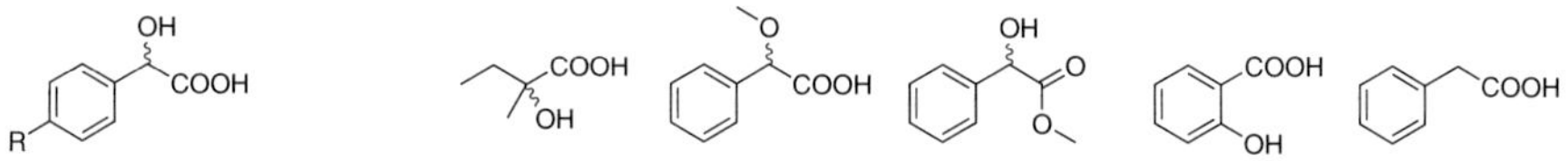

20	**21**	**22**	**23**	**24**	**25**	**26**	**27**
R = H	R = OMe	R = CF$_3$	~5% conv./24 h	no conv./26 h	no conv./26 h	~12% conv./12 h	no conv./26 h
>99% conv./22 h	>99% conv./19 h	>99% conv./16 h	(pK_a 4.05)	(pK_a 3.10)	(pK_a not available)	(pK_a 3.00)	(pK_a 4.28)
(pK_a 3.37)	(pK_a 3.43)	(pK_a 3.01)					

Figure 6.6 Structurally diverse Brønsted acids (1 mol% loading) screened in the cooperative Brønsted acid-type organocatalytic system (1 mol% thiourea **9**) utilizing the ethanolysis (12 equiv. EtOH) of styrene oxide as model reaction.

hydrolysis resulting in the removal of unsaturated toxic organic compounds (epoxidation–hydrolyis sequence) [34, 170]. Based on the concept of "hydrophobic amplification" in the epoxide aminolysis [153] (Scheme 6.8), Schreiner et al. developed a cooperative [18, 171, 172] Brønsted acid-type organocatalytic system for epoxide alcoholysis [173, 174]. Initial experiments performed for the ethanolysis of styrene oxide utilizing privileged thiourea catalyst **9** (1 mol%) in combination with mandelic acid **20** (1 mol%) (pK_a = 3.37) showed 99% conversion (GC/MS) after 22 h reaction time at room temperature and with ethanol (12 equiv. to avoid byproduct formation by nucleophilic attack of the product) as the solvent and the nucleophile. Further Brønsted acid screening in the model alcoholysis under otherwise identical conditions revealed that only aromatic acids incorporating a second coordination center in the α-position (hydroxy or carbonyl) induced appreciable conversions as shown for representative acids **20–27** in Figure 6.6. The removal or blocking of the α-coordination center (**24**; **27**) or removal of the aromatic system (**23**) drastically reduced the conversion rates. Aqueous acidity (pK_a) found to be not suitable to predict the catalytic activity (**22**; **26**). Performing the model alcoholysis in various solvents revealed a remarkable solvent effect; reac-

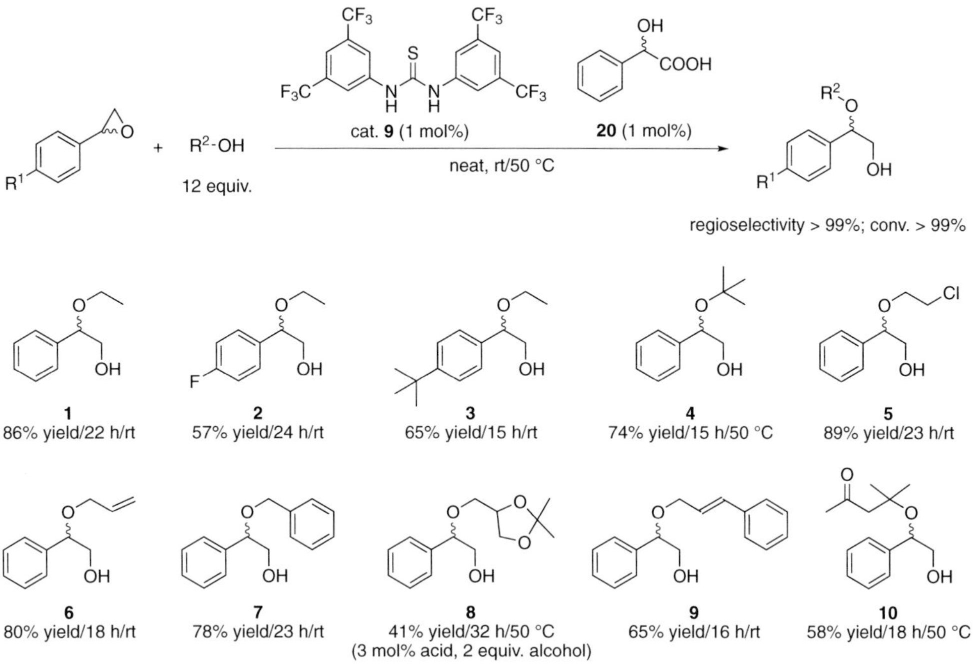

6.2 Synthetic Applications of Hydrogen-Bonding (Thio)urea Organocatalysts

Scheme 6.27 Typical β-alkoxy alcohols obtained from the highly regioselective alcoholysis of styrene oxides catalyzed by thiourea **9** and mandelic acid **20** in a cooperative organocatalytic system.

Compound data (products 1–10):

- **1** 86% yield/22 h/rt
- **2** 57% yield/24 h/rt
- **3** 65% yield/15 h/rt
- **4** 74% yield/15 h/50 °C
- **5** 89% yield/23 h/rt
- **6** 80% yield/18 h/rt
- **7** 78% yield/23 h/rt
- **8** 41% yield/32 h/50 °C (3 mol% acid, 2 equiv. alcohol)
- **9** 65% yield/16 h/rt
- **10** 58% yield/18 h/50 °C

Conditions: cat. **9** (1 mol%), **20** (1 mol%), neat, rt/50 °C, R²-OH 12 equiv.; regioselectivity > 99%; conv. > 99%.

tions in ethanol (99% conv./22 h/rt) were found to be two times faster than those in nonpolar or aprotic solvents (e.g., n-hexane: 99% conv./48 h/rt; THF: 99% conv./48 h/rt; CH_3CN: no conv./48 h/rt).

Under optimized conditions regarding the choice of Brønsted acid (mandelic acid **20**), stoichiometry (1 : 1 ratio **9** and mandelic acid **20**), solvent (the respective alcohol; neat conditions), temperature (rt or 50 °C), and catalyst loading (1 mol% **9** and 1 mol% mandelic acid **20**) electron-rich and electron-deficient styrene oxides underwent alcoholysis with simple aliphatic, sterically demanding as well as unsaturated and acid-labile alcohols. The completely regioselective (>99%) alcoholysis was reported to produce the corresponding β-alkoxy alcohols **1–10** in moderate (41%) to good (89%) yields without noticeable decomposition or polymerization reactions of acid-labile substrates (Scheme 6.27). Notably, all uncatalyzed reference experiments showed no conversion even after two weeks under otherwise identical conditions.

These experimental results suggested a hydrogen-bonding mediated cooperative Brønsted acid catalysis mechanism (Scheme 6.28). Thiourea cocatalyst **9** is viewed to coordinate to mandelic acid **20** through double hydrogen-bonding, stabilizes the acid in the chelate-like cis-hydroxy conformation, and acidifies the α-OH proton via an

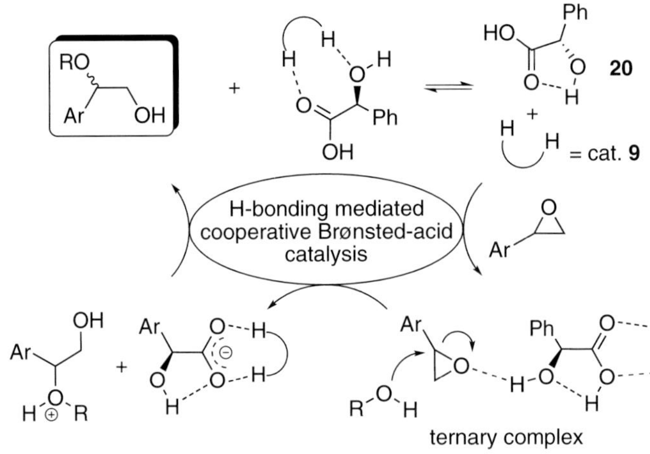

Scheme 6.28 Mechanistic proposal for the regioselective alcoholysis of styrene oxides utilizing a cooperative Brønsted acid-type organocatalytic system comprised of thiourea **9** and mandelic acid **20**.

additional intramolecular hydrogen bond. The epoxide was suggested to be activated by a single-point hydrogen bond that facilitates regioselective nucleophilic attack of the alcohol at the benzylic position similar to the monodentate binding reported for diol catalysts [175]. The incipient oxonium ion reprotonates the mandelate and provides the observed β-alkoxy alcohol products (Scheme 6.28). This mechanistic proposal, originating from the concept of cooperativity of two catalysts, was supported by DFT computations indicating that the ternary complex between styrene oxide, **9**, and mandelic acid **20** has an overall binding energy of 20.0 kcal mol^{-1}, while the possible binary complexes are less stable and disfavored (styrene oxide and **9**: 9.2 kcal mol^{-1}; mandelic acid **20** and **9**: 11.9 kcal mol^{-1}; styrene oxide and mandelic acid **20**: 5.7 kcal mol^{-1}).

6.2.1.2 Miscellaneous Nonstereoselective (Thio)urea Organocatalysts

In 2004, Connon and Maher identified simple symmetrically substituted achiral N,N'-bisphenyl (thio)urea derivatives as efficient, stable, and recyclable DABCO-compatible hydrogen-bonding promoters for the Morita–Baylis–Hillman (MBH) reaction [176, 177] for a range of aromatic aldehydes and methyl acrylate [178]. For catalyst screening, the pseudo-first-order rate constants of the reaction between methyl acrylate (10 equiv.) and benzaldehyde catalyzed by both DABCO (100 mol%) and (thio)ureas **9**, **15**, **16**, **26**, and **27** (each 20 mol%) were determined by ^{1}H NMR kinetic experiments. This revealed that urea analog **16** of catalyst **9** was the most efficient in accelerating this MBH reaction under solvent-free conditions at room temperature (Figure 6.7).

This result represented an uncommon example for the inferiority of thioureas as compared to ureas [1]. Under optimized conditions, the DABCO (100 mol%)-

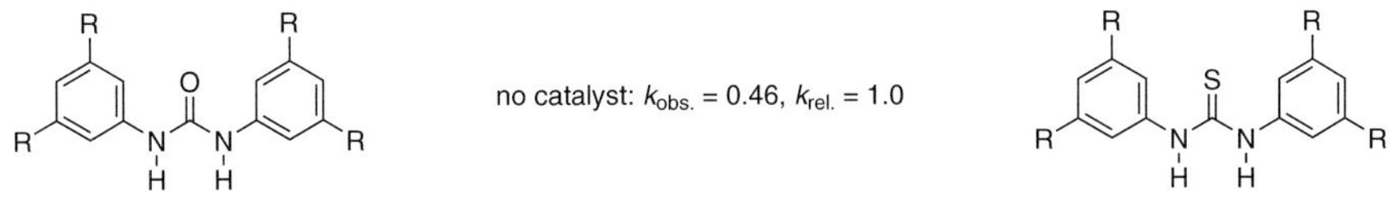

6.2 Synthetic Applications of Hydrogen-Bonding (Thio)urea Organocatalysts | 175

no catalyst: k_{obs} = 0.46, k_{rel} = 1.0

28: R = F; k_{obs} = 2.50; k_{rel} = 5.4 (20 mol%)
16: R = CF$_3$; k_{obs} = 3.06; k_{rel} = 6.7 (20 mol%)

15: R = H; k_{obs} = 0.76; k_{rel} = 1.7 (20 mol%)
29: R = F; k_{obs} = 2.63; k_{rel} = 5.7 (20 mol%)
9: R = CF$_3$; k_{obs} = 1.73; k_{rel} = 3.7 (20 mol%)

Figure 6.7 Hydrogen-bonding (thio)ureas screened in the DABCO-promoted MBH reaction between benzaldehyde and methyl acrylate; the pseudo-first-order rate constants relative to the uncatalyzed reaction are given in h^{-1}.

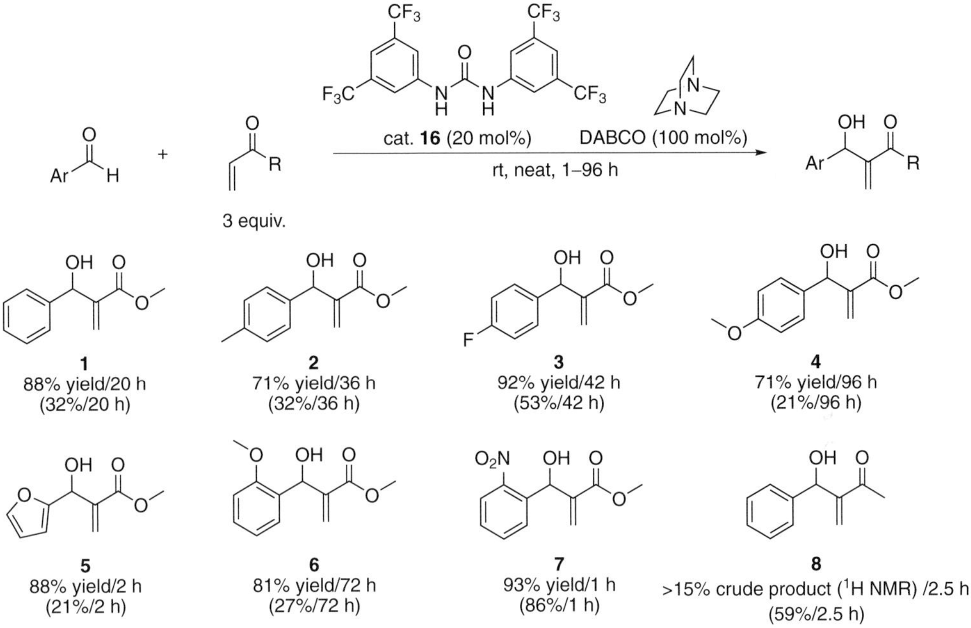

cat. **16** (20 mol%)   DABCO (100 mol%)
rt, neat, 1–96 h

3 equiv.

1
88% yield/20 h
(32%/20 h)

2
71% yield/36 h
(32%/36 h)

3
92% yield/42 h
(53%/42 h)

4
71% yield/96 h
(21%/96 h)

5
88% yield/2 h
(21%/2 h)

6
81% yield/72 h
(27%/72 h)

7
93% yield/1 h
(86%/1 h)

8
>15% crude product (^{1}H NMR) /2.5 h
(59%/2.5 h)

Scheme 6.29 Range of products for the DABCO-promoted MBH reaction utilizing urea derivative **16** as hydrogen-bonding organocatalyst. The results of the uncatalyzed reference reactions are given in parentheses.

promoted MBH reaction of various electron-deficient and electron-rich aromatic aldehydes (1 equiv.) with methyl acrylate (3 equiv.) was accelerated in the presence of urea catalysts **16** (20 mol%) and furnished the respective adducts **1–8** in moderate to very good yields (71–93%) within 2–42 h reaction time. Only the challenging deactivated anisaldehydes proved to require longer reaction times producing the MBH adducts **4** (96 h) and **6** (72 h), respectively (Scheme 6.29). The catalytic effect of urea catalyst **16** was illustrated by the poor product yields obtained in the

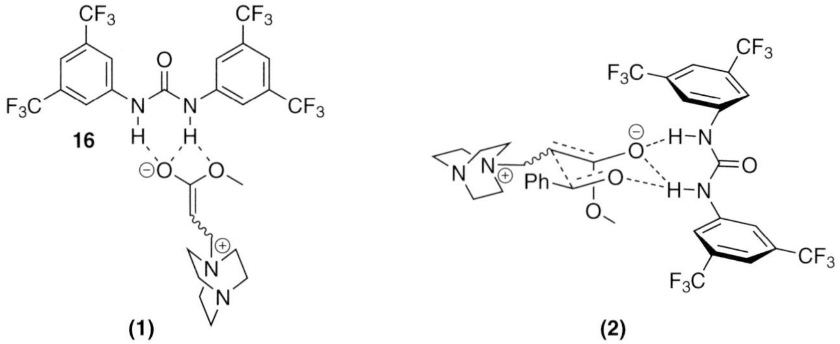

Figure 6.8 Proposed modes of action of hydrogen-bonding catalyst **16**: Bidentate hydrogen bonding coordination of the zwitterion derived from Michael-type DABCO attack to methyl acrylate **(1)** and Zimmerman–Traxler transition state for the reaction of methyl acrylate with benzaldehyde **(2)**.

urea-free control reactions involving deactivated substrates over the same time period (Scheme 6.29).

In all cases, urea catalyst **16** could be recovered unchanged after the reaction by column chromatography in good to excellent yield (82–95%) and could be reused without loss of activity. This protocol turned out to be not applicable to MBH reactions involving acrolein or methyl vinyl ketone, because catalyst **16** promoted the rapid decomposition of the Michael acceptor resulting in poor yields (e.g., MBH adduct **8** in Scheme 6.29: 15% yield/2.5 h). The screening reaction was also accelerated when replacing the (thio)ureas shown in Figure 6.7 with 40 mol% of the strong hydrogen-bond donors methanol (k_{obs} = 1.15 h^{-1}) or water (k_{obs} = 0.72 h^{-1}) under otherwise identical conditions, but the accelerating effect was reduced in comparison to the use of catalyst **16**. The authors suggested that **16** preferably coordinates through explicit hydrogen-bonding to the more basic zwitterionic MBH key intermediate (**(1)**; Figure 6.8) or possibly initiated and stabilized a Zimmerman–Traxler type transition state (**(2)**; Figure 6.8) for the addition of the zwitterion to benzaldehyde. This mechanistic proposal was supported by the experimental finding that **16** was completely deactivated in the presence of an equimolar amount of the strong hydrogen-bond acceptor tetrabutylammonium acetate (TBAA), which competed with the zwitterion for the urea N—H bonds [178].

To combine the efficiency of hydrogen-bonding (thio)urea and nucleophilic tertiary amine into one structure, the hybride compounds **30** and **31** were prepared by analogy to the known efficient catalyst 3-hydroxyquinuclidine. These compounds were evaluated in the MBH reaction between methyl acrylate and o-chlorobenzaldehyde (Figure 6.9). 3-Amino quinuclidine derivatives **30** and **31** (10 mol%) proved to be poor bifunctional organocatalysts (**30**: 31% yield/20 h; **31**: 23% yield/20 h) and these were less efficient than DABCO (86% yield/20 h). This result correlated with the findings by Aggarwal *et al.* that in quinuclidine derivative-catalyzed

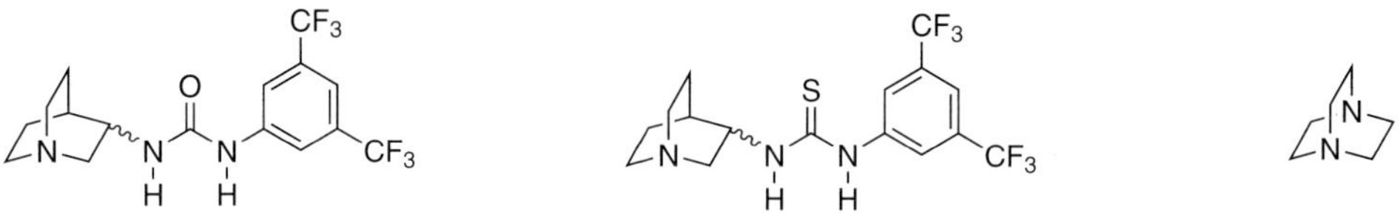

30
10 mol%
31% yield/20 h

31
10 mol%
23% yield/20 h

DABCO
10 mol%
86% yield/20 h

Figure 6.9 Bifunctional 3-amino quinuclidine derivatives **30** and **31** and DABCO probed in the MBH reaction between methyl acrylate and *o*-chlorobenzaldehyde.

Baylis–Hillman reactions, the protonated amine pK_a was the governing factor in determining catalyst efficiency, thus making quinuclidine itself a better catalyst than 3-heteroatom substituted analogs, which are of reduced basicity/nucleophilicity and consequently give lower reaction rates.

In 2008, the Connon group published a protocol for the organocatalyzed Corey–Chaykovsky epoxidation utilizing electron-deficient *N,N'*-3,5-bis(trifluoromethyl) phenyl substituted urea **16** as hydrogen-bonding catalyst (5 mol% loading) and trimethylsulfonium iodide (1.0 equiv.) as the precursor of sulfonium methylide generated *in situ* under biphasic conditions through deprotonation with aqueous NaOH [179]. The established reaction parameters (CH_2Cl_2, 50% $NaOH_{aq}$, rt) tolerated a broad substrate scope including electron-rich and electron-deficient aromatic aldehydes as well as cyclohexane carbaldehyde that were converted to the corresponding epoxides **1–6** in yields ranging from 57 to 96% (Scheme 6.30). Aldehydes bearing a α-proton proved to be incompatible with the chosen basic conditions and showed competitive aldol reactions. Thiourea derivatives such as **9** proved to be catalytically less effective than the urea counterparts presumably due to the higher N—H acidity of thioureas compared to the sulfonium methylide (pK_a in DMSO: *N,N'*-diphenyl urea: 19.55; *N,N'*-diphenyl thiourea: 13.4; $S(CH_3)_3I$: 18.2). Mechanistically, the epoxide formation is initiated with a nucleophilic attack of the *in situ* formed ylide to the aldehyde. This rate-determining addition step was suggested to be accelerated through hydrogen-bond-mediated stabilization of the developing negative charge on the carbonyl oxygen in the proposed transition state resulting in a zwitterionic intermediate; fast ring closure produced the observed epoxide product (Scheme 6.31). Based on the epoxidation protocol, the authors developed a tandem organocatalytic epoxidation–transition metal catalyzed ring-opening process employing the Cu(II) ion catalyzed (25 mol% $Cu(BF_3)_2$) Meinwald rearrangement [180] to reach a homologation of aldehydes via the respective epoxide. This straightforward procedure was exemplified for 2-methyl benzaldehyde that was epoxidized in the presence of **16** and subsequently chain extended to 2-methyl phenyl acetaldehyde in 69% overall yield.

Urea **32**, the bis-(mono-trifluoromethyl)phenyl derivative of urea catalyst **16** [178], was reported to operate as double hydrogen-bonding organocatalyst in the diastereoselective synthesis of γ-butenolide products substituted at the γ-position

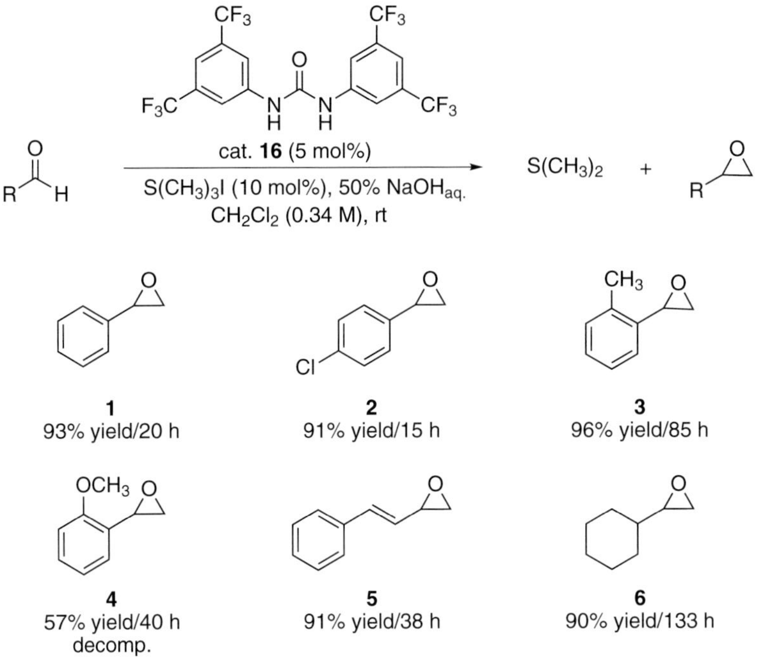

Scheme 6.30 Typical epoxides obtained from the Corey–Chaykovsky epoxidation of aldehydes catalyzed by urea **16**.

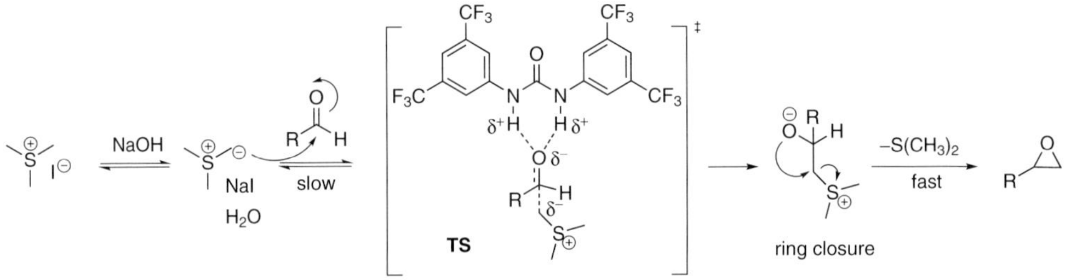

Scheme 6.31 Proposed mechanism for the Corey–Chaykovsky epoxidation of aldehydes catalyzed by urea **16**.

by a hydroxyl-functionalized alkyl chain [181]. This structural moiety is an important subunit in diverse natural products and biologically active compounds. Soriente and co-workers, in 2006, estimated the catalyzing efficiency of urea derivative **32** (10 mol% loading) in the exploratory addition experiment of commercially available 2-trimethylsilyloxyfuran (TMSOF) to benzaldehyde (5 equiv.) under variable conditions [181]. The corresponding model γ-butenolide **1** was obtained after 24 h at room temperature in 81% yield with a *dr* 66 : 34 (Scheme 6.32). Using only 1 equiv. benzaldehyde at room temperature or performing the

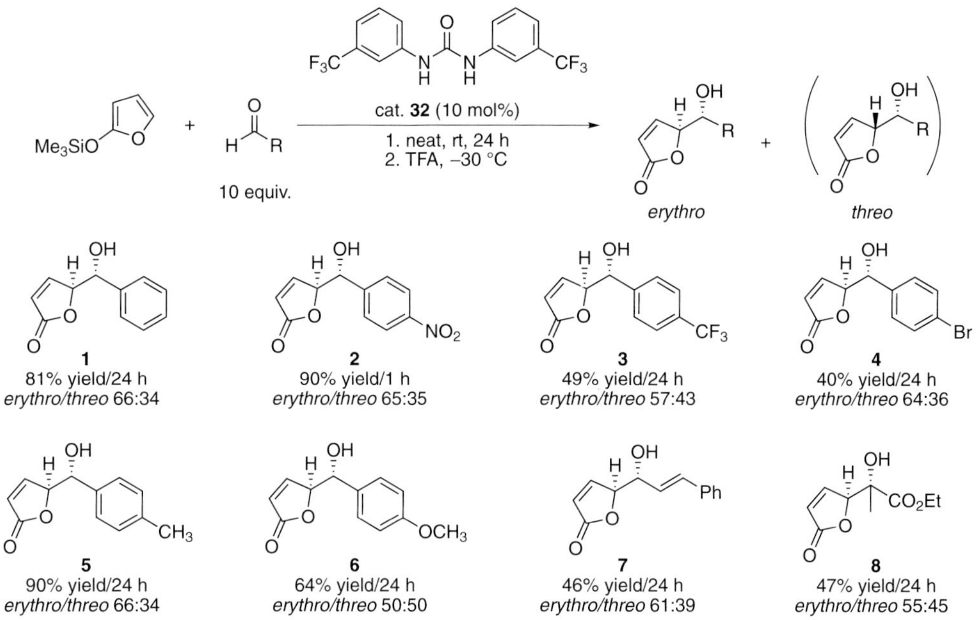

Scheme 6.32 γ-Butenolides obtained from diastereoselective aldol addition of 2-trimethylsilyloxyfuran to aldehydes catalyzed by urea **32**.

reaction with 5 equiv. at −20 °C affected only the yield of product **1** and gave 67% (24 h) and 73% (72 h), respectively; the diastereoselectivity remained constant at lower temperature (*dr* 66 : 34) or decreased marginally (*dr* 61 : 39). Under optimized conditions, the urea derivative **32** (10 mol%) catalyzed the addition of electron-rich and electron-deficient benzaldehydes to TMSOF followed by a TFA-mediated desilylation step giving the desired γ-butenolide adducts **1–7** in moderate to good yields (40–90%) in 1–24 h reaction time and with *dr* values up to 66 : 34 in favor of the *erythro* isomer (Scheme 6.32). The addition to a ketone was exemplified with the formation of product **8** (47% yield/24 h) from the addition of TMSOF to the activated ketone group of ethyl 2-oxopropanoate (Scheme 6.32).

The authors interpreted the observed stereochemical preference of the *anti* isomer in this addition by the synclinal arrangement of the reagents in the open-chain transition state (**TS2**) depicted in Scheme 6.33. The *anti* diastereoselectivity was ascribed to the unfavorable steric interaction between the R group of the aldehyde and the furan ring of TMSOF (**TS1**). An antiperiplanar arrangement of the reagents demonstrated steric congestion in both the conformers (**TS3** and **TS4**). In particular transition state **TS4**, which also furnished the *erythro* aldol adduct via **TS2**, possessed the same steric crowding as **TS1**. The examination of **TS3** also revealed steric repulsion between the bulky trimethylsilyl group and the trifluoromethyl-phenyl ring of the catalyst. Hence, the **32**-catalyzed reaction may

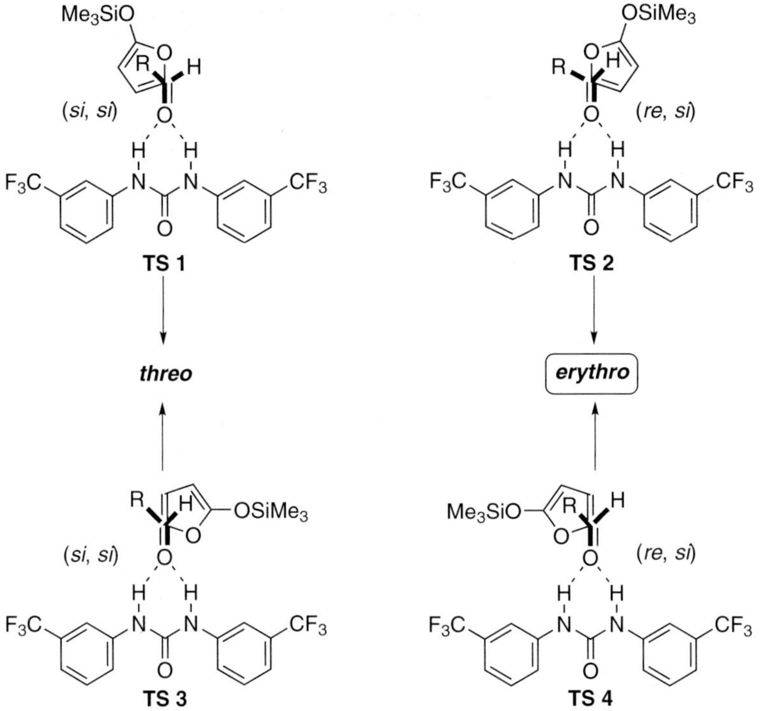

Scheme 6.33 Transitions states to explain the diastereoselectivity of the γ-butenolide formation from TMSOF and aldehydes in the presence of urea catalyst **32**.

proceed via the proposed transition state **TS2** and produces the observed *erythro* adduct (Scheme 6.33).

In 2007, Costero *et al.* reported on the basis of voltametric studies the cathodic reduction of aromatic carboxylates to give the corresponding dicarbonyl compounds (Scheme 6.34) [182]. This transformation was catalyzed by hydrogen-bonding biphenylthiourea derivatives **33**, **34**, and **35**, which coordinate with the carboxylate ion and facilitate the single electron transfer reducing step. The catalyst can be recovered unchanged after the reduction. Surprisingly, the authors assumed an activating hydrogen-bonding interaction of the anion with the thiourea derivative via a single amide proton as depicted in Scheme 6.34.

Connon and Procuranti, in 2008, introduced new bifunctional organocatalysts classified as reductase-mimicking (thio)urea catalysts incorporating both a substrate-activating hydrogen-bonding (thio)urea moiety and a covalently bound NADH analog moiety as hydride donor [183]. Thiourea derivative *rac*-**36** was found to catalyze the chemoselective reduction of benzil to benzoin in a biphasic aqueous/organic solvent (Et$_2$O). In contrast to a binary catalytic system employing Hantzsch ester or a stoichiometric Lewis-acidic additive such as Mg^{2+} in combination with an organocatalyst, this methodology was reported to operate under acid-free

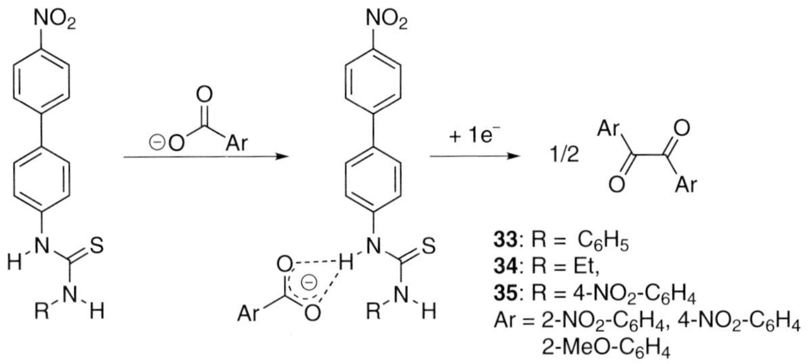
+ 1e⁻
1/2
33: R = C6H5
34: R = Et,
35: R = 4-NO2-C6H4
Ar = 2-NO2-C6H4, 4-NO2-C6H4
2-MeO-C6H4

Scheme 6.34 Electrochemical reduction of aromatic carboxylates to benzils in the presence of thiourea catalyst **33**, **34**, or **35**.

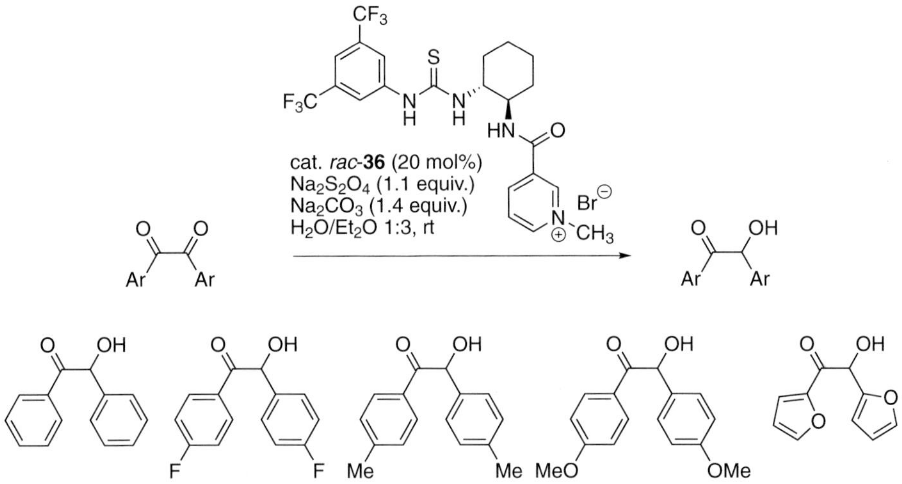
cat. rac-36 (20 mol%)
Na2S2O4 (1.1 equiv.)
Na2CO3 (1.4 equiv.)
H2O/Et2O 1:3, rt

1	**2**	**3**	**4**	**5**
96% yield/48 h	77% yield/48 h	60% yield/72 h	65% yield/120 h	72% yield/96 h

Scheme 6.35 Benzoins obtained from the reduction of benzils in the presence of thiourea derivative *rac*-**36** and sodium dithionate.

conditions and required only inexpensive sodium dithionate (1.1 equiv.) as coreductant that generated and recycled *in situ* the catalytically active hydropyridine species. To explain the experimental findings, the authors suggested a bifunctional reduction mechanism in which an intramolecular hydride donation from the *in situ* generated hydropyridine moiety to the hydrogen-bonded and activated 1,2-diketone occurred and not the thiourea-catalyzed reduction of the diketone by dithionate. Under optimized conditions, electron-rich and electron-deficient benzils were transformed to the target benzoins (2-hydroxyketones) **1–5** in good to excellent yields (60–96%) as shown in Scheme 6.35. The protocol was also applicable in the synthesis of unsymmetrical benzoins where it was difficult to have useful yields from the benzoin condensation of the corresponding aldehydes.

Rozas, Connon, and co-workers, in 2008, published a computational study guided catalyst design strategy (DFT computations) applied in combination with experimental structure-efficiency studies that led to the identification of acidic N-tosyl urea derivative 37 [184]. This novel thiourea catalyst turned out to be capable of accelerating the highly regioselective opening of styrene oxides with predominantly 1,2-dimethylindols as well as with anilines. Under optimized reaction conditions (CH_2Cl_2; MS 5 Å; rt) catalyst 37 (10 mol% loading) promoted the addition of various methylindols to electron-rich and electron-deficient styrene oxides resulting in desired adducts 1–6 in consistently high yields (89–98%) after 26 h–7.8 d as depicted in Scheme 6.36.

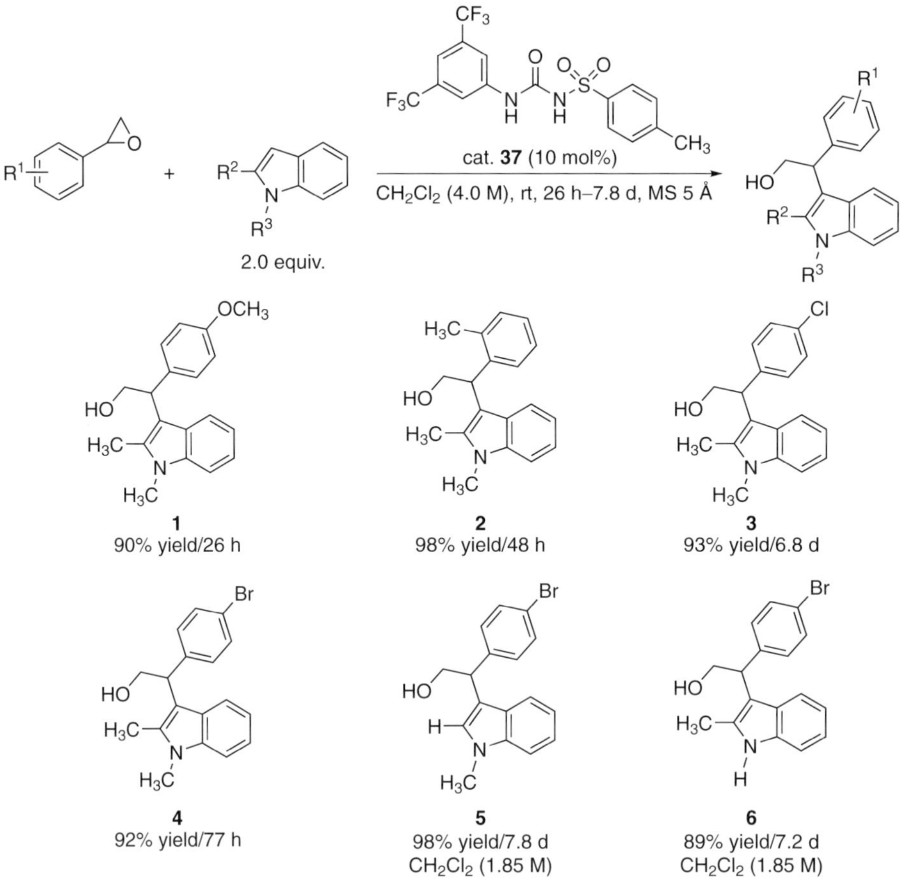

Scheme 6.36 Products obtained from the addition of indols to various styrene oxides in the presence of N-tosyl urea catalyst 37.

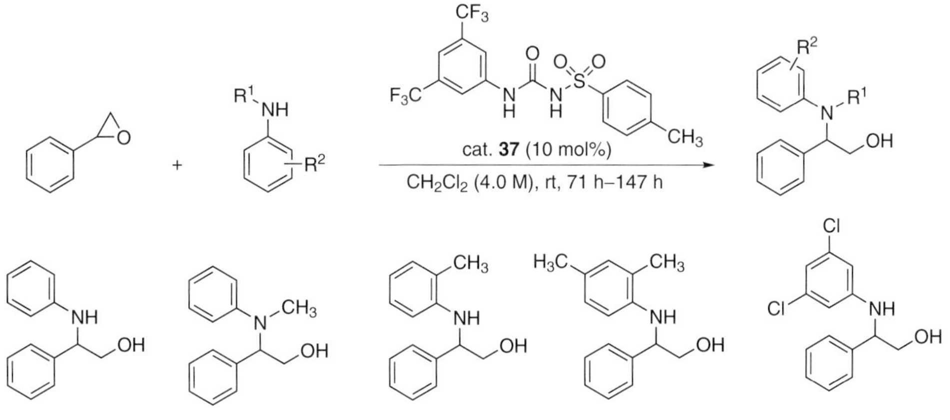

1	2	3	4	5
86% yield/91 h *dr* 94:6	78% yield/91 h	75% yield/90 h *dr* 92:8	92% yield/147 h (20 mol % **37**)	84% yield/71 h

Scheme 6.37 Product range of the styrene oxide opening with anilines promoted by *N*-tosyl urea catalyst **37**.

The addition of anilines to styrene oxide was reported to also proceed in the presence of 10 mol% **37** affording the corresponding β-amino alcohols **1–5** in yields ranging from 75% to 92% (Scheme 6.37). Additionally, urea derivative **37** (20 mol% loading) was found to catalyze the addition of aniline (2.0 equiv.) to (*E*)-stilbene oxide (92% yield; 5.9 d; 30 °C), the addition of thiophenol (2.0 equiv.) to 2-methoxy styrene oxide (85%; 20 h; rt), and the alcoholysis of 4-methoxy styrene oxide with benzyl alcohol (2.0 equiv.) affording the respective β-alkoxy alcohol (82%; 20 h; rt).

Ema, Sakai, and co-workers, in 2008, introduced trifunctional (thio)urea derivatives **38** and **39** mimicking the active site of serine hydrolases [185] in the biomimetic organocatalysis of the acetyl-transfer reaction (transesterification) [186, 187] from vinyl trifluoroacetate as a reactive acetyl donor to methanol and 2-propanol, respectively [188]. In analogy to the active site of serine hydrolases such as lipases, esterases, and serine proteases, the designed trifunctional catalysts **38** and **39** incorporated a hydroxy functionality expected to operate as a nucleophile, a pyridine moiety as a base, and a (thio)urea unit as an oxyanion hole [163] stabilizing the oxyanion in the transition state (Figure 6.10); the privileged 3,5-bis(trifluoromethyl)phenyl moiety [1, 116] was utilized to enhance the hydrogen-bond donor ability and thus the substrate binding of the (thio) urea unit. Proof-of-principle studies employing ^{19}F NMR kinetic experiments (100 mM vinyl trifluoroacetate; 500 mM MeOH) revealed urea catalyst **37** to be catalytically more active (100% conv.; 30 min; rt) than the thiourea counterpart **38** (100% conv.; 1 h; rt) at 1 mol% catalyst loading in $CDCl_3$ as the solvent. In the presence of 0.1 mol% catalyst, the acetylation was completed after 16 h (**38**) and 4 h (**39**). Control compounds lacking either the hydroxy, pyridine, or

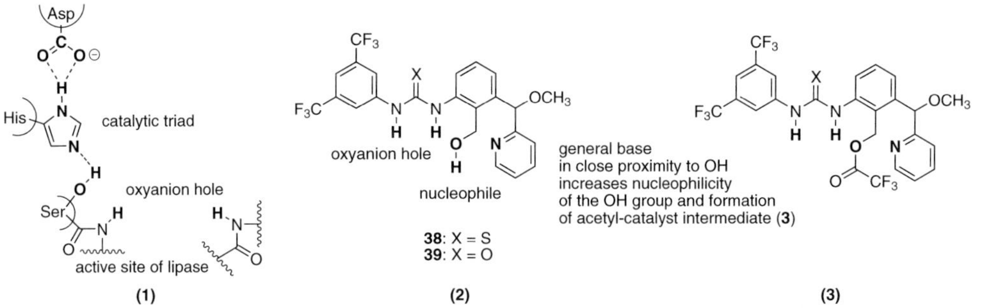

Figure 6.10 Active sites of lipase (1), trifunctional (thio)urea derivatives (38; 39) mimicking the acive site of serine hydrolases (2), and acetyl-catalyst intermediate of the biomimetic transesterification between vinyl trifluoroacetate methanol and 2-propanol, respectively (3).

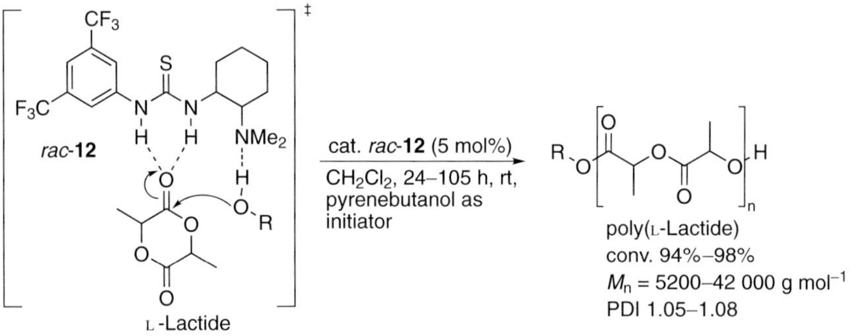

Scheme 6.38 Ring-opening polymerization of L-lactide catalyzed by tertiary amine-functionalized thiourea rac-12.

the (thio)urea moiety displayed little or no conversions under otherwise identical conditions indicating that the three functional groups operate cooperatively to effectively catalyze the model acetylation reactions with up to 3 700 000-fold accelerations (for the acetylation of 2-propanol) and with high turnover. The OH group of the catalyst was suggested to be acetylated in a fast step to an acetyl-catalyst intermediate, which was deacetylated in the rate-determining step resutling in the acetylation of the respective alcohol (Figure 6.10).

In 2005, Hedrick, Waymouth, and co-workers reported racemic tertiary amine-functionalized thiourea derivative rac-12 (5 mol% loading) to be catalytically active in supramolecular recognition for the living polymerization (evidenced by the linear correlation between ^{1}H NMR detected M_n and monomer conversion; PDI < 1.07) of L-lactide affording desired poly(L-Lactide) (Scheme 6.38) [189]. Employing pyrenebutanol as initiator this ring-opening polymerization (ROP)

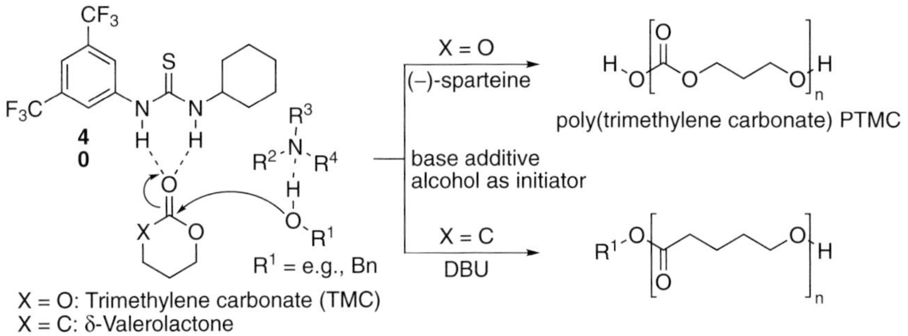

Scheme 6.39 Ring-opening polymerization of trimethylene carbonate and δ-valerolactone, respectively, catalyzed by a thiourea derivative **40** in the presence of a base.

proceeded in high conversions (94–98%; 24–105 h) at room temperature with minimal transesterification of the polymer chain due to the higher binding affinity of the thiourea catalyst to the cyclic ester monomer than to the linear ester polymer (examined by ^{1}H NMR titrations). As illustrated in Scheme 6.38, this specific bifunctional activation of the monomer favored the ROP instead of the competing transesterification [75, 190].

Hedrick, Waymouth, and co-workers modified the ROP methodology applied to the L-lactide monomer such that monofunctional electron-deficient N-bis(trifluoromethyl)phenyl N′-phenyl thiourea **40** was utilized in combination with a strong base [(−)-sparteine; DBU] [191, 192]. Under these conditions, the ROP of trimethylene carbonate (TMC) and, e.g., the cyclic ester δ-valerolactone proceeded to give the target polymers as depicted in Scheme 6.39. The authors assumed dual catalysis in which the base activated the alcohol that initiated the ROP by nucleophilic attack to the double hydrogen bonded carbonyl monomer. The growing research field of organocatalytic ROP including catalysis through hydrogen-bonding thiourea derivatives was reviewed by Hedrick, Waymouth, and co-workers in 2007 [75].

6.2.2
Stereoselective (Thio)urea Organocatalysts

6.2.2.1 (Thio)ureas Derived From Trans-1,2-Diaminocyclohexane and Related Chiral Primary Diamines

Trans-1,2-diaminocyclohexane, the most popular representative of the primary diamines and its derivatives, has been widely utilized in the field of stereoselective organometallic catalysis as chiral reagents, scaffolds, and ligands [137]. This C_2 symmetrical chiral diamine first reported in 1926 by Wieland and co-workers [193] is commercially available, since it is a component in a byproduct amine stream generated during the purification of 1,6-hexanediamine that is used in the industrial Nylon 66 production. The racemic mixture of this diamine can be easily resolved with D- or L-tartaric acid to provide the enantiopure (R,R)- or (S,S)-isomer, respectively [194]. Due to this cost-efficient availability of enantiopure *trans*-1,2-diaminocyclohexane and the multiple options of derivatization of the primary amine functionality, in particular the straightforward symmetrical or unsymmetrical incorporation of the urea or thiourea moiety via addition to (a)chiral isocyanates or isothiocyanates, respectively, this diamine has gained importance as an attractive chiral-building block for the design and synthesis of a broad spectrum of structurally diverse (thio)urea derivatives operating as stereoselective (bifunctional) hydrogen-bonding organocatalysts. This section is dedicated to the conceptually important class of diamine-derived (thio)ureas ranging from monofunctional Schiff base catalysts that were actually the first enantioselective (thio)urea catalysts, to bifunctional tertiary, secondary, and primary amine functionalized as well as pyrrole and bisthiourea catalysts. The few catalysts structures based on related chiral diamines such as 1,2-diphenylethylene diamine and their applications are also considered herein.

In 1998, Sigman and Jacobsen identified and optimized the first enantioselective (thio)urea organocatalysts from synthetic libraries of polystyrene-bound tridentate Schiff bases and their high-throughput screening [56, 71, 195] (HTS) in the asymmetric Strecker reaction of *N*-allyl benzaldimine with TBSCN as the cyanide source [122]. The basic catalyst design concept resulted from the original research goal to develop a novel tridentate chiral ligand for the efficient chirality transfer in an organometallic catalytic system as well as from the methodical requirement of solid-phase synthesis and systematic structure-optimizing variations to obtain high diversity of potential catalysts. Considering these aspects, the typical core structure of tridentate ligands (chiral amino alcohol, salicylaldehyde derivative, and metal ion) was modified such that the amino alcohol was replaced with a chiral 1,2-diamine ((R,R)-1,2-diaminocyclohexane or (R,R)-diphenyl-1,2-ethylenediamine) that could be attached with a linker to the solid support; additionally an amino acid was incorporated as a chiral diversity element located between the caproic acid and the (thio)urea linker (Figure 6.11).

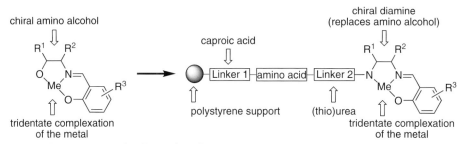

Figure 6.11 Typical tridentate ligand structure incorporating a chiral amino alcohol and modified diamine-based tridentate ligand structure attached to the solid support for parallel catalyst library strategy.

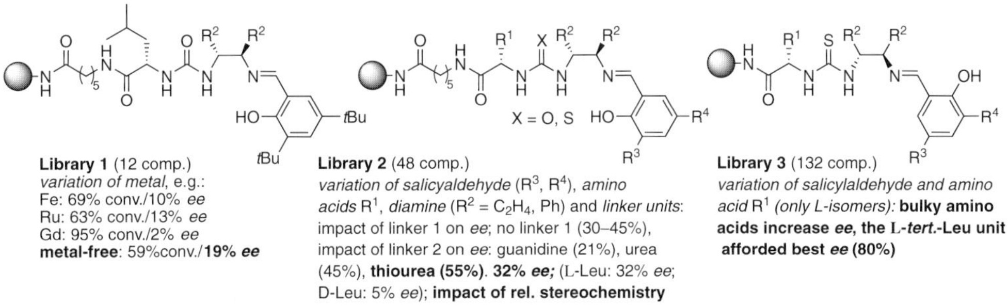

Figure 6.12 Polystyrene-bound Schiff base (thio)urea catalysts HTS-optimized in the asymmetric Strecker reaction between N-allyl-protected benzaldimine and TBSCN; key results obtained from the different libraries.

Library 1 (12 comp.) variation of metal, e.g.:
Fe: 69% conv./10% ee
Ru: 63% conv./13% ee
Gd: 95% conv./2% ee
metal-free: 59%conv./**19% ee**

Library 2 (48 comp.) variation of salicylaldehyde (R^3, R^4), amino acids R^1, diamine ($R^2 = C_2H_4$, Ph) and linker units: **impact of linker 1 on ee**; no linker 1 (30–45%), **impact of linker 2 on ee**: guanidine (21%), urea (45%), **thiourea (55%). 32% ee;** (L-Leu: 32% ee; D-Leu: 5% ee); **impact of rel. stereochemistry**

Library 3 (132 comp.) variation of salicylaldehyde and amino acid R^1 (only L-isomers): **bulky amino acids increase ee, the L-tert.-Leu unit afforded best ee (80%)**

On the basis of this initial ligand target structure, iterative HTS-optimization of the Strecker reaction enantioselectivity (GC analysis) was performed simultaneously with an autosampler in three stages with three libraries consisting of 12 (library 1), 48 (library 2), and 132 (library 3) polymer-bound catalyst candidates (Figure 6.12). For the screening, hydrocyanation reaction of N-allyl benzaldimine followed by trifluoroacetylation of the Strecker adduct library 1 (variation of the metal ion) demonstrated in this first screening stage that the metal-free, organocatalytic system, was more enantioselective (19% *ee*) than the 11 metal-ion containing alternatives (e.g., 13% *ee* with Ru) examined under the same conditions. Library 2 revealed that the amino acid unit, the relative stereochemistry of the diamine versus the amino acid and the salicylaldehyde derivative had a noticeable impact on the stereoinduction; linker 1 (caproic acid) was found to be responsible for background side reactions and was removed for improved enantioselectivity, while the amino acid group of the catalyst was directly attached to the polystyrene support. Linker 2, however, turned out to be crucial for stereoinduction: thiourea derivatives (55% *ee*) turned out to be superior to both urea (45% *ee*) and guanidine (21% *ee*)-based systems. The more focused library 3 consisting of 132 thiourea derivatives incorporating exclusively nonpolar L-amino acids and 3-*tert*-butyl-substituted salicylaldehydes supported the stereodifferentiating influence of the amino acid unit: the bulkiest thiourea derivatives bearing L-*tert*-Leu, cyclohexylglycine, and isoleucine produced the best *ee* values (up to 80%) (Figure 6.12) [56, 71].

Schiff base thiourea[5]) derivative **11** incorporating the (1*R*,2*R*)-diaminocyclohexane unit as chiral backbone was synthesized independently in solution on the basis

---

5) The Jacobsen group introduced the term "Schiff base catalyst" [122] to demonstrate that the structure of this novel catalyst class originates from Schiff base ligands and incorporates a Schiff base moiety; notably, this term does not indicate that these catalysts operate as bases, but their high catalytic efficiencies result from explicit double hydrogen-bonding as shown in Section 6.2.2.1. In the following the scientifically established term "Schiff base (thio)urea" is used to describe this class of organocatalysts.

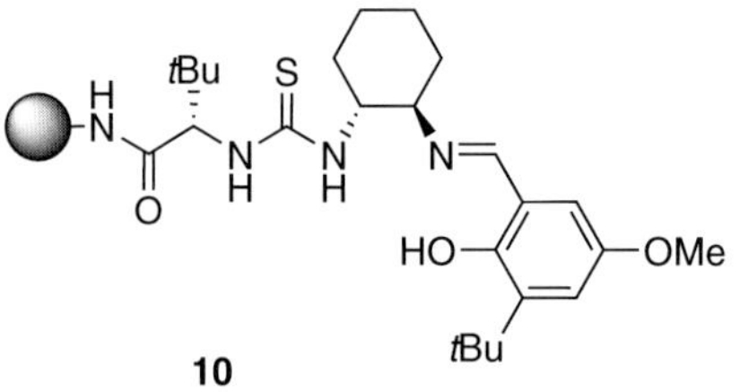

Figure 6.13 Diaminocyclohexane-derived Schiff base catalyst **10** representing the optimized structure identified from parallel catalyst screening and its polymer-free (solution-phase) counterpart **11** prepared for application in asymmetric Strecker reactions.

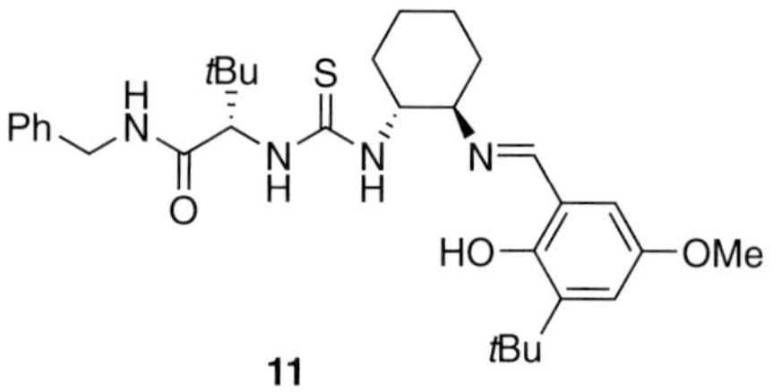

of the structural features of polystyrene-bound **10**, which was identified by parallel screening to be the most efficient in terms of enantioselectivity (Figure 6.13).

In the presence of Schiff base catalyst[6] **11** (2 mol%) and HCN as the cyanide source at −78 °C in toluene, the respective chiral trifluoroacetylated Strecker adducts **1–6** of aromatic and also aliphatic N-allyl aldimines were formed in yields ranging from 65 to 92% and ee values ranging from 70 to 91% as shown in Scheme 6.40. On the basis of the core structure illustrated in Figure 6.14, various slightly modified Schiff base (thio)urea derivatives have been developed and efficiently applied predominantly to the asymmetric Strecker reaction [157] of N-protected aldimines and ketimines as described in detail below. Although this new type of organocatalyst displayed high catalytic activity, stereoselectivity, and synthetic utility, the authors did not focus on the mode of action of the Schiff base catalysts until 2002, when they suggested that double hydrogen-bonding interactions between the (thio)urea amid protons and the imine lone pair activated the carbonyl analog electrophile for the product-forming nucleophilic attack by the cyanide ion [124]. This mechanistic proposal was supported by the theoretical and experimental findings toward double hydrogen-bonding activation of carbonyl-dienophiles through thiourea derivatives reported earlier by Schreiner and Wittkopp, in 2000 [112] and 2002 [116], and this ascribed the catalytically essential function of substrate binding and activation to the (thio)urea moiety originally incorporated as simple linker unit in the Schiff base structure.

On the basis of the core structure of catalyst **11** (Figure 6.14), the Jacobsen group constructed a new optimization parallel library of 70 Schiff base compounds incorporating seven amino acids with bulky α-substituents and 10 new salicylaldehyde derivatives [196]. Each library member was evaluated for enantioselectivity

---

6) The Jacobsen group introduced the term "Schiff base catalyst" [122] to demonstrate that the structure of this novel catalyst class originates from Schiff base ligands and incorporates a Schiff base moiety; notably, this term does not indicate that these catalysts operate as bases, but their high catalytic efficiencies result from explicit double hydrogen-bonding as shown in Section 6.2.2.1. In the following the scientifically established term "Schiff base (thio)urea" is used to describe this class of organocatalysts.

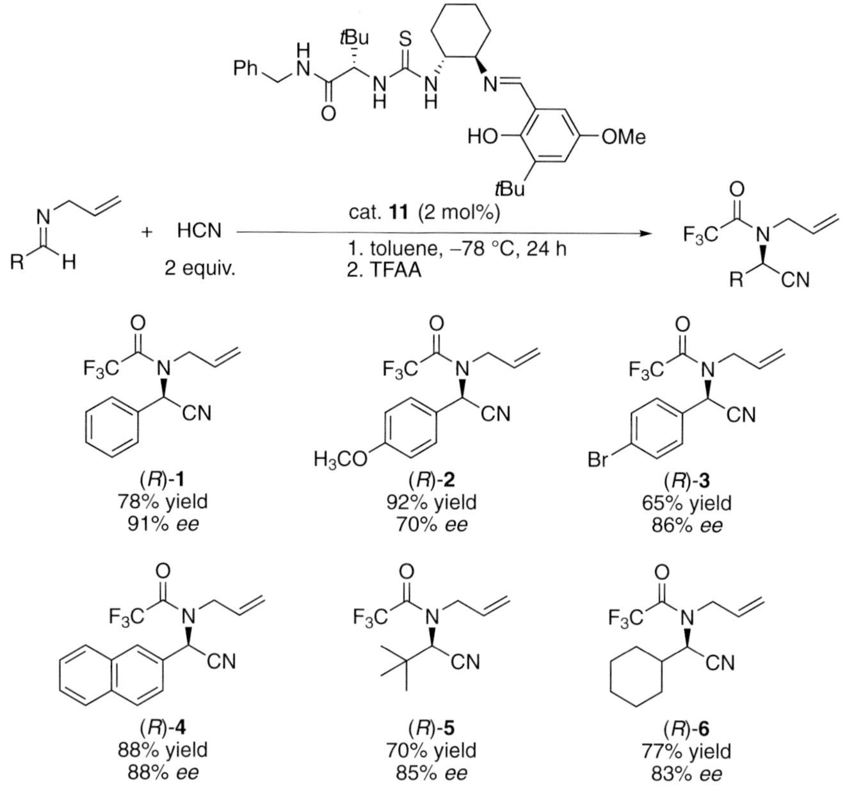

6.2 Synthetic Applications of Hydrogen-Bonding (Thio)urea Organocatalysts

Scheme 6.40 Product range of the **11**-catalyzed asymmetric Strecker reaction of aromatic and aliphatic N-allyl-protected aldimines.

(R)-**1** 78% yield, 91% ee
(R)-**2** 92% yield, 70% ee
(R)-**3** 65% yield, 86% ee
(R)-**4** 88% yield, 88% ee
(R)-**5** 70% yield, 85% ee
(R)-**6** 77% yield, 83% ee

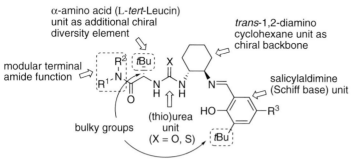

Figure 6.14 The key units of Jacobsen's Schiff base (thio)urea organocatalysts offering access to various structural modifications.

- α-amino acid (L-*tert*-Leucin) unit as additional chiral diversity element
- *trans*-1,2-diamino cyclohexane unit as chiral backbone
- modular terminal amide function
- salicylaldimine (Schiff base) unit
- bulky groups
- (thio)urea unit (X = O, S)

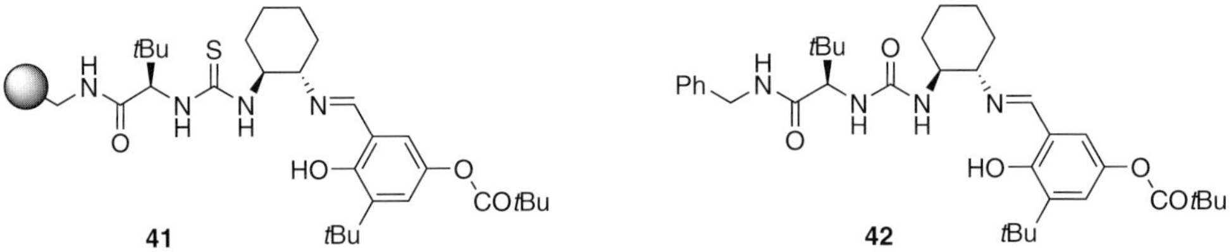

Figure 6.15 Polymer-bound Schiff base thiourea catalyst **41** bearing 5-pivaloyl-substitution and its nonimmobilized urea analog **42** optimized for the asymmetric Strecker reaction of aromatic and aliphatic aldimines.

in the HCN reaction [197] by adding the N-allyl imine of 2,2-dimethyl-propional-dehyde at 23 °C revealing polymer-bound 5-pivaloyl-substituted Schiff base **41** to be the most efficient catalyst. For further intensive studies with respect to scope and limitations, the soluble nonimmobilized urea-analogue **42** was employed for the asymmetric Strecker reaction of aldimines (Figure 6.15); the urea moiety was preferred due to an easier preparation at comparable catalyst efficiency.

In the presence of **42** (2 mol% loading), aliphatic and aromatic N-allyl as well as N-benzyl aldimines were efficiently converted after 20 h at −70 °C in toluene to the respective Strecker adducts and subsequently trifluoroacetylated to obtain the products **1–10** in good to excellent yields (65–99%) and *ee* values (77–97%) (Scheme 6.41). It turned out that N-benzyl imines could be used as substrates without significant difference in comparison to analogous N-allyl imines (e.g., N-benzyl adduct **8**: 85% yield, 87% *ee*; N-allyl adduct **9**: 88% yield, 86% *ee*; Scheme 6.41).

While all of the aryl imine substrates examined for this Strecker methodology existed predominantly or exclusively as the E-isomers, this did not appear to be a requirement for high enantioselectivity as demonstrated in the asymmetric **42**-catalyzed (2 mol% loading) hydrocyanation of the cyclic Z-imine 3,4-dihydroisoquinoline, which was converted to the corresponding adduct (88% yield, 91% *ee*) with the same sense of stereoinduction with respect to the benzylic stereogenic center as the examined acyclic E-imines (Schemes 6.41 and 6.42) [196].

To reveal the practical potential of polymer-bound Schiff base thiourea catalyst **41**, the Jacobsen group presented a recycling study by using this catalyst in 4 mol% loading at −78 °C for the hydrocyanation of N-benzyl pivalaldimine under preparative conditions (6.1 mmol scale) (Scheme 6.43). Clean hydrocyanation of the model imine was observed and the target Strecker adduct was obtained with only a slight reduction in enantioselectivity (**42**: 96% *ee*; **41** 93% *ee*). The product was isolated in nearly quantitative yield after removing the catalyst from the reaction mixture by simple filtration and no loss of catalyst activity or product enantioselectivity was detected after 10 reaction cycles (Scheme 6.43) [61, 62]. This simple work-up procedure, the easy catalysts recovery, and the reusability made **41** a practical alterna-

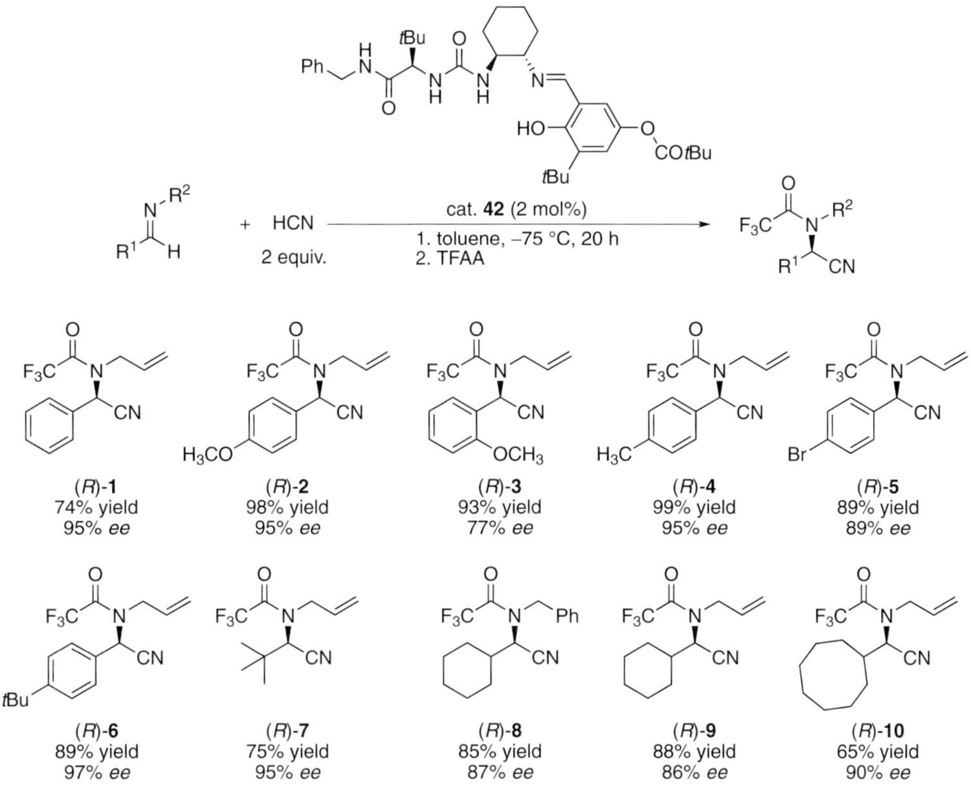

Scheme 6.41 Typical products obtained from the asymmetric Strecker reaction of aliphatic and aromatic aldimines catalyzed by urea **42**.

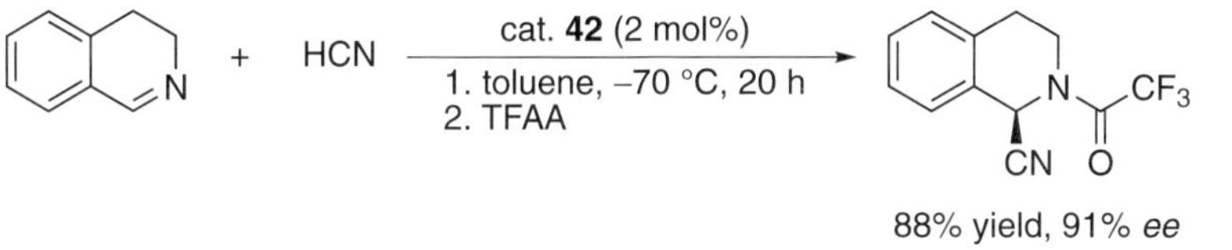

Scheme 6.42 In the presence of **42** the asymmetric Strecker reaction of the cyclic Z-imine 3,4-dihydroisoquinoline afforded the (R)-adduct.

tive to its soluble analog **42** that produced 2–4% higher *ee* values (loading 2 mol%) for the respective Strecker products, but reduced yields (e.g., for N-benzyl pivalaldimine: 88%) due to the required chromatographic product purification and catalyst separation.

The synthetic utility and importance of the obtained Strecker adducts as precursors for access to enantiomerically pure α-amino acids was examplified for the

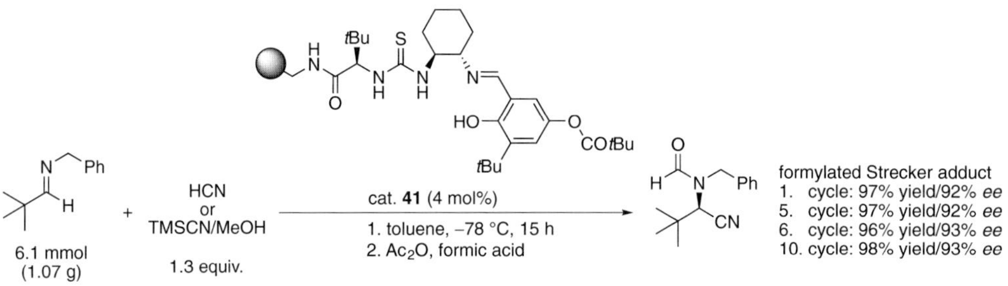

Scheme 6.43 Recycling study: Polymer-bound Schiff-base thiourea **41** catalyzed the Strecker reaction of pivalaldimine without loss of activity or enantioselectivity even after 10 catalytic cycles.

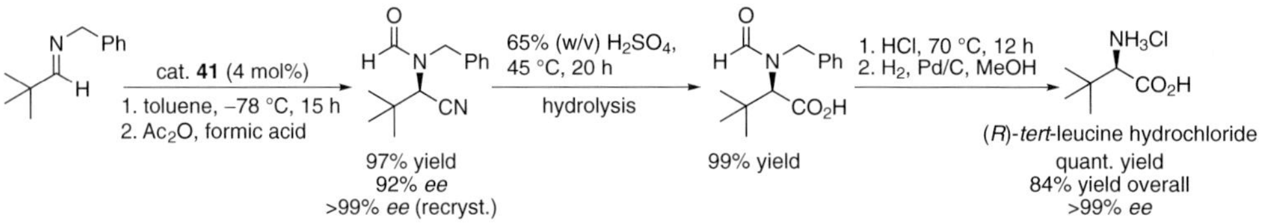

formylated Strecker adduct
1. cycle: 97% yield/92% *ee*
5. cycle: 97% yield/92% *ee*
6. cycle: 96% yield/93% *ee*
10. cycle: 98% yield/93% *ee*

Scheme 6.44 Reaction sequence for the synthesis of enantiopure (*R*)-*tert*-leucine hydrochloride starting from the pivalaldimine Strecker adduct obtained under catalysis with polymer-bound thiourea **41**.

synthesis of (*R*)-*tert*-leucine hydrochloride starting from *N*-benzyl pivalaldimine, which was enantioselectively converted in the presence of 4 mol% **41** to the corresponding α-amino nitrile. Since direct acidic hydrolysis of the (*R*)-amino nitrile failed and led to considerable decomposition due to harsh reaction conditions, the amino nitrile was first *N*-formylated (97% yield, 92% *ee*; 99% *ee* after recrystalization) before the racemization-free hydrolysis to the respective *N*-protected (*R*)-amino acid proceeded in 99% yield. The final steps included deformylation and removal of the *N*-benzyl group by using H_2/Pd/C producing the desired enantiopure (*R*)-*tert*-leucine hydrochloride (quant. yield, >99% *ee*) in high overall yield of 84% based on pivalaldimine (Scheme 6.44).

In addition to aldimines (Scheme 6.41), the Jacobsen group applied urea catalyst **42** to the asymmetric Strecker reaction of *N*-allyl and, in particular, *N*-benzyl ketimines resulting in α-amino nitriles bearing an (*R*)-configured stereogenic quaternary carbon center [198]. (*R*)-amino nitriles are suitable precursors for the synthesis of enantiopure α-quaternary α-amino acids, which are key intermediates for the synthesis of a broad variety of pharmaceutically important compounds. Although the resin-bound Schiff base catalyst **41**, which proved efficient in the hydrocyanation of aldimines (Scheme 6.44) also turned out to promote the model Strecker reaction of *N*-allyl acetophenone imine, long reaction times (180 h) were required to give the desired Strecker adduct (**41**: 95% yield; 85% *ee* at 4 mol% loading; −75 °C in toluene). Catalyst **42** (2 mol% loading), however, displayed higher catalytic activity under unchanged conditions (97% yield/30 h; 85% *ee*) and was therefore utilized for the enantioselective hydrocyanation of various *N*-benzyl-protected ketimines that were found to be the substrates of choice due to higher stability of their Strecker adducts under either acidic or basic conditions (no retro-Strecker decomposition) and due to slightly increased *ee* values (e.g., *N*-allyl acetophenone imine: 85% *ee*; *N*-benzyl: 90% *ee*; at −75 °C in toluene) compared to *N*-allyl-protected ketimines. Under optimized reaction conditions (−75 °C; 2 equiv. HCN, toluene) **42** (2 mol%) converted a variety of *N*-benzyl-protected ketimins bearing both electron-withdrawing and electron-donating aromatic substituents to their α-amino nitriles **1–8** in mostly excellent yields (45%, 97–100%) and *ee* values ranging from 42 to 95% (Scheme 6.45). Some of the Strecker adducts

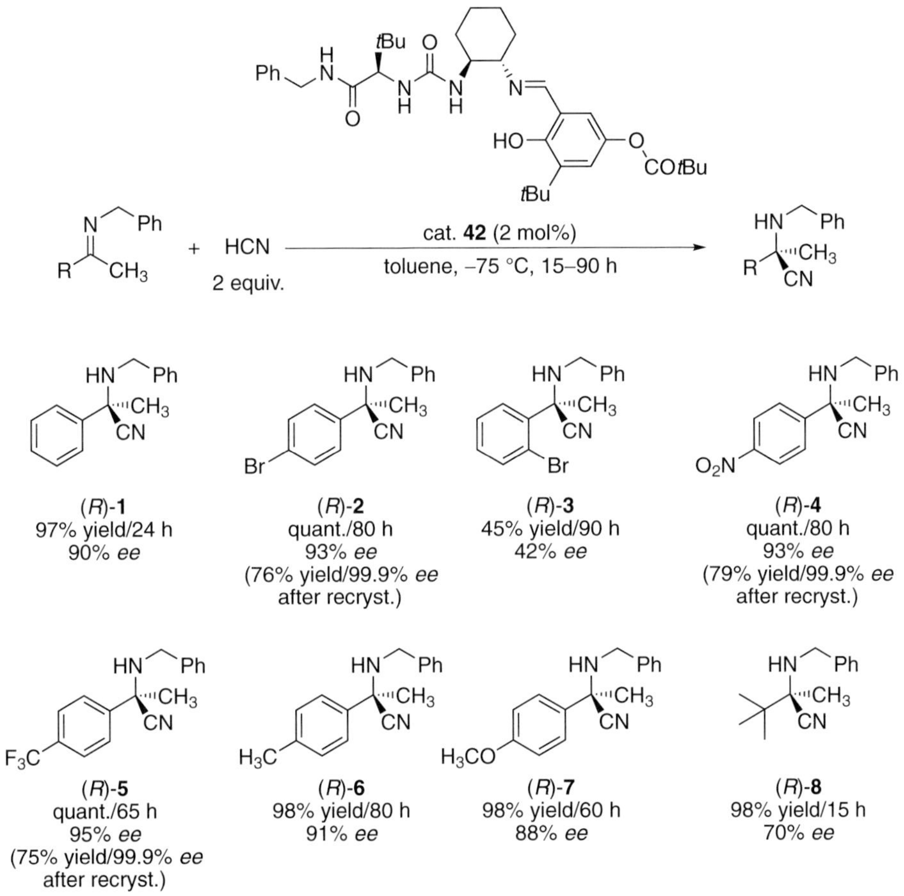

Scheme 6.45 Typical products of the asymmetric Strecker reaction of ketimines catalyzed by urea **42**.

(**2**, **4**, and **5**; Scheme 6.45) were isolated as crystalline compounds affording enantiomerically pure products (99.9% *ee*; 75–79% overall yield) after recrystallization from hexanes. Since catalysts **42** was soluble in hexanes, no chromatographic product purification was necessary as the catalysts remained in the mother liquors and could be recovered, purified, and reused in the Strecker reaction with results similar to that reached through freshly prepared catalyst. Employing the typical Strecker adducts of *N*-benzyl and *N*-4-bromobenzyl acetophenone imine (prepared in 40 h: 95% yield; 92% *ee*/75% yield, 99.9% *ee* after recryst.), the authors presented a three-step reaction sequence that furnished the desired enantiopure α-methyl phenylglycine in 92% overall yield (91% *ee*) for the *N*-benzyl and 93% overall yield (99.9% *ee*) for the *N*-4-bromobenzyl Strecker adduct (see also Scheme 6.44) [198].

In 2002, Vachal and Jacobsen utilized a number of approaches such as systematic structural modifications, NMR methods, kinetic, and computational studies to investigate and elucidate, in detail, the mode of action of Schiff-base type urea catalyst **42** [196, 198, 199] in the mechanism of highly enantioselective Strecker reaction between the model substrate *N*-allyl-4-methoxybenzaldimine and HCN as the cyanide source [124]. It was found through NOE and ROESY NMR experiments that **42** adopted a well-defined secondary structure in solution despite its relatively small size (fw = 621 g mol^{-1}). The hydrocyanation reaction of the model imine followed an enzyme-like Michaelis–Menten kinetic model with a first-order dependence on catalyst **42** and HCN, and saturation kinetics with respect to the imine substrate implicating reversible formation of a hydrogen-bonded imine–catalyst complex. To identify and locate the relevant proton(s) involved in these hydrogen-bond interactions, a series of analogs of **42** were prepared and evaluated as catalysts for catalytic activity and enantioselectivity in the model Strecker reaction. Deletion of the secondary amide proton, the phenolic proton, and the imine functionality did not suppress catalytic activity; in contrast, alkylation of either of the urea nitrogens or replacement with a carbamate group led to dramatic loss of activity and enantioselectivity. These findings suggested that exclusively the two urea amide protons interacted with the imine substrate resulting in the observed catalytic efficiency. NMR titrations of a solution of representative ketoimine *N-p*-methoxybenzyl acetophenone imine ($E:Z = 20:1$) with **42** resulted in a downfield shift of the *Z*-imine methyl resonance exclusively revealing the preferred binding of **42** to the *Z*-isomer instead to the *E*-isomer. This reaction occurred in quantitative yield and in 89% *ee*, while cyclic imines restricted to *E*-configurations such as 6-phenyl-2,3,4,5-tetrahydropyridine underwent no reaction under the same conditions even under extended reaction times, in contrast to the cyclic *Z*-imine 3,4-dihydroisoquinoline (Scheme 6.42). A detailed 3D structure model of the imine–catalyst complex, based on molecular modeling studies and experimentally on multiple NOE interactions between **42** and various *Z*-imines, mirrored the complex geometry resulting from the relative orientation of the imine substrate relative to the catalyst **42**. The imine substrate was placed in the bridging mode for an explicit double hydrogen-bonding interaction between the two urea amide protons of **42**, while the Strecker product–catalyst complex showed only single hydrogen-bonding. This provided the probable origin of the catalyst TON and TOF, since gas-phase calculations for model a (thio)urea and imine showed that the double hydrogen-bonding interaction (8.5 kcal mol^{-1} urea; 10.0 kcal mol^{-1} thiourea) was stronger than the single hydrogen-bonding interaction (5.0 kcal mol^{-1} urea; 6.3 kcal mol^{-1} thiourea). The model of the imine substrate–catalyst complex provided important rationalizations for the observed scope and stereoselectivity of the **42**-catalyzed Strecker reaction [196, 198, 199] and clues for further catalyst structure optimization: (1) The large group on the imine carbon was directed away from the catalyst into the solvent allowing **42**-catalyzed hydrocyanation of most aldimines with high *ee*, regardless of the steric and electronic properties of the substrate. (2) The small group (H for aldimines, Me for methylketoimines) was directed toward catalyst; ketoimines bearing larger substituents turned out to be poor substrates for the reaction, presumably because they could not be accommodated within the less hindered optimal geometry. (3) The *N*-substituent was also directed away from the catalyst. However, its size was restricted as a result of the requirement to access the *Z*-isomer of the imine. (4) On the basis of the observed stereoinduction trend, the addition of HCN took place over the diaminocyclohexane portion of the catalyst away from the amino acid and amide unit. The last hypothesis led to the prediction that a more sterically demanding amino acid or amide unit (Figure 6.14) could additionally favor the cyanide attack compared to the less bulky diaminocyclohexane unit and thus making the Schiff base catalyst more enantioselective in Strecker reactions of aldimines and ketimines. To evaluate this perspective, the authors performed a model- (mechanism-) driven systematic structure optimizations by stepwise modification of the amide, the amino acid, and the (thio)urea unit of catalyst **42** and examined these derivatives of **42** (1 mol% loading) in the model Strecker reaction (toluene; −78 °C; HCN) of *N*-benzyl-protected 2-methylpropionaldehyde imine (Figure 6.16). It was found that both the replacement of the secondary amide unit with a bulkier tertiary amide and the incorporation of a thiourea moiety instead of the urea unit resulted in a significant improvement in stereoinduction (from initial 80% *ee* obtained with **42** to 97% *ee*). This led to the identification of hydrogen-bonding Schiff base thiourea catalyst **47**, while the urea derivatives **43–46** gave lower *ee* values (Figure 6.16).

This tertiary amide-functionalized Schiff base thiourea was found to efficiently catalyze the asymmetric Strecker reaction [157] of *N*-benzyl-protected aldimines and also one ketimine in high enantioselectivities (86–99% *ee*) and proved superior to **42** examined under the same conditions (1 mol% loading, toluene, −78 °C, HCN) (Scheme 6.46) [198].

List and co-workers reported the **47**-catalyzed (1 mol% loading) asymmetric acetylcyanation of *N*-benzyl-protected aliphatic and aromatic aldimines by using commercially available liquid acetyl cyanide as the cyanide source instead of HCN [161]. Under optimized reaction parameters (toluene, −40 °C) the procedure resulted in the desired *N*-protected α-amino nitriles **1–5** in yields ranging from 62

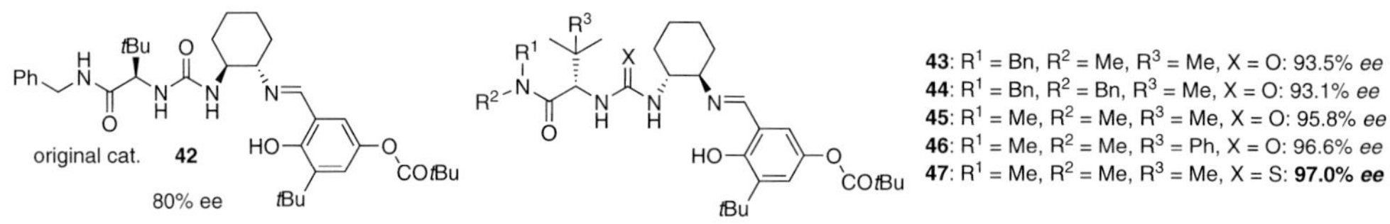

196 | *6 (Thio)urea Organocatalysts*

Figure 6.16 Structure optimization of **42** in the asymmetric Strecker reaction of N-benzyl-protected 2-methylpropionaldehyde imine identified tertiary amide-functionalized Schiff base thiourea **47** as the most enantioselective catalyst structure.

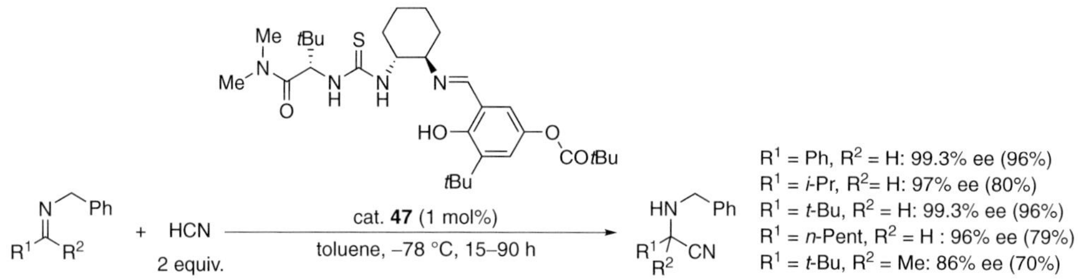

original cat. **42** 80% ee

43: R^1 = Bn, R^2 = Me, R^3 = Me, X = O: 93.5% ee
44: R^1 = Bn, R^2 = Bn, R^3 = Me, X = O: 93.1% ee
45: R^1 = Me, R^2 = Me, R^3 = Me, X = O: 95.8% ee
46: R^1 = Me, R^2 = Me, R^3 = Ph, X = O: 96.6% ee
47: R^1 = Me, R^2 = Me, R^3 = Me, X = S: **97.0% ee**

R^1 = Ph, R^2 = H: 99.3% ee (96%)
R^1 = *i*-Pr, R^2 = H: 97% ee (80%)
R^1 = *t*-Bu, R^2 = H: 99.3% ee (96%)
R^1 = *n*-Pent, R^2 = H: 96% ee (79%)
R^1 = *t*-Bu, R^2 = Me: 86% ee (70%)

Scheme 6.46 Products of the **47**-catalyzed asymmetric Strecker reaction; ee values obtained with urea **42** are given in parentheses.

to 95% and in high enantiomeric ratios (*er* up to 99:1) as shown for some typical products in Scheme 6.47.

Hydrogen-bonding Schiff base thiourea **47**, originally developed for the asymmetric HCN addition to N-protected imines [124] (Strecker reaction) [157] (Figure 6.16), was identified by Joly and Jacobsen to also promote the asymmetric addition of di(2-nitrobenzyl)phosphate to aliphatic and (hetero)aromatic N-benzyl-protected aldimines [200]. This **47**-catalyzed (10 mol% loading) enantioselective hydrophosphonylation generated the respective (R)-α-amino phosphonates **1–8** in yields in the range of 52–90% and in excellent *ee* values (92–99%) (Scheme 6.48). It is notable that the absolute configuration of the isolated adducts was found to be consistent with the stereoinduction observed for the asymmetric Strecker [122, 124, 196, 198] and Mannich reactions [72, 201]. For selected examples of the obtained (R)-α-amino phosphonates, the authors demonstrated the synthesis of the corresponding α-amino phosphonic acids via a selective hydrogenolysis strategy (87–93% yield; 96–98% *ee*).

Wenzel and Jacobsen, in 2002, identified Schiff base thiourea derivative **48** as catalyst for the asymmetric Mannich addition [72] of *tert*-butyldimethylsilyl ketene acetals to N-Boc-protected (hetero)aromatic aldimines (Scheme 6.49) [201]. The optimized structure of **48** was found through the construction of a small, parallel

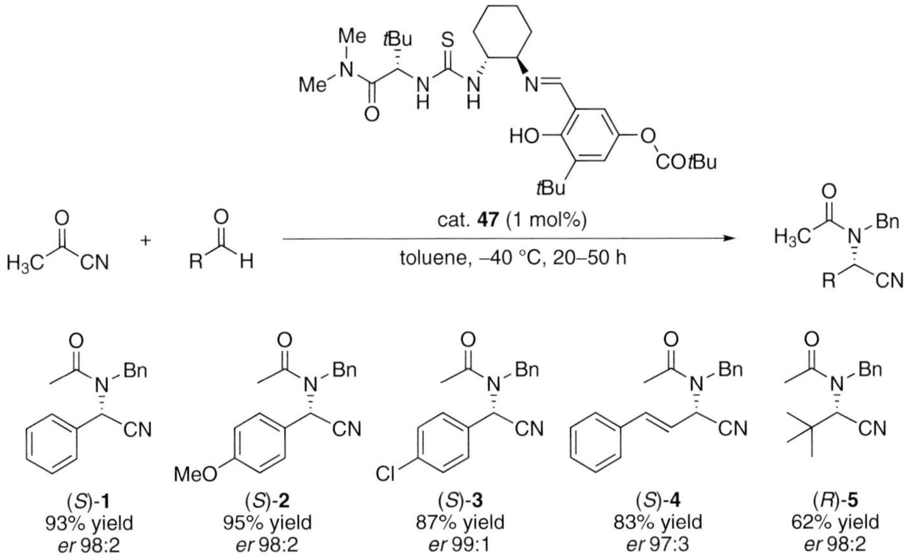

Scheme 6.47 Strecker products obtained from the **47**-catalyzed asymmetric acetylcyanation using acetyl cyanide as cyanide source.

(S)-**1**
93% yield
er 98:2

(S)-**2**
95% yield
er 98:2

(S)-**3**
87% yield
er 99:1

(S)-**4**
83% yield
er 97:3

(R)-**5**
62% yield
er 98:2

library of 22 catalyst candidates with systematic variation of the salicylaldimine, diamine, amino acid, and amide unit (Figure 6.14). Since the *para* substituent turned out to have no impact on the stereoinduction of the initially explored Mannich test reaction, the commercially available di-*tert*-butylsalicylaldehyde was used for the synthesis of subsequent catalyst candidates. As already shown in the case of catalyst **47** (Figure 6.16), the presence of a tertiary amide unit improved the stereoinduction and, additionally, prevented undesired formation of thiohydantoin byproducts during the preparation of the respective Schiff base catalyst. Under optimized conditions concerning the choice of catalyst (5 mol% **48**), the silyl ketene acetal (TBS-group, *iso*-propyloxy substituent), reaction temperature (at −40 °C or −30 °C the racemic background reaction was suppressed), and toluene as the solvent the Mannich reaction proceeded highly enantioselectively (86–98% ee) and gave the target (R)-configured Mannich adducts **1–6** in high yields (87–99%) in a practical reaction time (48 h) (Scheme 6.49). The N-Boc-protected β-amino acid derivatives were readily deprotected under mild acidic conditions to yield enantiopure β-aryl-β-amino acids suitable for peptide synthesis.

In 2004, Taylor and Jacobsen suggested a procedure for the enantioselective acetyl-Pictet–Spengler reaction, that is the cyclization of electron-rich aryl or heteroaryl groups onto N-acyliminium ion enabling access to substituted tetrahydro-β-carbolines and tetrahydroisoquinolines that are core structure elements in natural and synthetic organic compounds [202, 203]. Screening various thiourea catalyst candidates such as **47** in the formation of model product $N_β$-acetyl-

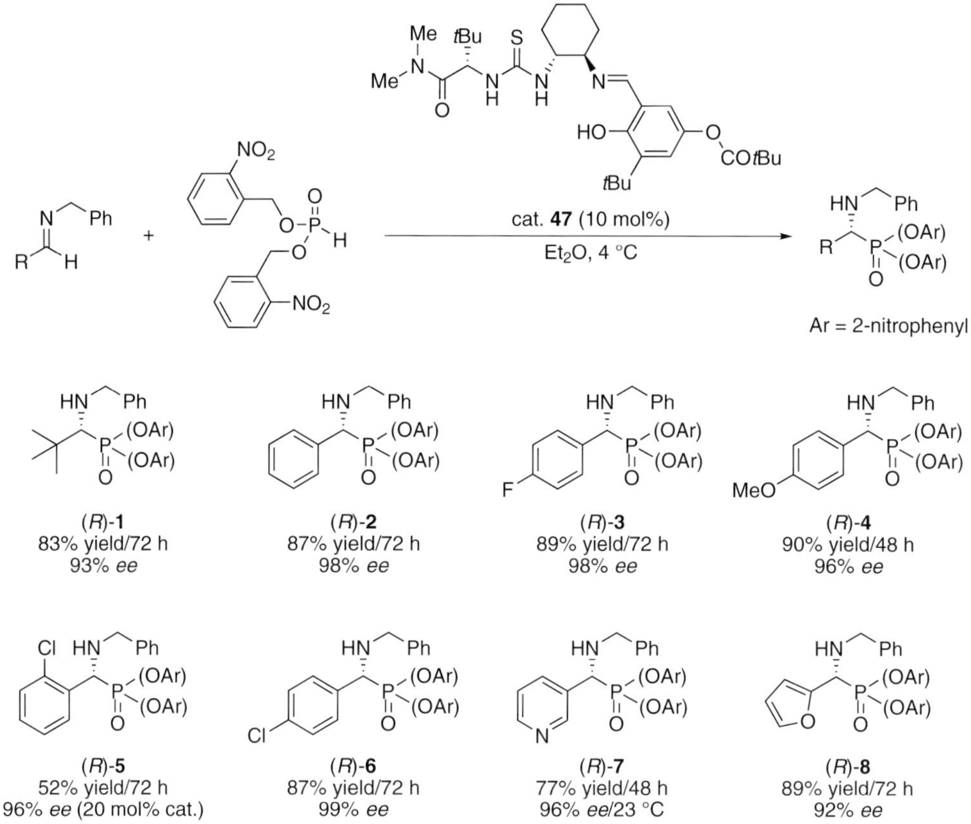

Scheme 6.48 Product range of the asymmetric hydrophosphonylation of N-benzylated aldimines promoted by thiourea derivative 47.

tetrahydro-β-carboline 1 (Scheme 6.50) from the corresponding tryptamine-derived imine starting material failed and showed no conversion even at high temperatures (Pictet–Spengler conditions). Performing the same test reaction, however, in the presence of acetyl chloride (1.0 equiv.) serving as the acetylating reagent, which activated the imine function through formation of the corresponding N-acetyliminim ion, the desired product-forming intramolecular Friedel–Crafts cyclization occurred and the target tetrahydro-β-carboline was isolated in 65% yield and 59% ee (47: 10 mol%; 2,6-lutidine, Et$_2$O; at −30 °C) [201]. The authors modulated the Schiff base structure of 47 such that the salicylaldimine moiety was replaced with the bulky N-pivaloyl amide unit resulting in catalytically more efficient novel thiourea derivatives 49–52 (Figure 6.17). Variation of the pyrrole substituents and fine-tuning of the amide unit demonstrated that 2-methyl-5-phenylpyrrole thiourea 52 bearing the N,N-diisobutyl amide group was the best catalyst in terms of activity (70% yield) and enantioselectivity (93% ee) (Figure 6.17).

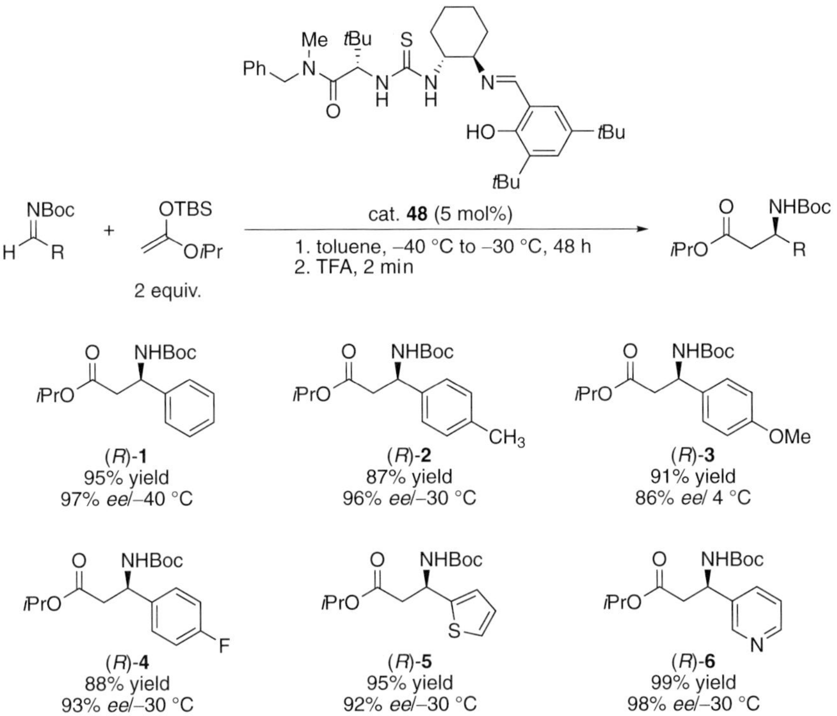

6.2 Synthetic Applications of Hydrogen-Bonding (Thio)urea Organocatalysts

Scheme 6.49 Typical Boc-protected β-amino acid derivatives obtained from **48**-catalyzed Mannich reactions of N-Boc-protected aldimines with silyl ketene acetal.

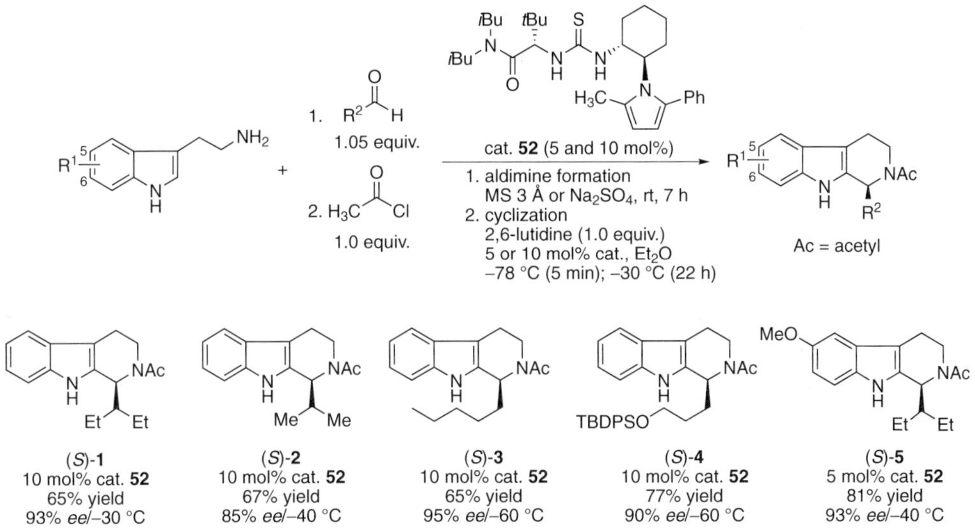

(R)-1
95% yield
97% ee/−40 °C

(R)-2
87% yield
96% ee/−30 °C

(R)-3
91% yield
86% ee/ 4 °C

(R)-4
88% yield
93% ee/−30 °C

(R)-5
95% yield
92% ee/−30 °C

(R)-6
99% yield
98% ee/−30 °C

Scheme 6.50 Typical tetrahydro-β-carbolines prepared with the **52**-catalyzed enantioselective acetyl-Pictet–Spengler reaction.

(S)-1
10 mol% cat. **52**
65% yield
93% ee/−30 °C

(S)-2
10 mol% cat. **52**
67% yield
85% ee/−40 °C

(S)-3
10 mol% cat. **52**
65% yield
95% ee/−60 °C

(S)-4
10 mol% cat. **52**
77% yield
90% ee/−60 °C

(S)-5
5 mol% cat. **52**
81% yield
93% ee/−40 °C

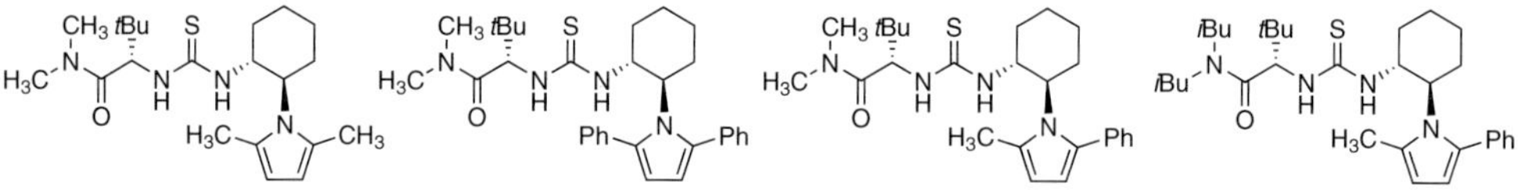

49
(10 mol%)
65% yield
77% ee

50
(10 mol%)
55% yield
71% ee

51
(10 mol%)
70% yield
93% ee

52
(10 mol%)
70% yield
93% ee

Figure 6.17 Pyrrole thiourea derivatives evaluated for catalytic activity and selectivity in the asymmetric acetyl-Pictet–Spengler reaction.

With thiourea **52** as the catalyst (5–10 mol%), a range of substituted S-configured tetrahydro-β-carbolines **1–5** were accessible in good yields (65–81%) and high enantioselectivities (85–95%) as shown in Scheme 6.50. The imine substrates of this two-step procedure were generated *in situ* by condensation of the tryptamine with the respective aldehyde (1.05 equiv.) and were directly used without further purification (Scheme 6.50). Recovered (by chromatography) pyrrole thiourea catalyst **52** could be reused without the loss of activity or selectivity.

Hydrogen-bonding pyrrole thiourea **52** proved capable of weakly activating Lewis-acid N-acetyliminium ions toward an enantioselective cyclization step in the acetyl-Pictet–Spengler reaction as reported by the Jacobsen group [201]. On the basis of this novel hydrogen-bonding activation, the same group applied catalyst **52** (10 mol%) to acyl-Mannich reactions [72] of substituted isoquinoline substrates offering access to a number of dihydroisoquinolines **1–5** in good yields (67–86%) and *ee* values (60–92%) (Scheme 6.51) [204]. The reaction provided the best results in terms of product yield and stereoinduction, when using *tert*-butyldimethylsilyl ketene acetal (2.0 equiv.) derived from isopropyl acetate as nucleophilic enolate equivalent and 2,2,2-trichloroethyl chloroformate (TrocCl) as the acylating reagent (1.1 equiv.). Similar to the acyl-Pictet–Spengler reaction (Scheme 6.50), the **52**-catalyzed acyl-Mannich reaction exhibited a strong dependence on the substitution pattern of the pyrrole moiety (e.g., 2,5-dimethyl-pyrrole: 60% yield; 30% *ee*; 2,5-diphenyl-pyrrole: 55% yield; 78% *ee* in then acyl-Mannich reaction of isoquinoline). The crystal structure geometry of catalyst **52** suggested an explanation of this observation. The 2-methyl-5-phenylpyrrole structural motif located the phenyl group in a position to interact closely with any incoming substrate that underwent hydrogen-bonding interactions with the thiourea amide protons [204]. To emphasize the synthetic importance of the obtained dihydroisoquinolines, the authors developed a two-step conversion to the corresponding 1-substituted tetrahydroisoquinoline derivatives (e.g., 83% yield; 86% *ee*) that are important chiral building blocks for, e.g., alkaloid synthesis.

In 2007, Jacobsen and co-workers reported the enantioselective Pictet–Spengler-type cyclization of β-indolyl ethyl hydroxylactams affording highly enantioenriched indolizidinones and quinolizidinones with fully substituted stereogenic centers [205]. The hydroxylactam substrates prepared either by imide reduction using

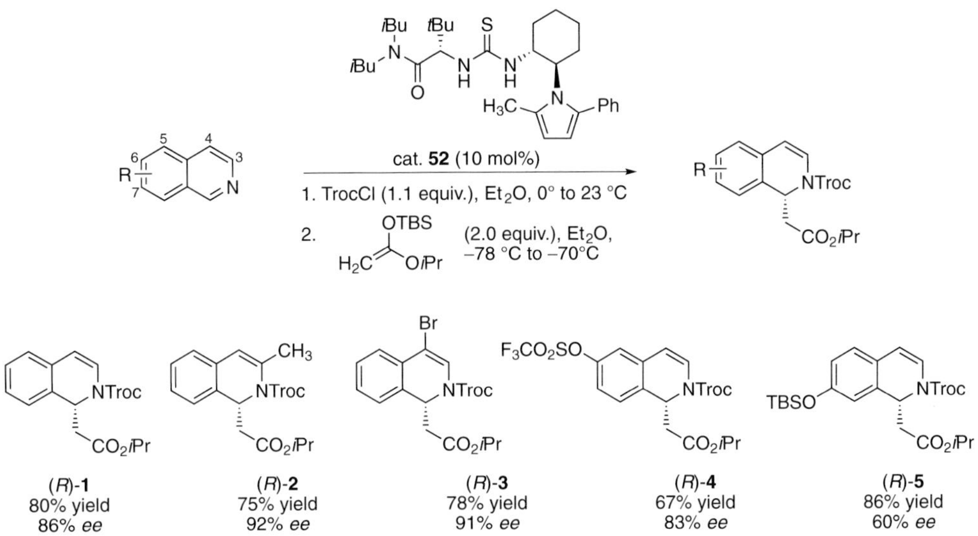

Scheme 6.51 Dihydroisoquinolines obtained form asymmetric acyl-Mannich reaction of substituted isoquinolines promoted by **52**.

NaBH$_4$ or by imide alkylation with organolithium reagents underwent conversion to the desired cyclization products **1–4** in yields ranging from 51 to 92% and in excellent *ee* values (90–99%) (Scheme 6.52). This methodology utilized pyrrole 2-methyl-5-phenylpyrrole catalyst **53** (10 mol%), the *N*-methylpentyl amide derivative of **52** (Schemes 6.50 and 6.51), in the presence of TMSCl (2 equiv.) that served as dehydrating agent to generate the mechanistically essential *N*-acyliminium ion [76] from the hydroxylactam (Scheme 6.53). Based on substituent, counterion, solvent effect, and variable temperature ^{1}H NMR studies the mechanistic proposal started with the TMSCl-induced irreversible and rapid formation of a chlorolactam. Catalysis and enantioinduction resulted from the **53**-promoted initial abstraction of a chloride counterion (S$_N$1-type rate-determining step) affording a chiral *N*-acyliminium chloride–thiourea complex that subsequently underwent product-forming cyclization mediated by anion-bound thiourea pyrrole catalyst **53** (Scheme 6.53). This methodology and the mechanistic proposal are of conceptual importance as for the first time a hydrogen-bonding organocatalyst is suggested to bind an anion (recognition) in an enantioselective reaction. An important parallel, e.g., represents the **9**-assisted orthoester heterolysis and subsequent nonstereoselective acetalization reported by Kotke and Schreiner (Scheme 6.11) [118, 146].

Pyrrole-containing thiourea derivatives **52** and **53** were developed and optimized for hydrogen-bonding activation of *N*-acyliminium ions [76] in the acyl-Pictet–Spengler [202, 205] (Schemes 6.50 and 52) and acyl-Mannich reaction [204] (Scheme 6.51). List *et al.* extended the applicability of this thiourea type to

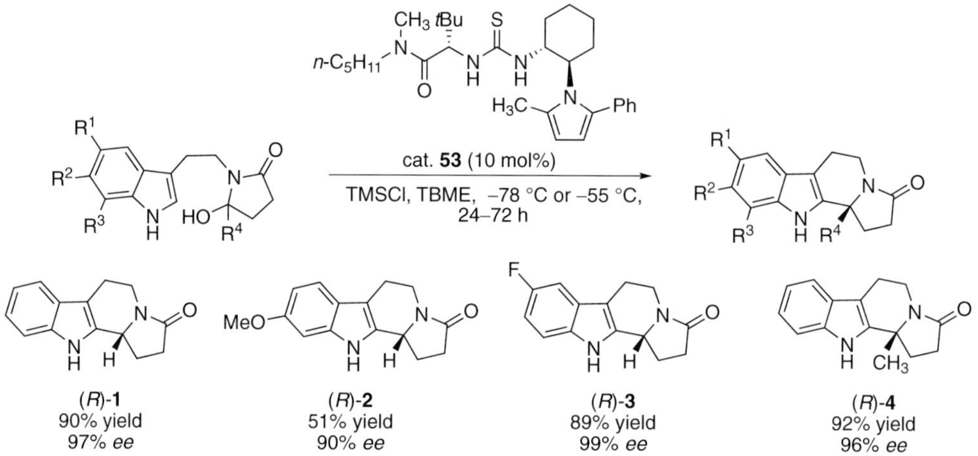

Scheme 6.52 Products of the asymmetric Pictet–Spengler-type cyclization of β-indolyl ethyl hydroxylactams catalyzed by **53**.

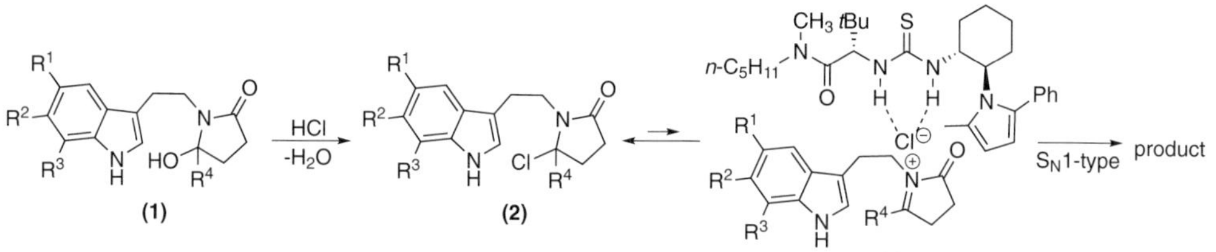

Scheme 6.53 Proposed mechanism for the **53**-catalyzed asymmetric Pictet–Spengler-type cyclization of β-indolyl ethyl hydroxylactams: Hydroxylactam (**1**) forms chlorolactam (**2**) followed by chiral N-acyliminium chloride-thiourea complex (**3**) and the observed product generated by intramolecular cyclization; catalysis and enantioinduction result from chloride abstraction and anion binding.

the asymmetric transfer hydrogenation of nitroalkenes and prepared a range of (S)-configured nitroalkanes such as **1–6** utilizing catalyst **54** (5 mol%) in the presence of Hantzsch ester [165] **55** (1.1 equiv.) [206]. The protocol was reported to be high yielding (82–99%) and enantioselective (up to er 97 : 3) for various nitroalkene substrates as shown in Scheme 6.54.

In 2003, Takemoto and co-workers introduced the first tertiary amine-functionalized thiourea catalyst [129]. This new type of stereoselective thiourea catalyst incorporating both (R,R)-1,2-diaminocyclohexane as the chiral scaffold and the privileged 3,5-bis(trifluoromethyl)phenyl thiourea motif for strong hydrogen-bonding substrate binding, marked the introduction of the concept of bifunctional-

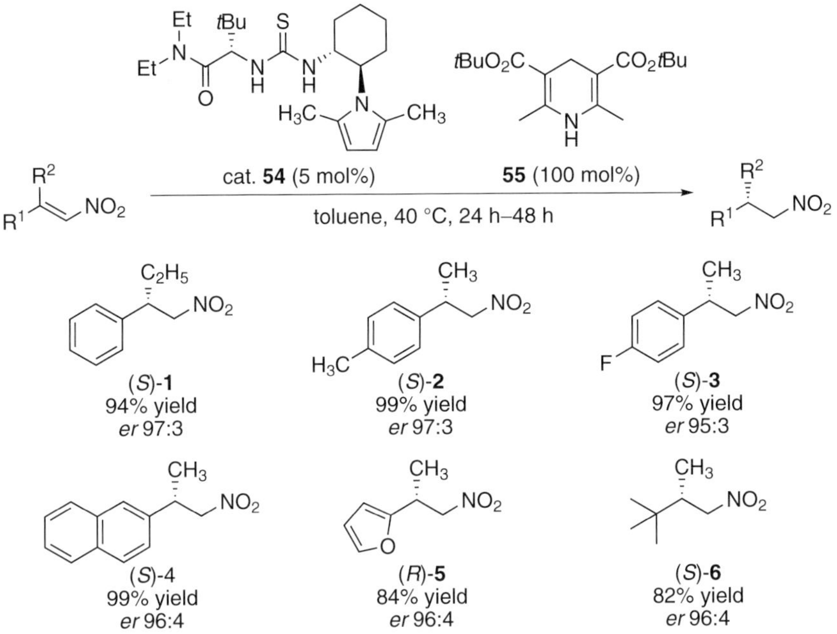

Scheme 6.54 Chiral nitroalkanes provided from the 54-catalyzed asymmetric transfer hydrogenation of nitroalkenes in the presence of **55**.

ity in synthetic (thio)urea catalysis systems. Bifunctionality is a structural principle orginating from highly efficient natural catalytic systems such as enzymes (Scheme 6.1) and has been mimicked by various synthetic catalytic systems to enable a catalyst to employ its Lewis/Brønsted acidic and Lewis/Brønsted basic functionality synergistically to activate both the nucleophilic and electrophilic reaction components simultaneously (Scheme 6.55). This bifunctional activation often results in much higher rate enhancements and stereoinductions than achieved by comparable monofunctional synthetic catalysts.[7]

The Takemoto group synthesized a series of diaminocyclohexane-based thiourea derivatives (e.g., **12**, **40**, **57**, and **58**) for catalysis of the Michael addition [149–152] of malonates to trans-β-nitrostyrenes (Figure 6.18) [129, 207]. In the model, Michael addition of diethyl malonate to trans-β-nitrostyrene at room temperature and in toluene as the solvent tertiary amine-functionalized thiourea **12** (10 mol% loading) was identified to be the most efficient catalyst in terms of catalytic activity (86%

---

7) In contrast to monofunctional (thio)urea organocatalysts, bifunctional catalyst structures enable simultaneous coordination, activation, and suitable relative orientation of both reaction components (the electrophile and the nucleophile) resulting in high catalytic efficiencies. Notably, the key principle of bifunctional catalysis is usually interpreted in terms of formation of ternary complexes, although this mechanistic model appears to be entropically disfavored. Refined models have to be developed.

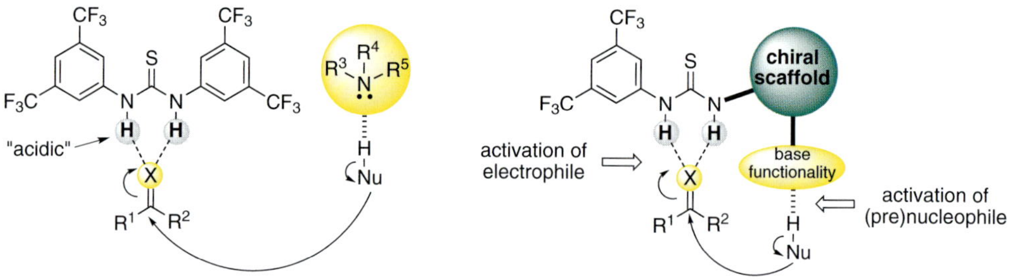

Scheme 6.55 Design principle of amine-functionalized bifunctional thiourea organocatalysts derived from privileged monofunctional thiourea **9** cooperating with an amine base additive (**A**) and basic bifunctional mode of action of chiral amine thioureas (**B**), simultaneous activation of both the electrophile and (pre)nucleophile (triple-collision scenario) through partial protonation (H-bonding) and (partial) deprotonation, respectively.

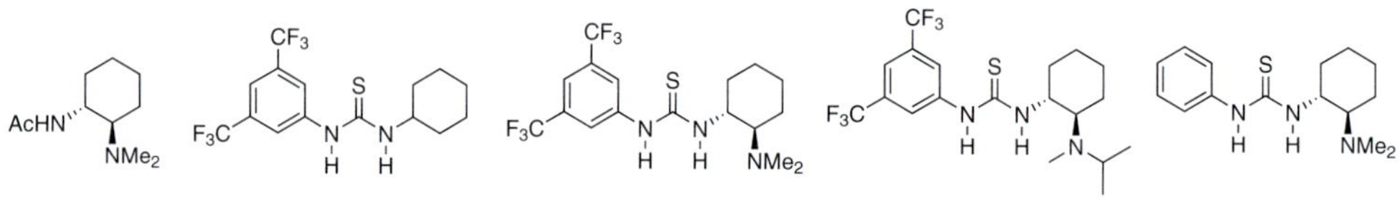

56
14% yield/24 h
35% ee/rt

40
TEA as additive
57% yield/24 h/rt

12
86% yield/24 h
93% ee/rt

57
76% yield/48 h
87% ee/rt

58
58% yield/48 h
80% ee/rt

Figure 6.18 Chiral amine **56** and thiourea derivatives (10 mol% loading) screened in the asymmetric Michael addition of diethyl malonate to trans-β-nitrostyrene in toluene.

yield/24 h) and enantioinduction (93% ee/rt). Performing the same reaction under otherwise identical conditions, in the presence of 10 mol% of chiral amine **56** lacking the thiourea moiety, the desired Michael adduct resulted in only 14% yield (24 h) and 35% ee; with triethylamine (TEA) as base additive (10 mol%) and achiral thiourea **40** (10 mol%) lacking the tertiary amine group the target Michael adduct was provided in only 57% yield after 24 h (Figure 6.18). These experimental findings indicated that for this type of reaction high yields and enantioselectivities require a bifunctional catalyst structure such that the catalyst incorporates both a thiourea moiety and a tertiary amine functionality in a suitable relative arrangement to each other (Scheme 6.55).

Apart from the increased catalytic efficiency, this structure design produced two positive side effects. In contrast to monofunctional (thio)ureas, which exhibit low solubility in nonpolar solvents due to intermolecular hydrogen-bonding association, tertiary amine thioureas of type **12** revealed intramolecular hydrogen bonding between the amine group and the amide protons making these (thio)ureas soluble in nonpolar reaction media such as toluene. The analysis of the X-ray crystal-

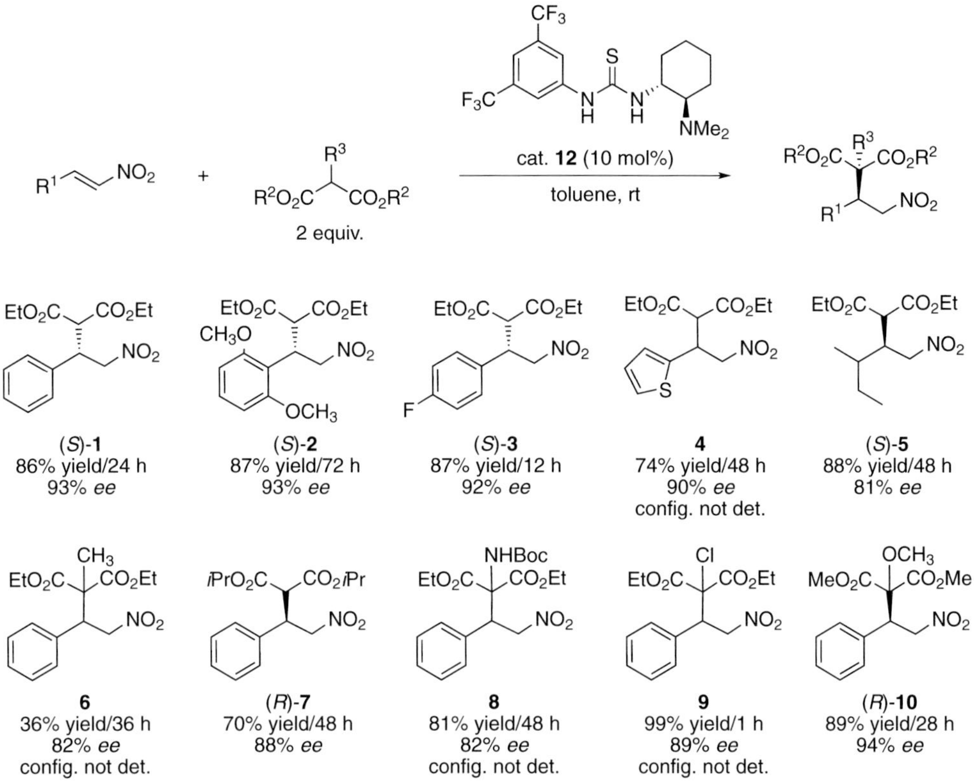

6.2 Synthetic Applications of Hydrogen-Bonding (Thio)urea Organocatalysts

Scheme 6.56 Typical products of the asymmetric Michael addition of dialkyl malonates to trans-β-nitrostyrenes in the presence of **12**.

lographic structure of rac-**12** supported this intramolecular hydrogen-bonding interaction [207]. Furthermore, the efficient application of bifunctional amine thiourea catalyst **12** without the practical necessity of a base additive such as TEA, DBU, or DABCO led to milder reaction procedures tolerating base-sensitive substrates (e.g., trans-β-nitrostyrene decomposes with DBU) [207] as well as acid-labile products due to a more facile, acid-free work-up. Under optimized reaction conditions (rt, toluene) bifunctional thiourea **12** (10 mol%) catalyzes the asymmetric Michael addition of various dimethyl-, diisopropyl-, and predominantly diethyl malonates to trans-β-nitrostyrene Michael acceptors. The protocol was reported to furnish the desired adducts **1–10** in yields ranging from 36 to 99% and in ee values ranging from 81 to 94% as depicted in Scheme 6.56 [129, 207].

At reduced reaction temperatures (−50 °C to rt) bifunctional thiourea **12** (10 mol% loading) catalyzes the stereoselective Michael addition [149–152] of prochiral α-substituted β-ketoesters to trans-β-nitrostyrene affording the respective adducts **1–5** in yields ranging from 76 to 97%, in high enantioselectivities (85–95%), and

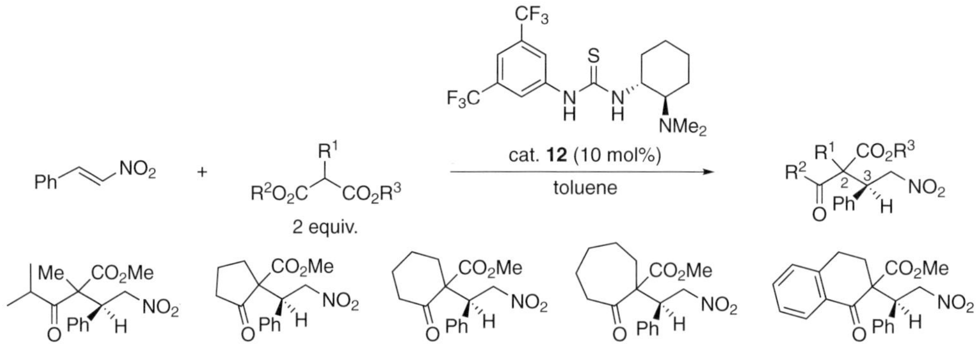

1
87% yield/48 h
95% *ee*/rt
dr (2*R*/2*S*) 36:64

(2
96% yield/24 h
93% *ee*/–50 °C
dr (2*R*/2*S*) 93:7

(3
93% yield/36 h
85% *ee*/–20 °C
dr (2*R*/2*S*) 96:4

4
76% yield/24 h
89% *ee*/–40 °C
dr (2*R*/2*S*) 83:17

5
97% yield/2 h
90% *ee*/rt
dr (2*R*/2*S*) 95:5

Scheme 6.57 Representative products of the **12**-catalyzed enantio- and diastereoselective Michael addition of α-substituted β-ketoesters to *trans*-β-nitrostyrene.

diastereoselectivities (up to *dr* 93 : 7) (Scheme 6.57) [207]. The absolute configuration of the constructed stereogenic center at the C3 position was determined to be (*R*) suggesting that the addition of symmetric (dialkyl malonates) and unsymmetrical (α-substituted β-ketoesters) 1,3-dicarbonyl compounds followed the same mechanism (Scheme 6.57).

The authors demonstrated the synthetic utility of the **12**-catalyzed enantioselective Michael reaction in the total synthesis of (*R*)-(–)-baclofen, an antispastic agent and lipophilic analog of GABA (γ-amino butyric acid) playing an important role as an inhibitory neurotransmitter in the central nervous system of mammals. Starting from 4-chlorobenzaldehyde and nitromethane (Henry reaction), the obtained alcohol underwent dehydration to give 4-chloro nitrostyrene that served as Michael acceptor for the **12**-catalyzed asymmetric addition of diethyl malonate producing the desired adduct in 80% yield and 94% *ee* in 24 h at rt. A single recrystallization of the adduct from *n*-hexane/ethyl acetate increased the *ee* value to 99%. Subsequent reduction of the nitro group, imide formation, hydrolysis of the remaining ester group, decarboxylation and final hydrolysis of the imide in the presence of hydrochloric acid provided the target γ-amino acid product (*R*)-(–)-baclofen as ammonium salt in 38% yield overall (Scheme 6.58) [207].

In the presence of 10 mol% of bifunctional thiourea **12** the intermolecular Michael addition [149–152] of γ,δ-unsaturated β-ketoesters as CH-acidic 1,3-dicarbonyl substrates to *trans*-β-nitrostyrene was catalyzed and after subsequent 1,1,3,3-tetramethylguanidine (TMG)-mediated intramolecular Michael cyclization 4-nitrocyclohexanone derivatives such as **1–4** were obtained (Scheme 6.59) [208]. This thiourea **12**/TMG-catalyzed double Michael reaction constructed three contiguous stereogenic centers and gave yields ranging from 71 to 93%, good enan-

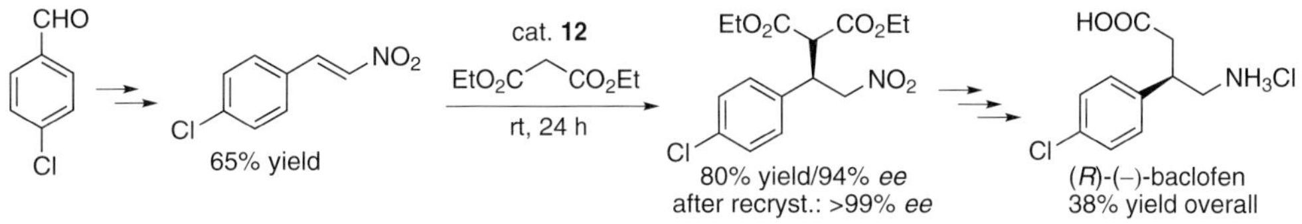

Scheme 6.58 Total synthesis of (R)-(–)-baclofen including an asymmetric Michael addition step promoted by bifunctional thiourea **12**.

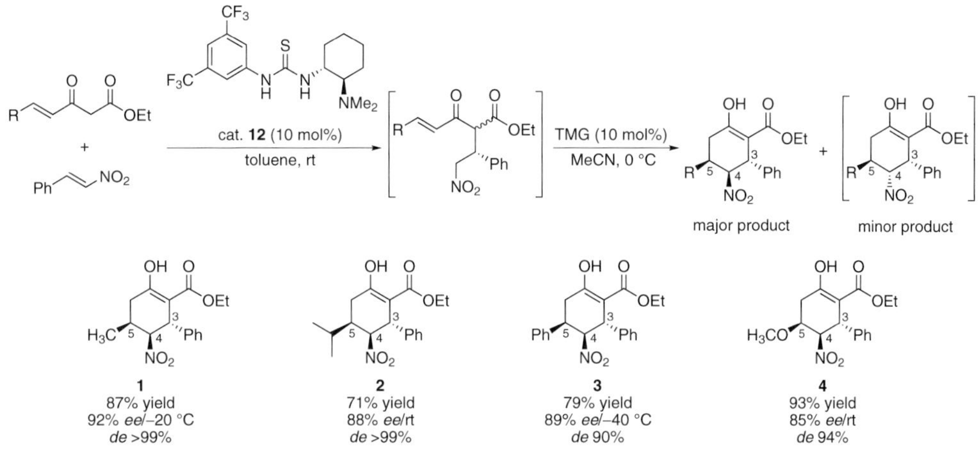

Scheme 6.59 Typical 4-nitrocyclohexanone derivatives prepared from the thiourea **12**/TMG-catalyzed enantio- and diastereoselective double Michael addition of γ,δ-unsaturated β-ketoesters to *trans*-β-nitrostyrene.

tioselectivities (85–92%), and diastereoselectivities (up to *de* 99%). The **12**-catalyzed asymmetric addition of a γ,δ-unsaturated β-ketoester to *trans*-β-nitrostyrene was successfully applied to the total synthesis of the alkaloid (–)-epibatidine as illustrated in Scheme 6.60 [208, 209].

The Takemoto group interpreted the catalytic activity and the observed stereochemical outcome of the Michael addition of CH-acidic 1,3-dicarbonyl compounds such as dialkyl malonates to *trans*-β-nitrostyrene on the basis of experimental data derived from the investigation of solvent effects and variable catalyst loadings as well as from NMR kinetic studies and NMR titrations [207]. As visualized by the reaction cascade **A**, **B**, and **C** in Scheme 6.61 diethyl malonate was enolized and adopted the six-membered enol form stabilized through interaction with the tertiary amine group of catalyst **12** that facilitated the enolization (**A**). The incoming Michael acceptor *trans*-β-nitrostyrene was coordinated through hydrogen-bonding interaction with the thiourea moiety resutling in ternary complex **B**, which allowed

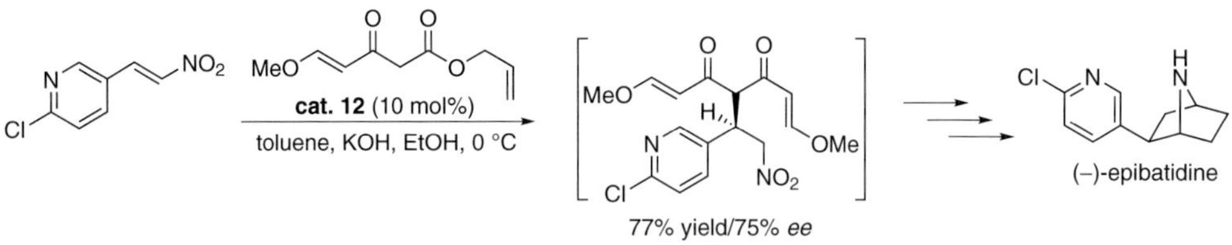

Scheme 6.60 Total synthesis of (−)-epibatidine including a 12-catalyzed Michael addition as enantioselective key step.

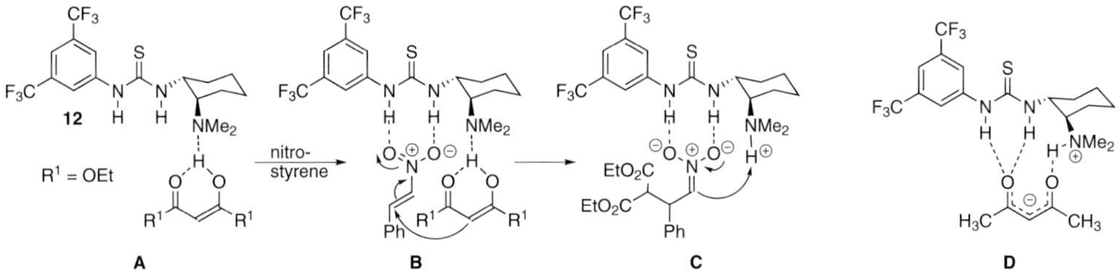

Scheme 6.61 Mechanistic proposals of the 12-catalyzed asymmetric Michael addition of diethyl malonate to trans-β-nitrostyrene proposed by the Takemoto group (**A**, **B**, and **C**) and initial enolate complex (**D**) with the ammonium group as additional hydrogen-bond donor initiating an alternative mechanism suggested by Soós, Pápai, and coworkers.

proper steric organization and relative orientation of the reaction partners such that the nucleophilic addition step could occur in an (R)-favored mode to give complex **C**; final protonation of the nitronate provided the desired Michael adduct [207]. DFT computations performed by Liu and co-workers suggested that the enantioselectivity of thiourea 12-catalyzed Michael reaction was controlled by the C–C bond-formation step (**B**), while the rate-determining step was identified to be the proton transfer from the amine group of the catalyst to the α-carbon of the nitronate (**C**) [210]. In contrast to the mechanistic picture shown in **A**, **B**, and **C**, Soós, Pápai, and co-workers favored an alternative mechanism based on theoretical investigations for the asymmetric Michael addition of acetylacetone to trans-β-nitrostyrene [211]. The amine thiourea catalyst 12 deprotonated the 1,3-dicarbonyl compound to form the corresponding enolate that was triple hydrogen bonded and stabilized through both the thiourea moiety (double hydrogen bonding) and the generated ammonium group (single hydrogen bonding) as visualized in **D** (Scheme 6.61). The Michael acceptor trans-β-nitrostyrene was found to be preferentially coordinated and activated through single hydrogen-bonding interaction with the ammonium hydrogen-bond donor, while the enolate was proposed to be hydrogen bonded via each oxygen atoms to one amide proton of the thiourea moiety. The enantioinduction was expected to originate from the extensive hydrogen-bond network provided by the three acidic N—H groups of protonated catalyst **12** that determined the preferential relative orientation of the approaching substrates.

In 2006, Takemoto and co-workers reported case studies toward the immobilization of bifunctional thiourea **12** on polymer support using an ester moiety as the linker group [212]. While soluble ester-functionalized thiourea derivative **59** (10 mol% loading) turned out to catalyze the model Michael addition of diethyl malonate to *trans*-β-nitrostyrene in 88% yield and in 91% enantioselectivity after 48 h in toluene, the insoluble crosslinked polystyrene-bound thiourea derivatives **60** and **61** exhibited a drastically reduced catalytic activity and gave the desired (*S*)-configured Michael adduct after 6 d at room temperature in 37% (87% *ee*) and 4% yield (88% *ee*) (Figure 6.19). In contrast, soluble poly(ethylene glycol) (PEG)-bound thiourea **62** could be used in dichloromethane under homogeneous conditions and at increased catalytic activity affording the model Michael adduct in 71% yield and in 86% *ee* (rt) after 6 d reaction time (Figure 6.19). The catalyst was readily recovered by filtration after addition of diethyl ether to the reaction mixture and could be reused without further purification to give the same adduct in comparable yield (74%) and *ee* value (90%) in a second run. As shown in Scheme 6.62, PEG-bound thiourea **62** also catalyzed the enantioselective double Michael addition of a γ,δ-unsaturated β-ketoester to *trans*-β-nitrostyrene resulting in the desired 4-nitrocyclohexanone derivative (63% yield; 76% *ee*; rt, 6 d). The recovery and reusability of the catalyst [61, 62] turned out to have no impact on catalytic activity and enantioselectivity (second run: 64% yield; 76% *ee* third run: 63% yield; 79% *ee*) in this test reaction.

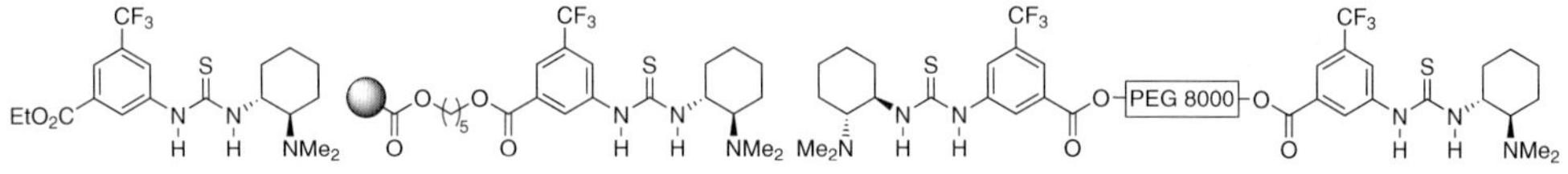

59: 88% yield/2 d, 91% *ee*/rt

60 = carboxypolystyrene HL resin, 37% yield/6 d, 87% *ee*/rt
61 = TentalGel carboxy resin, 4% yield/6 d, 88% *ee*/rt

62 (ca. 80% purity)
71% yield/6 d, 86% *ee*/rt (first run)
74% yield/6 d, 90% *ee*/rt (second run)

Figure 6.19 Ester-functionalized thiourea **59**, insoluble crosslinked polymer-bound thioureas (**60**; **61**) and soluble PEG-bound thiourea **62** screened in the asymmetric Michael addition of diethyl malonate to *trans*-β-nitrostyrene at rt in toluene (**59**, **60**, and **61**) and dichloromethane (**62**).

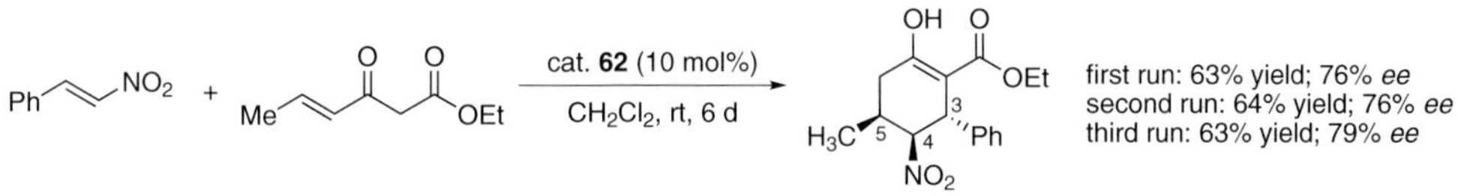

Scheme 6.62 Double Michael addition of a γ,δ-unsaturated β-ketoester to *trans*-β-nitrostyrene catalyzed by PEG-bound thiourea **62**.

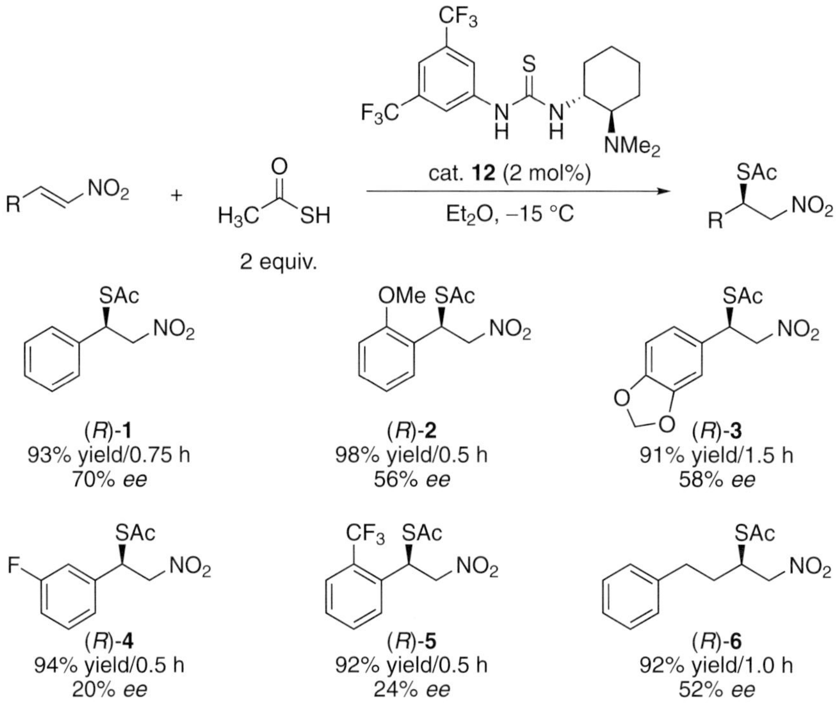

Scheme 6.63 Typical products obtained from the **12**-catalyzed asymmetric Michael addition of thioacetic acid to nitroalkenes.

Wang et al. identified bifunctional thiourea **12** to catalyze the enantioselective Michael addition [149–152] of thioacetic acid to a range of ortho-, meta-, and para-substituted trans-β-nitrostyrenes [213]. In the presence of 2 mol% **12** in diethyl ether as the solvent and at −15 °C reaction temperature, the transformations required short reaction times (0.5–1.5 h) to furnish the corresponding (R)-configured adducts **1–6**, precursors of synthetically useful thiols, in high yields (91–95%), but low (20% ee) to moderate (70% ee) enantioselectivities (Scheme 6.63). Nitrostyrenes bearing electron-donating substituents at the phenyl ring were found to provide Michael adducts with higher enantioselectivities (e.g., 56% and 58% ee) than nitrostyrene Michael acceptors bearing electron-withdrawing substituents (e.g., 20 and 24% ee) (Scheme 6.63). Mechanistically, bifunctional catalyst **12** deprotonated thioacetic acid generating nucleophilic thioacetate that selectively attacked the hydrogen-bonded nitroalkene. Final protonation of the primary adduct gave the observed target product and regenerated catalyst **12** as shown in Scheme 6.65.

The same group utilized thiourea **12** (10 mol% loading) for the catalysis of the enantioselective Michael addition of thioacetic acid to various chalcones [214]. At room temperature and otherwise unchanged conditions, in comparison to the

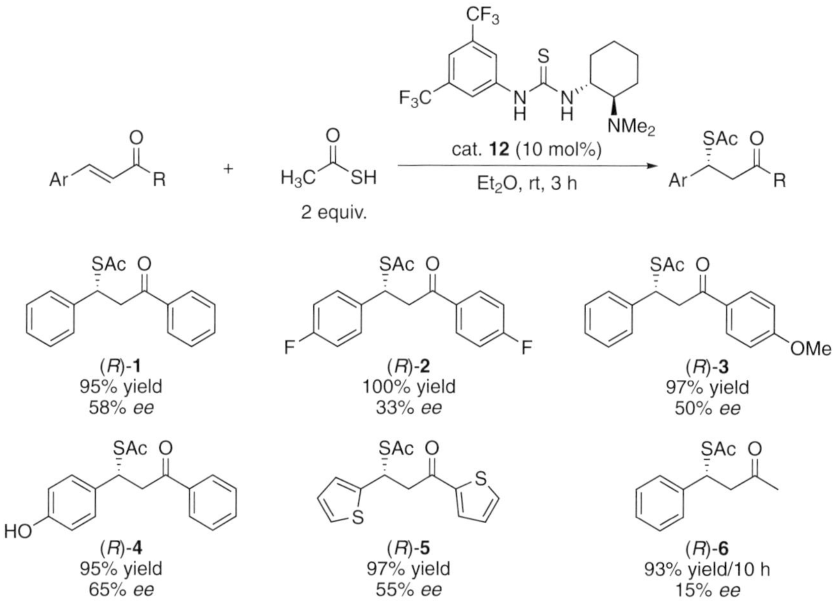

6.2 Synthetic Applications of Hydrogen-Bonding (Thio)urea Organocatalysts

Scheme 6.64 Michael adducts provided from the **12**-catalyzed asymmetric addition of thioacetic acid to various chalcones.

protocol described in Scheme 6.63, electron-rich and electron-deficient chalcones served as Michael acceptors producing the desired adducts **1–6** in high yields (93–100%) and low (15% ee) to moderate (65% ee) enantioselectivities (Scheme 6.64). The mechanistic picture of this reaction was described to be similar to that proposed for the Michael addition of thioacetic acid to nitroalkenes (Scheme 6.65) [213]. Catalyst **12** coordinated and activated the Michael acceptor chalcone through explicit double hydrogen-bonding interaction and facilitated the selective nucleophilic attack of the *in situ* generated thioacetate resulting in the observed products (Scheme 6.65).

Chen and co-workers described the **12**-catalyzed asymmetric Michael addition [149–152] of α-alkyl cyanoacetates to vinyl sulfone resulting in the respective adducts **1–4** in 52–98% yield and in ee values ranging from 72 to 99% after 48 h reaction time at −40 °C in toluene as the solvent (Scheme 6.66) [215]. After single recrystallization from 2-propanol/n-hexane one adduct example (R = Bn) could be purified to increase the ee value from 72% (98% yield) to 99% (72% yield). A one-pot strategy employing Raney-nickel hydrogenation at 50 psi H_2 pressure in the presence of Boc_2O exemplarily converted this adduct to the corresponding N-Boc-protected $\beta^{2,2}$-amino acid derivative (95% ee) as shown in Scheme 6.66.

The same group reported that bifunctional thiourea **12** catalyzed the enantioselective Michael addition [149–152] of α-alkyl and also α-aryl cyanoacetates to alkyl vinyl ketones and aryl vinyl ketones, respectively, to give the desired

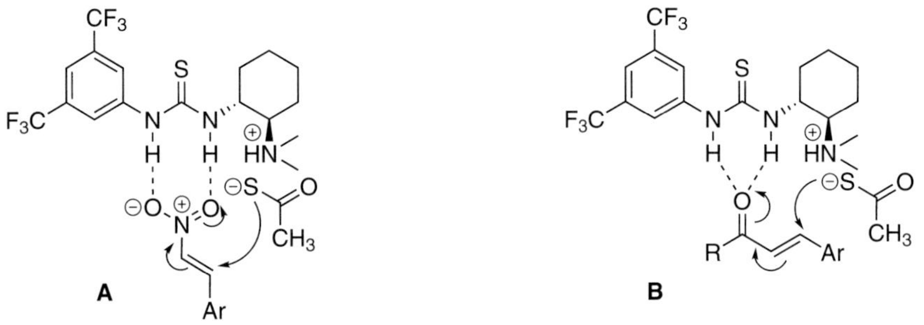

Scheme 6.65 Mechanistic proposals for the bifunctional mode of action of catalyst **12** in the Michael addition of thioacetic acid to nitroalkenes (**A**) and to chalcones (**B**).

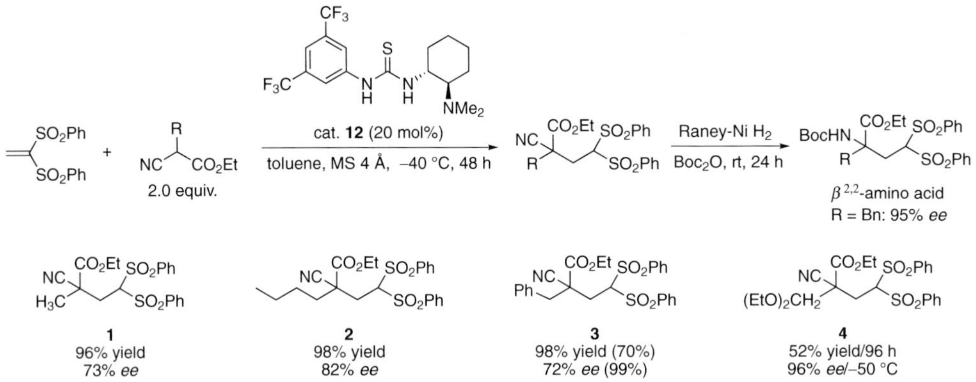

Scheme 6.66 Products of the **12**-catalyzed asymmetric Michael addition of α-alkyl cyanoacetates to vinyl sulfone and exemplary conversion of one adduct to the respective β^{2,2}-amino acid. The values in parentheses were obtained after single recrystallization; the absolute configurations of the products were not determined.

multifunctional adducts **1–8** in yields ranging from 61 to 99% and in enantioselectivities in the range of 73–97% (Scheme 6.67) [216]. The protocol was described to have broad substrate scope, when utilizing catalyst **12** (10 mol% loading) at −60 °C reaction temperature in toluene as the solvent and in the presence of molecular sieve 4 Å. Notably, the addition of α-alkyl cyanoacetates was found to be limited to aryl vinyl ketone Michael acceptors such as phenyl vinyl ketone. The authors demonstrated the synthetic versatility of the products in the conversion to various *N*-Boc-protected β^{2,2}-amino acid ethylesters (91–96% *ee*) using Raney-Ni-catalyzed hydrogenation in the presence of Boc$_2$O. Semiempirical computations indicated multiple hydrogen-bonding interactions between catalyst **12**, the enolized cyanoacetate nucleophile, and the vinyl ketone electrophile, although this

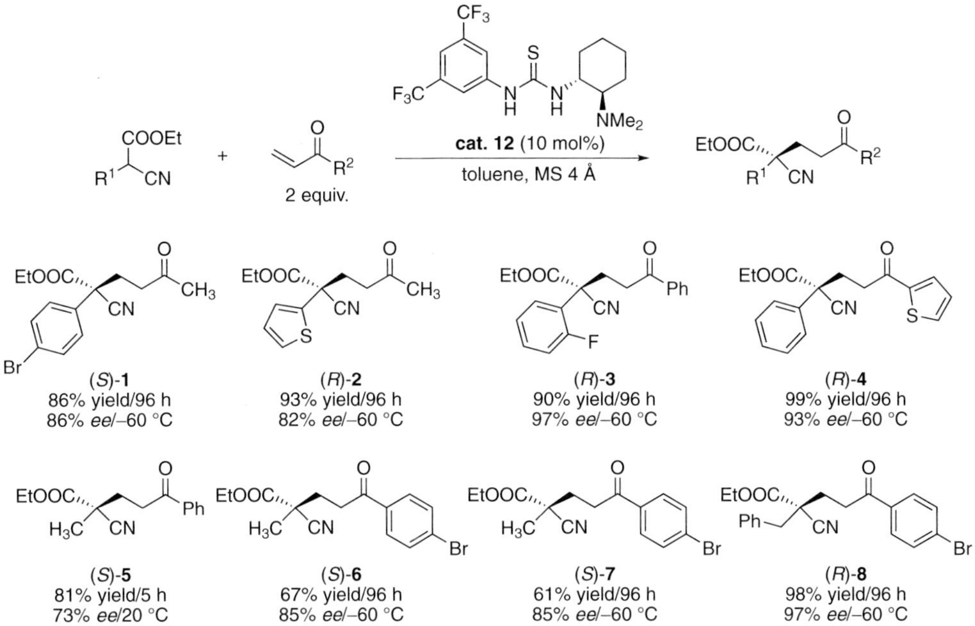

6.2 Synthetic Applications of Hydrogen-Bonding (Thio)urea Organocatalysts

Scheme 6.67 Products of the **12**-catalyzed enantioselective Michael addition of α-alkyl and α-aryl cyanoacetates to alkyl vinyl ketones and aryl vinyl ketones.

Products shown:
- (S)-**1**: 86% yield/96 h, 86% ee/−60 °C
- (R)-**2**: 93% yield/96 h, 82% ee/−60 °C
- (R)-**3**: 90% yield/96 h, 97% ee/−60 °C
- (R)-**4**: 99% yield/96 h, 93% ee/−60 °C
- (S)-**5**: 81% yield/5 h, 73% ee/20 °C
- (S)-**6**: 67% yield/96 h, 85% ee/−60 °C
- (S)-**7**: 61% yield/96 h, 85% ee/−60 °C
- (R)-**8**: 98% yield/96 h, 97% ee/−60 °C

theoretical approach is generally not suitable to describe hydrogen-bonding. Apart from a single hydrogen bond between the hydroxy group of the ester enol and the tertiary amine group of catalyst **12**, the computations located an additional weak hydrogen-bonding interaction between the ethoxy group of the enol and one amide proton of the thiourea moiety, which was proposed to be crucial to adjust the enol in a more rigid conformation resulting in improved enantiocontrol. The vinyl ketone was simultaneously activated and sterically oriented through typical double hydrogen bonding via the thiourea moiety facilitating the re-face attack of the Z-enolate to produce the observed (S)-configured Michael adduct. In contrast, the si-face attack of the E-enolate would be sterically disfavored resulting in the (R)-configured product [216].

To extend the applicability of bifunctional thiourea catalyst **12**, Takemoto et al. performed substrate screening experiments to evaluate the potenial of α,β-unsaturated imides to react as suitable acceptors in the Michael addition [149–152] of malononitrile [217]. In the presence of **12** (10 mol% loading) at room temperature and in toluene as the solvent 1-(3-phenyl-acryloyl)-pyrrolidin-2-one was found to give the highest yield (93%/60 h) and ee value (87%) for the desired malononitrile adduct, e.g., in comparison to the Michael acceptors 3-(3-phenyl-acryloyl)-oxazolidin-2-one (89 yield/96 h; 83% ee) and 1-(3-phenyl-acryloyl)-piperidin-2-one (42% yield/140 h, 59% ee). Under optimized reactions conditions (toluene 0.5 M,

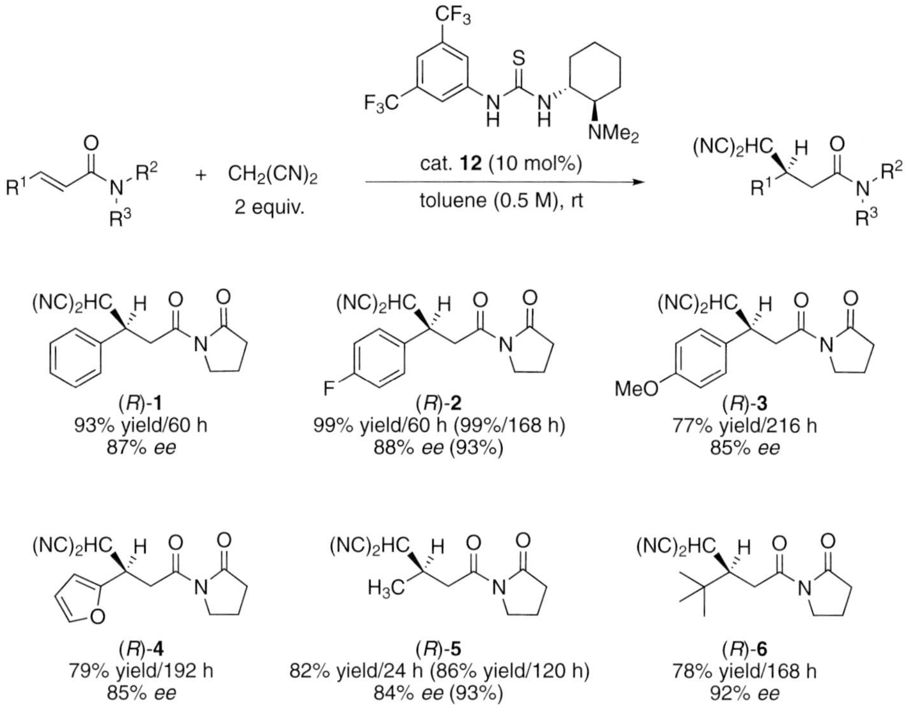

Scheme 6.68 Typical adducts obtained from the 12-catalyzed asymmetric Michael reaction between malononitrile and α,β-unsaturated N-acyl pyrrolidinones (cyclic imides); the values in parentheses refer to reactions at lower concentration (0.1 M).

rt) thiourea **12** (10 mol% loading) enantioselectively catalyzed the Michael addition of CH-acidic malononitrile to various β-aryl and alkyl substituted N-acyl pyrrolidinones resulting in yields ranging from 77 to 99% and in good *ee* values (84–92%) for the desired products **1–6** (Scheme 6.68).

As visualized in Scheme 6.70, a ternary complex (**A**) among bifunctional catalyst **12**, the imide, and malononitrile was suggested as transition state to explain the predominant formation of the (*R*)-configured adducts. After deprotonation of malononitrile, the corresponding nucleophilic anion is hydrogen bonded and sterically positioned through the tertiary ammonium group, while the 1,3-dicarbonyl groups of the imide coordinated and are activated in a bidentate mode through hydrogen-bonds provided by the thiourea moiety. This proposal based on ^{1}H NMR spectroscopic chemical shift experiments aiming at the imide–catalyst complex (1:1 mixture of imide and catalyst **12** in toluene-d$_8$) was supported by DFT computations performed for the addition of malononitrile to 1-but-2-enoyl-pyrrolidin-2-one published by Zhang and co-workers in 2008 [218]. The computed data obtained for the (*R*)- and (*S*)-reaction channel affording the (*R*)- and (*S*)-

adduct, respectively, revealed that this Michael addition underwent three elementary steps: the protonation of catalyst **12** at the tertiary amine group, the C—C bond formation, and the final deprotonation of the catalyst. The (*R*)-channel turned out to be energetically more favorable than the competing (*S*)-channel through each elementary step and led to the preferred formation of the observed (*R*)-configured adduct [218]. Again, further computations are desirable because current DFT implementations neglect dispersion forces [219–221].

Further investigations performed by the Takemoto group focussed on various *N*-cinnamoylbenzamide derivatives as Michael acceptors and demonstrated that the protocol developed for the **12**-catalyzed enantioselective Michael addition of malononitrile to α,β-unsaturated cyclic imides (Scheme 6.68) could be also successfully applied to acyclic imides [222]. Since *N*-cinnamoyl-2-methoxy benzamide exhibited the highest reactivity (95% yield/14 h) toward the Michael addition of malononitrile and gave the best *ee* value of the corresponding adduct (91%; rt) in comparison to *N*-cinnamoylbenzamide derivatives lacking a single 2-methoxy group (e.g., adducts **1**, **2**, and **3**; Scheme 6.69), the optimized protocol utilized α,β-unsaturated *N*-aryl substituted 2-methoxybenzamides ("2-methoxybenzimides") as preferred substrates. In the presence of 10 mol% loading of thiourea **12**, the respective (*R*)-configured adducts **4**, **5**, and **6** could be isolated in excellent yields (92–99%) and very good enantioselectivities (90–92%) as shown in Scheme

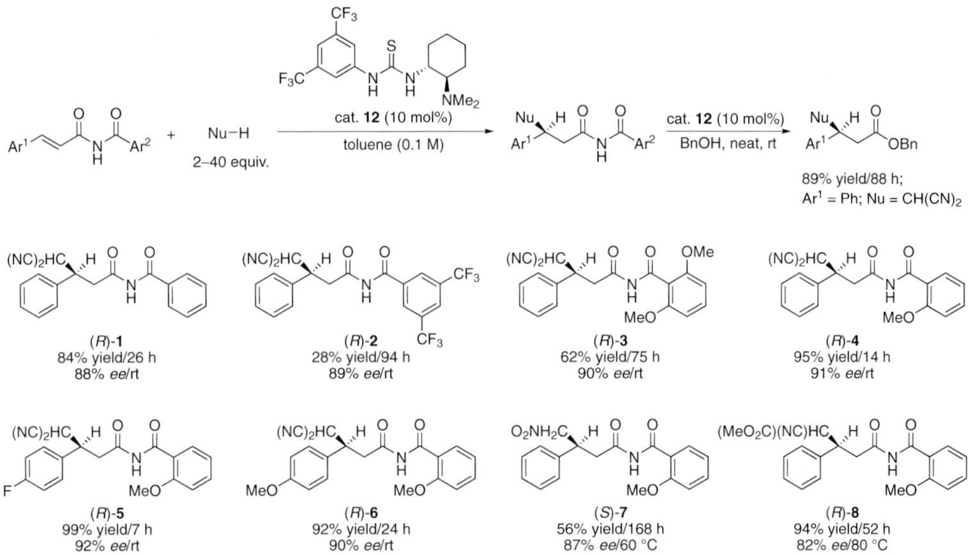

Scheme 6.69 Products obtained from the **12**-catalyzed asymmetric Michael addition of malononitrile, nitromethane, and methyl α-cyanoacetate to *N*-cinnamoylbenzamide derivatives (acyclic imides) and **12**-catalyzed derivatization of the Michael adduct.

6.69. The addition of nitromethane (56% yield/168 h; 87% ee) or methyl α-cyanoacetate (94% yield/52 h; 82% ee) as alternative CH-acidic methylene compounds required increased reaction temperatures (60 to 80 °C) to furnish the adducts **7** and **8**. As exemplarily depicted in Scheme 6.69 for benzylic alcohol thiourea **12** catalyzes the transformation of the obtained malononitrile Michael products to the respective carboxylic acid derivatives (89% yield/88 h). This method of derivatization also described for methanol (87% yield/24 h; rt), benzyl amine (77% yield/3 h; rt), and N,O-dimethylhydroxyamine (75% yield/20 h; 60 °C) as nucleophiles was reported to be feasible as a one-pot strategy without isolation of the initially formed Michael adduct [222].

IR and ^{1}H NMR experiments indicated that the observed higher reactivity of N-aryl substituted 2-methoxybenzamides in comparison to N-cinnamoylbenzamide probably originated from an intramolecular hydrogen-bonding interaction between the NH imide moiety and the 2-methoxy group increasing the electrophilicity of the α,β-unsaturated carbonyl moiety and thus the reactivity toward a Michael-type nucleophilic attack. Additionally, this interaction was proposed to be responsible for a coplanar orientation of the 2-methoxybenzamide moiety facilitating efficient bidentate hydrogen-bonding interactions with catalyst **12** as suggested by NMR titrations between N-cinnamoyl-2-methoxybenzamide (Figure 6.20). With an increase in the ratio of catalyst **12** to the imide substrate, the chemical shift of the imide N—H was gradually shifted downfield to obtain the maximum shift for a 1:1 mixture (from 10.22 to 10.24 ppm). The spectroscopic data and ^{1}H NMR kinetic studies under pseudo-first-order conditions supported a mechanistic picture consistent with the proposal for the Michael reaction between malononitrile and cyclic imides (Scheme 6.70). The bifunctional catalyst **12** simultaneously activated malononitrile through deprotonation and the bidendate imide through explicit hydrogen bonding forming a ternary complex with hydrogen-bonded reaction partners. The predominant formation of the (R)-configured adducts resulted

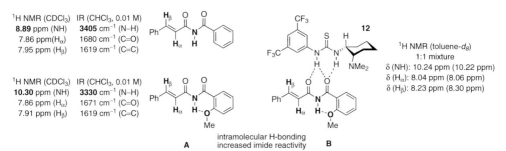

Figure 6.20 IR and ^{1}H NMR spectroscopic data suggesting intramolecular hydrogen-bonding responsible for increased reactivity of 2-methoxybenzimides (**A**) and proposed hydrogen-bonding pattern of the catalyst–imide complex supported by NMR titration (**B**); the chemical shift values in parentheses refer to the imide in the absence of **12**.

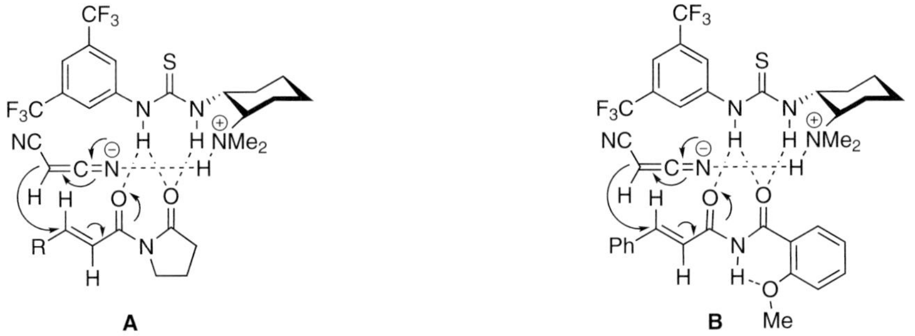

6.2 Synthetic Applications of Hydrogen-Bonding (Thio)urea Organocatalysts | 217

Scheme 6.70 Mechanistic proposal for the **12**-catalyzed asymmetric Michael addition of malononitrile to β-aryl and alkyl substituted N-acyl pyrrolidinones (cyclic imide) (**A**) and to α,β-unsaturated N-aryl substituted 2-methoxybenzamides such as N-cinnamoyl-2-methoxybenzamide (acyclic imide) (**B**).

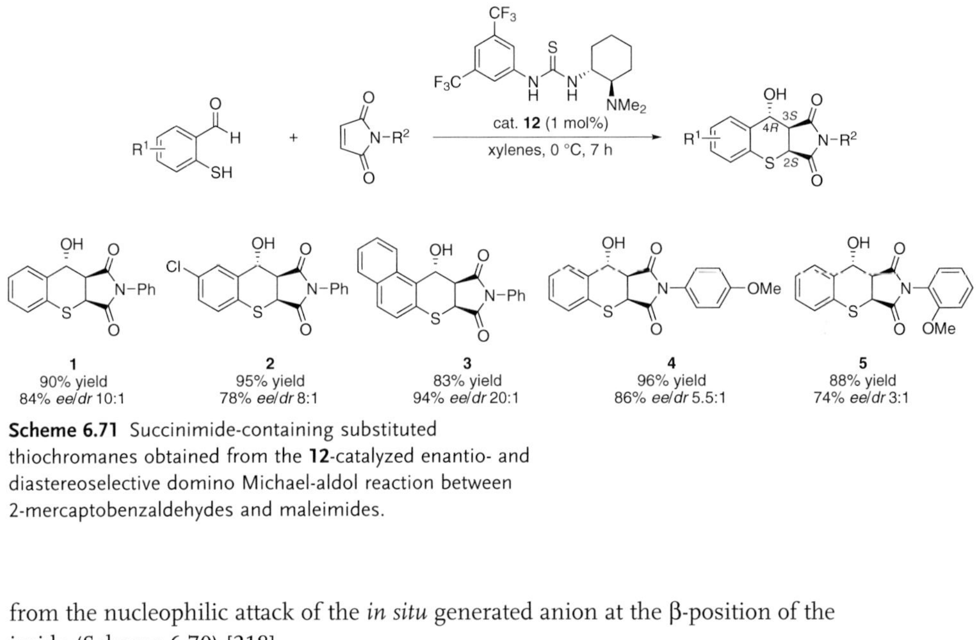

Scheme 6.71 Succinimide-containing substituted thiochromanes obtained from the **12**-catalyzed enantio- and diastereoselective domino Michael-aldol reaction between 2-mercaptobenzaldehydes and maleimides.

from the nucleophilic attack of the *in situ* generated anion at the β-position of the imide (Scheme 6.70) [218].

In 2007, Wang and co-workers published a protocol for an enantio- and diastereoselective domino Michael-aldol reaction using electron-rich and electron-deficient 2-mercaptobenzaldehydes and maleimides as substrates [223]. The conversion was described to proceed smoothly in the presence of bifunctional catalyst **12** (1 mol% loading) in xylenes at 0 °C reaction temperature producing the desired chiral succinimide-containing substituted thiochromanes **1–5** in high yields (83–96%), in synthetically useful *ee* values (74–94%), and diastereoselectivities (up to *dr* 20 : 1) in 7 h reaction time (Scheme 6.71).

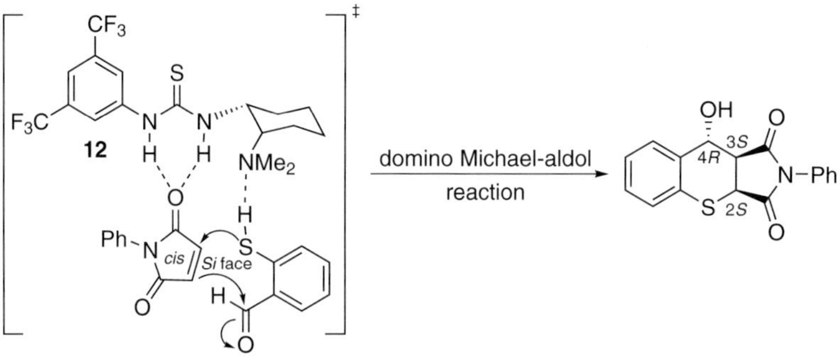

Scheme 6.72 Mechanistic proposal for the **12**-catalyzed enantio- and diastereoselective domino Michael-aldol reaction between N-phenyl maleimides and 2-mercaptobenzaldehyde; bifunctional mode of action of **12**.

The stereoconfigurations of the thiochromane products were determined by crystal structure analysis to be (2S, 3S) corresponding to *cis* stereochemistry. This stereochemical outcome was interpreted with a transition state in which the *cis*-cyclic maleimide was double hydrogen-bonded through the thiourea moiety of bifunctional catalyst **12** and thus properly activated and sterically organized for the initial Michael-addition step of the activated thiol group of 2-mercaptobenzaldehyde (Scheme 6.72). This *Si* face attack gave the (2S)-configured stereogenic center and was followed by the ring-forming aldol reaction resulting in the construction of the (3S)- and (4R)-configured stereogenic centers of the product. An alternative mechanism employing the Diels–Alder reaction between maleimide (dienophile) and enolized 2-mercaptobenzaldehyde (diene) would provide (2R, 3R, 4S)-configured products being not consistent with the detected stereochemistry (2S, 3S, 4R).

Takemoto *et al.* discovered N-phosphinoyl-protected aldimines as suitable electrophilic substrates for the enantioselective aza-Henry [224] (nitro-Mannich) reaction [72] with nitromethane, when utilizing thiourea **12** (10 mol%) as the catalyst in dichloromethane at room temperature [225]. The (S)-favored 1,2-addition of nitromethane to the electron-deficient C=N double bond allowed access to various β-aryl substituted N-phosphinoyl-protected adducts **1–5** in consistently moderate to good yields (72–87%) and moderate enantioselectivities (63–76%) as depicted in Scheme 6.73. Employing nitroethane under unchanged reaction conditions gave adduct **6** as a mixture of diastereomers (*dr* 73:27) at an *ee* value of 67% (83% yield) of the major isomer (Scheme 6.73).

N-Boc-protected (hetero)aromatic aldimines bearing both electron-donating or electron-withdrawing substituents were reported by the Takemoto group to undergo a **12**-catalyzed enantio- and diastereoselective aza-Henry reaction with various aliphatic nitroalkanes including nitromethane, nitroethane, and also nitroalcohols such as 2-nitroethanol [226]. Performing the conversion under optimized reaction parameters in the presence of 10 mol% of catalyst **12**

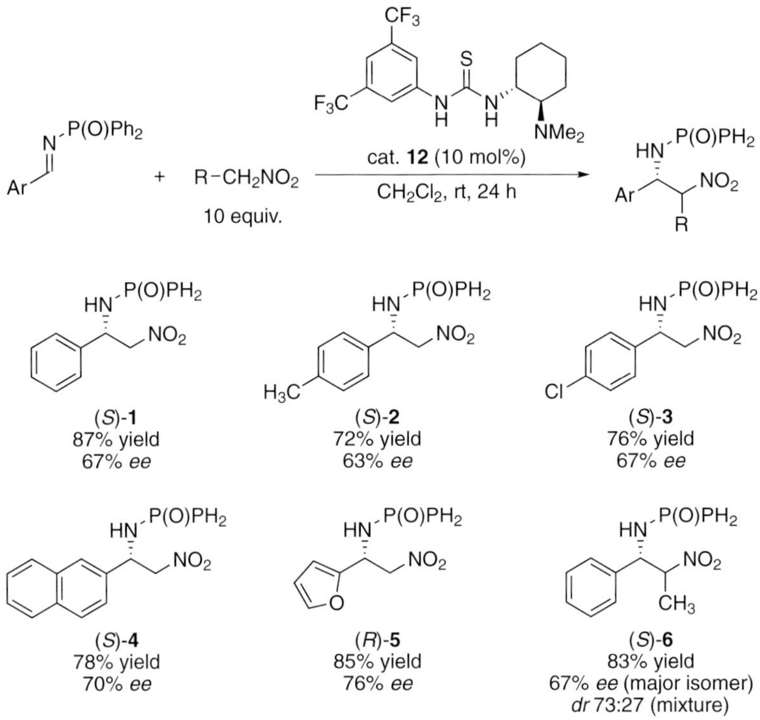

Scheme 6.73 Typical products of the enantioselective aza-Henry (nitro-Mannich) reaction between nitroalkanes and N-phosphinoylimines proceeding in the presence of catalyst **12**.

produced the respective aza-Henry adducts **1–8** in consistently high enantioselectivities (90–99% *ee* at −20 °C in CH$_2$Cl$_2$) and in a broad range of yields (71–94%; 24 h–72 h) (Scheme 6.74). In contrast to the aza-Henry reaction of N-phosphinoyl-protected aldimines [225] (Scheme 6.73), N-Boc-protected aldimines were found to give the (R)-configured adducts in preferred *syn*-diastereoselectivity (up to *dr* 97:3). Mechanistically, this outcome was explained through the ternary model complex visualized in Scheme 6.75.

The nucleophilic nitronate generated *in situ* by deprotonation of the corresponding nitroalkane was singly hydrogen-bonded through the tertiary ammonium group and properly sterically arranged to attack the hydrogen-bonded and activated electrophilic N-Boc-protected aldimine such that the *syn*-adduct was observed as the major diastereomer. Increasing the reaction temperature from −20 to 20 °C for 48 h in the presence of 10 mol% catalyst **12** decreased the diastereoselectivity indicating that **12** catalyzed the epimerization of the *syn*-adduct in favor of the thermodynamically not stable *anti*-adduct; the retro aza-Henry reaction, however, seemed not to occur keeping the *ee* value of the adduct unchanged at 20 °C (Scheme 6.75).

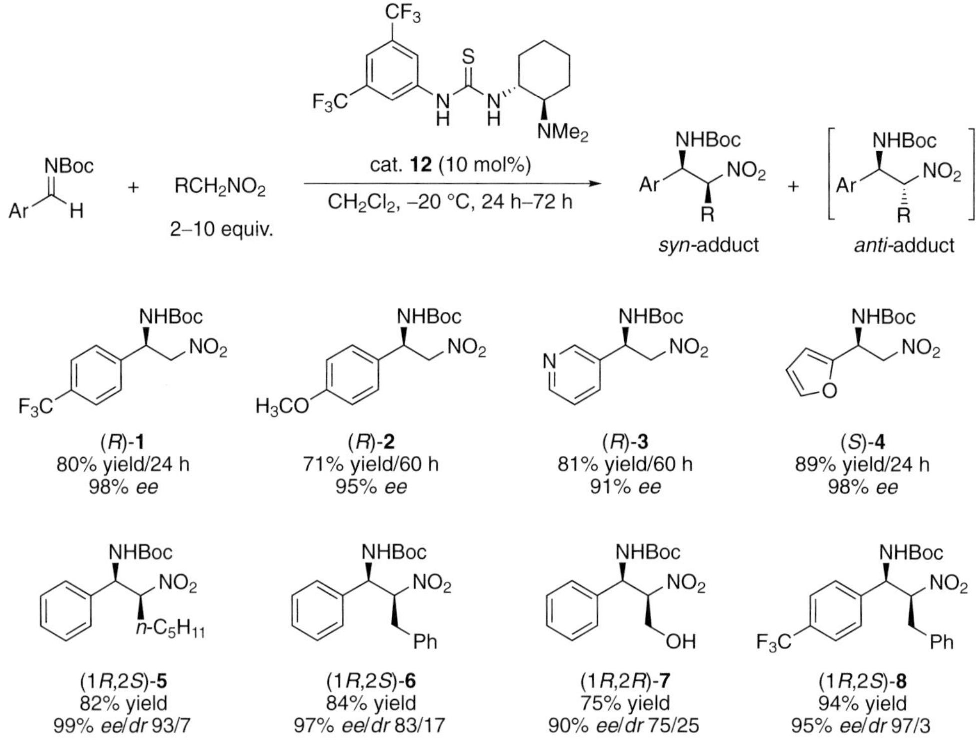

CF3
S
F3C
N H
N H
NMe2
NBoc
Ar H
+
RCH2NO2
2–10 equiv.
cat. 12 (10 mol%)
CH2Cl2, –20 °C, 24 h–72 h
NHBoc
Ar NO2
R
syn-adduct
+
anti-adduct
(R)-1
80% yield/24 h
98% ee
(R)-2
71% yield/60 h
95% ee
(R)-3
81% yield/60 h
91% ee
(S)-4
89% yield/24 h
98% ee
n-C5H11
Ph
OH
(1R,2S)-5
82% yield
99% ee/dr 93/7
(1R,2S)-6
84% yield
97% ee/dr 83/17
(1R,2R)-7
75% yield
90% ee/dr 75/25
(1R,2S)-8
94% yield
95% ee/dr 97/3

Scheme 6.74 Typical *N*-Boc-protected *syn*-β-nitroamines obtained from the enantio- and diastereoselective aza-Henry (nitro-Mannich) reaction between *N*-Boc-protected (hetero) aromatic aldimines and nitroalkanes in the presence of bifunctional thiourea catalyst **12**.

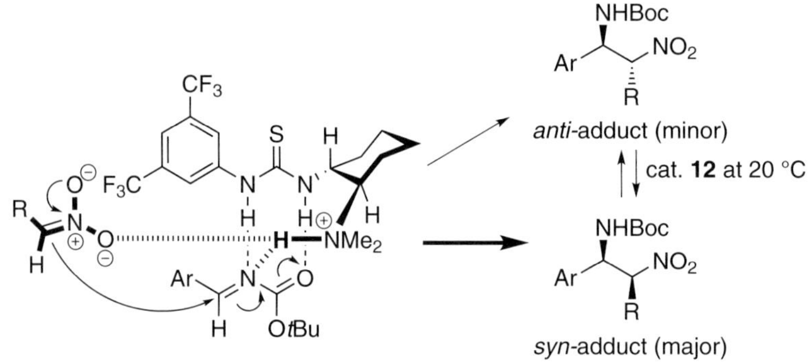

NHBoc
Ar NO2
R
anti-adduct (minor)
cat. 12 at 20 °C
syn-adduct (major)
CF3
F3C
NMe2
OtBu

Scheme 6.75 Proposed mechanism of the enantio- and diastereoselective aza-Henry reaction between *N*-Boc-protected aldimines and nitroalkanes in the presence of bifunctional catalyst **12** and catalyzed epimerization of the *syn*-adduct at increased temperature.

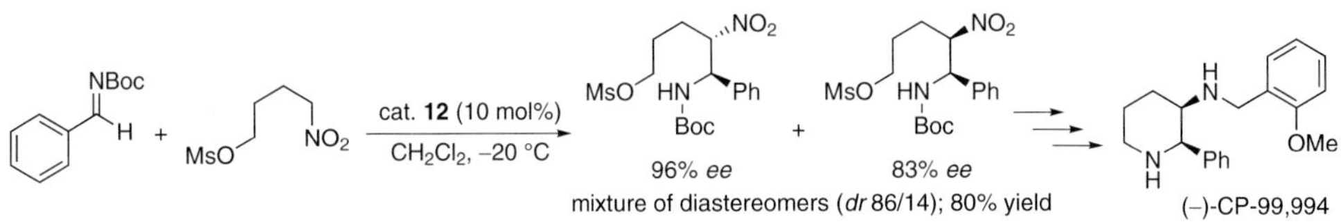

Scheme 6.76 Total synthesis of NK-1 receptor antagonist (−)-CP-99,994 utilizing the **12**-catalyzed enantio- and diastereoselective aza-Henry methodology.

This stereoselective aza-Henry methodology allowed access to physiologically important 2,3,6-trisubstituted and 2,3-disubstituted piperidines such as neurokinin-1 (NK-1) receptor antagonist (−)-CP-99,994 as shown for the key steps of the total synthesis in Scheme 6.76 [227]. Starting from N-Boc-protected benzaldimine, the enantio- and diastereoselective aza-Henry addition [224] of mesylated 4-nitrobutanol was catalyzed by **12** (at −20 °C in CH$_2$Cl$_2$) producing a mixture of diastereomers (dr 86/14; 96% ee and 83% ee, 80% yield), which could be directly used for the subsequent steps including the removal of the Boc group, cyclization, epimerization, reduction, imine formation with 2-anisaldehyde, and final reduction to yield the target compound spectroscopically consistent with the literature data [228].

Takemoto and co-workers could identify N-Boc-protected benzaldimine to react with prochiral cyclic 1,3-dicarbonyl compounds such as β-keto methylesters [229]. These stereoselective Mannich reactions [72] proceed in the presence of 10 mol% thiourea **12** in dichloromethane as the solvent at −78 °C to −20 °C giving the desired adducts **1–6** in good (81%) to excellent yields (98%), ee values ranging from 56 to 92%, and diastereoselectivities up to 99:1 (Scheme 6.77). The necessity and potential of bifunctional catalysis became evident from the results obtained in the model Mannich reaction (formation of adduct **1**; Scheme 6.77) using DBU (86% yield/8 h; dr 66:34), TEA (93% yield/8 h; dr 60:40), and bifunctional catalyst **12** (98% yield/9 h; dr 91:9; 82% ee) under identical conditions (10 mol% loading at rt in CH$_2$Cl$_2$).

The stereoselective direct vinylogous Mannich reaction (γ-aminoalkylation of α,β-unsaturated carbonyl compounds) is a variant of the traditional Mannich reaction [72] and offers facile access to highly functionalized δ-amino compounds. In 2007, Chen and co-workers described the development of a bifunctional thiourea catalyzed protocol for the regio- and enantioselective vinylogous Mannich reaction between C–H acidic α,α-dicyanoalkenes using the addition of 2-thiochroman-4-ylidene-malononitrile to N-Boc-benzaldimine as a model reaction resulting in Mannich adduct **1** (Scheme 6.78) [230]. At room temperature, in toluene, and at 10 mol% loading thiourea **12** (adduct **1**: 99% yield/6 h; 89% ee) proved to be less catalytically efficient than its N-cyclohexyl-substituted derivative **63** (adduct **1**: 99% yield/6 h; 99% ee) introduced earlier by the Berkessel group for the DKR of azlactones [231, 232] and the KR of oxazinones [233]. While this bifunctional aliphatic thiourea catalyst revealed excellent catalytic activity and enantioinduction even at

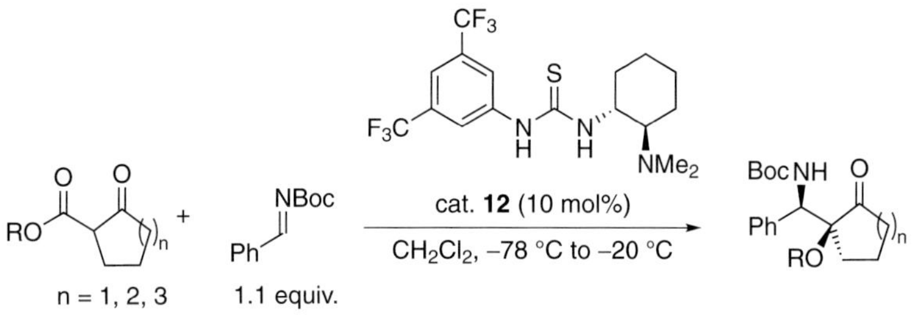

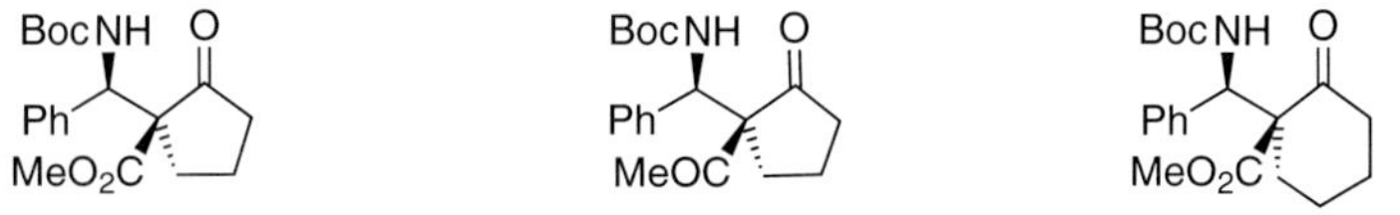

6 (Thio)urea Organocatalysts

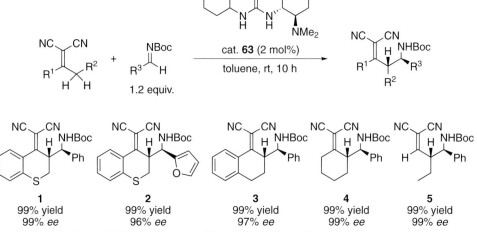

Scheme 6.77 Product range of the **12**-catalyzed enantio- and diastereoselective Mannich addition of prochiral cyclic 1,3-dicarbonyl compounds to N-Boc-protected benzaldimine.

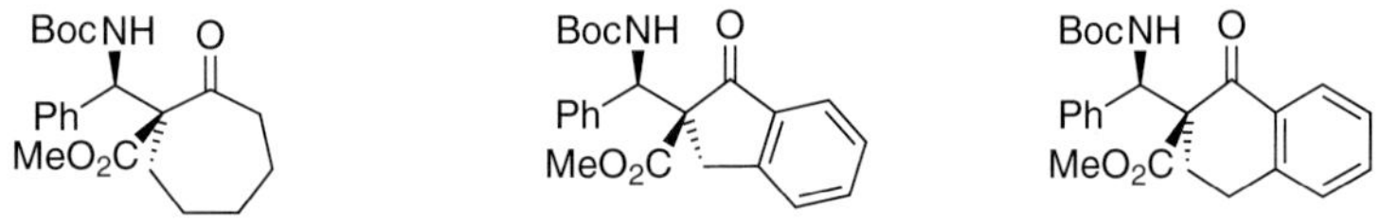

Scheme 6.78 Typical products of the vinylogous Mannich addition of α,α-dicyanoalkenes to N-Boc-protected aldimines catalyzed by **63**.

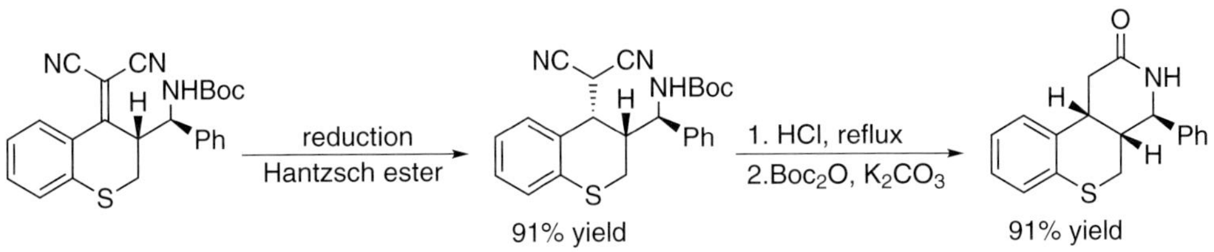

Scheme 6.79 Conversion of one vinylogous Mannich adduct obtained from the **63**-catalyzed Mannich addition of α,α-dicyanoalkenes to N-Boc-protected aldimines to the corresponding chiral δ-lactam.

0.5 mol% (adduct **1**: 99% yield/16 h; 99% ee) and 0.1 mol% loading (adduct **1**: 98% yield/24 h; 98% ee), the authors decided to evaluate the scope of the optimized protocol at the more practical 2 mol% catalyst loading. Various aliphatic and aromatic α,α-dicyanoalkenes reacted with N-Boc-protected aldimines to give the respective vinylogous Mannich adducts such as **1–5** in consistently nearly quantitative yields (99%) and in excellent enantioselectivities (96–99%) (Scheme 6.78). Adduct **1** in Scheme 6.78 was exemplarily converted to the corresponding chiral δ-lactam utilizing a high-yielding reduction-hydrolysis-cyclization sequence as outlined in Scheme 6.79.

The enantioselective addition of CH acidic α-substituted β-keto esters to N=N bonds incorporated in electrophilic azodicarboxylates such as di-*tert*-butyl azodicarboxylate was reported by the Takemoto group in 2006 [234]. This α-hydrazination of cyclic 1,3-dicarbonyl compounds (Diels amination) [235, 236] offered access to highly functionalized precursors of nonnatural cyclic α,α-disubstituted α-amino acid derivatives that represented valuable chiral building blocks for, e.g., peptide synthesis [237] or the development of neuroactive glutamate analogs [238]. Initial experiments revealed that thiourea catalyst **12** exhibited catalytic activity (91% yield/1 h) and enantioselectivity (75% ee/rt) in the formation of model adduct **1** (Scheme 6.80), but turned out to be unstable under the hydrazination conditions (10 mol% loading, toluene, rt) at longer reaction times; the urea analog **64** of thiourea **12**, however, was identified to be compatible to the reaction conditions and catalyzed the formation of model adduct **1** in 84% yield and in 60% ee after 0.5 h at room temperature. In the presence of **64** (10 mol%) representative five-, six-, and seven-membered monocyclic and in one example bicyclic β-keto methyl-, *iso*-propyl-, and *tert*-butyl esters underwent the enantioselective addition to di-*tert*-butyl azodicarboxylate providing the (S)-configured adducts **1–6** in predominantly high yields (52%; 90–99%) and ee values (87–91%) as shown in Scheme 6.80. For one adduct, the authors demonstrated the straightforward three-step transformation to the corresponding oxazolidinone amino acid derivative (32% yield overall) (Scheme 6.81).

The catalytic efficiency of bifunctional thiourea **12** turned out to be mainly limited to (pre)nucleophiles such as CH-acidic 1,3-dicarbonyl compounds and

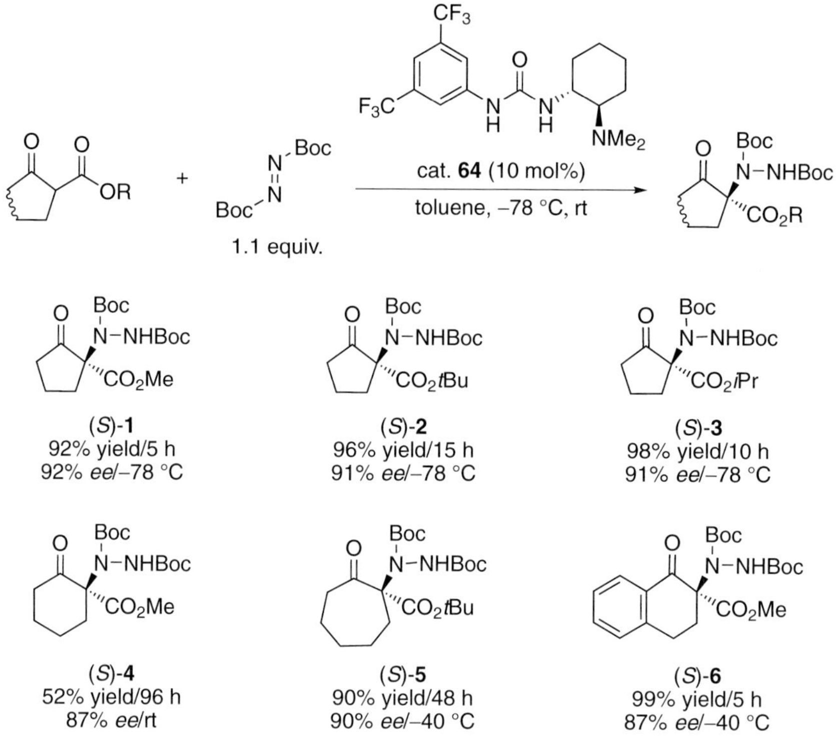

(S)-1
92% yield/5 h
92% *ee*/–78 °C

(S)-2
96% yield/15 h
91% *ee*/–78 °C

(S)-3
98% yield/10 h
91% *ee*/–78 °C

(S)-4
52% yield/96 h
87% *ee*/rt

(S)-5
90% yield/48 h
90% *ee*/–40 °C

(S)-6
99% yield/5 h
87% *ee*/–40 °C

Scheme 6.80 Typical products obtained from the **64**-catalyzed enantioselective addition of α-substituted β-keto esters to di-*tert*-butyl azodicarboxylate (α-hydrazination).

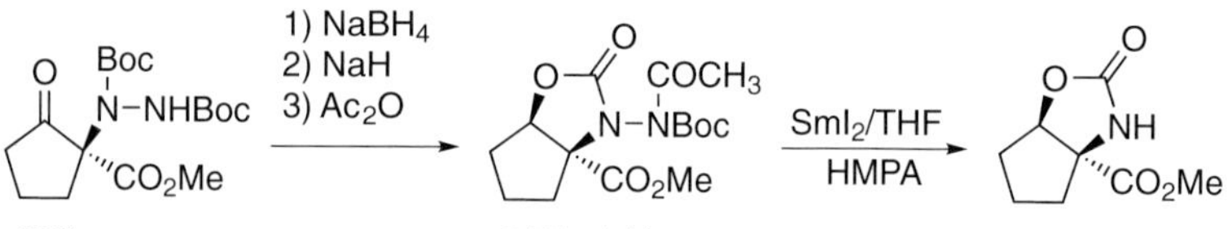

83% ee 61% yield 53% yield

Scheme 6.81 Transformation of one adduct prepared from the **64**-catalyzed asymmetric addition of α-substituted β-keto esters to di-*tert*-butyl azodicarboxylate (α-hydrazination) into the corresponding oxazolidinone amino acid derivative.

nitroalkanes that could be deprotonated by the tertiary amine functionality to be activated for the product-forming nucleophilic attack to the hydrogen-bonded electrophile. On the basis of the suggested key complex (**A**) of the Petasis reaction (Petasis boronic acid-Mannich reaction) utilizing α-hydroxy aldehydes, amines, and organic boronic acids for the vinylation of *in situ* generated iminium ions [76]

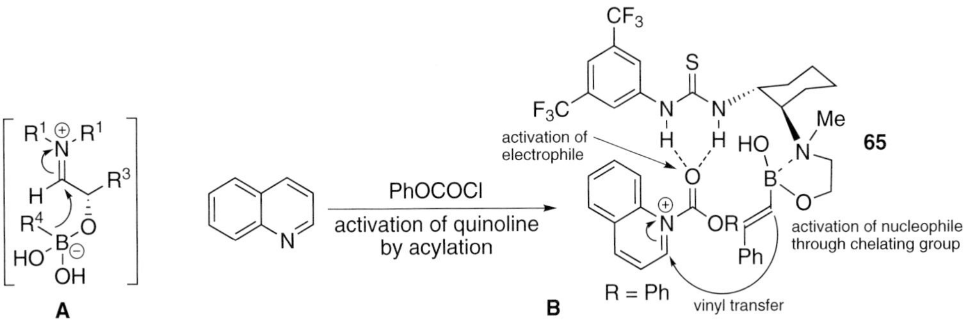

Scheme 6.82 Proposed reactive complex of the Petasis reaction utilizing α-hydroxy aldehydes, amines, and organic boronic acids (**A**) and bifunctional mode of action of chelating thiourea catalyst **65** in the enantioselective Petasis-type 2-vinylation of N-acetylated quinolinium ions (**B**).

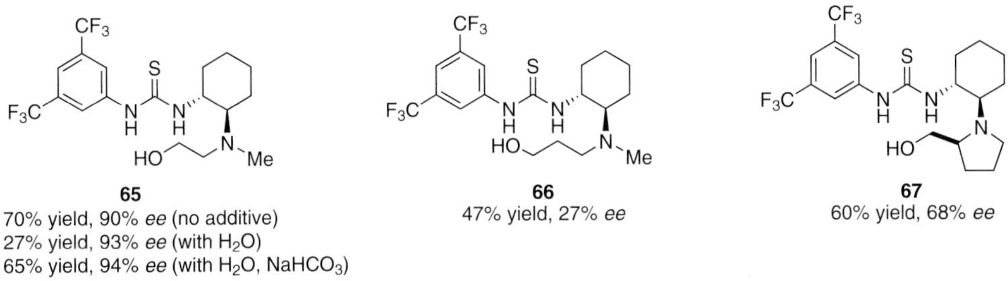

65
70% yield, 90% ee (no additive)
27% yield, 93% ee (with H₂O)
65% yield, 94% ee (with H₂O, NaHCO₃)

66
47% yield, 27% ee

67
60% yield, 68% ee

Figure 6.21 Chelating thiourea derivatives screened in the Petasis-type 2-vinylation of the N-acetylated quinolinium ion at −65 °C.

(Scheme 6.82), the Takemoto group modified thiourea catalyst **12** through formal mono-hydroxy alkylation of the tertiary amine group resulting in thiourea derivatives **65**–**67** (Figure 6.21) [239]. These newly designed thiourea derivatives were proposed to operate bifunctionally in an enantioselective Petasis-type 2-vinylation of N-acetylated quinolinium ions such that the chelating hydroxy group activated the vinylboronic acid inducing the nucleophilic addition of the alkyl vinyl unit to the iminium moiety of the hydrogen-bonded electrophile as visualized in complex (**B**) (Scheme 6.82). The catalyst screening (10 mol% loading) was performed in the reaction among quinoline, phenylvinyl boronic acid, and phenyl chloroformate (2 equiv.) at −65 °C in dichloromethane. After 24 h reaction time, thiourea derivative **65** (70% yield; 90% ee) bearing a 1,2-amino alcohol functionality proved to be more efficient in both catalytic activity and enantioselectivity than the derivative **66** (47% yield; 27% ee) bearing a 1,3-amino alcohol group and thiourea **67** incorporating the hydroxy methyl pyrrolidine moiety, respectively (Figure 6.21). Under identical

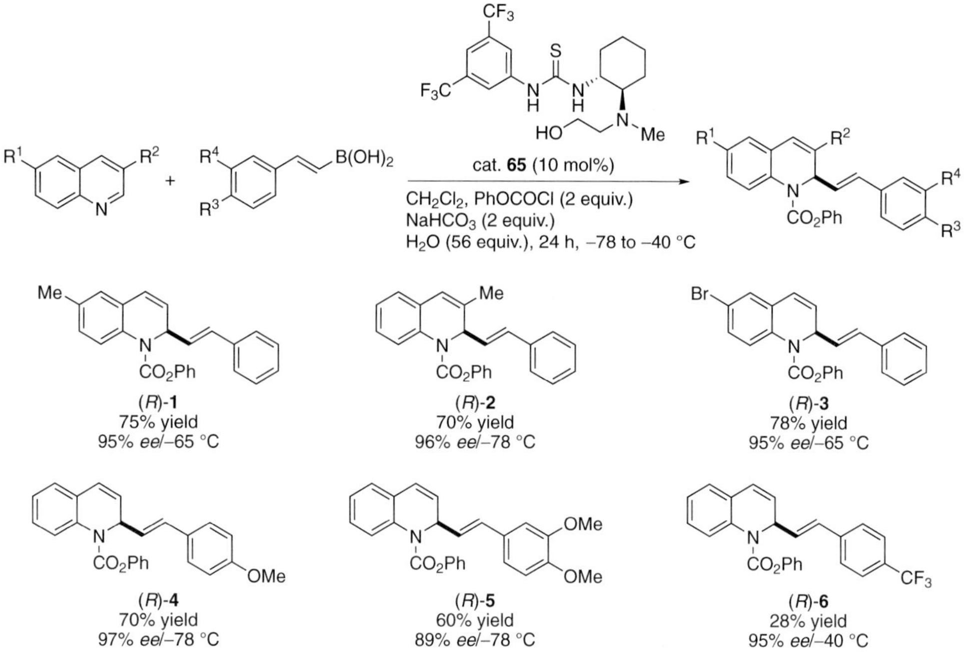

Scheme 6.83 Product range of the **65**-catalyzed asymmetric Petasis-type 2-vinylation of N-phenoxycarbonyl quinolinium salts.

conditions, tertiary amine catalyst **12** gave only 34% yield of the model adduct and exhibited no stereoinduction producing a racemic product mixture. The addition of water (56 equiv.) as proton source decreased the catalytic activity (27% yield) but slightly increased the enantioselectivity (93% *ee*) of thiourea **65** in the model Petasis-type reaction, while the combination of the additives water and $NaHCO_3$ (2 equiv.) improved the yield and led to optimized results (65% yield; 94% *ee*) (Figure 6.21). The remarkable impact of water and $NaHCO_3$ on this reaction was ascribed to the promoted regeneration of the active catalyst structure through the proton source and the removal of the resulting boronic acid side product through the base.

Employing chelating thiourea catalyst **65** (10 mol% loading) along with variable boronic acids, phenyl chloroformate, and water/$NaHCO_3$ as additives in dichloromethane as the reaction medium, numerous N-phenoxycarbonyl quinolinium salts were regioselectively (no 1,4-addition) and enantioselectively converted to the corresponding Petasis-type (*R*)-configured 1,2-adducts such as **1–6** (Scheme 6.83). The yields of this reaction ranged from 28 to 78% and the adducts were isolated with high *ee* values (89–97%) [239].

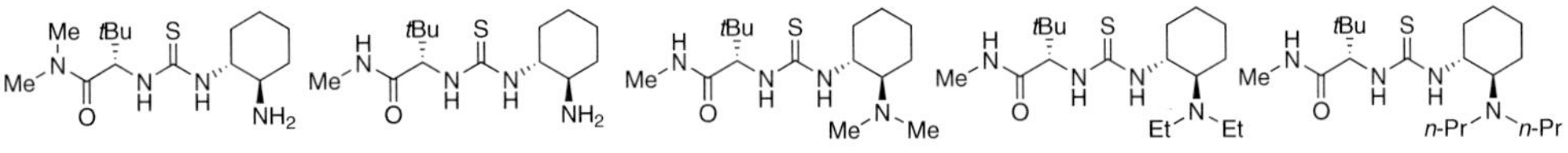

6.2 Synthetic Applications of Hydrogen-Bonding (Thio)urea Organocatalysts | 227

Figure 6.22 Primary and tertiary amine thioureas evaluated for catalytic efficiency in the cyanosilylation of acetophenone.

	68	69	70	70	71	72
	(10 mol%); toluene	(10 mol%); toluene	(10 mol%); toluene	(10 mol%); toluene	(10 mol%); CH$_2$Cl$_2$	(10 mol%); CH$_2$Cl$_2$
	100% conv./3 h	100% conv./3 h	0% conv./24 h	80% conv./24 h	100% conv./12 h	100% conv./12 h
	25% ee/−40 °C	55% ee/−40 °C	no ee/−40 °C	90% ee/−40 °C	94% ee/−78 °C	97% ee/−78 °C
				CF$_3$CH$_2$OH	CF$_3$CH$_2$OH	CF$_3$CH$_2$OH

The Jacobsen group, in 2005, described the systematic structure optimization and identification of a bifunctional tertiary amine-functionalized thiourea derivative operating as a bifunctional catalyst in the enantioselective cyanosilylation of ketones [240]. While the established Schiff base thiourea catalyst **47** (Figure 6.16; Schemes 6.46–6.48) displayed no detectable catalytic activity in the model cyanosilylation (with TMSCN) of acetophenone primary amine **68** (10 mol%), the immediate synthetic precursor of **47**, was found to be highly active (100% conv./3 h). Further structure modulation revealed that less bulky amide derivatives such as secondary methyl amide catalyst **69** produced the best *ee* values (55%/−40 °C). N,N-Dimethylation of the primary amine function in **69** led to catalytically inactive N,N-dimethyl tertiary amine thiourea **70**; the addition of trifluoroethanol for *in situ* generation of HCN as the active nucleophile, however, restored catalytic activity (80% conv./24 h) and enantioselectivity (90% *ee*). The crucial role of the amine substituent on both catalyst activity and selectivity became evident upon comparison of N,N′-dimethyl (**70**), N,N′-diethyl (**71**), and N,N′-di-n-propyl (**72**) thiourea derivatives, when the most sterically demanding catalyst **72** was identified to be the most efficient in the cyanosilylation of acetophenone (Figure 6.22).

The optimized protocol utilizing thiourea derivative **72** as the catalysts (5 mol%), trifluoroethanol (1 equiv.) as the additive, and dichloromethane as the solvent was applied to the asymmetric cyanosilylation of a broad spectrum of ketone substrates (e.g., **1–10**) including various alkyl aryl ketones, heteroaromatic ketones, α,β-unsaturated ketones, and also to the aldehydes benzaldehyde (96% *ee*/2 h; 0.05 mol% **72**) and *trans*-cinnamaldehyde (93% *ee*/2 h; 0.05 mol% **72**). The silylated cyanohydrins, important precursors of α-hydroxy acids, β-amino alcohols, and other valuable chiral building blocks [241], were isolated in high yields (81–98%) and *ee* values (86–97%) (Scheme 6.84). A combination of experimental (e.g., kinetic analysis, structure-efficiency studies) and theoretical methods (DFT computations) utilized for the elucidation of the **72**-catalyzed cyanosilylation of ketones suggested that this 1,2-carbonyl addition followed a cooperative mechanism in which both the thiourea and the tertiary amine of the catalyst were involved productively in the rate-limiting addition step [242]. DFT transition-state analyses distinguished between two mechanistic pathways and thus two transition states (**TS 1** and **TS 2**, Figure 6.23) involving thiourea hydrogen-bonding activation of the ketone (**TS 1**) or of cyanide (**TS 2**). Computed transition-state energies and the

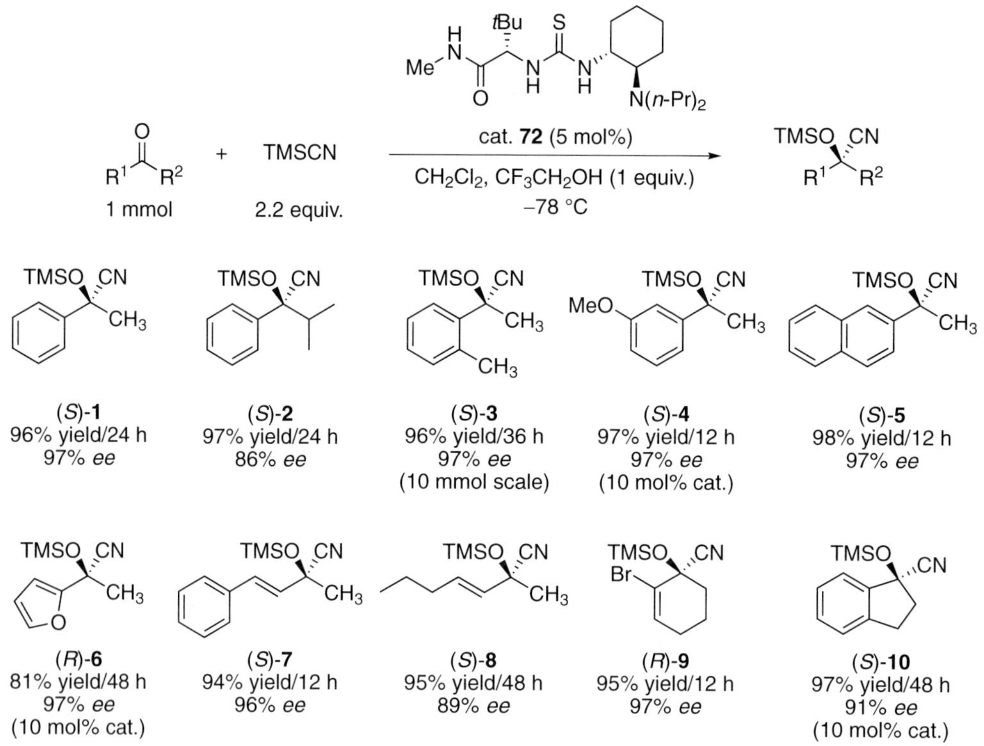

Scheme 6.84 Typical silylated cyanohydrins prepared from various ketones under asymmetric **72**-catalysis (cyanosilylation).

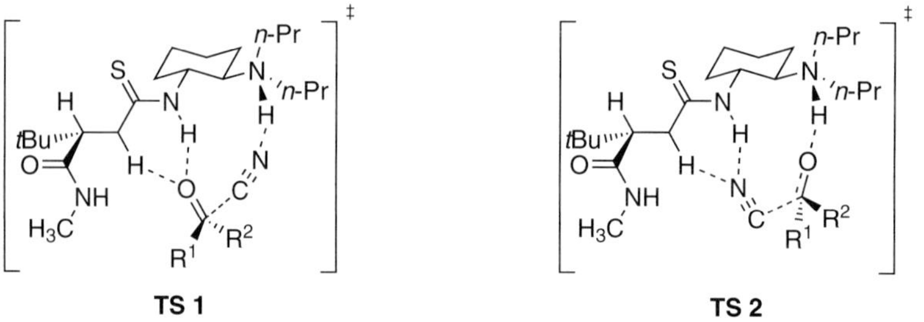

Figure 6.23 Proposed transition states of the asymmetric **72**-catalyzed cyanosilylation of ketones describe two alternative mechanistic pathways for cooperative catalysis: Addition via thiourea-bound ketone (**TS 1**, preferred) and addition via thiourea-bound cyanide (**TS 2**).

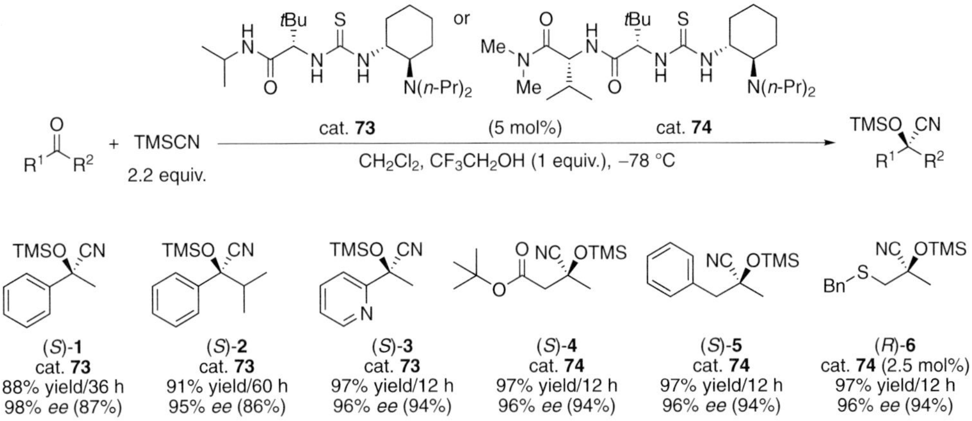

Scheme 6.85 Product range of the **73**- and **74**-catalyzed asymmetric cyanosilylation of ketones.

strong correlation between the experimentally observed sense and degree of enantioinduction for various catalysts and ketone substrates favored **TS 1**, in which the tertiary amine group activated the nucleophile HNC (isonitrile-form of HCN) generated from tautomerization of HCN and the thiourea moiety of the ketone through explicit double hydrogen-bonding interaction with both ketone lone pairs, respectively [242].

The computations suggested that the enantioselectivity of the cyanosilylation arose from direct interactions between the ketone substrate and the amino-acid derived unit of the catalyst type represented by thiourea **72**. On the basis of this insight, the Jacobsen group designed thiourea catalysts **73** and dipeptide thiourea catalyst **74** [67]. These optimized catalysts gave access to a broader spectrum of silylated cyanohydrins (e.g., **1–6**) and proved to be more active (88–97% yield) and more enantioselective (98–98% ee) than **72** (Scheme 6.85) [242].

Based on the structure of highly efficient enantioselective *tert*-leucine-derived thiourea catalysts bearing a tertiary amine functionality such as **72**, **73**, and **74** (Schemes 6.84 and 6.85), Jacobsen and Fang developed a new class of chiral bifunctional phosphinothioureas derived from readily accessible *trans*-2-amino-1-(diphenylphosphino)cyclohexane [243, 244]. Structure optimization studies including the variation of the amide group and the amino acid unit demonstrated alanine-derived phosphinothiourea **75** to be the most efficient in the asymmetric model [3 + 2] cycloaddition reaction between diphenylphosphinoyl-(DPP) protected benzaldehyde imine and buta-2,3-dienoic acid ethyl ester as the allene component (adduct **1**; Scheme 6.86). In the presence of Et_3N (5 mol%) and H_2O (20 mol%), this thiourea catalyst (10 or 20 mol% loading) exhibited an increased accelerating effect and promoted the enantioselective [3 + 2] cycloaddition between the model allene and various DPP-protected (hetero)aromatic imines in 48 h reaction time resulting in the target 2-aryl-2,5-dihydropyrrole derivatives **1–5** in good yields (68–90%) and with excellent ee values (94–98%) (Scheme 6.86).

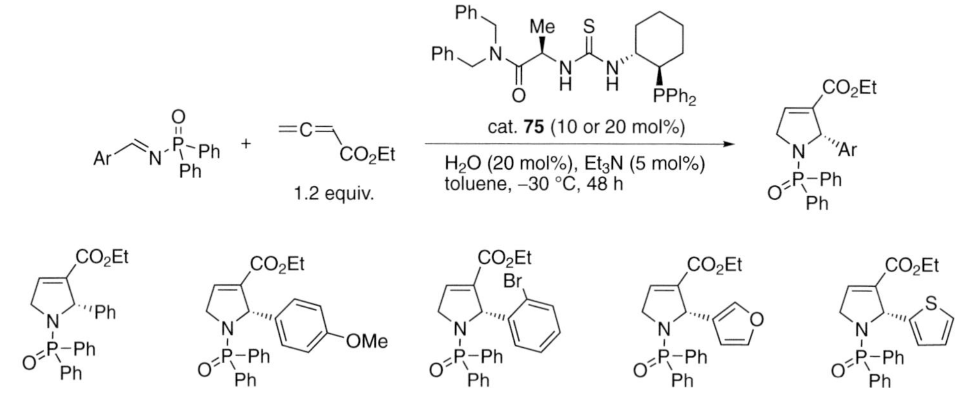

230 | 6 (Thio)urea Organocatalysts

Scheme 6.86 Typical 2-aryl-2,5-dihydropyrrole derivatives prepared with the asymmetric [3 + 2] cycloaddition between buta-2,3-dienoic acid ethyl ester and various DPP-protected (hetero)aromatic imines catalyzed by phosphinothiourea **75**.

(S)-1
(10 mol% cat. **75**)
84% yield
98% ee

(S)-2
(20 mol% cat. **75**)
80% yield
97% ee

(R)-3
(10 mol% cat. **75**)
90% yield
95% ee

(S)-4
(20 mol% cat. **75**)
68% yield
94% ee

(R)-5
(20 mol% cat. **75**)
77% yield
97% ee

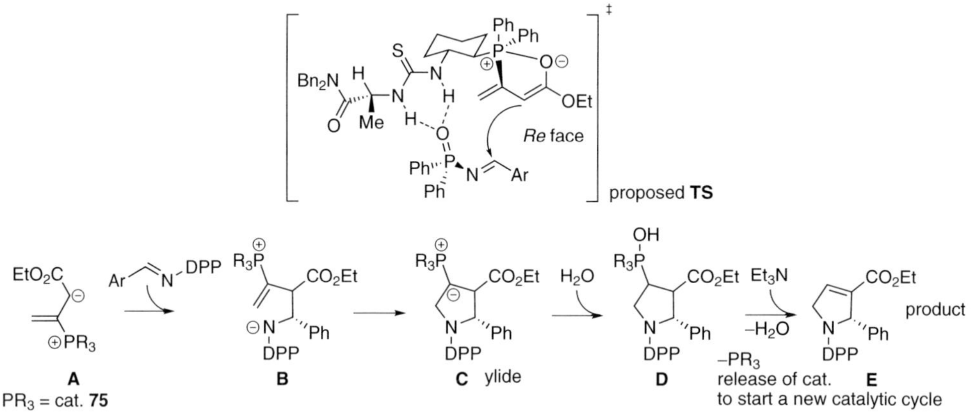

Scheme 6.87 Mechanistic proposal for the asymmetric [3 + 2] cycloaddition between buta-2,3-dienoic acid ethyl ester and various DPP-protected imines catalyzed by phosphinothiourea **75**.

In the mechanistic scenario visualized in Scheme 6.87, the allene is activated through the nucleophilic attack of the phosphino group providing zwitterion (**A**) that takes part in the proposed transition state for asymmetric addition on the hydrogen-bonded DPP-protected imine (**B**). The beneficial effect of Et$_3$N and H$_2$O on the reaction rate without affecting the enantioinduction suggests that these additives are not involved in the rate-determining step(s). Most likely H$_2$O affects

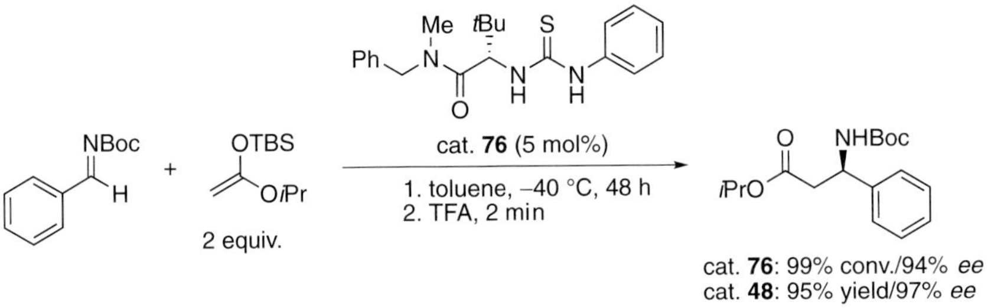

Scheme 6.88 Asymmetric Mannich reaction of N-Boc-protected aldimines catalyzed by simplified thiourea **76**.

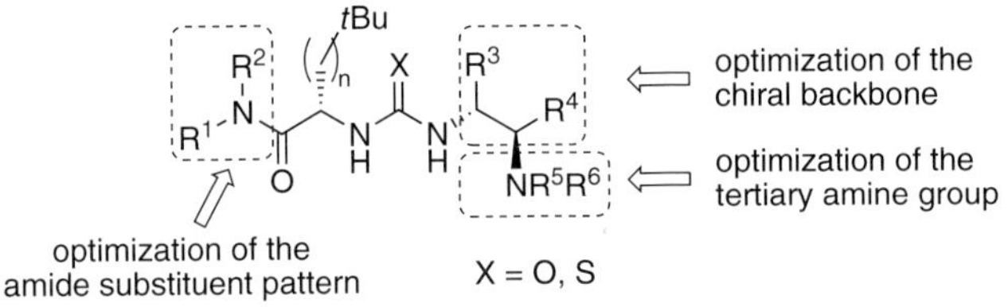

Figure 6.24 Systematic structure modification led to "second generation" catalyst **78** optimized for the asymmetric DKR of azlactones.

the formation of (**D**) through deprotonation of (**C**) and Et$_3$N promotes the elimination and the release of catalyst **75** via either E$_2$ or E$_1$cb mechanisms affording the product (**E**). Secondary interactions (π–π stacking or C=O···Ar) between the amide portion of the catalyst and the diphenyl portion of the imine were suggested to additionally stabilize the lowest energy **TS** leading to the observed high enantioselectivities.

Systematic investigations of the catalyst structure–enantioselectivity profile in the Mannich reaction [72] led to significantly simplified thiourea catalyst **76** lacking both the Schiff base unit and the chiral diaminocyclohexane backbone (Figure 6.14; Scheme 6.88). Yet, catalyst **76** displayed comparable catalytic activity (99% conv.) and enantioselectivity (94% *ee*) to the Schiff base catalyst **48** in the asymmetric Mannich reaction of N-Boc-protected aldimines (Schemes 6.49 and 6.88) [245]. This confirmed the enantioinductive function of the amino acid–thiourea side chain unit, which also appeared responsible for high enantioselectivities obtained with catalysts **72**, **73**, and **74**, respectively, in the cyanosilylation of ketones (Schemes 6.84 and 6.85) [240, 242].

Berkessel and co-workers synthesized a library of structurally diverse tertiary amine-functionalized catalyst candidates incorporating a chiral 1,2- or 1,4-diamine chiral backbone [231, 232, 246]. Structure-efficiency studies through sequential modification of the diamine backbone, the tertiary amine functionality, the (thio)urea N-substituents as well as of the amide substituent pattern, exemplarily illustrated a Jacobsen-type 1,2-diamine-based structure (Figure 6.24), identified

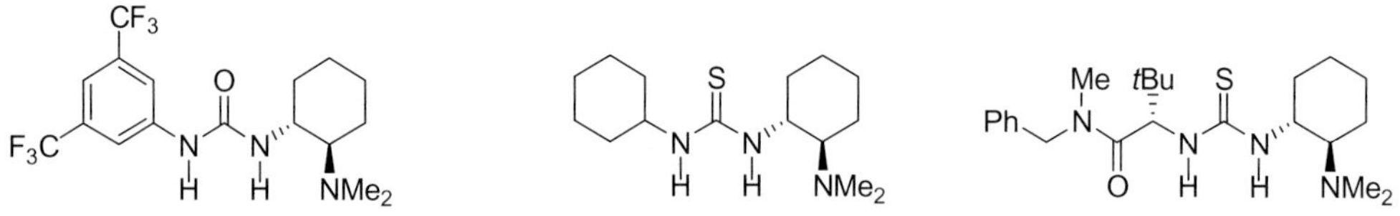

64
96% conv./48 h
72% ee/rt

77
96% conv./24 h
75% ee/rt

78
59% conv./24 h
78% ee/rt

Figure 6.25 The most efficient (thio)urea derivatives in the asymmetric DKR of phenylalanine-derived azlactone (R = Bn; Scheme 6.90).

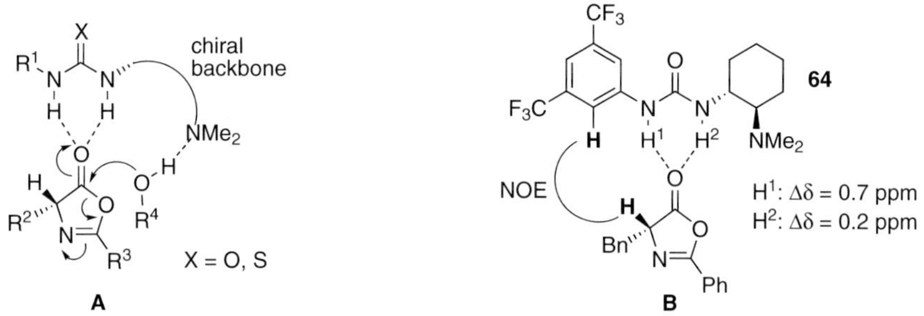

H^1: Δδ = 0.7 ppm
H^2: Δδ = 0.2 ppm

Scheme 6.89 Proposed mechanistic picture for the asymmetric alcoholytic DKR of racemic azlactones promoted by bifunctional (thio)urea catalysts **64**, **77**, and **78** (**A**); hydrogen-bonded azlactone-**64** complex supported by NMR methods (**B**).

dimethylated amine (*R,R*)-diaminocyclohexane-(thio)ureas **64**, **77**, and **78** (5 mol% loading) to be the most active and stereoselective catalysts in the model asymmetric alcoholytic dynamic kinetic resolution (DKR) [247, 248] of phenylalanine-(R = Bn; Scheme 6.90) and *tert*-leucine-derived (R = *t*Bu; Scheme 6.90) racemic oxazol-5(4*H*)-ones ("azlactones") with allyl alcohol (Figure 6.25). Azlactones are readily accessible α-amino acid derivatives prepared by the Erlenmeyer azlactone synthesis or from *N*-acylated (e.g., with benzoyl chloride) racemic α-amino acids through cyclodehydration in the presence of a condensation agent (e.g., acetic anhydride) [249]. Owing to the acidic hydrogen atom (pK_a = 8.9), azlactones are configurationally labile substrates that undergo base-catalyzed or autocatalytic racemization through enol formation.

The authors assumed that **64**, **77**, and also "second generation" catalyst **78** structurally closely related to **72** (Figure 6.22) operated in a bifunctional mode such that the azlactone carbonyl was activated through double hydrogen-bonding interaction with the (thio)urea moiety, while the tertiary amine group increased the nucleophilicity of the attacking alcohol (Scheme 6.89). This proposal for the hydrogen-bonded azlactone–catalyst complex was supported by NMR spectroscopic experiments using **64** as the catalyst. Upon addition of the racemic azlactone substrate to a solution of catalyst **64** in d_8-toluene (NMR titration), downfield shifts of $\Delta\delta = 0.2$ ppm and $\Delta\delta = 0.7$ ppm were observed for the urea N–H hydrogen atoms of the catalyst (Scheme 6.89). Furthermore, an intermolecular NOE of the azlactone 4-H resonance was observed upon irradiation at the resonance frequency of the aromatic hydrogen atoms at the 2- and 6-positions of the catalyst. The ^{1}H and ^{13}C NMR spectra of this azlactone-**64** complex consisted of only one set of signals indicating the preferential formation of one of the two possible diastereomers of the catalyst–azlactone complex due to rapid interconversion of the azlactone enantiomers [246].

The chiral information of the *tert*-leucine amide motif introduced by the Jacobsen group (Figure 6.22) [240] was found to enhance the stereodifferentiation predominantly induced by the diaminocyclohexane moiety [231]. Employing catalyst **64** (5 mol%) or **78** (5 mol%) under optimized conditions (toluene, rt; 1.5 equiv. allyl alcohol) to the alcoholytic ring opening of selected racemic azlactones exhibited conversions of 28–96% (24–48 h) and enantioselectivities of 72–95% (at rt) for the formation of the desired enantiomerically enriched N-benzoyl-protected α-amino acid allyl esters **1–4** (Scheme 6.90) [231, 246], that could be readily converted to synthetically useful enantiopure α-amino acids. Additionally, this organocatalytic DKR methodology was utilized for clean stereoinversion of enantiopure natural or nonnatural L- (or D-) α-amino acids resulting in the corresponding N-benzoyl-D- (or L-) α-amino acids allyl esters [231].

Berkessel and co-workers extended the synthetic applicability of hydrogen-bonding thiourea catalyst **78** in the DKR of azlactones to the kinetic resolution (KR) of structurally related, but configurationally stable 4,5-dihydro-1,3-oxazine-6-

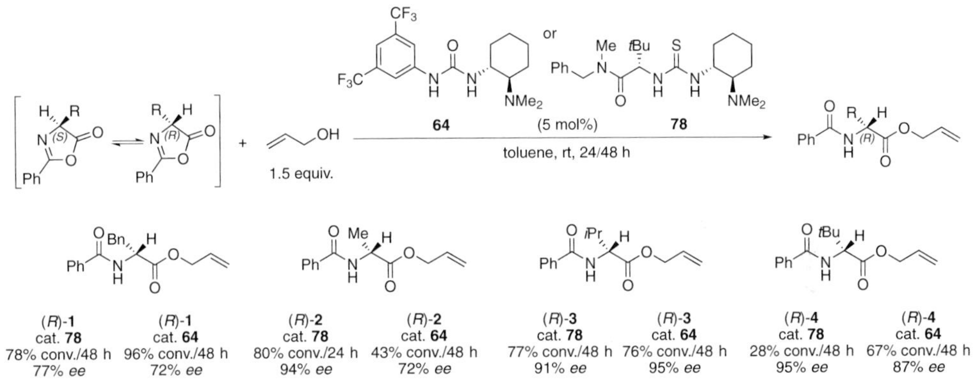

Scheme 6.90 Chiral N-benzoyl-protected α-amino acid allyl esters obtained from **64**- and **78**-catalyzed asymmetric DKR of racemic azlactones derived from racemic natural nonnatural α-amino acids.

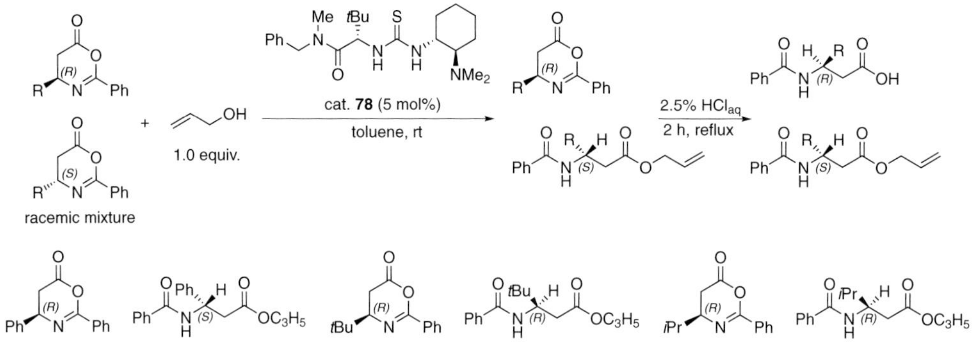

(R)-1a
57% conv./6.5 h
99% ee

(S)-2a
86% ee

(R)-3b
54% conv./48 h
97% ee

(S)-4b
82% ee

(R)-5c
53% conv./48 h
98% ee

(S)-6c
88% ee

Scheme 6.91 Typical enantioenriched (*R*)-oxazinones and (*S*)-configured *N*-benzoyl-protected β-amino acid allyl esters obtained from the **78**-catalyzed kinetic resolution of racemic oxazinone mixtures; subsequent isolation of the ester through (*R*)-oxazinone hydrolysis.

ones ("oxazinones") [233]. Oxazinones are six-membered cyclic derivatives of β-amino acids and can be synthesized similar to azlactones by cyclodehydration of the corresponding *N*-benzoyl-protected amino acids or, alternatively, by the one-step protocol reported by Tan and Weaver for the synthesis of racemic β-amino acids using aldehydes, malonic acid, and ammonium acetate as inexpensive starting materials [250]. The KR of racemic oxazinone mixtures based on the (*S*)-favored hydrogen-bonding activation through **78** and thus preferred alcoholytic (1.0 equiv. allyl alcohol) ring opening of the (*S*)-configured oxazinone in comparison to its (*R*)-counterpart resulting in both the formation of the corresponding enantiomerically enriched (*S*)-configured *N*-benzoyl-protected β-amino acid allyl ester and the remaining(*R*)-oxazinone (Scheme 6.91). Due to the difficulty in the accurate determination of the selectivity factor *S*, the authors used the conversion and *ee* values of the substrates and products, respectively, to describe the quality of the KR [251]. After practical reaction times (6.5–48 h at rt) and with 5 mol% loading of **78** the unreacted (*R*)-oxazinones **1a**, **3b**, and **5c** were obtained in high enantioselectivities (97–99%) and the (*S*)-configured *N*-benzoyl-protected β-amino acid allyl ester **2a**, **4b**, and **6c** in *ee* values ranging from 82 to 88% (Scheme 6.91). Performing the KR for oxazinone R = Ph in the presence of only 1 mol% catalyst **78** produced 61% conversion after 10 h, 98% *ee* for the (*R*)-**1a**, and 85% *ee* for the protected allyl ester (*S*)-**2a** (Scheme 6.91). Simple hydrolytic work-up of the product mixture with aqueous HCl converted the remaining (*R*)-oxazinone to the corresponding insoluble *N*-benzoyl-protected β-amino acid that was separable from the desired (*S*)-configured *N*-benzoyl-protected β-amino acid allyl ester through filtration (Scheme 6.91) [233].

Chen *et al.* identified (*R*,*R*)-1,2-diphenylethylenediamine-derived tertiary amine-functionalized thiourea **79** (20 mol% loading), an analog of Takemoto's bifunc-

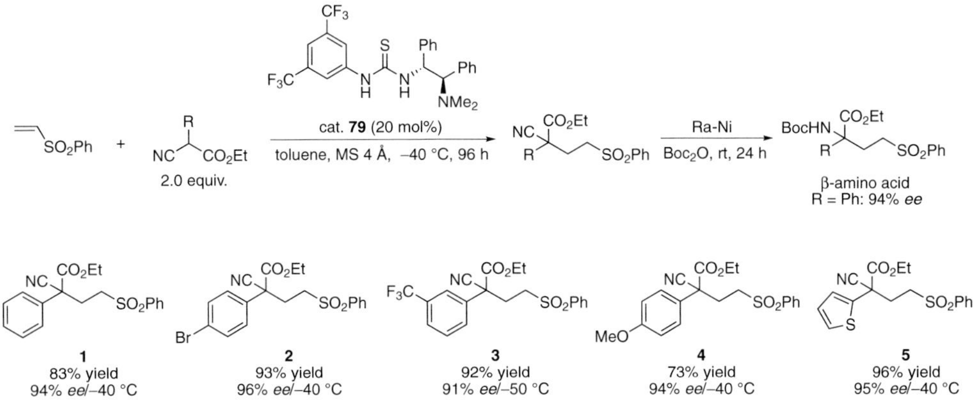

Scheme 6.92 Product range of the **79**-catalyzed Michael addition of α-aryl cyanoacetates to phenyl vinyl sulfone and conversion of one exemplary adduct (R = Ph) to the corresponding protected β-amino acid. The absolute configurations of the adducts were not determined.

tional thiourea catalyst **12** (Figure 6.18) [129], as enantioselective catalyst for the Michael addition [149–152] of α-aryl cyanoacetates to phenyl vinyl sulfone producing the desired adducts **1–5** in good (73%) to excellent (96%) yields (after 96 h) and in very good enantioselectivities (91–96% at −50 to −40 °C) (Scheme 6.92) [215]. The protocol using **79** was limited to aromatic cyanoacetates (R = aryl), while thiourea **12** catalyzed the addition of aliphatic (R = alkyl) α-alkyl cyanoacetates to phenyl vinyl sulfone and produced the adducts in yields ranging from 52 to 96% and ee values of 72–96% after 48 h at −40 °C [215]. The authors presented a one-pot protocol to convert the obtained Michael adducts efficiently into synthetically important protected β-amino acid precursors as shown in one example (R = Ph: 94% ee) in Scheme 6.92 [215]. In analogy to catalyst **12**, thiourea derivative **79** was classified as bifunctional catalyst activating both the phenyl vinyl sulfone through hydrogen bonding to the sulfone functionality and also the respective cyanoacetates through tertiary-amine mediated deprotonation resulting in the formation of the active nucleophile.

In 2008, Wu and co-workers introduced a small series of novel multiple hydrogen bonding tertiary amine-functionalized thiourea derivatives **80–83** for catalysis of the asymmetric Michael addition [149–152] of acetylacetone to aliphatic and aromatic nitroalkenes (Figure 6.26) [252]. The new concept of catalyst design was based on the working hypothesis that amine thioureas bearing multiple hydrogen-bonding donors could form more activating hydrogen bonds to the substrates and thus could display higher catalytic efficiency (at reduced catalyst loadings) compared to a thiourea catalyst bearing only one thiourea group capable of only one double hydrogen-bonding interaction. This strategy was already known from bis-thiourea catalysts **106** and **114** (Schemes 6.107 and 6.110) but was for the first time

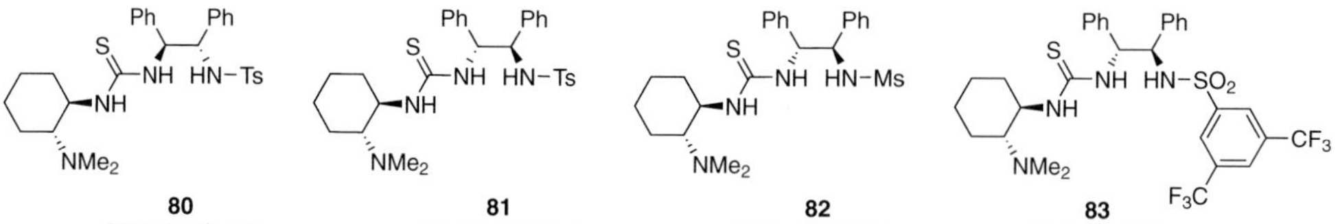

Figure 6.26 Multiple hydrogen-bonding tertiary amine-functionalized thioureas screened in the asymmetric Michael reaction between trans-β-nitrostyrene and acetylacetone at 10 mol% loading.

80 97% yield/0.5 h, 76% ee/rt/Et$_2$O
81 96% yield/0.5 h, 93% ee/rt/Et$_2$O
82 96% yield/0.5 h, 89% ee/rt/Et$_2$O
83 97% yield/1 h, 97% ee/rt/Et$_2$O

realized in chiral amine thioureas. Additionally, these thiourea derivatives incorporated both the 1,2-diphenylethylenediamine as well as the 1,2-diaminocyclohexane backbone closely connected via a bridging thiourea functionality to create an efficient chiral environment. In the model Michael reaction between trans-β-nitrostyrene and acetylacetone at room temperature, with diethyl ether as the solvent, and in the presence of **80**, **81**, **82**, or **83** (10 mol% loading) all catalyst candidates exhibited enantioinduction and consistently high catalytic activity (Figure 6.26). Thiourea **80**, however, combining the (R,R)-diaminocyclohexane with the (S,S)-diphenylethylenediamine unit furnished only a moderate ee value (76%), while the (R,R)-configuration in both diamine units proved to be the matched combination and registered increased enantioinduction (e.g., with **81**: 93% ee at rt). Replacing the tosyl (Ts) group of **81** with the less bulky mesyl (Ms) group led to sterically more flexible, but less enantioselective (89% ee) thiourea derivative **82**. The best results in terms of catalytic activity (97% yield/1 h) and enantioselectivity (97% ee/rt/Et$_2$O) were reached with electron-deficient thiourea derivative **83** again incorporating the 3,5-bis(trifluoromethyl)phenyl moiety. The **83**-catalyzed (10 mol%) model reaction demonstrated that polar protic solvents were strongly incompatible and drastically reduced both catalytic activity and enantioselectivity (e.g., MeOH: 69% yield/17 h, 18% ee/rt; DMSO: 72% yield/16 h; 0% ee), while aprotic nonpolar or less polar solvents such as DCM (97% yield/0.5 h/89% ee/rt) or MeCN (90% yield/10 h/86% ee/rt) gave better results. Diethyl ether was identified as the solvent of choice (97% yield, 1 h; 97% ee at rt). Reducing the reaction temperature from 0 to −20 °C did not affect the ee values. Lowering the catalyst loading from initial 10 mol% to 1 mol% catalyst **83** gave an unchanged yield (97%/1 h) and ee value (97% ee/rt) for the model Michael adduct, while 0.1 mol% loading resulted in 81% yield and 95% ee after 10 h at room temperature.

Under optimized conditions, the **83**-catalyzed (1 mol% loading) Michael addition of acetylacetone to various aryl nitroalkenes as well as alkyl nitroalkenes proceeded in good to excellent yields (80–97%) and enantioselectivities (82–99%) of the desired adducts **1–5** (Scheme 6.93). The authors also reported the successful enantioselective Michael addition of 1,3-diphenylpropane-1,3-dione (adduct in

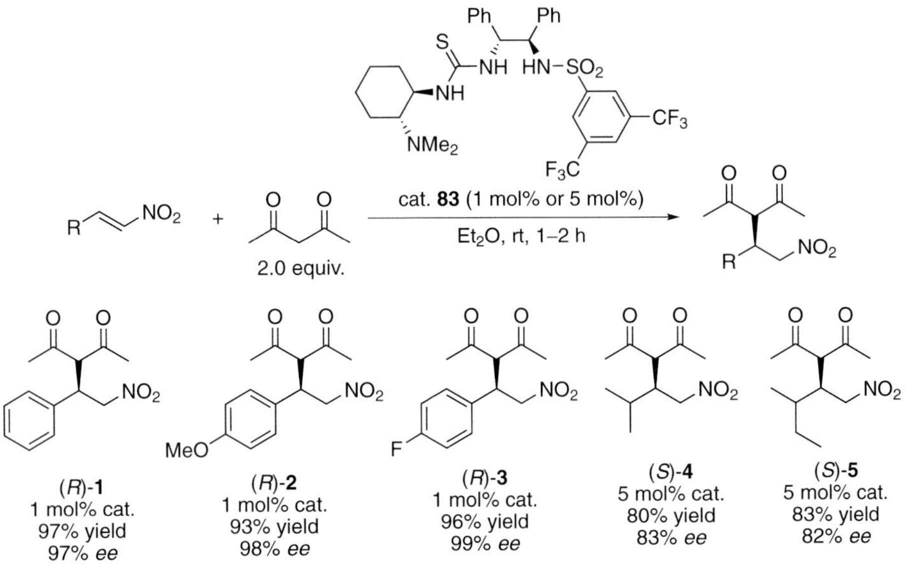

Scheme 6.93 Typical products of the asymmetric Michael addition of acetylacetone to various nitroalkenes catalyzed by **83**.

(R)-**1**
1 mol% cat.
97% yield
97% ee

(R)-**2**
1 mol% cat.
93% yield
98% ee

(R)-**3**
1 mol% cat.
96% yield
99% ee

(S)-**4**
5 mol% cat.
80% yield
83% ee

(S)-**5**
5 mol% cat.
83% yield
82% ee

95% yield/12 h; 85% ee/rt) and 2-acetylcyclopentanone (adduct in 92% yield/10 h; 96% ee/dr :85:15; rt) to trans-β-nitrostyrene. A derivative of **81** lacking the third amide proton due to methylation of the sulfonamide proved less effective in the model Michael reaction (80% yield/16 h; 68% ee); this observation indicated that the NH of the sulfonamide on the 1,2-diphenylethylenediamine moiety played a significant role in this Michael reaction [252].

Dixon and Richardson developed (R,R)-1,2-diaminocyclohexane-derived hydrogen-bonding thiourea derivatives incorporating the phthalimide (Phthal) and tetraphenylphthalimide (TPhP) unit, respectively, for enantioselective catalysis of the conjugate addition [149–152] of aryl methyl-ketone-derived morpholine enamines to various aromatic nitroalkenes [253]. The catalyst evaluation performed for the Michael reaction between acetophenone-derived morpholine enamine and trans-β-nitrostyrene (10 mol% catalyst loading; in toluene at −10 °C) produced the desired (S)-adduct after acidic hydrolytic work-up in very low ee values for thioureas bearing the phthalimide moiety such as **84** (8% ee) and **85** (12% ee) (Figure 6.27). In contrast, the tetraphenylphthalimide-containing counterparts showed under the same screening conditions enhanced and reversal enantioinduction resulting in 38% (with **86**) and 25% (with **87**) enantioselectivity, respectively, for the (R)-configured Michael adduct (Figure 6.27). Thiourea-derivative **86**, which was prepared in two steps (20% yield) from (R,R)-1,2-diaminocyclohexane and tetraphenylphthalic anhydride, turned out to be the most efficient catalyst and was utilized for the synthesis (10 mol% loading) of a broad range of Michael adducts.

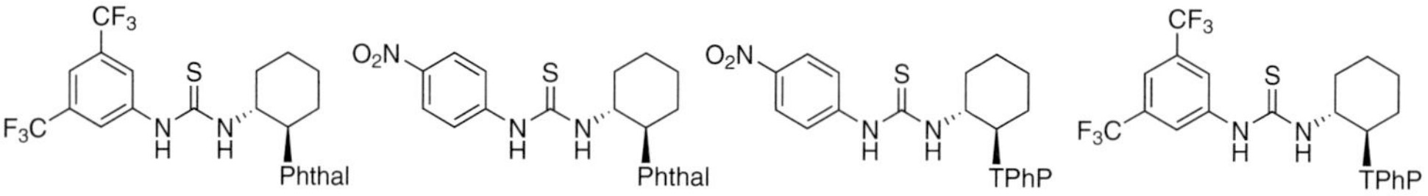

84
90% conv./2 h
8% ee (S)-adduct/−10 °C

85
35% conv./2 h
12% ee (S)-adduct/−10 °C

86
40% conv./2 h
38% ee (R)-adduct/−10 °C

87
73% conv./2 h
25% ee (R)-adduct/−10 °C

Figure 6.27 Representative (R,R)-1,2-diaminocyclohexane-derived thiourea derivatives incorporating a phthalimide (Phthal) and tetraphenylphthalimide (TPhP) moiety; catalyst screening was performed in the Michael addition of acetophenone-derived morpholine enamine to trans-β-nitrostyrene in toluene as the solvent.

The optimized protocol reached yields ranging from 67 to 98% and moderate ee values (37–65%) of the typical adducts **1–6** (Scheme 6.94). The enantioenriched products represented useful synthetic intermediates serving, e.g., as precursors of 1,4-dicarbonyl compounds and γ-amino acids [254].

The Roussel group introduced new atropoisomeric thiourea derivatives bearing the (R,R)-1,2-diaminocyclohexane backbone and a tertiary amine functionality to bifunctionally catalyze the enantioselective addition of TMSCN (cyanosilylation) to aromatic and aliphatic aldehydes [255]. Thiourea catalyst (aR/aR)-(R,R)-**88**, which was an inseparable mixture of the two diastereomers (aR)-(R,R)-**88** and the (aS)-(R,R) isomer, was prepared from the racemic mixture (aR/aS) of atropoisomeric N-(2-aminophenyl)-4-methyl-thiazoline-2-thione ($\Delta G_{rot} > 145$ kJ/mol) through isothiocyanation (1. CS_2, Et_3N, 1 h; 2. DCC, MeCN, 24 h) and subsequent coupling with enantiopure (R,R)-N,N-dimethyl-1,2-diaminocyclohexane. The single diastereomeric thioureas (aR)-(R,R)-**88** and (aR)-(S,S)-**88** were accessible, when the pure (aR)-atropoisomeric of the 2-amino-thiozoline-2-thione (obtained by semipreparative HPLC) was transformed to the corresponding atropoisomeric (aR)-isothiocyante followed by the addition of enantiopure (R,R)- or (S,S)- N,N-dimethyl-1,2-diaminocyclohexane (Figure 6.28). The cyanosilylation of the model substrate benzaldehyde at 10 mol% loading of the prepared thiourea derivatives demonstrated that the mixture of diastereomers (aR/aR)-(R,R)-**88** showed higher catalytic activity (100% yield/27 h) and enantioselectivity (66% ee/−20 °C) of the acetylated adduct than the single diastereomers (Figure 6.28). The authors proposed that this observation could originate from the involvement of more than one thiourea molecule in the transition state of this cyanosilylation so that the diastereomers could self-associate ("preorganization") to stabilize a catalytically more active thiourea conformation than a single diastereomer could adopt. Thiourea **58** lacking the atropoisomeric unit turned out to be less active (15% yield/27 h) and enantioselective (32% ee at −20 °C).

Utilizing the readily accessible diastereomeric atropoisomeric thioureas (aR/aR)-(R,R)-**88** as the catalyst (10 mol%) various (hetero)aromatic and aliphatic aldehydes could be cyanosilylated to the corresponding TMS-protected cyanohydrins

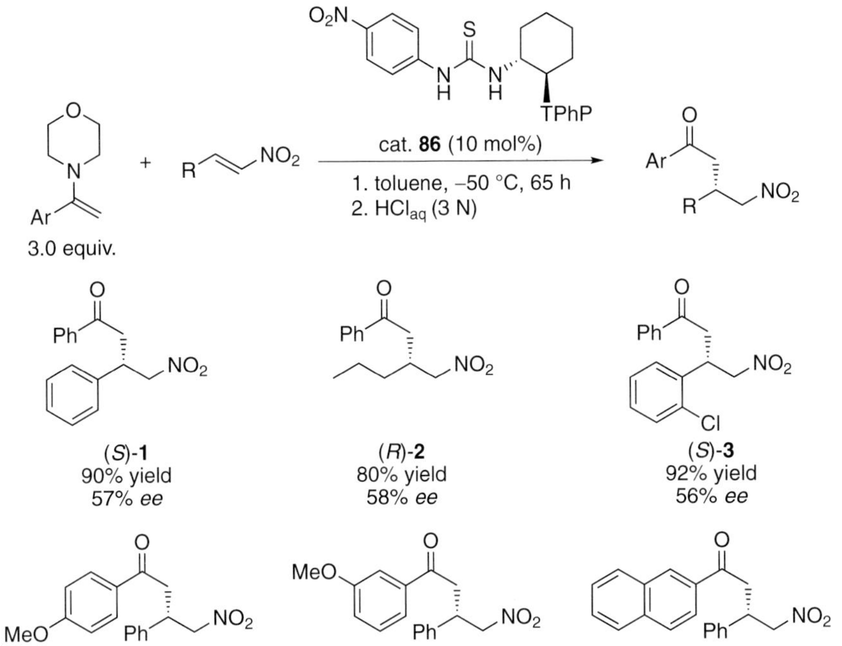

(*S*)-**1** 90% yield 57% *ee*

(*R*)-**2** 80% yield 58% *ee*

(*S*)-**3** 92% yield 56% *ee*

(*S*)-**4** 90% yield 48% *ee*

(*S*)-**5** 98% yield 65% *ee*

(*S*)-**6** 67% yield 37% *ee*

Scheme 6.94 Typical products obtained from the **86**-catalyzed Michael addition of aryl methyl-ketone-derived morpholine enamines to various aromatic nitroalkenes and subsequent acidic hydrolysis.

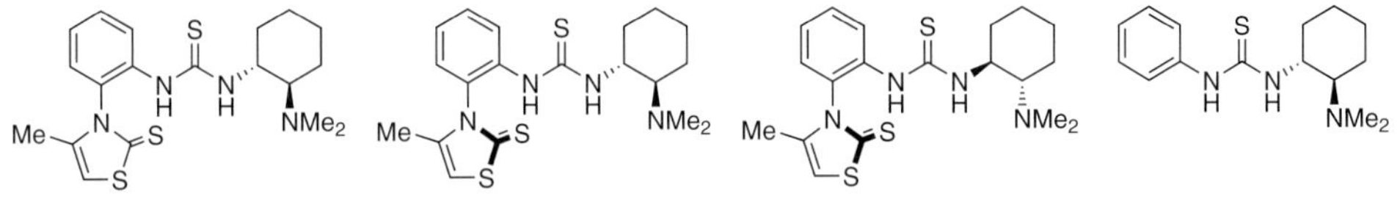

(*aR*/*aS*)-(*R,R*)-**88** 100% yield/27 h 66% *ee*/(*S*)-adduct)–20 °C

(*aR*)-(*R,R*)-**88** 30% yield/24 h 47% *ee*/(*S*)-adduct)/–20 °C

(*aR*)-(*S,S*)-**88** 100% yield/26 h 55% *ee* (*S*)-adduct)/–20 °C

58 15% yield/27 h 32% *ee* (*R*)-adduct)/–20 °C

Figure 6.28 Bifunctional atropoisomeric thioureas and **58** lacking axial chirality screened in the cyanosilylation of benzaldehyde.

1–6 and desilylated as well as acetylated to the respective acetates (Scheme 6.95). The yields of the TMSCN-adducts ranged from 43 to 100% and the *ee* values of the obtained acetates were moderate (45–69%).

Based on the modular structure of Schiff base catalysts such as "first-generation" Strecker urea catalyst **42** [196, 198] (Figure 6.15; Schemes 6.41 and 6.45), Yoon

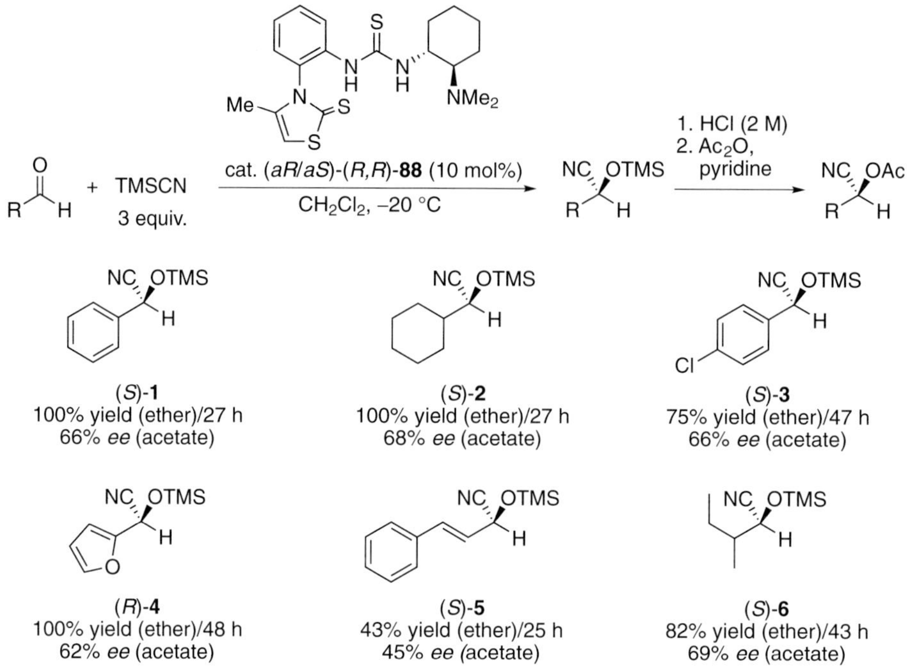

Scheme 6.95 TMS-protected cyanohydrins prepared from the cyanosilylation of aldehydes in the presence of atropoisomeric thiourea catalyst (aR/aR)-(R,R)-**88**. Desilylation and acetylation to the respective more stable acetates.

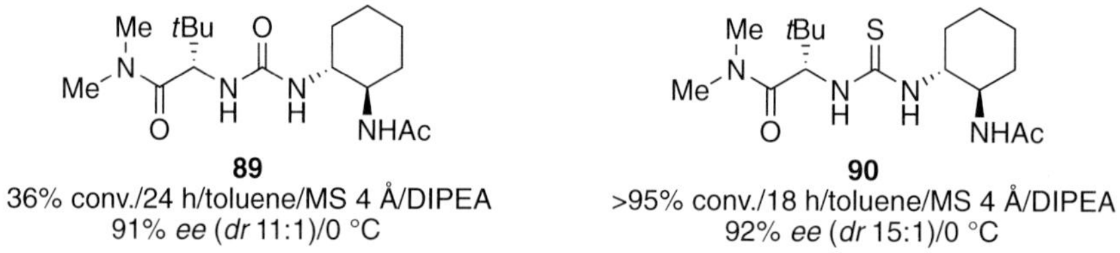

89
36% conv./24 h/toluene/MS 4 Å/DIPEA
91% ee (dr 11:1)/0 °C

90
>95% conv./18 h/toluene/MS 4 Å/DIPEA
92% ee (dr 15:1)/0 °C

Figure 6.29 Acetamide (thio)urea derivatives evaluated for catalytic efficiency in the nitro-Mannich reaction between N-Boc-protected benzaldimine and nitroethane.

and Jacobsen developed acetamide (thio)urea derivatives **89** and **90** for enantio- and diastereoselective catalysis of the nitro-Mannich reaction [72] between N-Boc-protected aldimines and nitroethane as well as nitropropane (Figure 6.29) [256]. Thiourea **90** (10 mol% loading) was found to accelerate the model Mannich reaction between N-Boc-protected benzaldimine and nitroethane (in toluene at 0 °C) more efficiently (95% conv./18 h) than the urea analog **89** (36% conv./

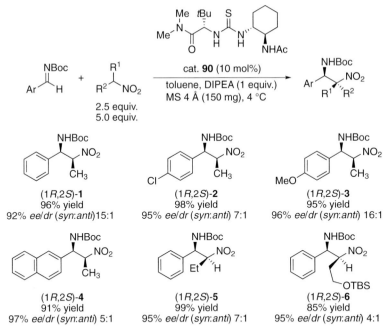

Scheme 6.96 Product range of the **90**-catalyzed nitro-Mannich reaction between N-Boc-protected aromatic aldimines and nitroalkanes.

24 h); the stereoinduction, however, turned out to be nearly identical (**89**: 91% *ee*; **90**: 92% *ee*). A screen of additives revealed that the addition of MS 4 Å improved both the reproducibility and the diastereoselectivity of the reaction, while the use of MS 5 Å and MS 3 Å had a detrimental effect on the reaction rate and on the enantioselectivity. This suggested a direct function of the MS in the reaction mechanism. Replacing initially used triethyl amine with diisopropylethylamine ("Hünig's base") as base additive increased the diastereoselectivity of the reaction.

In the presence of 10 mol% loading of **90**, the optimized reaction conditions allowed the high-yielding (85–99%) addition of nitroethane and nitropropane to a range of N-protected aromatic aldimines and furnished the respective nitro-Mannich adducts **1–6** in high enantioselectivities (92–97% *ee*) and *syn*-favored diastereoselectivities (up to *dr* 16 : 1) (Scheme 6.96). The acid labile TBS-group (adduct **6**; Scheme 6.96) underwent neither desilylation nor elimination and remained stable under the mild reaction conditions. Since thioureas are known to bind and modulate the reactivity of nitronate anions the catalytic function of thiourea **90** could be the activation of the nitroalkane component through hydrogen bonding or dual activation of both the aldimine and the nitroalkane. The sense of enantiofacial selectivity in this reaction was observed to be identical to that reported for the thiourea-catalyzed Strecker [122, 124, 196, 198], Mannich [201], and hydrophosphonylation reactions [200] suggesting a commonality in the mode of substrate activation and comparable mechanisms.

Tan and Jacobsen discovered urea derivative **91** to catalyze the enantioselective addition of *in situ* generated allylindium N-benzoyl-protected hydrazones derived from aromatic aldehydes [257]. The catalyst incorporated both a hydrogen-bonding urea group and a Lewis basic *tert*-butyl sulfinamide functionality properly positioned and in close proximity to enable a bifunctional mode of action. Crystallographic analysis of the catalyst solid structure revealed an interaction between these groups through an intramolecular hydrogen bond between the NH group of the sulfinamide unit and the oxygen of the urea carbonyl group. This interaction could serve to increase the hydrogen-bonding ability of the urea group and/or the rigidity of the catalyst structure to attain high catalytic activity and enantioselectivity. The **91**-catalyzed (10 mol%) allylation of N-benzoyl-protected hydrazones occurred in toluene at −40 °C in the presence of indium (0) powder (1.75 equiv.) and furnished the corresponding adducts **1–5** (Scheme 6.97) in yields ranging from 78 to 92% and in good (76%) to high (95%) *ee* values. N-benzoyl-protected hydrazones derived from aliphatic aldehydes were allylated with substantially lower *ee* values (>50%). During the studies, the authors observed that the batch of indium and the stirring rate affected the *in situ* generation of the organometallic allylindium species; a high stirring rate led to a rapid formation of allylindium and entire consumption of the indium powder resulting in increased yields, but slightly

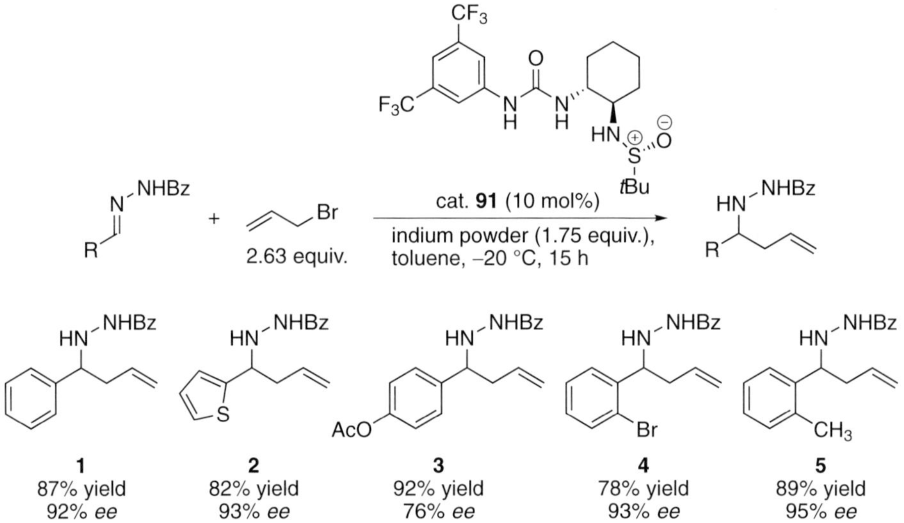

Scheme 6.97 Typical products obtained from the **91**-catalyzed asymmetric allylation of N-benzoyl-protected aromatic hydrazones. The product configurations were not determined.

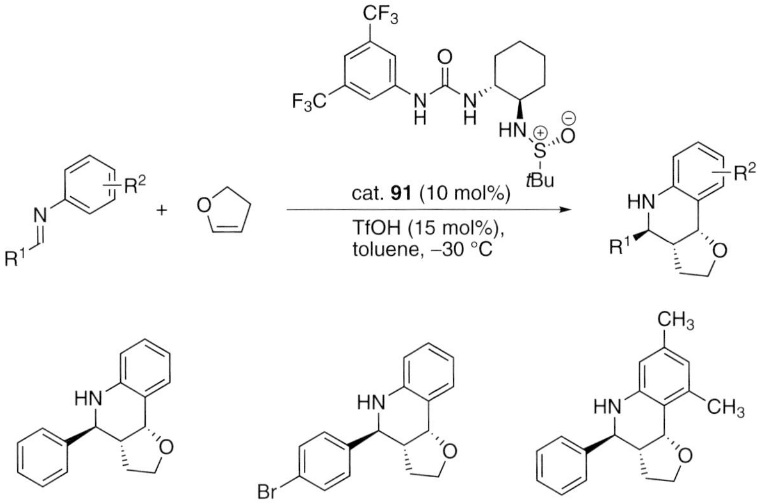

Scheme 6.98 Products of the enantioselective Povarov reaction of N-aryl imines with 2,3-dihydrofuran promoted by thiourea catalyst **91**.

1
86% yield
94% ee/ dr 4:1

2
64% yield
94% ee/ dr 4:1

3
96% yield
92% ee/ dr 5:1

lowered *ee* values (e.g., adduct 4; Scheme 6.97: high stirring rate gave 89% yield, 90% *ee*).

Woll and Jacobsen found that sulfinamide-functionalized thiourea **91** (10 mol%) catalyzed the asymmetric Povarov reaction between N-aryl imines and 2,3-dihydrofuran. At −30 °C in toluene, the resulting cyclic products **1–3** were obtained in yields ranging from 64 to 96% and in high enantioselectivities (92–94%) (Scheme 6.98) [7]. Mechanistically, an anion-binding model analogous to that proposed for the **53**-catalyzed acyl-Pictet–Spengler reaction was suggested (Scheme 6.53) [205]. In the case of the **91**-catalyzed Povarov reaction, the urea/strong acid (TfOH) system was proposed to generate an active electrophilic species consisting of a protioiminium [76] electrophile with a catalyst-bound sulfonate counterion.

Tsogoeva and co-workers explored the catalytic potential of pyridyl- and imidazoyl-containing thiourea derivatives (e.g., thiourea **92** and **93**) in the asymmetric model Strecker reactions [157] of N-benzyl- and benzhydryl-protected benzaldimine with HCN [258]. The observed enantioselectivities were consistently very low (4–14% *ee*) for all catalyst candidates and were far below synthetically useful levels, while imidazoyl-thiourea **93** was reported to be highly active and displayed 100% conversion (at 7% *ee*) of the N-benzhydryl-protected benzaldimine (Scheme 6.99). X-ray structure analysis of a pyridyl-thiourea revealed an intramolecular hydrogen-bond between the basic ring nitrogen and one amide proton. This could make this

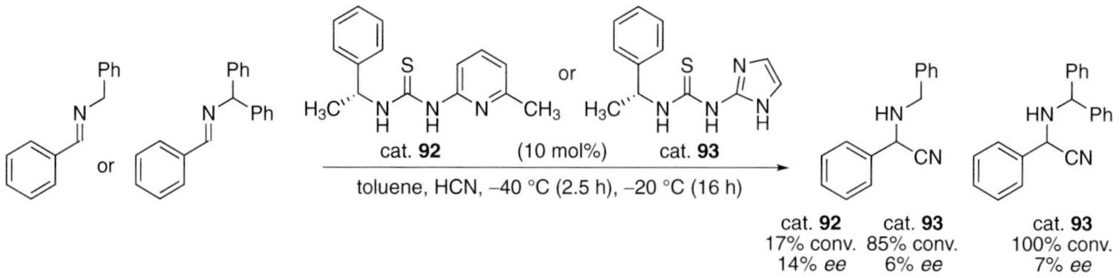

Scheme 6.99 Typical pyridyl- and imidazoyl-thioureas evaluated for bifunctional catalysis in the asymmetric Strecker reaction of aldimines.

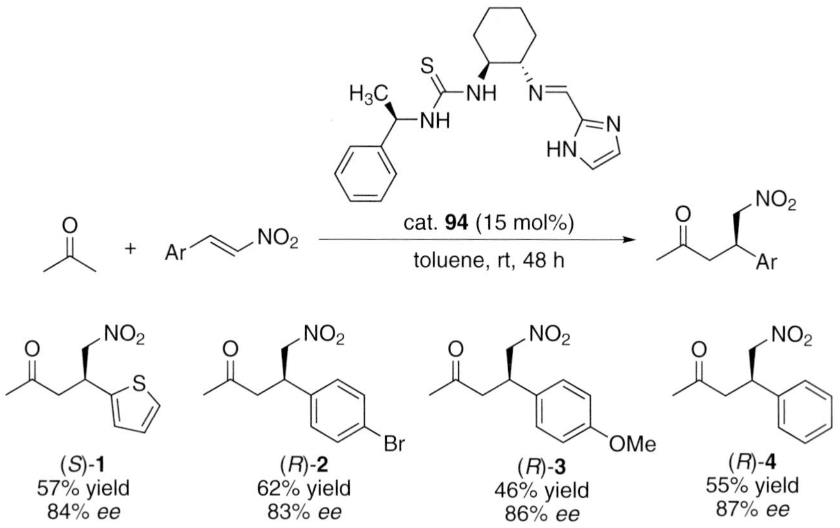

Scheme 6.100 Products of the asymmetric nitro-Michael addition of acetone to trans-β-nitrostyrenes catalyzed by thiourea **94**.

type of thiourea incapable of providing double hydrogen-bonding to the imine resulting in low enantioinduction.

The modification of thiourea catalyst **93** through incorporation of the (S,S)-diaminocyclohexane backbone as an additional chirality element and a Schiff base imidazoyl-moiety led to the bifunctional catalyst **94** that, in contrast to **93** in the Strecker reaction (Scheme 6.99), exhibited enantioinduction (83–87% ee) in the nitro-Michael addition of acetone to trans-β-nitrostyrenes. The desired adducts were isolated in moderate yields (46–62%) as depicted in Scheme 6.100) [259].

The Tsogoeva group, in 2006, reported the introduction of newly designed bifunctional secondary amine-functionalized proline-based thioureas (**95** and **96**) and the primary amine-functionalized thioureas (**97–99**) for catalysis of the asymmetric addition of ketones to trans-β-nitrostyrenes (Figure 6.30) [260, 261]. Using

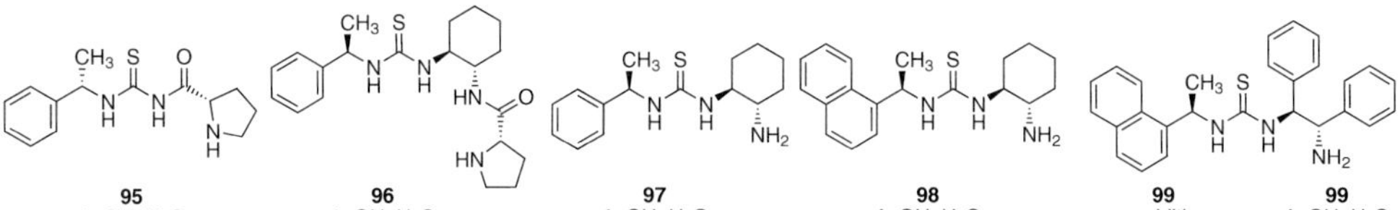

6.2 Synthetic Applications of Hydrogen-Bonding (Thio)urea Organocatalysts | 245

95	**96**	**97**	**98**	**99**	**99**
AcOH, H$_2$O	AcOH, H$_2$O	AcOH, H$_2$O	AcOH, H$_2$O	no additives	AcOH, H$_2$O
50% yield/72 h	55% yield/72 h	85% yield/16 h	97% yield/48 h	75% yield/72 h	98% yield/48 h
3% ee/rt	3% ee/rt	86% ee/rt	84% ee/rt	87% ee/rt	91% ee/rt

Figure 6.30 Secondary amine- and primary amine-functionalized bifunctional thiourea derivatives (15 mol% loading) screened in the model Michael addition of acetone to trans-β-nitrostyrene in toluene at rt.

the Michael reaction [149–152] between acetone and trans-β-nitrostyrene for catalyst screening (15 mol% loading; toluene; rt) and optimization of the reaction conditions demonstrated that primary amine-thioureas were superior to the proline-based candidates in both catalytic activity and enantioselectivity. The thioureas **97** (85% yield/16 h; 86% ee) and **99** (98% yield/48 h; 91% ee) turned out to be the most efficient when the reaction was performed in the presence of water (2.0 equiv.) and acetic acid (0.15 equiv.) as additives that facilitated the reversible formation of an enamine intermediate. In the absence of these additives, the catalytic activity as well as the enantioinduction of the catalysts decreased as exemplarily shown for **99** (75% yield/72 h; 87% ee) in Figure 6.30.

Catalyst **97** derived from (S,S)-diaminocyclohexane and **99** derived from (S,S)-diphenylethylenediamine promoted the enantioselective Michael addition of various ketones to electron-rich and electron-deficient trans-β-nitrostyrenes, respectively, to afford predominantly the respective (R)-adducts **1–6** in good to excellent yields (84–99%) and ee values (84–99%) (Scheme 6.101). Notably, the addition of methyl ethyl ketone occurred anti-diastereoselectively (adduct **5**: dr syn:anti 14 : 86), while the addition of tetrahydrothiopyran-4-one favored the formation of the opposite diastereomer (adduct **6**: dr syn:anti 83 : 17). This stereochemical outcome was interpreted on the basis of the proposed transition states **A** and **B** visualized in Scheme 6.102. An acyclic ketone formed a Z-configured enamine intermediate [55, 58, 77] with the primary amine group and attacked the hydrogen-bonded nitrostyrene in a mode that the anti-adduct was preferred; a cyclic ketone, however, reacted from the E-configured enamine to give the syn-diastereomer. Computational studies performed for catalyst **97** suggested an activation through double hydrogen bonding to only a single oxygen of the nitro group (Scheme 6.102) [260].

The Jacobsen group independently focussed on the development of primary amine-functionalized thiourea derivatives and published, in 2006, the thioureas **100–103** incorporating the established tert-leucine (amide) motif (Figure 6.14) and the diaminocyclohexane or diphenylethylenediamine chiral backbone, respectively (Figure 6.31) [262]. The catalyst screening was carried out in the asymmetric Michael addition [149–152] of 2-phenylpropionaldehyde, an α,α-disubstituted aldehyde, to 1-nitrohex-1-ene (at 20 mol% loading, DCM, rt, variable equiv. of H$_2$O)

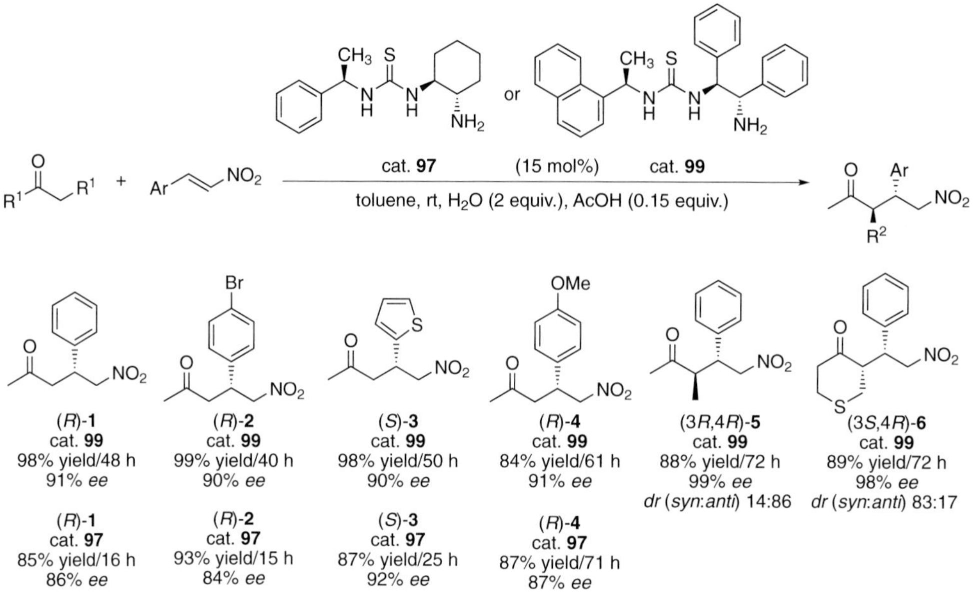

Scheme 6.101 Typical Michael products obtained from the 97- and 99-catalyzed addition of ketones to *trans*-β-nitrostyrenes.

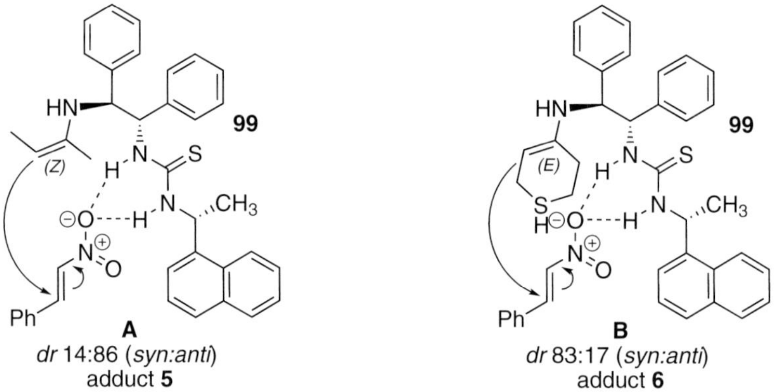

Scheme 6.102 Bifunctional catalysis with primary amine thiourea **99**: Proposed transition states to explain the *anti*-diastereoselectivity (**A**) and the *syn*-diastereoselectivity (**B**) of the Michael addition of both acyclic and cyclic ketones to *trans*-β-nitrostyrene.

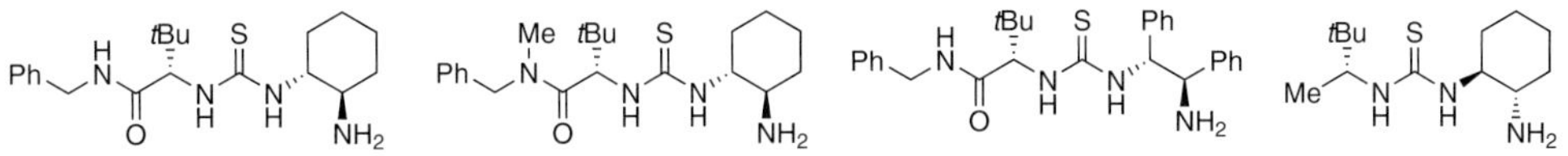

Figure 6.31 Primary amine-functionalized thioureas screened in the asymmetric Michael addition of 2-phenylpropionaldehyde to 1-nitrohex-1-ene using DCM as the solvent.

100
no H$_2$O
34% yield/24 h
96% ee
dr (syn:anti) 10:1

100
H$_2$O (5 equiv.)
64% yield/24 h
96% ee
dr (syn:anti) 10:1

101
H$_2$O (5 equiv.)
31% yield/24
96% ee
dr (syn:anti) 10:1

102
H$_2$O (5 equiv.)
100% yield/24 h
99% ee
dr (syn:anti) 10:1

103
H$_2$O (5 equiv.)
<5% yield/24 h

and identified primary amine thioureas **100** (64% yield/24 h; 96% ee) and **102** (100% yield/24 h; 99% ee) bearing a secondary amide functionality to be the most catalytically active and stereoselective catalysts (Figure 6.31). In the presence of 5.0 equiv. water, the more applicable diaminocyclohexane-derived thiourea **100** was utilized instead of **102** to catalyze the enantio- and diastereoselective Michael addition of various α,α-disubstituted aldehydes to aliphatic and aromatic nitroalkenes. The protocol tolerates a broad substrate scope and provides the corresponding Michael adducts such as **1–6** in yields ranging from 54 to 94%, in excellent enantioselectivities (96–99%), and in *syn*-favored diastereoselectivity (up to dr 28:1) (Scheme 6.103).

To explain the mode of action of bifunctional thiourea catalyst **100** in the studied Michael reactions, the authors proposed a catalytic cycle in which **100** initially formed an imine (**A**) through condensation with the aldehyde substrate (Scheme 6.104). Tautomerization of the imine led to the preferred formation of the thermodynamically favored *E*-enamine (**B**), which was proposed to be responsible for the observed *syn*-diastereoselectivities. The double hydrogen-bonding activation of the nitroalkene via only one oxygen atom [260] allowed the enamine to attain sufficiently close proximity for the carbon–carbon bond forming nucleophilic attack (**C**) resulting in zwitterionic intermediate (**D**) typical for the conjugate addition of enamines to nitroalkenes. Intramolecular proton transfer followed by imine hydrolysis yields the desired Michael adduct and regenerates catalyst **100** to start a new cyclus (Scheme 6.104). The beneficial impact of water on catalyst turnover and yield, respectively, was especially ascribed to the acceleration of this final imine hydrolysis as well as to the initial imine formation (**A**) affecting the subsequent imine–enamine equilibrium. Too much water reduced the enamine formation and thus the product-forming steps as shown in the model Michael reaction (10 equiv. H$_2$O: 54% yield); in contrast, too less water affected the release of the product and the catalyst (2.0 equiv. H$_2$O: 54% yield), while the ideal amount of H$_2$O (5 equiv.) gave 64% yield. The enantioselectivity (96% ee) and the diastereoselectivity (dr 10:1 syn/anti) of this reaction appeared to be independent of the amount of water [262].

The same group reported primary amine-functionalized diphenylethylenethiourea derivatives **104** and **105** to catalyze the enantioselective and diastereose-

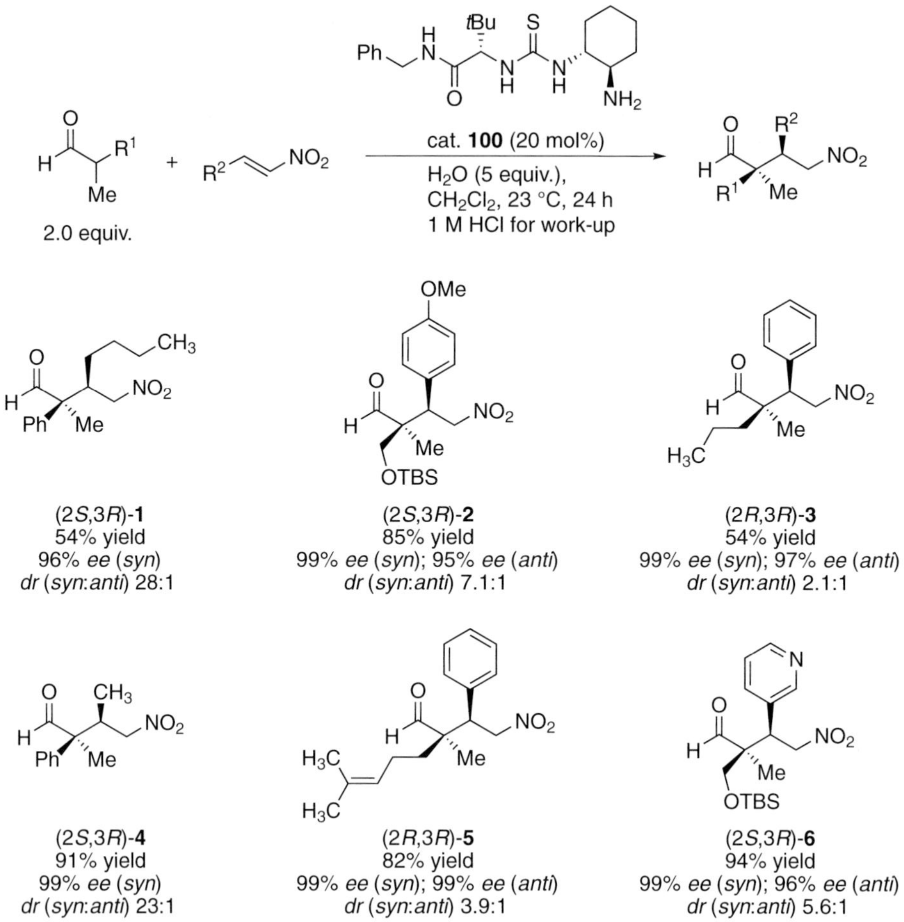
cat. 100 (20 mol%)
H2O (5 equiv.),
CH2Cl2, 23 °C, 24 h
1 M HCl for work-up
2.0 equiv.
(2S,3R)-1
54% yield
96% ee (syn)
dr (syn:anti) 28:1
(2S,3R)-2
85% yield
99% ee (syn); 95% ee (anti)
dr (syn:anti) 7.1:1
(2R,3R)-3
54% yield
99% ee (syn); 97% ee (anti)
dr (syn:anti) 2.1:1
(2S,3R)-4
91% yield
99% ee (syn)
dr (syn:anti) 23:1
(2R,3R)-5
82% yield
99% ee (syn); 99% ee (anti)
dr (syn:anti) 3.9:1
(2S,3R)-6
94% yield
99% ee (syn); 96% ee (anti)
dr (syn:anti) 5.6:1

Scheme 6.103 Representative products provided from the **100**-catalyzed asymmetric Michael addition of α,α-disubstituted aldehydes to aliphatic and aromatic nitroalkenes.

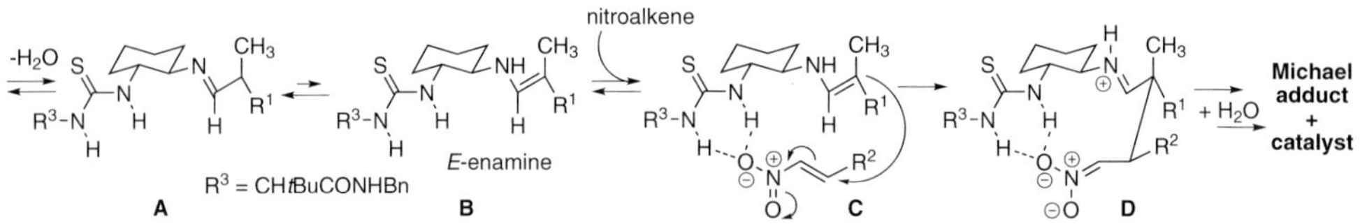
nitroalkene
-H2O
E-enamine
R3 = CHtBuCONHBn
A
B
C
D
+ H2O
Michael adduct
+
catalyst

Scheme 6.104 Key intermediates of the proposed catalytic cycle for the **100**-catalyzed Michael addition of α,α-disubstituted aldehydes to aliphatic and aromatic nitroalkenes: Formation of imine (**A**) and *E*-enamine (**B**), double hydrogen-bonding activation of the nitroalkene and nucleophilic enamine attack (**C**), zwitterionic structure (**D**), product-forming proton transfer, and hydrolysis.

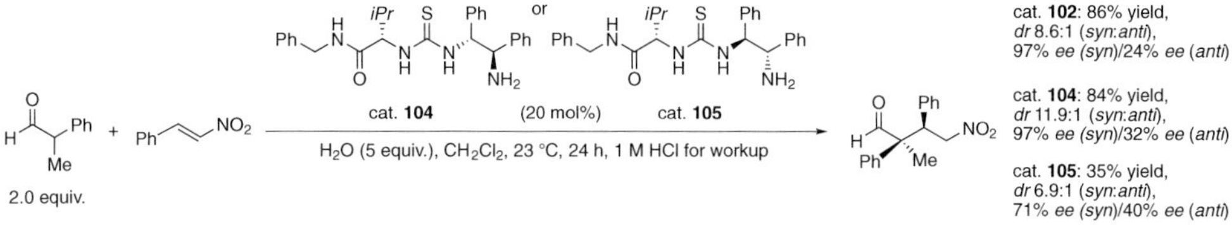
cat. 104
(20 mol%)
cat. 105
2.0 equiv.
H2O (5 equiv.), CH2Cl2, 23 °C, 24 h, 1 M HCl for workup
cat. 102: 86% yield, dr 8.6:1 (syn:anti), 97% ee (syn)/24% ee (anti)
cat. 104: 84% yield, dr 11.9:1 (syn:anti), 97% ee (syn)/32% ee (anti)
cat. 105: 35% yield, dr 6.9:1 (syn:anti), 71% ee (syn)/40% ee (anti)

Scheme 6.105 Asymmetric Michael addition of phenylpropionaldehyde to *trans*-β-nitrostyrene catalyzed by primary amine thioureas **102**, **104**, and **105**.

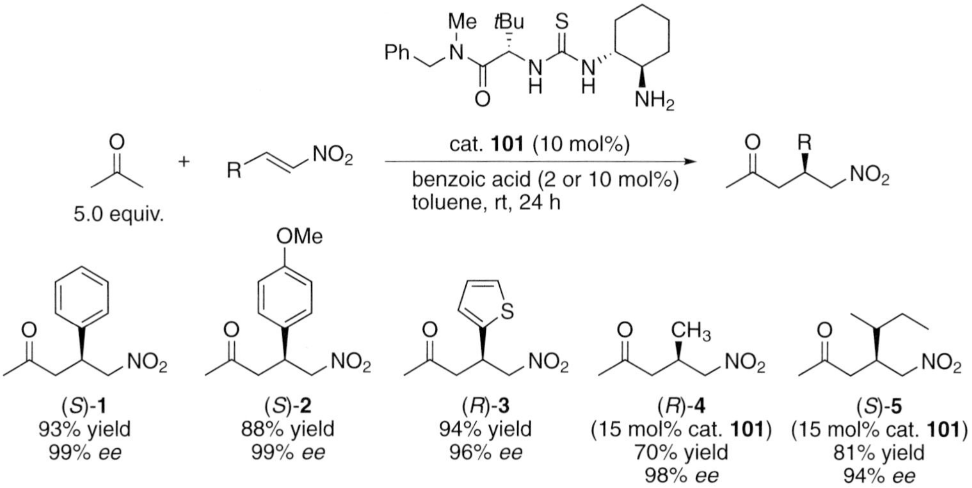
cat. 101 (10 mol%)
benzoic acid (2 or 10 mol%)
toluene, rt, 24 h
5.0 equiv.
(S)-1
93% yield
99% ee
(S)-2
88% yield
99% ee
(R)-3
94% yield
96% ee
(R)-4
(15 mol% cat. 101)
70% yield
98% ee
(S)-5
(15 mol% cat. 101)
81% yield
94% ee

Scheme 6.106 Typical product obtained from the Michael addition of acetone to *trans*-β-nitrostyrenes in the presence of catalyst **101**.

lective Michael addition of phenylpropionaldehyde to *trans*-β-nitrostyrene (Scheme 6.105).

Primary amine thiourea derivative **101** bearing a tertiary amide functionality was found by Huang and Jacobsen to catalyze the enantio- and diastereoselective Michael addition of ketones to *trans*-β-nitrostyrenes at 10 mol% standard catalyst loading and in the presence of benzoic acid (2–10 mol%) in toluene [263]. As shown in Scheme 6.106 for the addition of acetone to various nitroalkenes this protocol reached synthetically useful yields (70–94%) and excellent *ee* values (94–99%) of the corresponding adducts **1–5**. Catalyst **101** was identified to show a strong bias for the activation of ethyl ketones allowing highly regio- (up to *rr* 30 : 1) and *anti*-diastereoselective (up to *dr* 20.1) addition reactions of dialkyl ketones to β-alkyl and β-aryl nitroalkenes producing the respective Michael adducts in high enantioselectivities (86–99%) [263].

A bifunctional mechanism involving enamine catalysis [55, 58, 77] was clearly indicated in the Michael reactions promoted by catalyst **101**. The observed

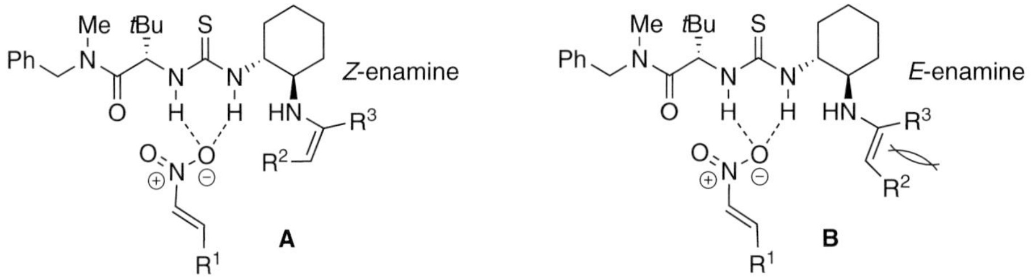

Figure 6.32 Proposed intermediates in the **100**-catalyzed Michael addition of ketones to nitroalkenes: favored Z-enamine (**A**) and disfavored E-enamine (**B**).

anti-diastereoselectivity suggested the participation of a Z-enamine intermediate (Figure 6.32) that provided the complementary diastereoselectivity obtained in analogous reactions involving E-enamines such as in **100**-catalyzed Michael reactions producing *syn*-adducts (Scheme 6.103) [262].

Nagasawa and co-worker, in 2004, introduced the first bis-thiourea-type catalyst **106** to accelerate the DMAP-catalyzed asymmetric MBH reaction [176, 177] between cyclohexenone and selected aliphatic and aromatic aldehydes [264]. The catalyst was reported to be readily accessible as a crystalline solid from (R,R)-diaminocyclohexane and 3,5-bis(trifluoromethyl)phenyl isothiocyanate (2 equiv.) in a high-yielding (94%) one-step procedure. At −5 °C reaction temperature and in the presence of 40 mol% DMAP as base additive bis-thiourea **106** (40 mol%) converted predominantly CF$_3$-substituted benzaldehydes as well as cyclic and acyclic aliphatic aldehydes with 2-cyclohexen-1-one into the corresponding (R)-configured MBH allyl alcohols **1–6** (Scheme 6.107). The obtained yields (38–99%/72 h) and enantioselectivities (19–90% *ee*) varied strongly with the aldehyde substrate. Due to the incorporation of two electron-deficient thiourea functionalities catalyst **106** was capable of providing twice double hydrogen-bonding interactions resulting in simultaneous activation of both electrophilic components of the aldehyde and the enone as depicted in Scheme 6.108. Since the aldehyde and the enone were coordinated in such a way that the organic residue R of the aldehyde was located on the opposite side of the thiourea group interacting with the enone the carbon–carbon bond forming nucleophilc attack of the base-activated enone preferably produced the (R)-MBH adduct (Scheme 6.108). The importance of bifunctionality in this tranformations was demonstrated by the use of monofunctional thiourea **107** that displayed low catalytic activity (**107**: 20% yield; **106**: 88%) in the MBH reaction between benzaldehyde and 2-cyclohexen-1-one (Scheme 6.108) [264].

β-Amino carbonyl compounds containing an α-alkylidene group are densely functionalized materials, which are widely applied in the synthesis of medicinal reagents and natural products [265]. These products are usually prepared through the classic aza-Morita–Baylis–Hillman reaction [176, 177] of activated imines and electron-deficient alkenes catalyzed by tertiary amines or phosphines. Chen and co-workers, in 2008, identified bis-thiourea **106** as a suitable catalyst for the

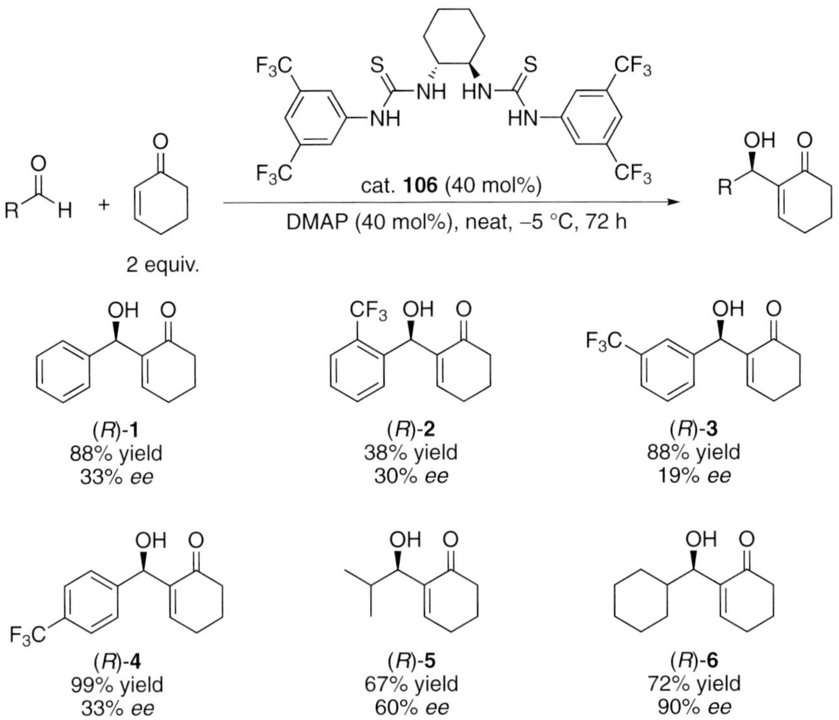

6.2 Synthetic Applications of Hydrogen-Bonding (Thio)urea Organocatalysts

Scheme 6.107 Product range of the **106**-catalyzed asymmetric MBH reaction between aldehydes and 2-cyclohexen-1-one.

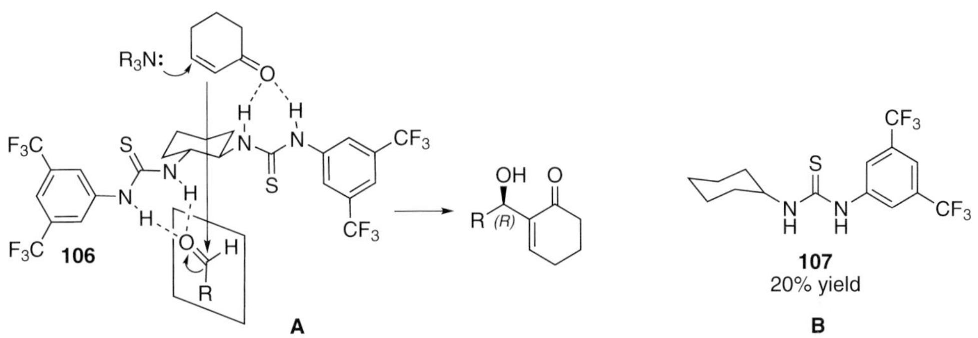

(R)-1, 88% yield, 33% ee
(R)-2, 38% yield, 30% ee
(R)-3, 88% yield, 19% ee
(R)-4, 99% yield, 33% ee
(R)-5, 67% yield, 60% ee
(R)-6, 72% yield, 90% ee

Scheme 6.108 Proposed mechanistic picture for the **106**-catalyzed MBH reaction affording (R)-adducts (**A**) and monofunctional thiourea **107** displaying low catalytic activity (**B**).

107 20% yield

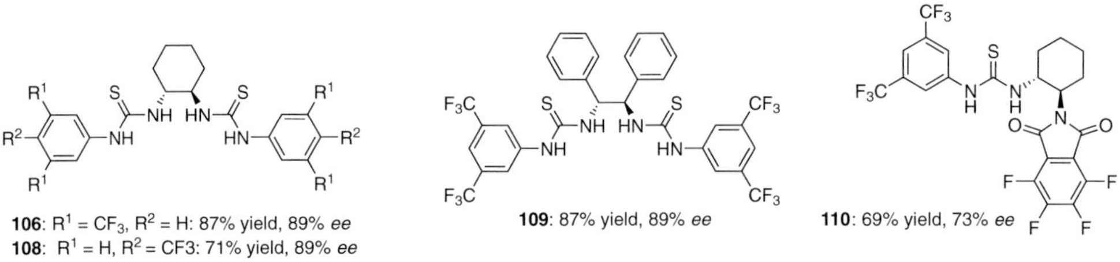

106: R¹ = CF₃, R² = H: 87% yield, 89% ee
108: R¹ = H, R² = CF3: 71% yield, 89% ee

109: 87% yield, 89% ee

110: 69% yield, 73% ee

Figure 6.33 Thiourea derivatives evaluated for catalytic efficiency in the Mannich addition of P-ylides to N-Boc-protected benzaldimine.

asymmetric Mannich addition reaction [72] of stabilized phosphorus ylides to activated N-Boc-protected aliphatic and (hetero)aromatic aldimines [266]. Alternative thiourea catalysts such as bis-thioureas **108** and **109** as well as the imide-functionalized thiourea **110** turned out to be less efficient in the screening Mannich reaction with N-Boc-protected benzaldimine (Figure 6.33). Under optimized conditions (m-xylene, MS 4Å; −20 °C; rt), bis-thiourea **106** promoted the nucleophilc P-ylide attack to various N-Boc-protected aldimines (Mannich reaction). The subsequent treatment of the Mannich adducts with formaldehyde at rt (Wittig reaction) gave access to the corresponding N-Boc-protected β-amino-α-methylene ethyl esters **1–6** in yields ranging from 37 to 85% and in ee values ranging from 57 to 96% (Scheme 6.109).

The Berkessel group, in 2006, introduced novel bis-(thio)urea catalysts **111–114** derived from isophosphoronediamine [3-(aminomethyl)-3,5,5-trimethyl cyclohexylamine, IPDA] for the enantioselective catalysis of the Morita–Baylis–Hillman reaction [176, 177] between aldehydes and 2-cyclohexen-1-one as well as 2-cylopenten-1-one [267]. IPDA is a readily available 1,4-diamine produced industrially on a multiton scale. IPDA and its derivative isophoronediisocyanate [5-isocyanato-1-(isocyanatomethyl)-1,3,3-trimethylcyclohexane, IPDI] are used as monomers for urethane and epoxy resins [268]. **111–114** were prepared in yields ranging from 60 to 85% (ureas) to 100% (thioureas) from IPDA and the respective iso(thio)cyanates in a straightforward one-step procedure. Catalyst evaluation (20 mol% loading) in the model MHB reaction between cyclohexanecarbaldehyde and 2-cyclohexen-1-one in the presence of DABCO as the nucleophilic promoter revealed bis-thiourea **114** bearing the privileged 3,5-bis(trifluoromethyl)phenyl thiourea moiety to be the most active (81% yield/72 h) and enantioselective (90% ee/10 °C) catalyst (Figure 6.34).

At 20 mol% loading of **114**, with DABCO (20 mol%), and under solvent-free conditions the desired MBH adducts **1–6** were obtained in very different yields (22–100%) and enantioselectivities (34–96%) (Scheme 6.110). Generally, aliphatic aldehydes gave higher ee values than aromatic aldehydes independent of the enone.

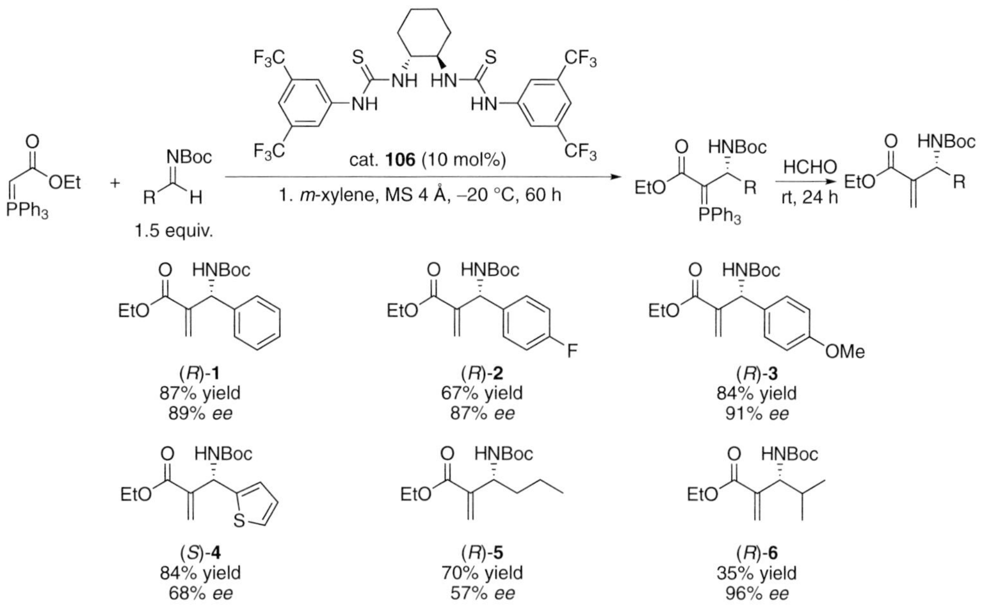

Scheme 6.109 Typical *N*-Boc-protected β-amino-α-methylene ethyl esters obtained from the **106**-catalyzed asymmetric Mannich reaction and subsequent Wittig reaction with formaldehyde.

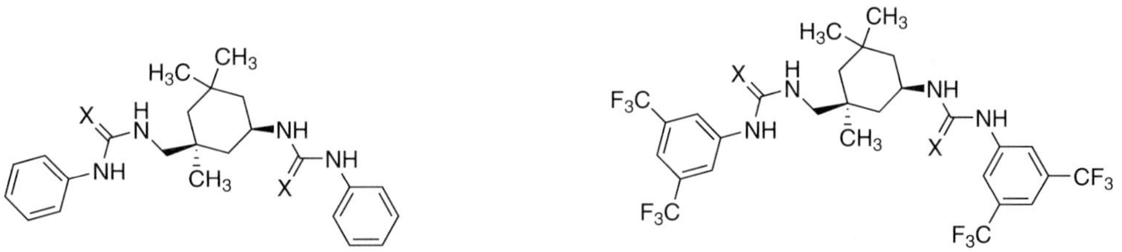

111 : X = O: 10% yield, 86% *ee*
112: X = S: 30% yield, 87% *ee*

113: X = O: 11% yield, 93% *ee*
114: X = S: 81% yield, 90% *ee*

Figure 6.34 Bis-(thio)ureas **111–114** derived from IPDA and results of the screening in the DABCO-promoted MBH reaction between cyclohexanecarbaldehyde and 2-cyclohexen-1-one under neat conditions at 10 °C.

6.2.2.2 **(Thio)ureas Derived from Cinchona Alkaloids**

Naturally occurring cinchona alkaloids and numerous analogs have been widely utilized in a broad spectrum of organic transformations as chiral auxiliaries, as ligands for transition-metal catalysis, as phase-transfer catalysts, and as organocatalysts (Figure 6.35) [139, 269]. Bredig and Fiske, in 1913, reported the first asymmetric organocatalytic reaction utilizing pseudoenantiomeric alkaloids

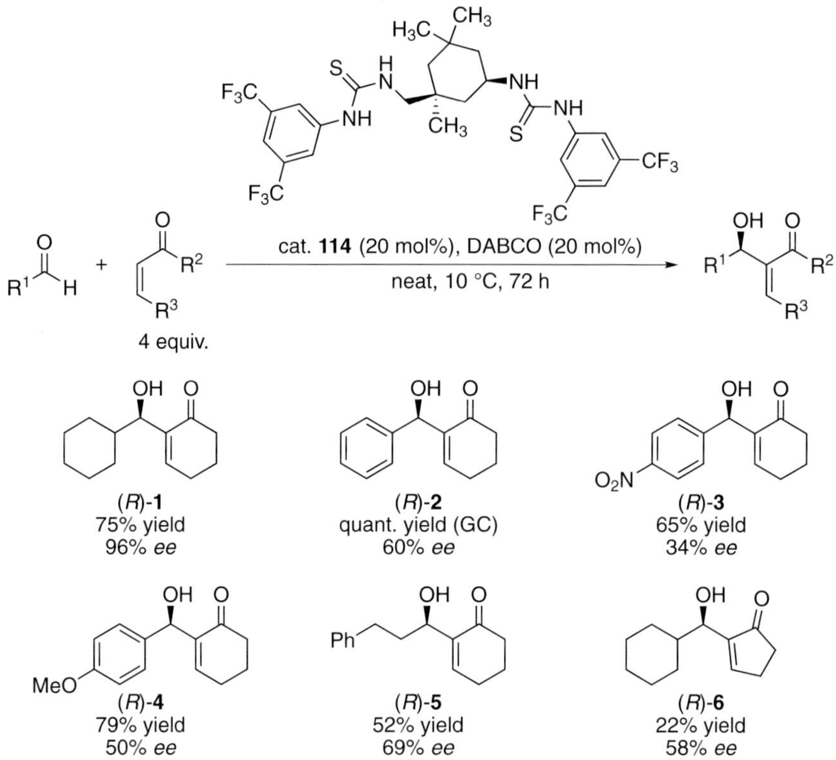

Scheme 6.110 Typical products of the **114**-catalyzed MBH reaction between various aldehydes and 2-cyclohexen-1-one as well as 2-cylopenten-1-one.

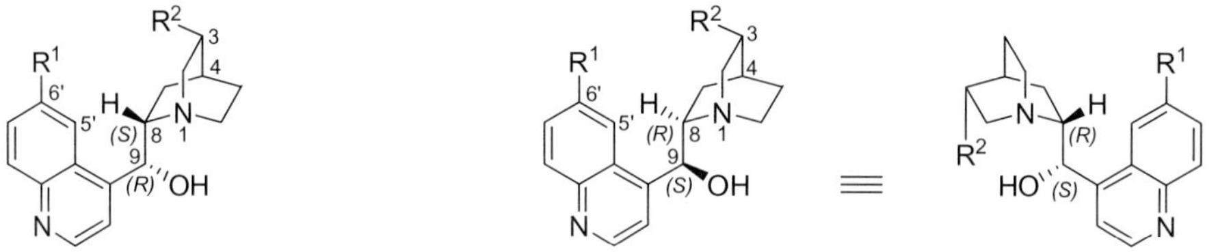

C3-(*R*), C4-(*S*) and **C8**-(*S*), **C9**-(*R*)

quinine (QN): R^1 = OMe, R^2 = vinyl
dihydroquinine (DHQ): R^1 = OMe, R^2 = Et
cinchonidine (CD): R^1 = H, R^2 = vinyl
cupreine (CPN): R^1 = OH, R^2 = vinyl

C3-(*R*), C4-(*S*) and **C8**-(*R*), **C9**-(*S*)

quinidine (QD): R^1 = OMe, R^2 = vinyl
dihydroquinidine (DHQD): R^1 = OMe, R^2 = Et
cinchonine (CN): R^1 = H, R^2 = vinyl
cupreidine (CPD): R^1 = OH, R^2 = vinyl

Figure 6.35 Starting materials for the synthesis of (thio)urea-functionalized cinchona alkaloids: The main cinchona alkaloids **QN**, **CD**, and their *pseudo*-enantiomers **QD** and **CN**, respectively, with opposite absolute configuration at the key stereogenic centers (N1, C8, C9) and identical absolute configuration in the quinuclidine fragment (C3, C4), the dihydrogenated derivatives **DHQ** and **DHQD** as well as the 6′-OH derivatives **CPN** and **CPD**.

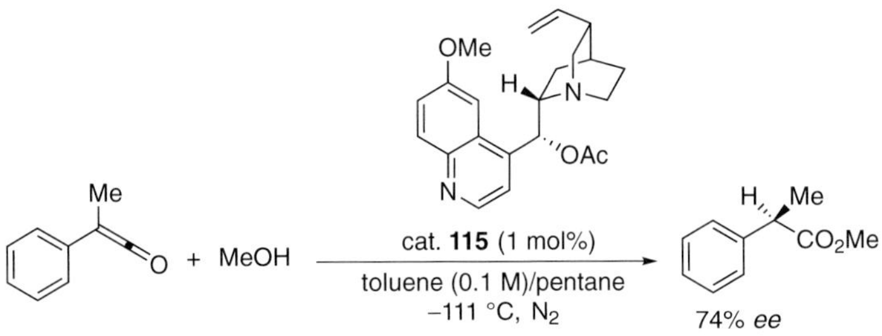

Scheme 6.111 Synthesis of (−)-α-phenyl methylpropionate in the presence of O-acetylquinine as catalyst.

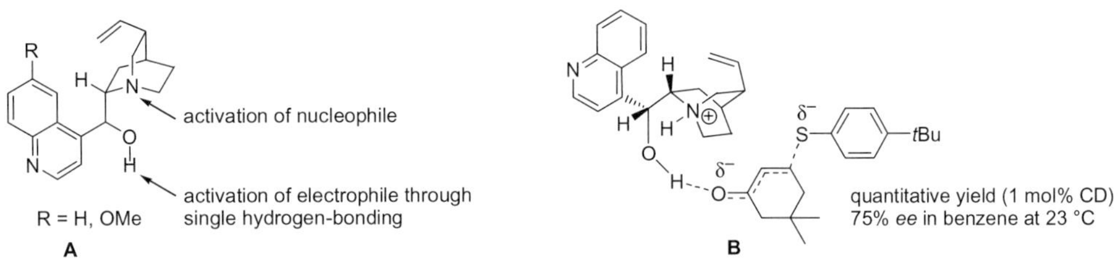

Figure 6.36 Bifunctionality of cinchona alkaloids (**A**) and Wynberg's proposal for the transition state of the cinchonidine-catalyzed Michael addition of 4-*tert*-butylthiophenol to 5,5-dimethyl-2-cyclohexenone (**B**).

quinine and quinidine as catalysts for the addition of HCN to benzaldehyde producing mandelonitrile in less than 10% *ee* [270].

The first synthetically useful levels of stereoinduction can be dated back to 1960, when Pracejus systematically studied the addition of various alcohols such as methanol to phenyl ketene. In the presence of O-acetylquinine **115** as catalyst, the respective (−)-α-phenyl methylpropionate was isolated in 74% *ee* in nearly quantitative yield (Scheme 6.111) [271].

Wynberg and Hiemstra, in 1981, published pioneering research results on cinchona alkaloids as chiral nucleophilic catalysts for the enantioselective 1,2- and 1,4-addition reactions to cycloalkenones [272]. Unmodified natural quinine, quinidine, cinchonine, and cinchonidine were found to promote the Michael addition reactions of aromatic thiols to, e.g., cyclohexenone in higher rates and enantioselectivities (up to 75% *ee*) than their derivatives acylated at the C9-OH group. These results led to the proposal that cinchona alkaloids are bifunctional catalysts operating through a mechanism in which the cyclic enone and the attacking thiol nucleophile are simultaneously activated through hydrogen-bonding (OH group) and deprotonation (quinuclidine nitrogen), respectively, as visualized in Figure 6.36 [64].

In 2005, various groups independently realized the potential of the easily available cinchona alkaloids as chiral templates for the synthesis of the new class

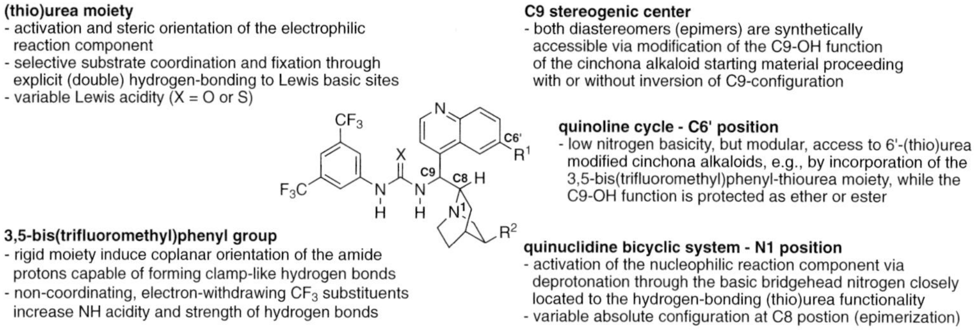

(thio)urea moiety
- activation and steric orientation of the electrophilic reaction component
- selective substrate coordination and fixation through explicit (double) hydrogen-bonding to Lewis basic sites
- variable Lewis acidity (X = O or S)

C9 stereogenic center
- both diastereomers (epimers) are synthetically accessible via modification of the C9-OH function of the cinchona alkaloid starting material proceeding with or without inversion of C9-configuration

quinoline cycle - C6' position
- low nitrogen basicity, but modular, access to 6'-(thio)urea modified cinchona alkaloids, e.g., by incorporation of the 3,5-bis(trifluoromethyl)phenyl-thiourea moiety, while the C9-OH function is protected as ether or ester

3,5-bis(trifluoromethyl)phenyl group
- rigid moiety induce coplanar orientation of the amide protons capable of forming clamp-like hydrogen bonds
- non-coordinating, electron-withdrawing CF_3 substituents increase NH acidity and strength of hydrogen bonds

quinuclidine bicyclic system - N1 position
- activation of the nucleophilic reaction component via deprotonation through the basic bridgehead nitrogen closely located to the hydrogen-bonding (thio)urea functionality
- variable absolute configuration at C8 postion (epimerization)

Figure 6.37 Design principles, functionalities, and characteristics of bifunctional H-bonding (thio)ureas derived from cinchona alkaloids.

bifunctional hydrogen-bonding (thio)urea organocatalysts utilizing cinchona alkaloids (Figure 6.35). The cinchona alkaloid backbone incorporates both a basic quinuclidine moiety and a secondary alcohol function in a well-defined chiral environment and offers easy modulation for an improvement of the bifunctional character. The basic catalyst design results from Wynberg's original proposal [272] that the C9-OH group of cinchonine and cinchonidine participate in electrophile activation hydrogen-bond donation and also from Takemoto's bifunctional hydrogen-bonding amine-thiourea catalyst **12** (see also Section 6.2.2.1) [129], which demonstrated the compatibility of the (thio)urea moiety with a Lewis basic site incorporated in one catalyst structure. Replacing the C9-OH group with a stronger hydrogen-bond donor moiety, such as the privileged 3,5-bis(trifluoromethyl)phenyl-thiourea group furnishes cinchona alkaloid-derived (thio)ureas exhibiting in the most cases higher activities as well as selectivities than the unmodified cinchona alkaloids (Figure 6.37). This section presents the catalytically efficient bifunctional hydrogen-bonding (thio)urea-alkaloids in the chronological order of the cited literature and separates principally into C9-(thio)ureas derived from cinchonidine, cinchonine, quinine, and quinidine; a few examples of alkaloids having the thiourea moiety at C6'-position are also considered (Figure 6.37).

The Chen group early in 2005 constituted the novel class of thiourea-functionalized cinchona alkaloids with the first reported synthesis and application of thioureas **116** (8R, 9S) and **117** (8R, 9R) prepared from cinchonidine and cinchonine in over 60% yield, respectively (Scheme 6.112) [273]. In the Michael addition of thiophenol to an α,β-unsaturated imide, the thioureas **116** and **117** displayed only poor stereoinduction (at rt **116**: 7% ee; **117**: 17% ee), but high catalytic activity (99% yield/2 h) (Scheme 6.112).

Dixon *et al.* screened cinchonine-derived thioureas **117–120** for their performance in the dimethyl malonate Michael addition to *trans*-β-nitrostyrene in dichloromethane at room temperature and at −20 °C [274]. As shown in Figure 6.38, all candidates revealed comparable activity, but monodentate hydrogen-bond donor **118** exhibited very low asymmetric induction producing the desired Michael

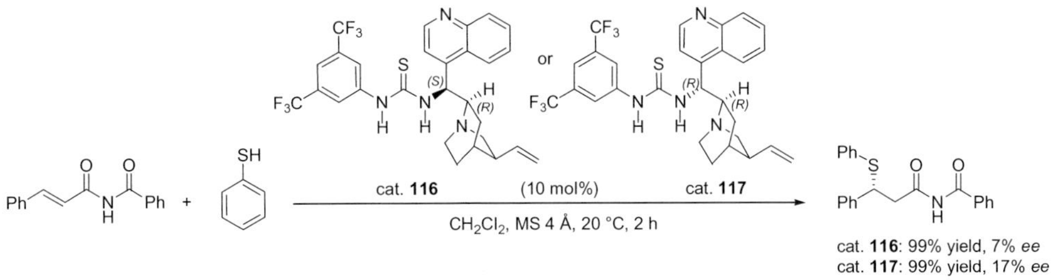

Scheme 6.112 Michael addition of thiophenol to an α,β-unsaturated imide catalyzed by cinchonidine-derived thiourea **116** and cinchonine-derived thiourea **117**, the first representatives of this class of bifunctional hydrogen-bonding cinchona alkaloid-thioureas.

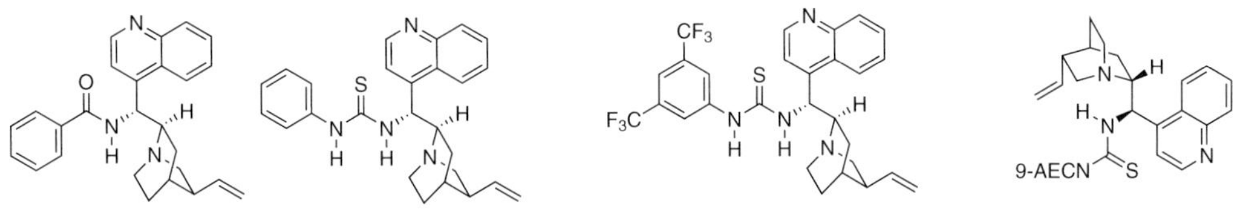

118	**119**	**117**	**117**	**120**
93% conv./72 h	96% conv./44 h	98% conv./22 h	98% conv./40 h	99% conv./44 h
8% *ee* (*R*)-adduct/rt	87% *ee* (*R*)-adduct/rt	87% *ee* (*R*)-adduct/rt	94% *ee* (*R*)-adduct/–20 °C	17% *ee* (*R*)-adduct/rt

Figure 6.38 Cinchonine-derived thioureas (10 mol% loading) screened in the Michael reaction of dimethyl malonate to *trans*-β-nitrostyrene.

adduct in only 8% *ee*. Derivative **117** was identified to be the most active (98% conv./40 h) and most selective (94% *ee*/–20 °C) catalyst.

The model reaction turned out to be independent of the choice of solvent, but due to solubilizing properties for substrates and products, the authors decided to study the Michael addition [149–152] of methyl malonates to various *trans*-β-nitroalkenes in dichloromethane. In the presence of 10 mol% of catalyst **117** at –20 °C, aromatic and heteroaromatic substrates were converted to the corresponding Michael adducts **1–8** in good to excellent yields (83–99%) and very good enantioselectivities ranging from 89 to 95% after practical reaction times (30–48 h) as shown in Scheme 6.113. Aliphatic nitroalkenes as exemplified by the formation of adduct **8** reacted more slowly and gave lower *ee* values; in the case of R = *t*Bu no conversion was detected (Scheme 6.113). The Dixon group extended the application of *epi*-cinchonine-thiourea **117** to the direct enantio- and diastereoselective Mannich reaction [72] of dialkylmalonates and β-ketoesters with *N*-Boc- and *N*-Cbz-protected aldimines [275]. The optimizing experiments were performed by adding acetylacetone to *N*-Boc benzaldimine using toluene as the solvent. With 10 mol% **117** at room temperature the Mannich adduct was isolated in quantitative yield with 37% *ee*. Reducing the reaction temperature to –78 °C for 72 h produced

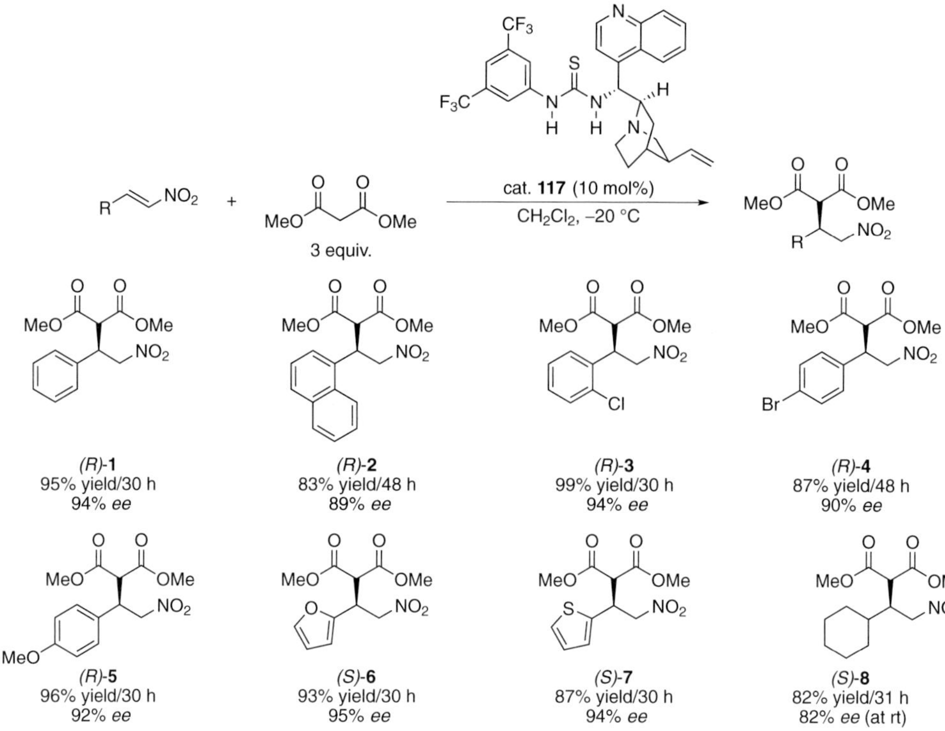

Scheme 6.113 Product range for the **117**-catalyzed Michael reaction of dimethyl malonate to various *trans*-β-nitrostyrenes.

the same adduct in 82% *ee*. *N*-Cbz-protected benzaldimine as model substrate was converted with acetylacetone to give the adduct in 73% yield and 86% *ee*. A range of malonates was investigated to determine potential nucleophiles; dimethyl malonate turned out to be the most reactive and produced 99% yield and 89% *ee* in the reaction with *N*-Boc-protected benzaldimine in toluene at −78 °C. In dichloromethane under otherwise identical conditions, the yield dropped to 76% (87% *ee*). The optimized protocol was applicable to a series of *N*-Boc- and *N*-Cbz-protected aromatic and heteroaromatic aldimines independent on the substituent pattern. Scheme 6.114 shows typical Mannich adducts **1–6** obtained after 3 d reaction time at −78 °C in the presence of catalyst **117** (10 mol%).

The construction of a quaternary α-stereocenters was demonstrated in the **117**-catalyzed Mannich addition [72] of methylcyclo-pentanone-2-carboxylate to *N*-Boc-protected aromatic aldimines and furnished the adducts **1–3** in 70–97% yield, with good *ee* values (85–87%) and diastereoselectivities (Scheme 6.115). The authors exemplified the synthetic utility of the protocol by a simple racemization-free decarboxylation of the Mannich adduct of *N*-Boc benzaldimine and dimethyl malonate to obtain the respective *N*-Boc-protected β-amino ester (68% yield/89%

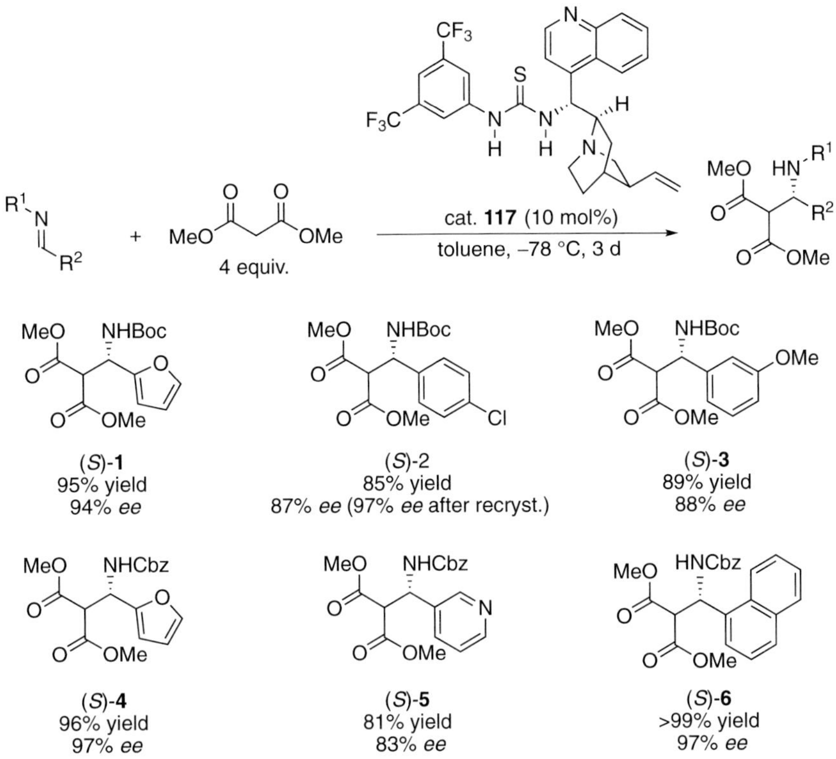

6.2 Synthetic Applications of Hydrogen-Bonding (Thio)urea Organocatalysts | 259

Scheme 6.114 Chiral Mannich adducts of the **117**-promoted reaction between dimethyl malonate and N-Boc as well as N-Cbz aldimines.

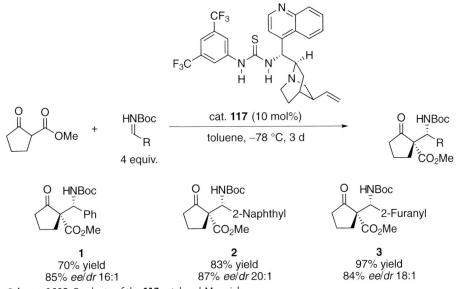

Scheme 6.115 Products of the **117**-catalyzed Mannich addition of methylcyclo-pentanone-2-carboxylate to N-Boc-protected aldimines.

ee/12 h at 160 °C), which are precursors of β-amino acids. Bifunctional hydrogen-bonding **117** also revealed catalytic efficiency in the enantio- and diastereoselective Michael addition reaction between 2,2-bis(trifluoromethyl)-substituted 5-aryl-1,3-dioxolan-4-ones, which are easily accessible from mandelic acid derivatives and hexafluoreacetone, and electron-rich as well as electron-deficient aromatic and heteroaromatic nitroalkenes [276]. The Dixon group identified this type of dioxolan-4-ones as new and efficient prenucleophiles that were deprotonated by the basic quinuclidine nitrogen at the acidic benzylic α-position and produced the enolate attacking the *trans*-β-nitroalkene in the product-forming step. The optimized protocol allowed the Michael addition [149–152] of various enolizable dioxolan-4-ones to *trans*-β-nitroalkenes at 0 °C and 5 mol% loading of **117** producing the Michael products **1–10** in moderate (58%) to very good yields (92%), with moderate to good enantioselectivities (60–89% ee), and very good diastereoselectivities (>93% de) (Scheme 6.116). The preferred (*R*)-configuration of the Michael

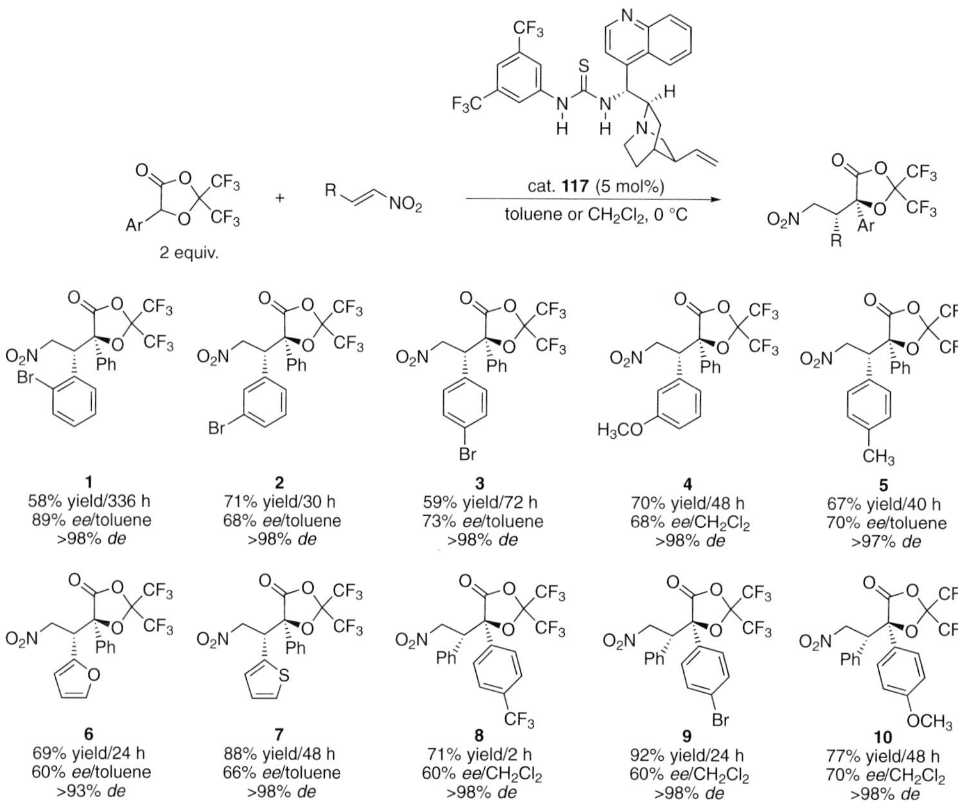

Scheme 6.116 Adducts of the **117**-catalyzed Michael reaction between 5-aryl-1,3-dioxolan-4-ones and various *trans*-β-nitroalkenes.

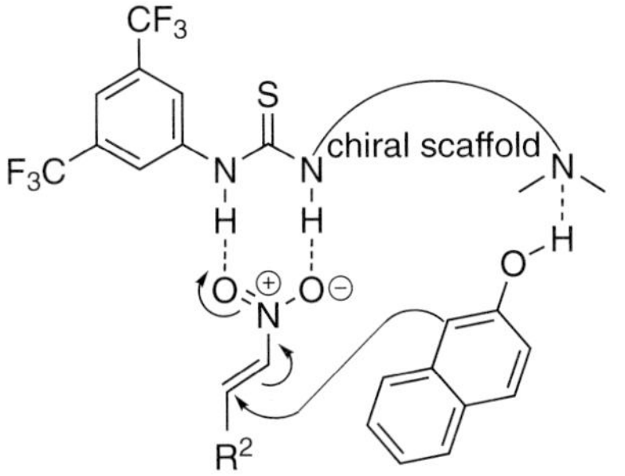

Scheme 6.117 Proposed bifunctional activation of the reactants through catalyst **117** in the asymmetric Michael-type Friedel–Crafts alkylation of 2-naphthols.

adducts, already observed for the malonate addition (Scheme 6.113) [274], was confirmed by an efficient multistep transformation of one example to its (R,R,R)-α-hydroxy acid derivative and subsequent X-ray analysis of the hydroxy acid crystals. This result supported the addition of the dioxolan-4-one nucleophile to the re-face of the trans-β-nitroalkene.

Chen and co-workers presented, in 2007, a Michael-type Friedel–Crafts reaction of 2-naphthols and trans-β-nitroalkenes utilizing the bifunctional activating mode of cinchonine-derived catalyst **117** [277]. The nitroalkene was activated and sterically orientated by double hydrogen bonding, while the tertiary amino group interacts with the naphthol hydroxy group to activate the naphthol for the nucleophilic β-attack at the Michael acceptor nitroalkene (Scheme 6.117).

Employing 10 mol% loading of **117** at −50 °C in toluene this Friedel–Crafts protocol allowed the synthesis of the products **1–8** in yields ranging from 69 to 83% with good to very good ee values (85–95%) (Scheme 6.118). The authors identified a dimeric tricyclic hydroxyl amine derivative as a side product (yield <10%), when conducting the reaction at optimized conditions (96 h reaction time). Performing the Friedel–Crafts alkylation at longer reaction time (144 h) under otherwise unchanged conditions, the side product became the major product in moderate yields (52–67%) with excellent stereochemical induction (99.5% ee; 99.5% dr).

The Soós group, in 2005, prepared the first thiourea derivatives from the cinchona alkaloids quinine **QN** (8S, 9R-**121**), dihydroquinidine **DHQD** (8S, 9S-**122**), C9-epi-**QN** (8S, 9R-**123**), and quinidine **QD** (8R, 9R-**124**) via an experimentally simple one-step protocol with epimerization at the C9-position of the alkaloid starting material (Figure 6.39) [278]. The catalytic efficiency of these new thiourea derivatives and also of unmodified **QN** and C9-epi-**QN** was evaluated in the enantioselective Michael addition [149–152] of nitromethane to the simple model chalcone 1,3-diphenyl-propenone resulting in adduct **1** in Scheme 6.119. After 99 h reaction time at 25 °C in toluene and at 10 mol% catalyst loading **QN** turned out to be a poor catalyst (4% yield/42% ee; (S)-adduct) and C9-epi-**QN** even failed to accelerate the screening reaction. In contrast, the C9-modified cinchona alkaloid

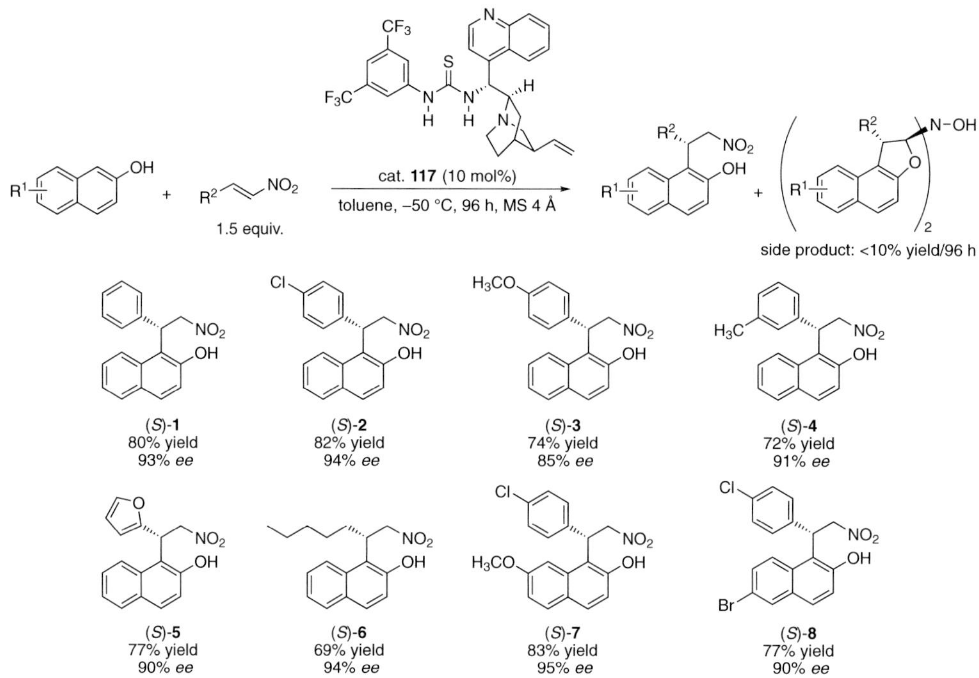

Scheme 6.118 Products resulting from the **117**-catalyzed Michael-type Friedel–Crafts alkylation of 2-naphthols and observed side products.

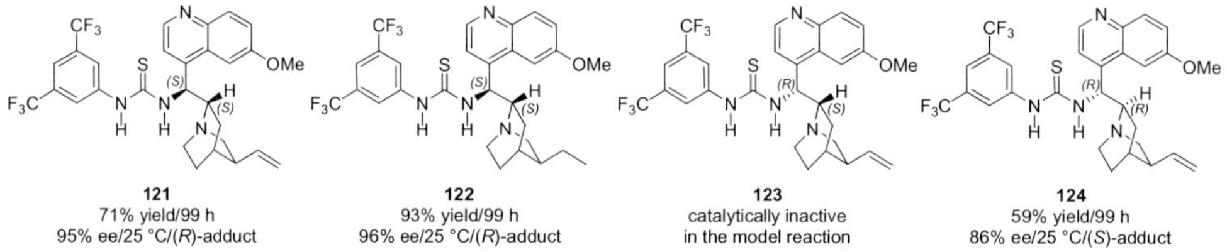

Figure 6.39 Cinchona alkaloid-thioureas prepared from quinine (**121**), dihydroquinine (**122**), C9-*epi*-quinine (**123**), and quinidine (**124**); catalytic efficiency evaluated in the Michael addition of nitromethane to *trans*-chalcone 1,3-diphenyl-propenone at 10 mol% loading and rt.

thiourea **121** (8*S*, 9*S*) and its pseudoenantiomer **124** (8*R*, 9*R*) revealed catalytic activity (**121**: 71% yield; **124**: 59% yield) and selectivity (**121**: 95% *ee*, (*R*)-adduct **1**; **124**: 86% *ee*, (*S*)-adduct **1**). **DHQ**-derived thiourea **122** (8*S*, 9*S*) was identified to be the most active (93% yield) and selective (96% *ee*, (*R*)-adduct **1**) candidate (Figure 6.39). It is remarkable that C9-*epi*-**QN**-derived thiourea **123** having the natural (8*S*, 9*R*)-configuration of the alkaloid turned out to be inactive in this reaction, while

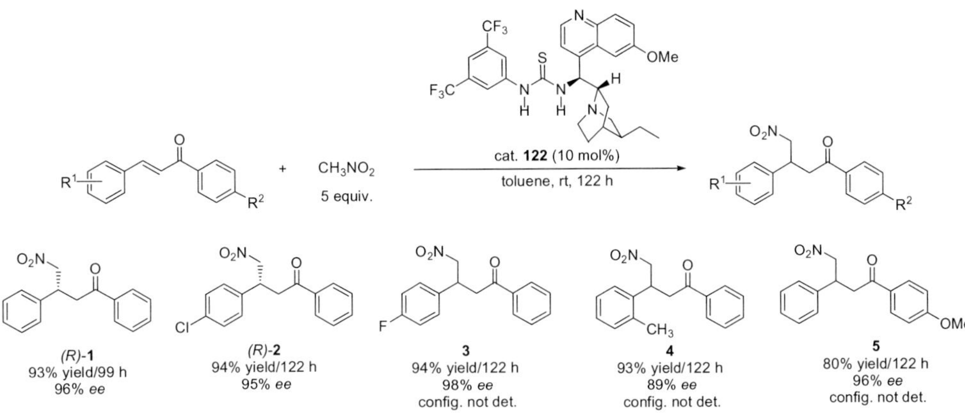

6.2 Synthetic Applications of Hydrogen-Bonding (Thio)urea Organocatalysts | 263

Scheme 6.119 Michael adducts from the asymmetric addition of nitromethane to *trans*-chalcones catalyzed by **122**.

Reaction: R¹-aryl chalcone + CH₃NO₂ (5 equiv.) → Michael adduct, cat. **122** (10 mol%), toluene, rt, 122 h.

Products:
- (R)-**1**: 93% yield/99 h, 96% ee
- (R)-**2** (Cl): 94% yield/122 h, 95% ee
- **3** (F): 94% yield/122 h, 98% ee, config. not det.
- **4** (CH₃): 93% yield/122 h, 89% ee, config. not det.
- **5** (OMe): 80% yield/122 h, 96% ee, config. not det.

the C9-modified candidates **121**, **122**, and **124** with the unnatural C9-stereochemistry (C9-epimerized) exhibited conversion as well as stereoinduction [279].

These findings strongly indicated that a well-defined relative stereochemical arrangement of the hydrogen-bonding thiourea moiety and the basic sites in cinchona alkaloid derived thioureas was an essential structural prerequisite to enable synergetic operation of these functionalities and thus for stereodifferentiating bifunctional catalysis. Additionally, it became evident from these results that the quinuclidine subunit of the cinchona alkaloids themselves was not able to facilitate the model reaction and that incorporation of the thiourea moiety drastically improved catalytic activity and selectivity. Under optimized reaction conditions (10 mol% loading; in toluene at 25 °C), thiourea **122** catalyzed the Michael additions of some electron-rich and electron-deficient chalcones providing the respective products **1–5** in synthetically useful yields (80–94%) and very good *ee* values (95–96%) as depicted in Scheme 6.119.

Connon and McCooey, shortly after the report published by the Soós laboratory [278], presented their independently performed investigations focusing on the catalytic efficiency of (thio)urea derivatives prepared from the dihydrogenated cinchona alkaloids dihydroquinine and dihydroquinidine (Figure 6.40) [279]. The catalyst evaluation in the Michael addition [149–152] of dimethyl malonate to *trans*-β-nitrostyrene demonstrated the effect of the relative stereochemistry at the C8- and C9-position of these materials on the catalyst performance. Neither epimerization of dihydroquinine (C9-*epi*-**DHQ**: 46% conv./144 h/18% *ee* at 5 mol% loading) nor the substitution with an N-arylurea moiety (**125**: 26% conv./25% *ee*/24 h at 5 mol% loading) increased catalyst activity which dropped from 98% conv. to 26% conv. with a small increase of the stereoinduction from 12% *ee* to 25% *ee* in comparison to unmodified **DHQ** (Figure 6.40). The combination of both cinchona alkaloid modifications, however, resulted in more active and selective catalysts; urea C9-*epi*-**126** showed 98% conv. (88% *ee*) after 24 h and thiourea

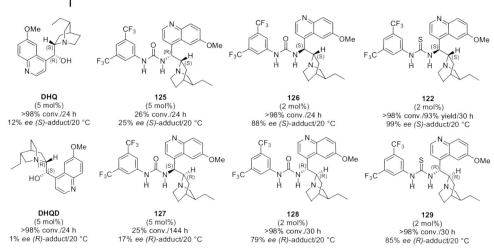

Figure 6.40 (Thio)urea catalysts derived from dihydroquinine and dihydroquinidine; screening results obtained from the asymmetric Michael addition of dimethyl malonate to *trans*-β-nitrostyrene.

C9-*epi*-**122** 98% conv. (99% *ee*) after 30 h, respectively (Figure 6.40). This structure–efficiency relationship supported the results already published by the Soós group for quinine- and quinidine-derived thioureas (Figure 6.39) [278]. C9-epimeric catalysts were found to be remarkably more efficient in terms of rate acceleration and stereoinduction than the analogs of natural cinchona alkaloid stereochemistry. This trend was also observed for the corresponding (thio)ureas derived from **DHQD** as shown by the experimental results in Figure 6.40 [279].

In the presence of thiourea catalyst **122**, the authors converted various (hetero) aromatic and aliphatic *trans*-β-nitroalkenes with dimethyl malonate to the desired (*S*)-configured Michael adducts **1–8**. The reaction occurred at low **122**-loading (2–5 mol%) in toluene at −20 to 20 °C and furnished very good yields (88–95%) and *ee* values (75–99%) for the respective products (Scheme 6.120). The dependency of the catalytic efficiency and selectivity on both the presence of the (thio)urea functionality and the relative stereochemistry at the key stereogenic centers C8/C9 suggested bifunctional catalysis, that is, a quinuclidine-moiety-assisted generation of the deprotonated malonate nucleophile and its asymmetric addition to the (thio)urea-bound nitroalkene Michael acceptor [279].

In 2006, the Connon group reported a one-pot methodology for the stereoselective synthesis of highly functionalized nitrocyclopropanes [280]. This protocol utilized **DHQ**-derived thiourea **122**, which was introduced by Soós *et al.* (Figure 6.39) [278], to catalyze the initial Michael addition [149–152] of dimethyl chloromalonate to various *trans*-β-nitroalkenes followed by a DBU-mediated cyclization in the presence of HMPA as the solvent. At −30 °C and 2 mol% loading of **122** the reaction sequence proceeded highly diastereoselectively (98% *de*) and gave moder-

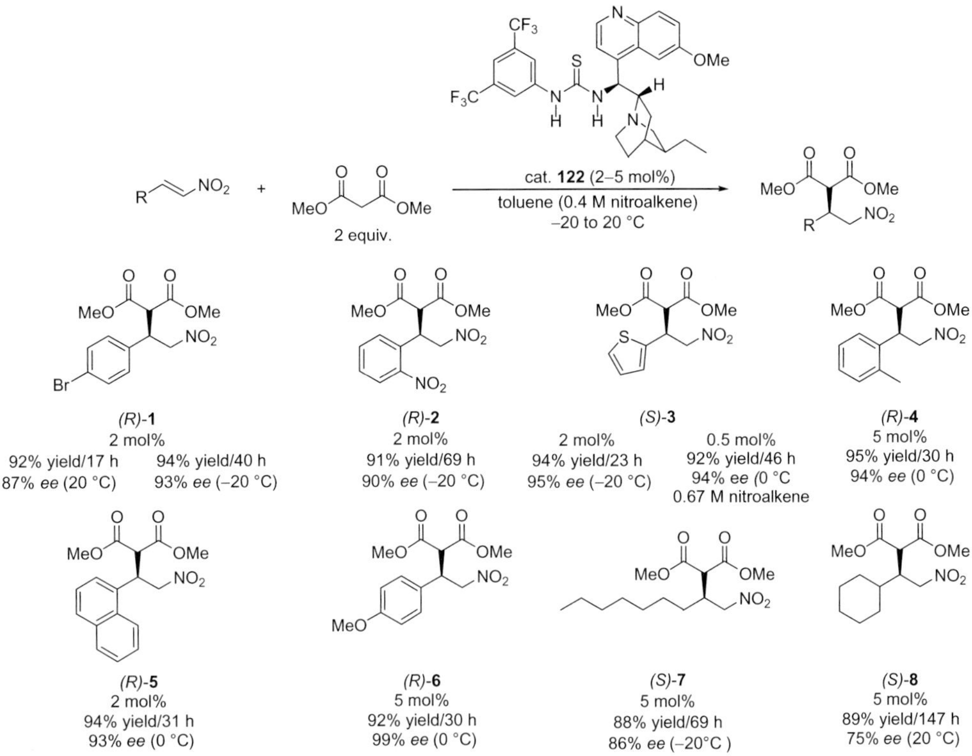

6.2 Synthetic Applications of Hydrogen-Bonding (Thio)urea Organocatalysts | 265

Scheme 6.120 Michael adducts prepared from the **122**-catalyzed asymmetric addition of dimethyl malonate to *trans*-β-nitroalkenes.

ate to good yields (64–69%) of the corresponding nitrocyclopropanes **1–8**; the *ee* values were low and ranged from 14–47% as depicted in Scheme 6.121.

In 2006, Schaus *et al.* identified acetyl-protected (hetero)aromatic aldimines as suitable substrates for the **122**-catalyzed asymmetric aza-Henry [224] (nitro-Mannich) additions [72] of nitromethane and nitroethane producing the corresponding β-nitroamines **1–10** in yields ranging from 60 to 98% with enantioselectivities of 90–98% and diastereoselectivities of 83–97% (Scheme 6.122) [281]. Replacing the nitroalkane with dimethyl malonate as alternative nucleophile, a range of acylated (hetero)aromatic aldimines were enantioselectively (86–94% *ee*) converted to the respective Mannich adducts **1–5** (65–98% yield) under thiourea **122** catalysis (Scheme 6.123). The authors also developed a new method for transforming the obtained Mannich adducts to the respective β-amino esters; as exemplified for adduct **1** in Scheme 6.123 microwave irradiation at 160 °C gave the corresponding ester in 80% yield and 85% *ee* after only 10 min reaction time.

The utility of C9-*epi*-**DHQ** thiourea **122** for the catalysis of asymmetric Michael reactions [149–152] was further demonstrated by the Wang group [282] in 2006.

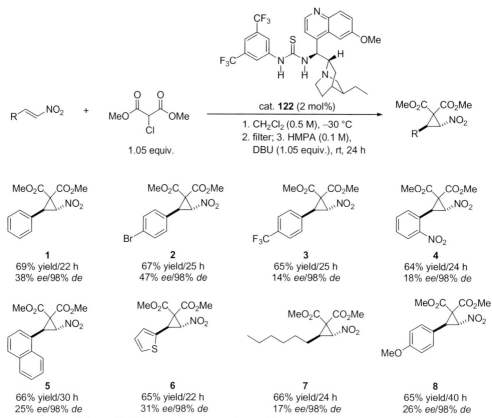

Scheme 6.121 Nitrocyclopropanes obtained from **122**-catalyzed Michael addition of dimethyl chloromalonate to trans-β-nitroalkenes and subsequent DBU-promoted cyclization.

The presented protocol tolerated a spectrum of enolizable species including dialkyl malonates, 1,3-diketone, ketoester, 1,3-dinitriles, and nitroesters to be added to the model trans-chalcone 1,3-diphenyl-propenone producing the desired Michael adducts in good to excellent yields (67–99%) and with attractive ee values (88–93%). Figure 6.41 shows some selected results of the nucleophile evaluation.

Diethyl malonate was found to be the most suitable nucleophile and was used to probe the scope and limitations of the catalyzed addition to various electron-rich and electron-deficient trans-chalcones. With optimized reaction parameters including the choice of bifunctional thiourea catalyst **122**, xylenes as reaction medium (e.g., THF: 40% yield; 81% ee; Et$_2$O: 35% yield; 53% ee; toluene: 86% yield; 90% ee), catalyst loading (10 mol%), and reaction temperature (rt) the reaction furnished the target adducts **1–8** in yields of 61–97% and ee values ranging from 85 to 98% (Scheme 6.124). Chalcones bearing a methyl group on the carbonyl unit proved

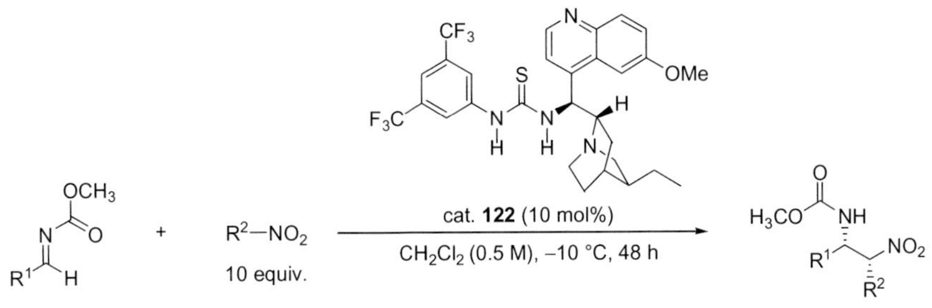

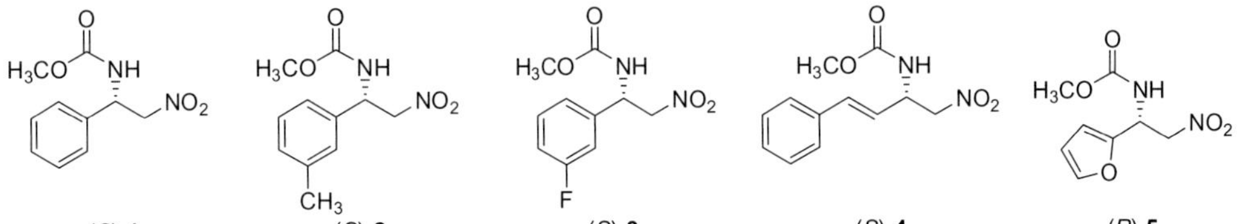

Scheme 6.122 Products of the **122**-catalyzed aza-Henry (nitro-Mannich) addition of nitromethane and nitroethane to acylated aldimines.

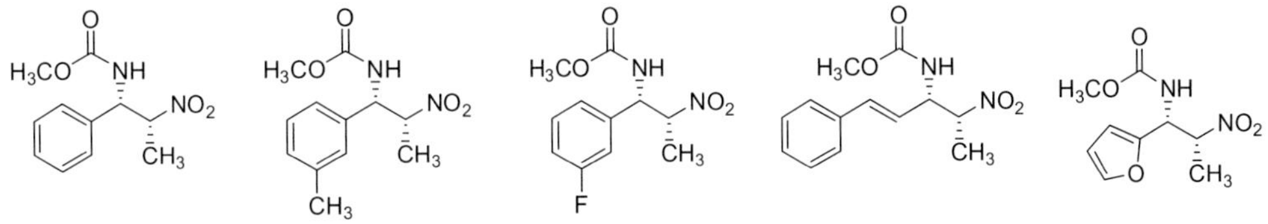

(S)-**1**
91% yield/24 h
93% ee

(S)-**2**
98% yield/48 h
91% ee

(S)-**3**
98% yield/48 h
98% ee

(S)-**4**
80% yield/48 h
90% ee

(R)-**5**
60% yield/48 h
92% ee

(1S,2R)-**6**
96% yield/24 h
94% ee/83% de

(1S,2R)-**7**
98% yield/48 h
97% ee/90% de

(1S,2R)-**8**
98% yield/48 h
91% ee/97% de

(1S,2R)-**9**
80% yield/48 h
90% ee/92% de

(1R,2R)-**10**
73% yield/48 h
97% ee/92% de

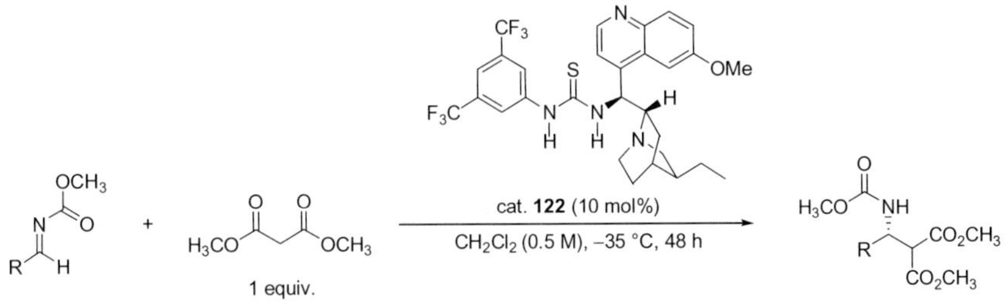

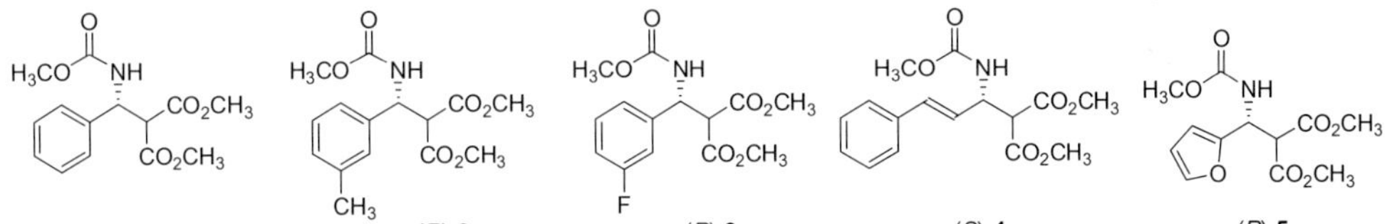

Scheme 6.123 Spectrum of adducts of the **122**-catalyzed asymmetric Mannich addition of dimethyl malonate to acylated aldimines.

(R)-**1**
98% yield/24 h
92% ee

(R)-**2**
97% yield/48 h
86% ee

(R)-**3**
98% yield/48 h
90% ee

(S)-**4**
97% yield/48 h
90% ee

(R)-**5**
65% yield/48 h
94% ee (−50 °C)

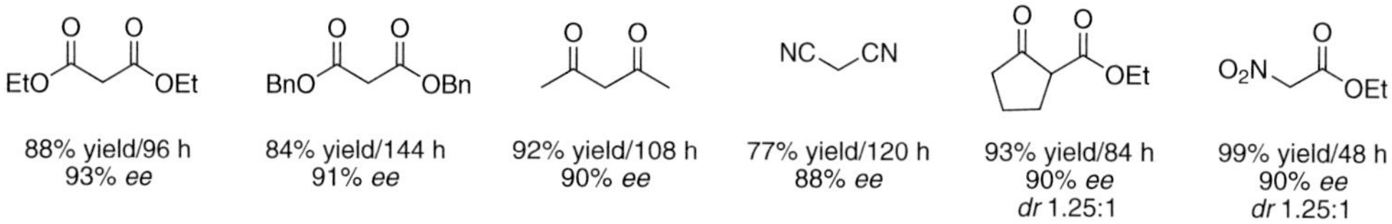

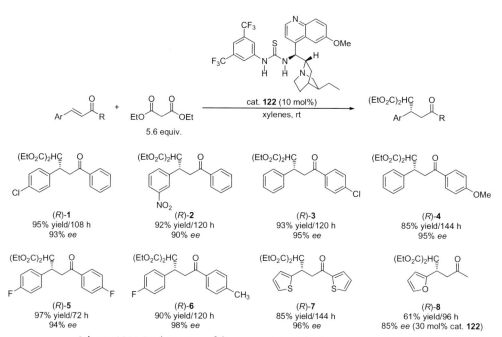

Figure 6.41 Enolizable compounds screened for the **122**-catalyzed asymmetric Michael addition to 1,3-diphenyl-propenone.

Scheme 6.124 Product range of the asymmetric Michael addition of diethyl malonate to various *trans*-chalcones promoted by **122**.

to be less reactive and desired 30 mol% loading under otherwise unchanged conditions to be converted to the respective Michael adduct in reasonable reaction times.

In 2007, Chen and co-workers reported the **122**-catalyzed (10 mol% loading) enantioselective Michael addition [149–152] of ethyl α-cyanoacetate to various electron-rich and electron-deficient *trans*-chalcones [283]. The reaction was performed for a broad spectrum of chalcones and gave the corresponding adducts in yields of 80–95% and in *ee* values of 83–95%, but at low *syn/anti*-diastereoselectivities as shown for representative products **1–8** in Scheme 6.125.

In 2006, Deng and co-workers identified **QN**-derived thiourea **121** and **QD**-derived thiourea **124**, which were already introduced by the Soós group for the

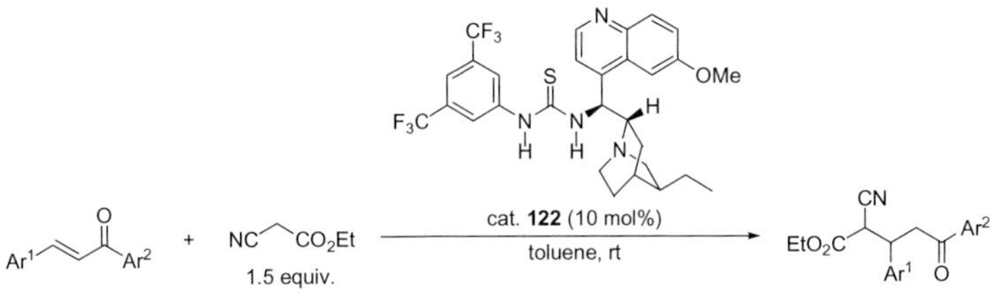

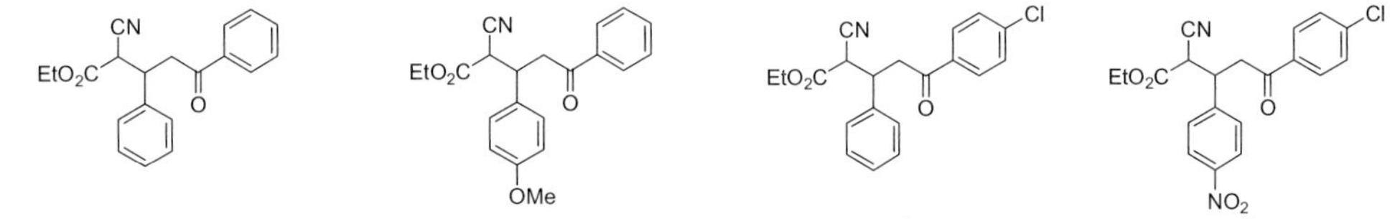

1
84% yield (*syn*/*anti* 63:37)/144 h
88:89% *ee* (*syn*/*anti*)

2
82% yield (*syn*/*anti* 59:41)/144 h
83:83% *ee* (*syn*/*anti*)

3
88% yield (*syn*/*anti* 62:38)/72 h
92:94% *ee* (*syn*/*anti*)

4
92% yield (*syn*/*anti* 60:40)/24 h
89:86% *ee* (*syn*/*anti*)

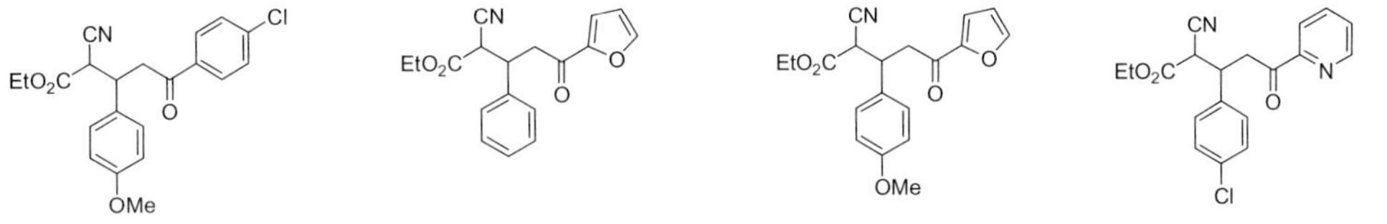

5
84% yield (*syn*/*anti* 59:41)/96 h
88:88% *ee* (*syn*/*anti*)

6
95% yield (*syn*/*anti* 67:33)/24 h
93:92% *ee* (*syn*/*anti*)

7
94% yield (*syn*/*anti* 61:39)/24 h
93:93% *ee* (*syn*/*anti*)
(99:99% *ee* after recryst.)

8
80% yield (*syn*/*anti* 60:40)/20 h
94:95% *ee* (*syn*/*anti*)

Scheme 6.125 Typical products of the **122**-catalyzed enantioselective Michael addition of ethyl α-cyanoacetate to various chalcones. The product configurations were not determined.

Michael addition [149–152] of nitromethane to *trans*-chalcones (Figure 6.39) [278], to efficiently catalyze the enantioselective Mannich reaction [72] of *N*-Boc-protected (hetero)aromatic and aliphatic aldimines [284]. The choice of acetone as a solvent at −60 °C reaction temperature resulted from the initial addition reaction of dimethyl malonate to the *N*-Boc-protected aldimine of 4-methyl-benzaldehyde. The optimized protocol utilizing thioureas **121** and **125** at 10 mol% loading was reported to be high yielding (81–99%, 36 h) and highly enantioselective (92–99%) in the case of (hetero)aromatic substrates producing the respective Mannich adducts **1–8** depicted in Scheme 6.126. To sustain a useful level of enantioselectivity for α-branched *N*-Boc-protected aldimine substrates, e.g., the *N*-Boc-protected aldimine of cyclohexane carbaldehyde, the catalyst loading of **124** was increased to 100 mol%; however, **124** could be recycled in more than 95% yield after the reaction (Scheme 6.126). Since both catalysts **121** and **124** tolerated malonates of different bulk (Scheme 6.126) and also various β-ketoester as nucleophiles for the asymmetric Mannich addition to *N*-Boc-protected aldimines, this protocol

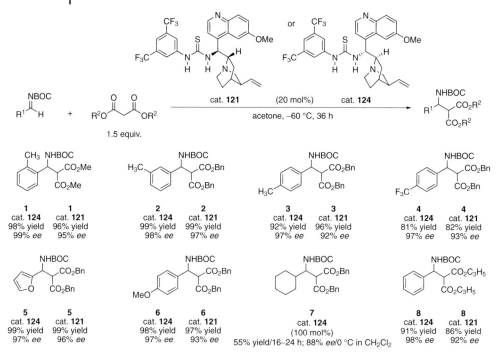

Scheme 6.126 Mannich adducts obtained from the **121**- and **124**-catalyzed asymmetric addition of dialkyl malonates to N-Boc aldimines. The product configurations were not determined.

provided access to a wide variety of optically active β-amino ketones and β-amino acids [284].

The same group developed a methodology for the enantioselective Friedel–Crafts reaction between indols and N-protected aldimines to construct the 3-indolyl methanamine structural motif [285]. In the presence of 10 mol% of thiourea **121** or **124**, various indols reacted with brosyl(Bs)- and tosyl(Ts)-N-protected aromatic as well as aliphatic aldimines to give the corresponding Friedel–Crafts adducts **1–8** (Scheme 6.127) in yields ranging from 53 to 98% and with high ee values (83–97%), although the standard reaction temperature was remarkably high (50 °C). The stereoinduction was found to be insensitive to the electronic properties of the indole ring and to the N-protective group of the alkylating aldimines. Due to the efficient selectivity of **121** and **124** even at increased temperature electron-deficient 6-bromo-1H-indole could be alkylated with an electron-rich N-tosyl aldimine giving the desired Friedel–Crafts adduct **6** in good yields (83%; 86%) and ee values (83%; 86%) (Scheme 6.127).

Ricci and co-workers published a protocol for the enantioselective aza-Henry reaction [224] of N-protected aldimines with nitromethane in the presence of C9-epi-quinine thiourea **121** [8]. The reaction was optimized for 20 mol% loading of

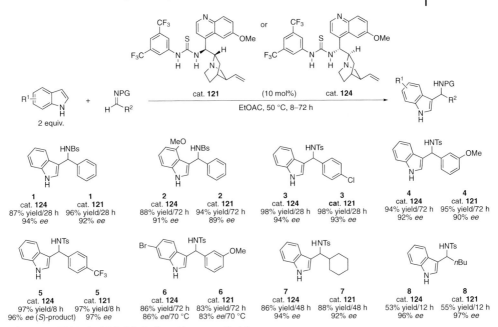

Scheme 6.127 Typical Friedel–Crafts adducts resulted from the **121**- and **124**-catalyzed alkylation of indols with N-protected aldimines. The configurations of the products were not determined.

121 at −24 °C in toluene and converted preferably N-Boc-protected (hetero) aromatic aldimines to the desired N-protected β-nitroamine adducts **1–6** in moderate to very good yields ranging from 50 to 95% and with synthetically useful levels of enantioselectivity (63–94%). 2-Naphthaldehyde-derived imine (adduct **1**; Scheme 6.127) and benzaldimine derivatives having both electron-donating and electron-withdrawing substituents proved to be suitable substrates and the tolerance of the Cbz-group was exemplified by the formation of adduct **3** (58% yield; 90% ee; 48 h). Among the aromatic heterocyclic aldimines, the 2-thiophenecarboxyaldehyde-derived imine (adduct **5**: 50% yield; 82% ee; 40 h) gave better results than its oxygenated analog (adduct **6**: 58% yield; 53% ee; 48 h) (Scheme 6.128).

QN-derived **121** and QD-derived **124** also proved to have catalytic efficiency as the bifunctional hydrogen-bonding organocatalyst in the diastereoselective and enantioselective conjugate addition [149–152] of α-chloroacrylonitrile and acrylonitrile to a variety to cyclic α-cyanoketones and acyclic α-substituted cyanoesters as reported by Deng and co-workers in 2007 [286]. The reaction occurred at room temperature in toluene and gave the Michael adducts **1–10** in high diastereoselectivities (9–25 : 1 dr), enantioselectivities (88–97% ee), and good (82%) to excellent (100%) yields (Scheme 6.129). This asymmetric tandem conjugate addition–protonation sequence allowed the stereoselective construction of 1,3-tertiary-quaternary stereocenters.

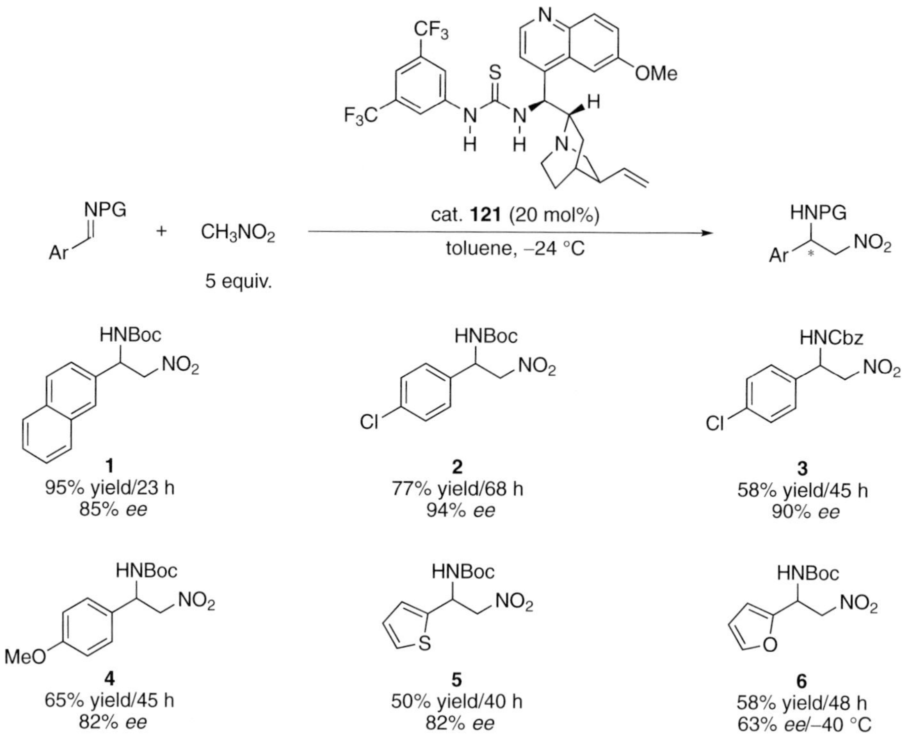

Scheme 6.128 Product range of **121**-catalyzed asymmetric aza-Henry reactions between N-protected aldimines and nitromethane. The configurations of the products were not determined.

To interpret the stereochemical outcome of this tandem reaction, the authors suggested a transition-state model, in which the bifunctional catalyst **121** interacted with the Michael donor and acceptor in a mode that the enolic Michael donor approached the acceptor from its *Si* face providing the observed adduct (Figure 6.42).

Chiral benzothiopyrans (thiochromanes) were prepared from 2-mercaptobenzaldehydes and α,β-unsaturated oxazolidinones via a one-pot method developed by the Wang group [287]. This asymmetric tandem thio-Michael-aldol process [149–152] utilized thiourea **121** as the highly efficient catalyst (1 mol% loading) that bifunctionally activated both the oxazolidinone Michael acceptor through hydrogen bonding and the 2-mercaptobenzaldehyde through deprotonation to induce the initial thio-Michael addition and the ring-forming aldol reaction (Scheme 6.130). It is noteworthy that the proposed hydrogen-bonding interactions with bidentate oxazolidinones are not in line with the theoretical and experimental findings published by Schreiner and Wittkopp [116].

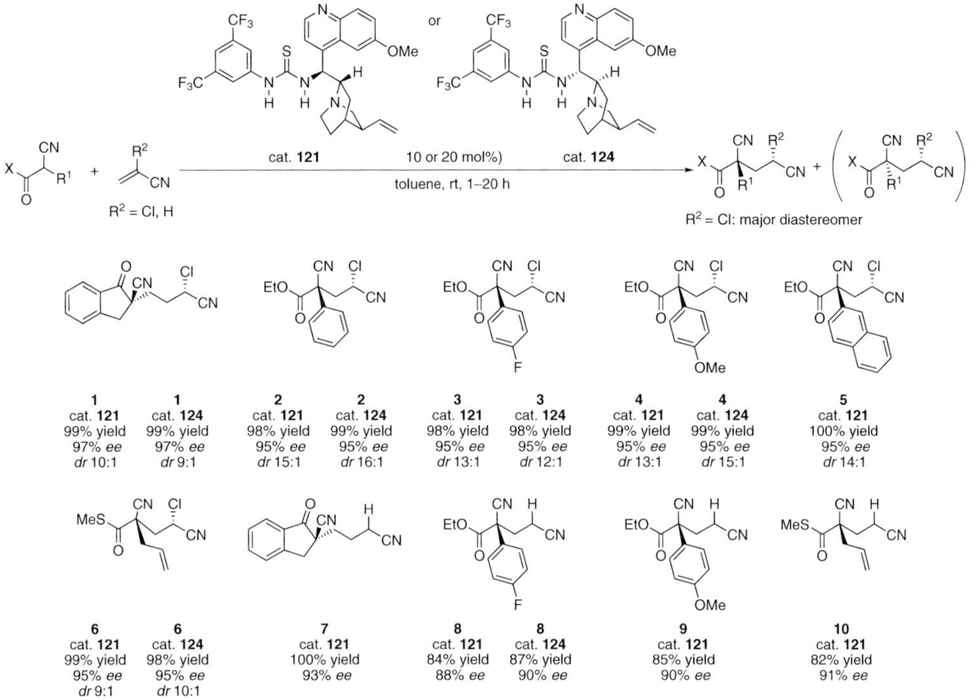

6.2 Synthetic Applications of Hydrogen-Bonding (Thio)urea Organocatalysts | 273

Scheme 6.129 Products prepared from the **121**- and **124**-catalyzed stereoselective Michael additions of α-chloroacrylonitrile and acrylonitrile to a variety to cyclic α-cyanoketones and acyclic α-substituted cyanoesters.

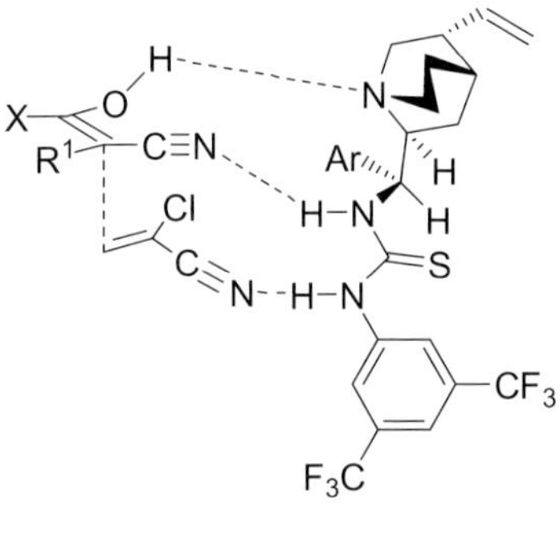

Figure 6.42 Transition-state model for the Michael addition of α-chloroacrylonitrile to cyclic α-cyanoketones or acyclic α-substituted cyanoesters; activation mode of quinine-derived thiourea **121**.

With established reaction parameters, this tandem process offered the facile synthesis of a range of substituted thiochromanes bearing three stereogenic centers. The reaction was reported to be high-yielding (75–97%) and produced the respective products **1–8** in excellent *ee* values (91–99%) as well as in very

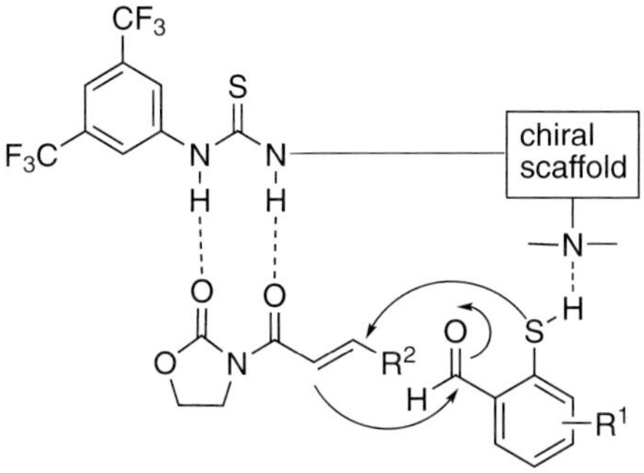

Scheme 6.130 Proposed mechanism of the **121**-catalyzed enantioselective thio-Michael-aldol tandem reaction of 2-mercaptobenzaldehydes with α,β-unsaturated oxazolidinones; bifunctional activation through thiourea **121**.

good diastereoselectivities (20:1 *dr*) after short reaction times (1–10 h) in 1,2-dichloroethane; the electronic properties of the 2-mercaptobenzaldehydes and the steric nature of the α,β-unsaturated oxazolidinones had only minimal impact on the yields and the stereoinduction of this **121**-catalyzed transformation (Scheme 6.131).

On the basis of the **121**- and **124**-catalyzed enantioselective direct Mannich additions [72] of dialkyl malonates to *N*-Boc-protected (hetero)aromatic and aliphatic aldimines [284] (Scheme 6.126), Deng and co-workers developed a modified Mannich protocol starting from readily accessible and stable *N*-Boc- and *N*-Cbz-protected α-amido sulfones instead of using often unstable *N*-protected aldimines [288]. In the presence of an inorganic base additive (0.1 M aqueous solution: Na_2CO_3 for aromatic, Cs_2CO_3 or CsOH for aliphatic aldimines), *N*-Boc- and *N*-Cbz-protected α-amido sulfones derived from (hetero)aromatic and aliphatic aldehydes **121** and **124** (5 and 10 mol%) catalyzed the asymmetric Mannich addition to the *in situ* generated *N*-Boc and *N*-Cbz aldimines. This one-pot strategy avoided the handling of Cbz-protected aldimines as unstable starting materials and furnished the desired Mannich adducts **1–8**, which are precursors for optically acive β-amino acids, in yields of 45–99% and *ee* values ranging from 85 to 96% (Scheme 6.132). The stereoinduction of the catalysts **121** and **124** proved to be complementary to each other and produced the respective opposite enantiomer [288].

The Deng group identified **QN**-derived thiourea **121** and **QD**-derived thiourea **124** to be also efficient promoters of enantio- and diastereoselective Diels–Alder reactions between the 2-pyrone diene 3-hydroxypyran-2-one and the dienophiles fumaronitrile, maleonitrile as well as acrylonitrile, while various C9-hydroxy acylated and alkylated (dihydro)cupreines and (dihydro)cupreidines failed for the same reactions under identical conditions (e.g., 97% yield, 15% *ee*, 64:36 *endo:exo*) [289]. Catalysts **121** and **124** (5 mol% loading), however, produced the corresponding Diels–Alder adducts **1–3** with synthetically useful enantioselectivities (85–

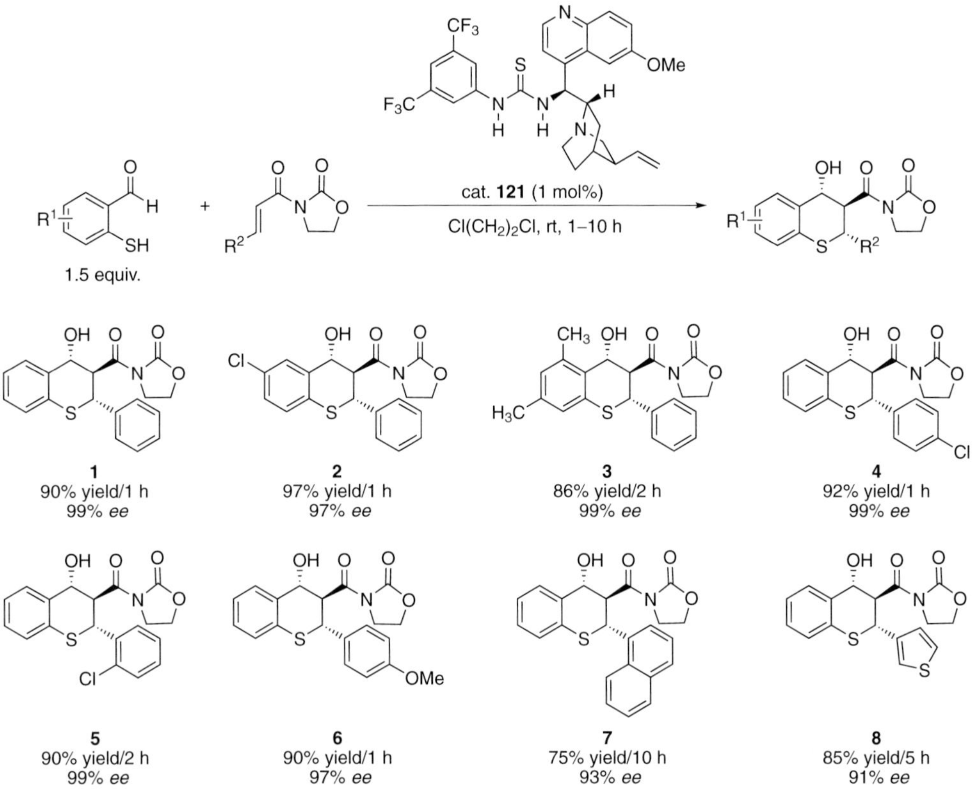

Scheme 6.131 Typical thiochromanes obtained from the 121-catalyzed asymmetric thio-Michael–aldol tandem reaction between various 2-mercaptobenzaldehydes and α,β-unsaturated oxazolidinones.

97%), diastereoselectivities (up to 96:4 *exo:endo*), and yields (87–91%), when performing the transformation at −20 °C in TBME (Scheme 6.133). The tolerance of dienophiles bearing either *E*- or *Z*-double bonds was illustrated by the successful conversion of fumaronitrile and maleonitrile. The *exo*-stereoselectivity of both C9-thiourea functionalized alkaloids **121** and **124** became evident from the Diels–Alder reaction between α-chloroacrylonitrile and 3-hydroxypyran-2-one, which produced the adducts in high yields (91%; 93%) and stereoselectivities (89% *ee*; *dr* 93:7 and 85% *ee*; *dr* 91:9) as depicted in Scheme 6.134. In contrast, various C9-modified **CPN** and **CPD** (5 mol% in Et$_2$O), respectively, were found to be complementary to the thiourea catalysts (**121**; **124**) and provided the respective *endo*-diastereomers in up to 90% yield, 85% *ee*, and *dr* 87:13.

The Rouden group utilized **121** and **124** as "organic bases" for the asymmetric decarboxylative protonation of cyclic, acyclic, and bicyclic *N*-acylated α-amino hemimalonates [290]. The introduced protocol suffered from high catalyst loading

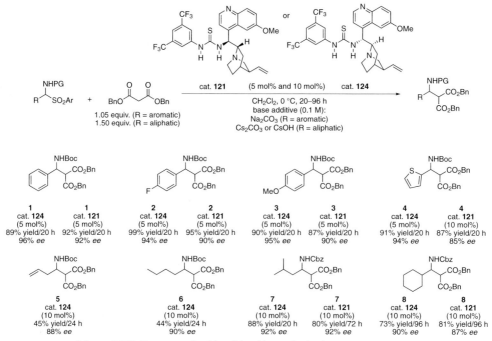

Scheme 6.132 Representative Mannich adducts obtained from the base-catalyzed *in situ* generation of *N*-protected aldimines from *N*-Boc- and *N*-Cbz-protected α-amido sulfones and subsequent **121**- and **124**-catalyzed addition of dibenzyl malonate. The product configurations were not determined.

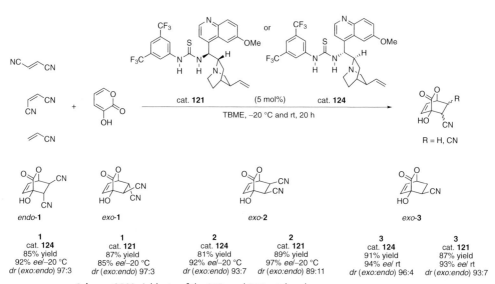

Scheme 6.133 Adducts of the **121**- and **124**-catalyzed stereoselective Diels–Alder reactions between the 3-hydroxypyran-2-one and the dienophiles fumaronitrile, maleonitrile, and acrylonitrile.

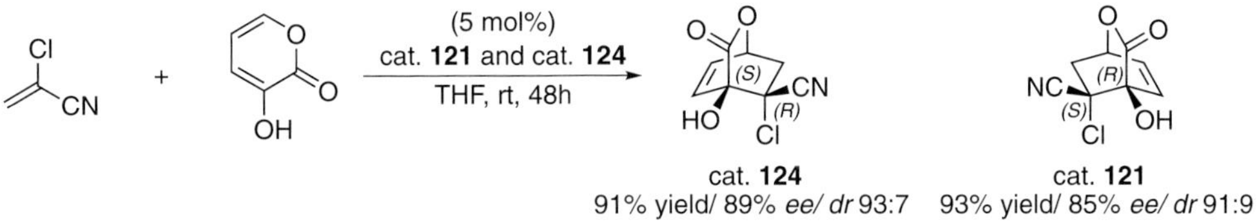

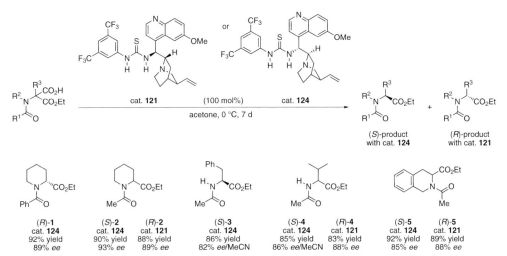

Scheme 6.134 *Exo*-selective **121**- and **124**-catalyzed Diels–Alder reaction between α-chloroacrylonitrile and 3-hydroxypyran-2-one.

Scheme 6.135 Typical products obtained from the **121**- and **124**-catalyzed asymmetric decarboxylative protonation of *N*-acylated α-amino hemimalonates.

(100 mol%; catalysts were recycled by acid–base work-up) and long reaction times (7 d), but it furnished the protected α-amino esters **1–5**, which represent precursors of α-amino acids, in good yields (83–92%) and *ee* values (82–93%) at low temperature (0 °C). The catalyst activity and selectivity turned out to be highly dependent on the reaction temperature (e.g., product **1** in Scheme 6.135: 85% yield/70% *ee* at 25 °C after 2 d; 25% yield/83% *ee* at −15 °C after 7 d), but appeared nearly insensitive to the choice of solvent; a wide range of polar aprotic (CH_3CN, acetone) and apolar aprotic (toluene, Et_2O, THF) was tolerated. The pseudoenantiomeric catalysts **121** and **124** provided access to both enantiomers of the desired *N*-acylated α-amino esters (Scheme 6.135).

Wang and co-workers developed a **121**-catalyzed enantioselective Michael addition [149–152] of 1*H*-benzotriazole to a variety of α,β-unsaturated ketones such as the model substrate 3-(4-chloro-phenyl)-1-phenyl-propenone affording the N-1 product **1** (Scheme 6.136) [291]. The evaluation of the reaction medium revealed

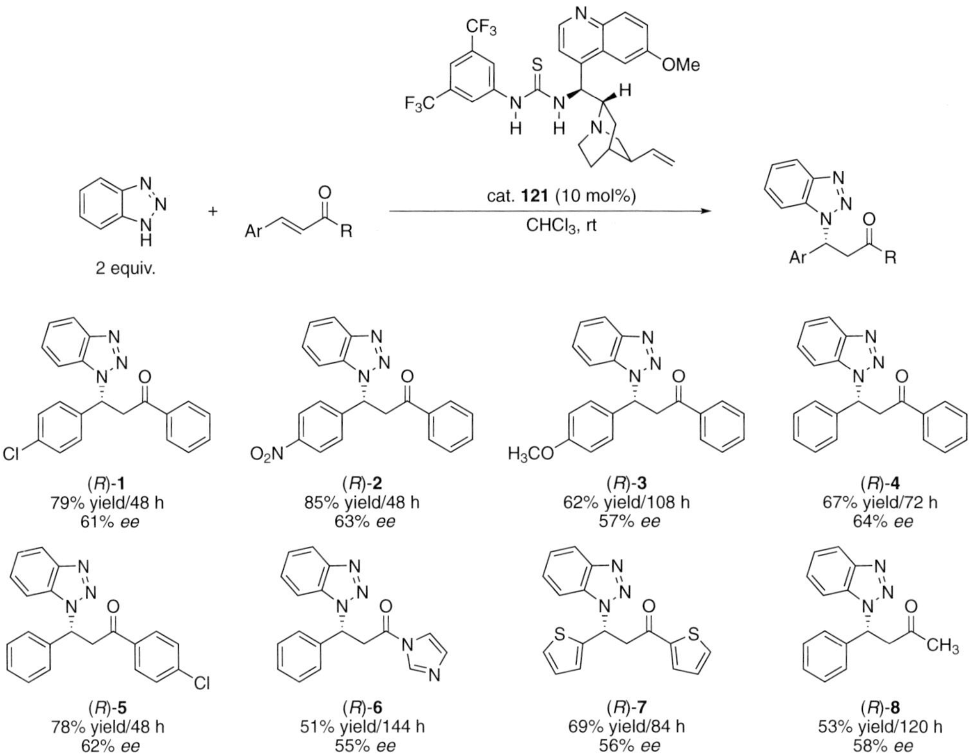

Scheme 6.136 Products of **121**-catalyzed asymmetric Michael additions of 1H-benzotriazole to a variety of α,β-unsaturated ketones.

that in polar and/or protic solvents (e.g., isopropyl alcohol, DMSO) no reaction occurred possibly due to competitive hydrogen-bonding interactions disturbing those between catalyst **121** and the substrate. Nonpolar and aprotic solvents, however, such as diethyl ether (70% yield; 43% ee), acetonitrile (77% yield, 46% ee), toluene (74% yield, 44% ee), and dichloromethane (81% yield; 52% ee) gave conversion as well as stereoinduction; the best result was obtained with chloroform (79% yield, 61% ee). With optimized reaction parameters including the choice of solvent (CHCl$_3$), catalyst loading (10 mol% **121**), and reaction temperature (rt) electron-rich and electron-deficient α,β-unsaturated ketones turned out to be suitable Michael acceptors for the addition of 1H-benzotriazole. After 48–144 h reaction time the desired products **1–8** were isolated in moderate to good yields (51–85%) and moderate ee values (55–64%) (Scheme 6.136). The authors proposed a synergistic activation of the substrates based on double hydrogen-bonding activation of the enone and on amine-group mediated deprotonation of 1H-benzotriazole facilitating the product-forming Michael addition step (Scheme 6.137).

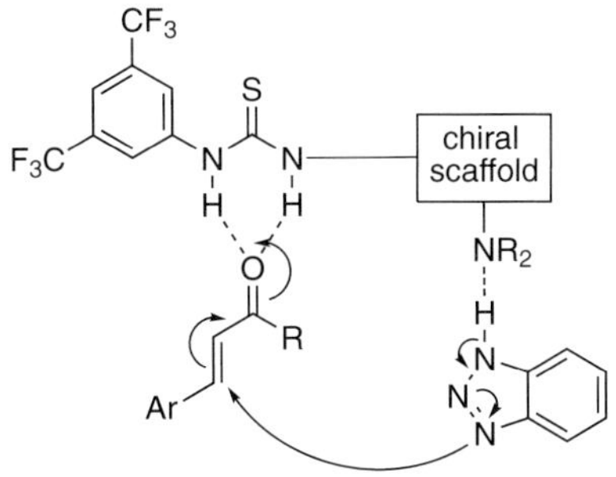

Scheme 6.137 Mechanistic proposal for **121**-catalyzed asymmetric Michael additions of 1H-benzotriazole to α,β-unsaturated enones.

The Jørgensen group extended the application of **QN**-derived thiourea catalyst **121** to the enantioselective β-hydroxylation of functionalized aliphatic nitroalkenes using predominantly ethyl glyoxylate oxime as oxygen source [292]. **CN**, **CD**, **QD**, and **QN** revealed poor selectivities (5–10% ee/–24 °C/73–82% conv. after 16 h/ toluene) in the catalyst-screening Michael reaction between (E)-1-nitrohept-1-ene and ethyl glyoxylate oxime (adduct **1**; Scheme 6.138), while **121** was found to be more efficient (95% conv./91% ee) even at 5 mol% loading at otherwise identical conditions. These experimental results represented a further example of improved bifunctional catalysis through incorporation of the privileged 3,5-bis(trifluoromethyl) phenyl-thiourea motif into the cinchona alkaloid backbone; catalyst **121** was suggested to activate both the oxime by single hydrogen bonding to the basic quinuclidine nitrogen atom and by double hydrogen-bonding of the C9-thiourea moiety to the nitroalkene Michael acceptor. Employing the optimized protocol (5 mol% **121**; toluene; –24 °C; 16 h) to the enantioselective Michael additions of oximes to various aliphatic nitroalkenes furnished the desired adducts **1–8** in 68–83% yields and 48–93% ee (Scheme 6.138). This β-hydroxylation tolerated variable functionalizations of the nitroalkane substrates such as a phenyl substitution, the presence of an isolated double bond, an ester group, and a thioether group as outlined in Scheme 6.138. To demonstrate the synthetic utility of the protocol and the prepared adducts, the authors presented selective reductions to transform the adduct of ethyl glyoxylate oxime to (2-nitro-vinyl)-cyclohexane (prepared in 82% yield/90% ee) either to its corresponding optically active N-Boc-protected amino alcohol (via hydrogenation: 1. H_2/Pd/C 2. (Boc)$_2$O) or to its (R)-configured β-hydroxy nitroalcohol (ZrCl$_4$, NaBH$_4$/THF) [292].

In the presence of thiourea **121** (20 mol% in toluene at 4 °C/20 °C), the aza-Michael addition [149–152] of O-benzylhyroxylamine to numerous trans-chalcones bearing electron-rich and electron-deficient (hetero)aromatic substituents as well as aliphatic side chains provided the respective β-keto hydroxylamines **1–8** in moderate to very good yields (35–94%) and low to moderate (30–60%) ee values

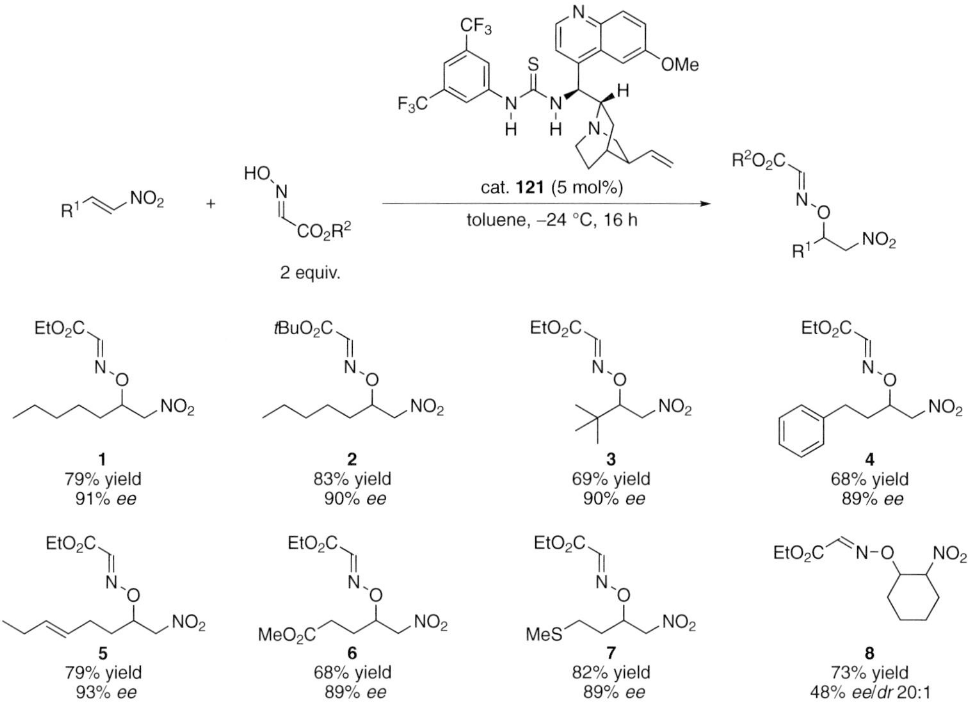

Scheme 6.138 Product range of the **121**-catalyzed asymmetric β-hydroxylating Michael addition of oximes to aliphatic nitroalkenes. The product configurations were not determined.

(Scheme 6.139) [293]. Ricci and co-workers explained the outcome of their aza-Michael reaction with the mechanistic picture visualized in Scheme 6.140; C9-*epi*-QN-derived thiourea **121** displayed a bifunctional mode of catalysis, which simultaneously activated both the chalcone Michael acceptor and the donor *O*-benzylhydroxylamine through explicit hydrogen bonding.

In 2008, Falck *et al.* reported the **121**-catalyzed enantioselective intramolecular oxy-Michael addition [149–152] of γ-hydroxy-α,β-enones and δ-hydroxy-α,β-enones in the presence of phenylboronic acid resulting in the respective β,γ-dihydroxy-enones and β,δ-dihydroxy-enones, respectively [294]. The β-hydroxylation of the hydroxy-enone substrates was described as an asymmetric intramolecular conjugate addition of an *in situ* generated boronic acid hemiester to the enone β-position forming dioxaborolane (*n* = 0) or dioxaborinane (*n* = 1) intermediates, respectively, which produced the chiral target diol after mild oxidative work-up (H_2O_2/Na_2CO_3) (Scheme 6.141). The authors suggested that **121** operated as a push/pull-type bifunctional hydrogen-bonding organocatalyst. As shown in Scheme 6.141, hydrogen-bonding coordination of the carbonyl group by the thiourea moiety (the pull) and complexation of the tertiary nitrogen to boron (the push) were expected to enhance simultaneously the nucleophilicity of the boronate oxygen as well as

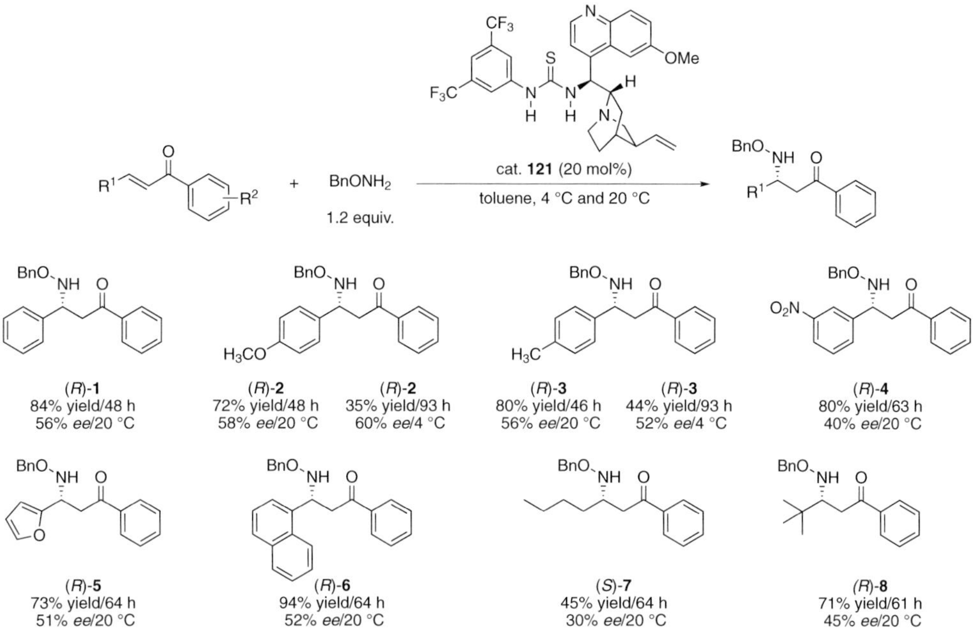

Scheme 6.139 Typical products of the aza-Michael addition of O-benzylhydroxylamine to *trans*-chalcones under bifunctional **121**-catalysis.

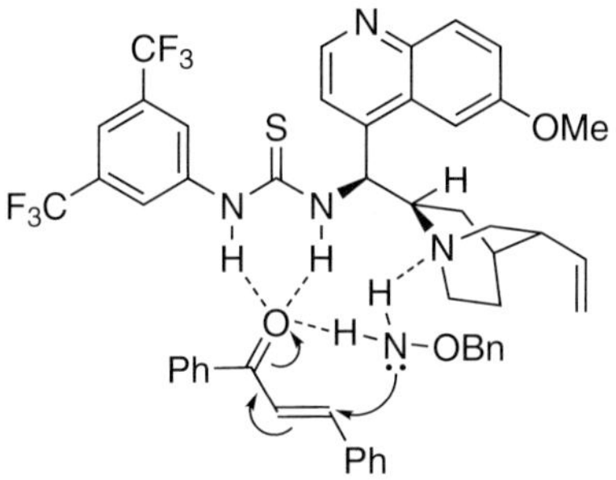

Scheme 6.140 Proposal for the role of catalyst **121** in the Michael reaction between O-benzylhydroxylamine and 1,3-diphenyl-propenone.

envelope the enone in a chiral environment giving rise to stereoinduction. In this mechanistic picture, the intermediate amine–boronate complex served as a chiral hydroxide surrogate or synthon. Notably, organic bases such as triethylamine (42% yield/48 h), diisopropylamine (86% yield/16 h), and DABCO (70%/48 h) in dichloromethane also catalyzed the intramolecular Michael addition of the model

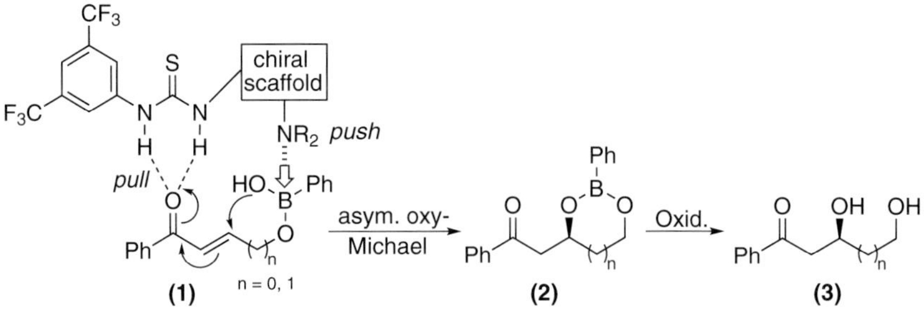

Scheme 6.141 Mechanistic proposal for the **121**-catalyzed asymmetric intramolecular Michael addition exemplified for the model substrates (E)-4-hydroxy-1-phenyl-2-buten-1-one (n = 0) and (E)-5-hydroxy-1-phenyl-2-buten-1-one (n = 1); **121** functions as push/pull-type bifunctional catalyst inducing the cyclization of boronic acid hemiester (**1**) to form intermediate (**2**); release of diol product (**3**) by oxidation.

substrate (E)-4-hydroxy-1-phenyl-2-buten-1-one, while inorganic bases Na_2CO_3 and $NaHCO_3$ incapable of forming an activating push-type boron complexation remained catalytically inactive under otherwise identical conditions.

Under optimized conditions, the substrate scope of the **121**-catalyzed (10 mol% loading) intramolecular oxy-Michael addition was reported to cover various γ-hydroxy-α,β-enones that could be converted at room temperature in dichloromethane or 1,2-dimethoxyethane to the desired (R)-configured adducts **1–8** in good to excellent yields (62–95%) and ee values (87–99%) (Scheme 6.142). Arylketones having strong electron-withdrawing substituents reacted faster than electron-rich systems, although the ee values of the latter were better. Aliphatic ketones, regardless of steric congestion adjacent to the carbonyl, had longer reaction times, yet giving excellent overall yields. The tolerance of the labile triethylsilyl (TES) ether, the additional substitution at the double bond, and the carbinol group testified the mildness of the reaction conditions and the level of structural substrate complexity, respectively (Scheme 6.142) [294].

The transformations of δ-hydroxy-α,β-enones to the corresponding internal Michael adducts were performed at 20 mol% loading of C9-*epi*-quinine-thiourea **121** in toluene at increased reaction temperature (50 °C) using 3,4,5-trimethoxyphenylboronic acid for aliphatic and phenylboronic acid for aromatic enones. Under these conditions, this protocol furnished the desired (R)-configured adducts **1–5** in yields ranging from 73 to 86% and ee values of 84–96% (Scheme 6.143) [294]. Product **5** in Scheme 6.143 was identical in all respects with (+)-(S)-streptenol A, one of four known streptenols produced by *Streptomyces luteogriseus* that has attracted attention as an immunostimulant as well as an inhibitor of cholesterol biosynthesis and tumor cells [295].

The **121**-catalyzed synthesis of thiochromanes starting from 2-mercaptobenzaldehydes and α,β-unsaturated oxazolidinones was developed by the Wang group

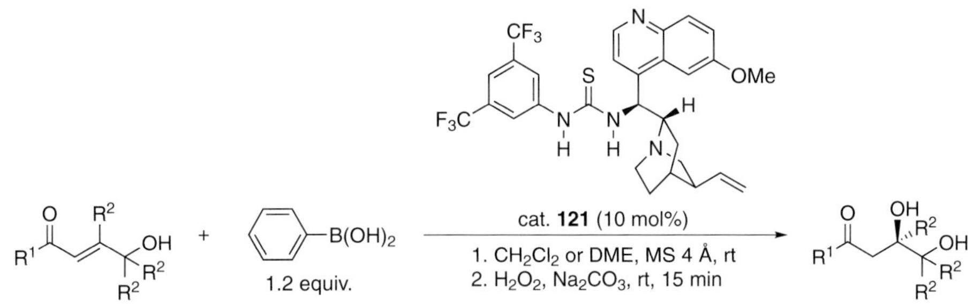

6.2 Synthetic Applications of Hydrogen-Bonding (Thio)urea Organocatalysts | 283

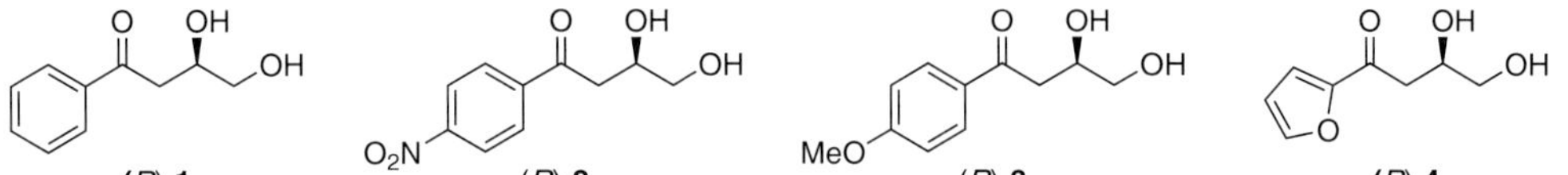

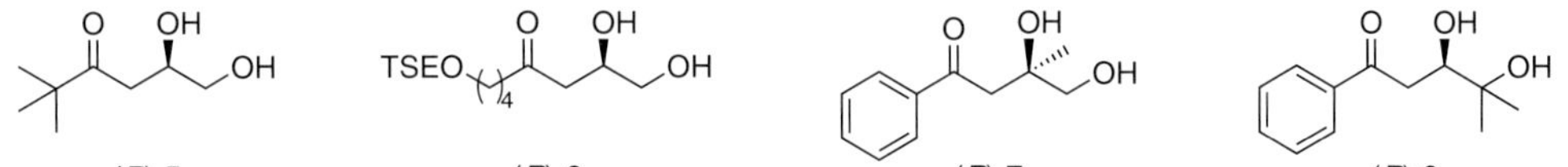

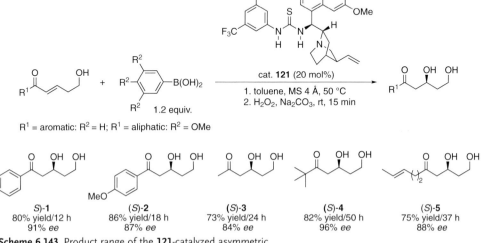

Scheme 6.142 Product range of the catalyzed asymmetric intramolecular Michael addition of γ-hydroxy-α,β-enones catalyzed by **121**.

Scheme 6.143 Product range of the **121**-catalyzed asymmetric intramolecular Michael addition of δ-hydroxy-α,β-enones.

and proceeded via a tandem Michael process [149–152] (Scheme 6.131) [287]. Zhao and co-workers, however, reported a tandem Michael–Knoevenagel approach toward tetra-substituted thiochromanes replacing the oxazolidinone component with benzylidenemalonates that were readily accessible through Knoevenagel condensation of malonates and aldeyhdes [296]. Utilizing catalyst **121** (5 mol% loading) in dichloromethane at −40 °C, various 2-mercaptobenzaldehydes and diethyl malonates were converted to the corresponding thiochromanes **1–6** in yields of 70–95% (after 2 h), *ee* values of 49–96%, and *dr* values up to 93:7 (Scheme 6.144). Steric and electronic effects turned out to have a strong impact on the stereochemical outcome of this reaction. Diastereoselectivity proved sensitive mainly toward

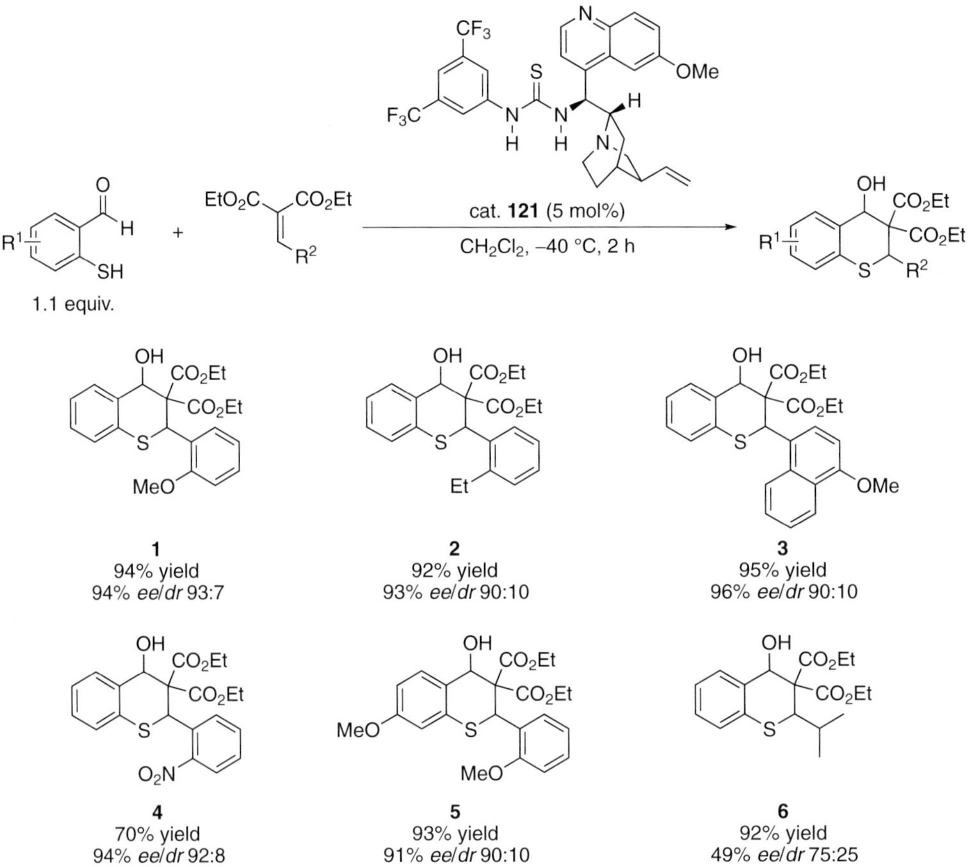

Scheme 6.144 Thiochromanes prepared from the **121**-catalyzed enantioselective tandem Michael–Knoevenagel reaction between 2-mercaptobenzaldehydes and diethyl methylenemalonates. The product configurations were not determined.

steric factors at the 2-position of the malonate phenyl ring, while enantioselectivity was influenced by the electronic effects of the substituent on the aromatic ring (=R^2). Substituents (=R^1) on the 2-mercaptobenzaldehyde ring had minimal impact on the reaction.

Chin, Song, and co-workers experimentally identified quinine-derived thiourea **121** (10 mol%) as an efficient bifunctional organocatalyst for the methanolic desymmetrization of bicyclic *meso*-anhydrides resulting in the respective chiral hemiesters **1–5** in excellent yields (92–96%) and good enantioselectivities ranging from 81 to 85% (Scheme 6.145) [297]. The evaluation of the reaction conditions performed for the **121**-catalyzed (10 mol%) asymmetric methanolysis of *cis*-1,2-cyclohexane dicarboxylic anhydride in THF (product **1**; Scheme 6.145) demonstrated the influence of concentration, temperature, and solvent effects on the level of stereoinduction. Dilution of the reaction mixture increased the *ee* value of the model hemiester **1** from 82% *ee* (2.5 ml THF) to 96% *ee* (80 ml THF) and a higher reaction temperature also registered an improved stereoinduction (77% *ee* at −20 °C; 82% *ee* at 25 °C in 2.5 ml THF) under otherwise unchanged conditions. Employing polar protic solvents such as methanol that accept and donate hydrogen bonds weakened the explicit hydrogen-bonding between catalysts **121** and the anhydride resulting in reduced stereoinduction (31% *ee*), while aprotic, hydrogen-bonding accepting solvents such as dioxane (97% *ee*) or THF (95% *ee*) appeared compatible with this approach. This observations suggested that catalyst **121** existed mainly in the dimeric (or higher order) form by hydrogen-bonded

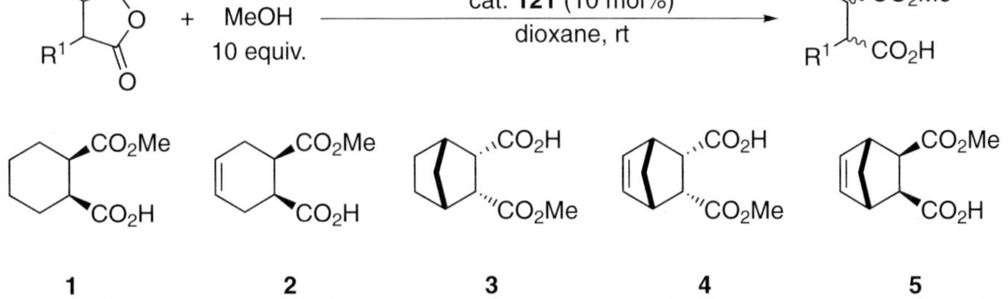

1	**2**	**3**	**4**	**5**
85% yield/10 h	84% yield/15 h	85% yield/28 h	81% yield/25 h	82% yield/25 h
97% *ee*	96% *ee*	93% *ee*	92% *ee*	95% *ee*

Scheme 6.145 Chiral hemiesters obtained from the **121**-catalyzed methanolic desymmetrization of cyclic *meso*-anhydrides.

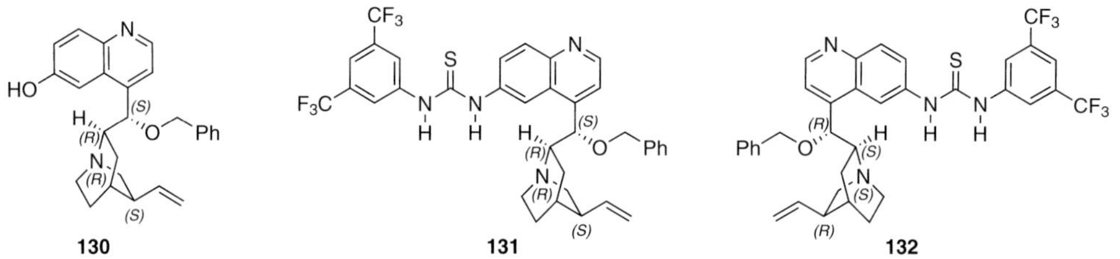

Figure 6.43 C9-OH protected cinchona alkaloid **130**, C6′-thiourea derivative **131** and its pseudoenantiomer **132**.

self-association at high concentrations, low temperature, and with noncoordinating solvents. On the other hand, at reduced concentration of the reaction mixture, at ambient temperature, and in the presence of coordinating solvents **121** existed mainly in the monomeric form that was responsible for high enantioselectivity. To confirm the hypothesis, ^{1}H NMR dilution experiments of **121** were carried out in d_8-toluene and concentration dependences were detected for the chemical shift of the —C(=S)N(H)—Ar proton; the chemical shift of this proton was downfield-shifted from 9.3 to 11.1 ppm upon concentration from 10 to 212 mM. This concentration dependency was consistent with the hydrogen-bonded self-association of **121**. The authors assumed that the efficiency of **121** originated from the simultaneous activation of the nucleophile (methanol) by the basic quinuclidine that functioned as a general base and the electrophile (the *meso*-anhydride) through double hydrogen-bonding interactions with the thiourea moiety. The rate-determining step for the methanolysis reaction was assumed to be a general base-catalyzed addition of methanol to the anhydride rather than the ring-opening step since carboxylate was a better leaving group than methoxide [297].

The novel class of C6′-thiourea functionalized cinchona alkaloids was constituted by Hiemstra and co-workers in 2005, when cupreidine-derived **131** and its cupreine-derived [64] pseudoenantiomer **132** were introduced as bifunctional hydrogen-bonding organocatalysts for the asymmetric Henry (nitroaldol) reaction [224] between nitromethane and various (hetero)aromatic aldehydes (Figure 6.43) [298]. Thiourea **131**, a bench-stable crystalline solid, was synthesized on a multigram scale from C9-hydroxy-benzylated **CPD** via a high-yielding reaction sequence [298]. Bifunctional cinchona alkaloid **130**, designed by Deng *et al.* [299–301], was previously utilized by the Hiemstra group to catalyze the asymmetric Henry addition of nitromethane to benzaldehydes producing yields of 70–92% and low *ee* values (6–35%) [302], while **131** revealed drastically improved catalytic activity and selectivity for an extended substrate spectrum.

For the model Henry reaction between benzaldehyde and nitromethane a solvent dependency of the enantioselectivity was detected (e.g., CH_2Cl_2: 6% *ee*; MeOH: 49% *ee*; THF: 62% *ee*; all at rt). Under optimized reaction conditions concerning catalyst loading (10 mol% of **131**), solvent (THF), and reaction temperature

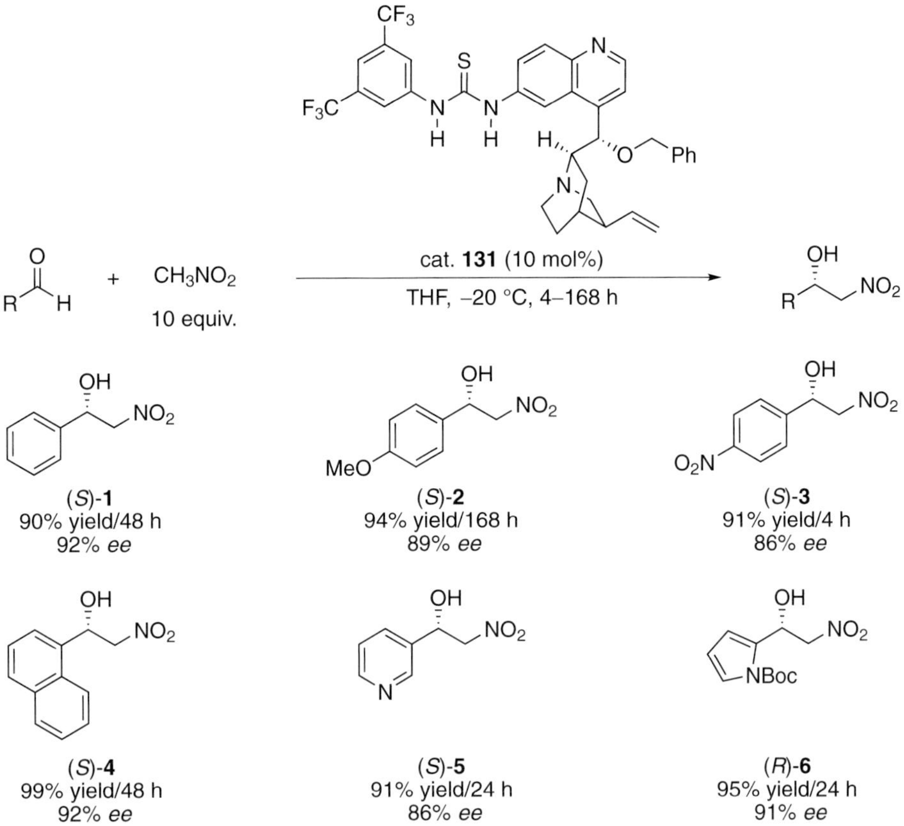

Scheme 6.146 Representative adducts obtained from the asymmetric Henry reaction between nitromethane and (hetero)aromatic aldehydes under bifunctional catalysis of C6′-thiourea-functionalized cinchona alkaloid **131**.

(−20 °C), various (hetero)aromatic aldehydes were transformed into nitroalcohols **1–6** in consistently high yields (90–99%) and *ee* values (86–92%) as shown in Scheme 6.146. The protocol failed for aliphatic aldehydes such as cyclohexanecarboxaldehyde and isobutyraldehyde that displayed incomplete conversion to the respective nitroalcohols even after 1 week reaction time and gave low *ee* values (<20%) of the adducts. Catalyst **132**, the pseudoenantiomer of **131**, gave access to nitroalcohols with the opposite configuration and comparable enantiomeric excess, as exemplified for three aldehydes (e.g., (*R*)-adduct **3**: 87% yield; 93% *ee*).

Scheme 6.147 visualizes two proposals for the mechanism of the **131**-catalyzed Henry addition of nitromethane to benzaldehyde. In (**A**), benzaldehyde is activated by the thiourea moiety through double hydrogen bonding to the carbonyl function, while the nitromethane is deprotonated and activated by the basic quinuclidine nitrogen [298]; proposal (**B**), however, based on detailed DFT computations

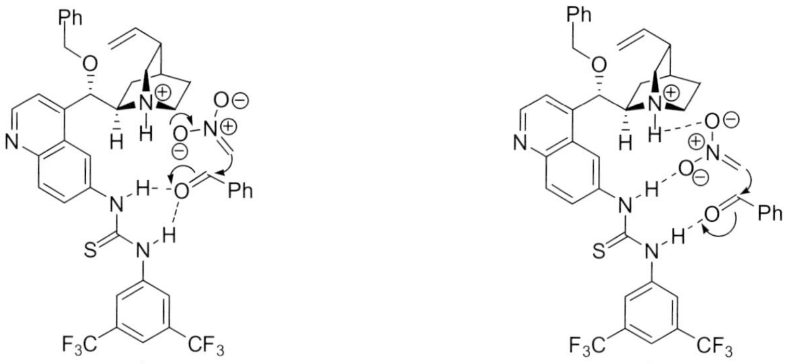

Scheme 6.147 Mechanistic proposals for the **131**-catalyzed asymmetric Henry addition of nitromethane to benzaldehyde. Preliminary model (**A**) involving double hydrogen bonding and DFT-based model (**B**) supporting single hydrogen bonding to benzaldehyde and nitronate.

favors single hydrogen-bonding to the nitronate and the carbonyl group facilitating C—C-bond formation through nucleophilc attack resulting in the preferred (S)-configured adduct [303].

6.2.2.3 (Thio)urea Catalysts Derived from Chiral Amino Alcohols

Chiral amino alcohols have been utilized as auxiliaries, ligands, and as important building blocks in organic synthesis [138]. Recently the chiral amino alcohols D- and L-prolinol **133** were utilized as organocatalysts for the asymmetric fluoroaldol reaction of aldehydes with fluoroacetone [304]; β-amino alcohols served as starting material for the synthesis of a series of bifunctional hydrogen-bonding L-prolinamides such as *(1S,2R)-cis*-1-amino-2-indanol-derived catalyst **134** applicable to asymmetric nitro-Michael additions [305], and for the synthesis of compounds **135** and **136** that catalyze enantioselective direct *syn*-aldol reactions of various aldehydes with ketones (Figure 6.44) [306].

Ricci and co-workers introduced a new class of amino- alcohol- based thiourea derivatives, which were easily accessible in a one-step coupling reaction in nearly quanitative yield from the commercially available chiral amino alcohols and 3,5-bis(trifluoromethyl)phenyl isothiocyanate or isocyanate, respectively (Figure 6.45) [307]. The screening of (thio)urea derivatives **137–140** in the enantioselective Friedel–Crafts reaction of indole with *trans*-β-nitrostyrene at 20 °C in toluene demonstrated (1*R*,2*S*)-*cis*-1-amino-2-indanol-derived thiourea **139** to be the most active catalyst regarding conversion (95% conv./60 h) as well as stereoinduction (35% *ee*), while the canditates **137**, **138**, and the urea derivative **140** displayed a lower accelerating effect and poorer asymmetric induction (Figure 6.45). The uncatalyzed reference reaction performed under otherwise identical conditions showed 17% conversion in 65 h reaction time.

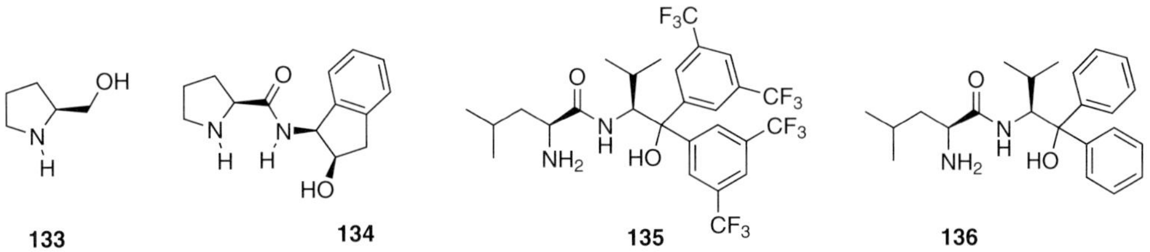

133 **134** **135** **136**

Figure 6.44 Stereoselective organocatalysts derived from amino alcohols.

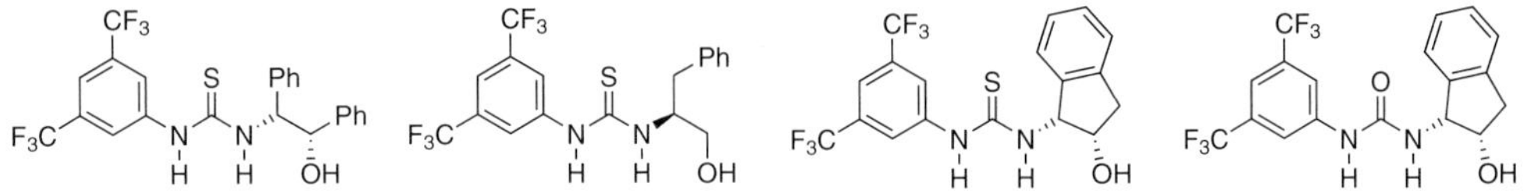

137
86% conv./ 64 h
13% ee

138
41% conv./45 h
13% ee

139
>95% conv./60 h
35% ee

140
23% conv./118 h
25% ee

Figure 6.45 Hydroxy-functionalized thiourea derivatives (20 mol% loading) screened in the enantioselective Friedel–Crafts reaction of indole with trans-β-nitrostyrene at 20 °C in toluene.

To examine the scope and limitations of this protocol, Friedel–Crafts reactions of various nitroalkenes and indole were carried out under optimized conditions utilizing thiourea catalyst **139** in dichloromethane at 20 mol% loading (Scheme 6.148). The transformations of indole and electron-rich indole derivatives such as 5-methoxyindole with aliphatic and aromatic nitroalkenes to the corresponding 2-indyl-1-nitro compounds proceeded in good enantioselectivities (71–89%) and in moderate to good yields (35–88%). Electron-poor 5-chloro indole substrate gave 71% yield at longer reaction time (142 h), but provided a significantly lower ee value (35%) similar to the ee value reported for the isopropyl-substituted nitroalkene substrate resulting in adduct **6** (37% ee) shown in Scheme 6.148. The yield of adduct **6** could be considerably increased without remarkable loss of enantioselectivity when performing the reaction at 0 °C (76% yield/96 h; 72% ee) instead of −24 °C (37% yield/96 h; 81% ee) (Scheme 6.148).

To elucidate the substrate–catalyst interactions the authors prepared derivatives of structure **139** and probed their efficiencies in the reactions of indole with trans-β-nitrostyrene. TMS-protected **141** and **142** lacking the hydroxyl functionality and the use of N-methylated indole (75%yield/6% ee) as a substrate gave lower conversions and enantioselectivities, indicating the presence of a weak single hydrogen-bonding interaction between the indole proton and the hydroxy oxygen in addition to the double hydrogen bonding activation of the nitroalkene by the thiourea amide protons. This assumed bifunctional catalysis induced the nucleophilic attack of the incoming indole on the *Si face* of the nitroalkene (Figure 6.46). The Ricci group demonstrated the synthetic versatility of the optically active Friedel–Crafts adducts

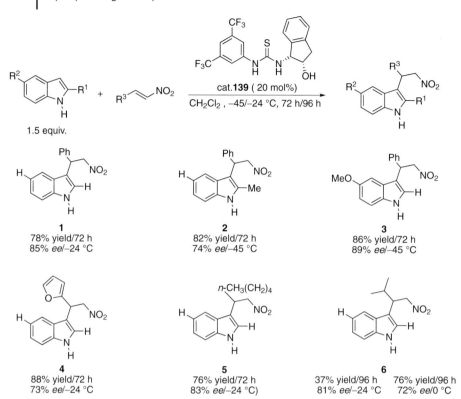

Scheme 6.148 Product range of the enantioselective 139-catalyzed Friedel–Crafts alkylations of various indols. The product configurations were not determined.

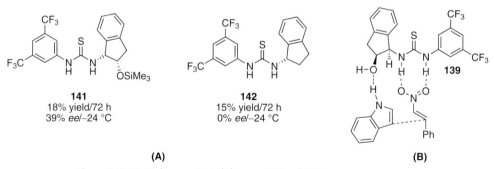

Figure 6.46 (A) Hydroxy-protected thiourea 141 and 142 lacking the hydroxy function and their catalytic efficiency in the Friedel–Crafts alkylation of indole with trans-β-nitrostyrene (139: 78% yield; 85% ee under identical conditions). (B) Proposal for the key hydrogen-bonding interactions between 139 and the model substrates.

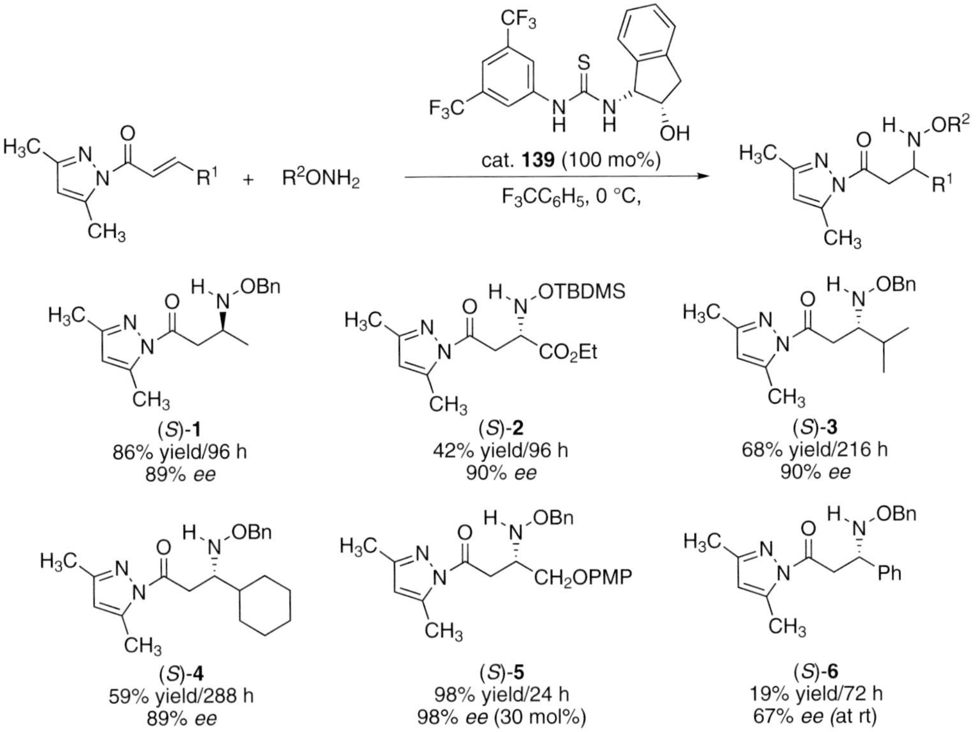

Scheme 6.149 Typical products obtained from the **139**-catalyzed asymmetric Michael addition of protected hydroxyl amines to α,β-unsaturated 2,4-dimethyl pyrazole substituted substrates.

through a two-step conversion of product **1** in Scheme 6.148 to tryptamine (74% yield; 85% *ee*) and 1,2,3,4,-tetrahydro-β-carboline (76% yield; 85% *ee*) [307].

To develop a metal-free synthetic access to precursors of β-amino acids, Sibi and Itoh studied the thiourea-mediated conjugate addition [149–152] of various amines to pyrazole-derived enoates [308]. The addition products are synthetically useful due to their simple conversion to β-amino acids [309]. The authors evaluated the optimal reaction conditions using a stoichiometric amount of **139** to catalyze the model conjugate addition of *O*-benzylhydroxylamine to 2,4-dimethyl pyrazole crotonate. The best results concerning product yield and asymmetric induction for model product **1** (Scheme 6.149) were obtained when the reaction proceeded in the nonhydrogen-bonding solvent trifluorotoluene at 0 °C reaction temperature and in the presence of powdered activated MS 4 Å (500 mg/1 mmol α,β-unsaturated amide). The MS was found to improve the yield (no MS: 76%/48 h; with MS: 75%/24 h; at rt), but had no impact on the *ee* value (unchanged 71% *ee* at rt); the function of this additive was not examined and remained ambiguous. The

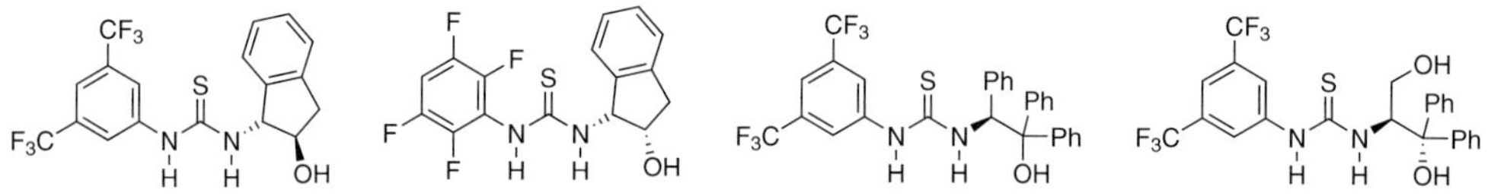

Figure 6.47 Various thiourea catalysts screened in the Michael addition of O-benzylhydroxylamine to 2,4-dimethyl pyrazole crotonate.

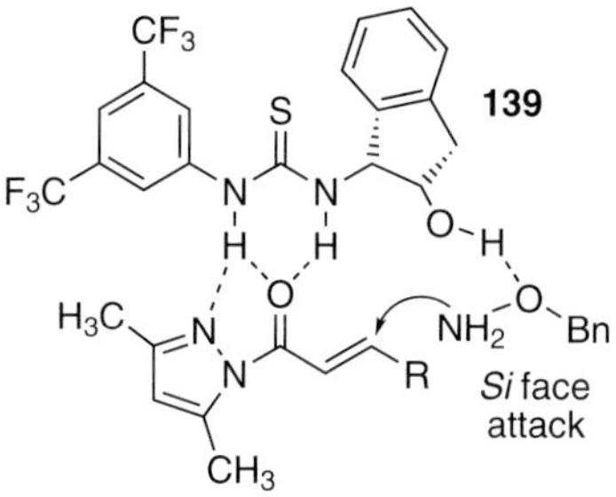

143
61% yield
9% ee/rt

144
92% yield
40% ee/rt

145
23% yield
7% ee/rt

146
44% yield
7% ee/rt

Scheme 6.150 Proposed mechanistic picture for (S)-favored enantioselective Michael addition of O-benzylhydroxylamine to 2,4-dimethyl pyrazole substituted α,β-unsaturated substrates in the presence of hydrogen-bonding thiourea catalyst **139**.

screening of amino alcohol derived (thio)ureas such as **137**, **140**, and **142–146** in their role as "chiral activators" (Figures 6.45–6.47) in the formation of model product **1** at room temperature revealed bifunctionality, rigidity, and the incorporation of the 3,5-bis(trifluoromethyl)phenyl thiourea functionality as catalyst structure key features (Figure 6.47) [308].

Lowering the catalyst loading from 100 to 30 mol% led to longer reaction times (up to 168 h at rt), while the *ee* value of model product **1** remained unchanged. Employing the optimized protocol (100 mol% loading of thiourea **139**; $F_3CC_6H_5$ as the solvent at 0 °C) and 2,4-dimethyl pyrazole substituted α,β-unsaturated substrates bearing an aliphatic substituent at the β-carbon the respective Michael adducts **1–5** were obtained in moderate to excellent product yields (42–98%; 1 d to >10 d reaction time) and in high *ee* values (89–98%) (Scheme 6.149). In contrast, β-phenyl substituted adduct **6** was isolated in only 19% yield (72 h) and 67% enantioselectivity (Scheme 6.149).

The authors suggested a double hydrogen-bonding coordination of the bidentate α,β-unsaturated 2,4-dimethyl pyrazole substituted substrates resulting in a hydrogen bond pattern examined by Schreiner and Wittkopp [1, 116]. This additional fixation of O-benzylhydroxylamine in favor of a *Si face* nucleophilic Michael addition was led to the observed stereochemical outcome of (S)-configured adducts (Scheme 6.150).

Thiourea catalyst **139** was also screened in the asymmetric Friedel–Crafts reaction between 2-naphthol *trans*-nitrostyrene (73% yield; 0% *ee*; 18 h in toluene at −20 °C and 10 mol%) [277], in the asymmetric aza-Michael reaction of *O*-benzylhydroxylamine to chalcone (72% conv.; 19% *ee*; 72 h in toluene at 20 °C and 20 mol% catalyst loading) [293], and in the asymmetric Morita–Baylis–Hillman [176, 177] reaction between cyclohexenecarbaldehyde and 2-cyclohexene-1-one (20% yield; 31% *ee*; 46 h at rt and 20 mol% DABCO and **139**) [310]. In all these transformations, thiourea **139** proved to be not competitive to the organocatalysts probed for these transformations under identical screening conditions and thus was not employed in the optimized protocols.

Lattanzi screened various amino alcolhol derived thioureas in the enantioselective Morita–Baylis–Hillman model reaction between cyclohexenecarbaldehyde and 2-cyclohexene-1-one (MBH adduct **1**; Scheme 6.151) under solvent-free

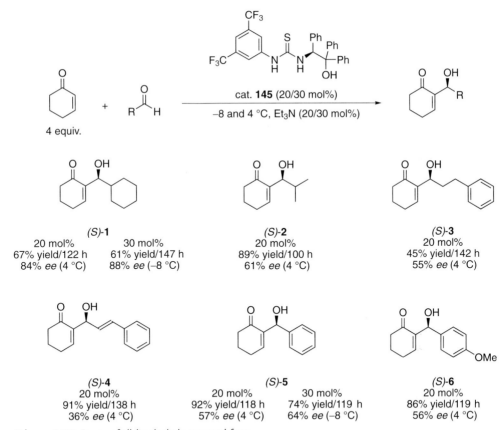

Scheme 6.151 Range of allylic alcohols prepared from **145**-catalyzed MBH reactions of 2-cyclohexene-1-one with various aldehydes.

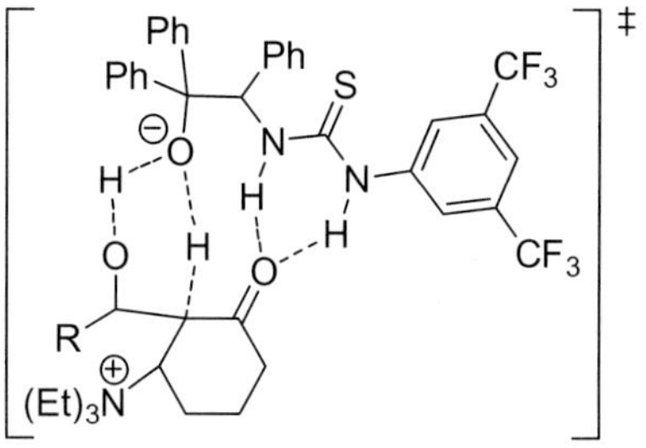

Figure 6.48 Proposed transition state for the proton-transfer step promoted by amino-alcohol-derived hydrogen-bonding thiourea **145**.

conditions and identified readily accessible thiourea **145** as the most effective catalyst [310]. Notably, thiourea **145** exhibited poor activity and enantioinduction in the Michael addition of O-benzylhydroxlyamine to α,β-unsaturated 2,4-dimethyl pyrazole substituted substrates (Figure 6.47).

To evaluate the scope and limitations of this protocol, various aldehydes were utilized as substrates to react with 2-cyclohexene-1-one under optimized conditions in the presence of catalyst **145** (20 and 30 mol% loading) and triethylamine (20/30 mol%) as base additive at 4 °C as the standard reaction temperature. After an average reaction time of approx. 120 h, the corresponding (S)-configured allylic alcohols **1–6** were provided in enantioselectivities ranging from 36 to 88% ee and yields ranging from 45 to 92%.(Scheme 6.151) The asymmetric induction decreased significantly when using aliphatic sterically unhindered aldehydes such as 2-methyl-propionaldehyde forming product (S)-**2** (Scheme 6.151). Running the reaction with benzaldehyde at −8 °C but at 30 mol% loading of **145** and triethylamine (instead of 20 mol% standard amount) increased the ee value of MBH adduct (S)-**5** from 57 to 64% (Scheme 6.151). The author suggested that bifunctional hydrogen-bonding thiourea catalyst **145** stabilized the zwitterionic intermediate and accelerated the product-forming proton transfer by the closely located catalyst hydroxyl group followed by the catalyst regeneration after base elimination (Figure 6.48).

In 2007, Fernández, Lassaletta, and co-workers reported an approach for the enantioselective conjugate addition [149–152] of formaldehyde hydrazone 1-methyleneaminopyrrolidine to aliphatic γ,β-unsaturated α-keto ethylesters as enoate surrogates providing the 1,4-dicarbonyl compounds **1–6** in moderate (60%) to good yields (82%) and ee values (58–80%) as depicted in Scheme 6.132 [311]. The reaction occurred in the presence of (1S,2R)-1-aminoindan-2-ol-derived thiourea catalyst **147** (10 mol% loading), the enantiomer of thiourea catalyst **139** introduced by the Ricci group (Scheme 6.152). In contrast to the umpolung strategies applied in the catalytic Stetter reaction that failed in the formaldehyde case due to oligomerizations [312] in this mild protocol 1-methyleneaminopyrrolidine turned

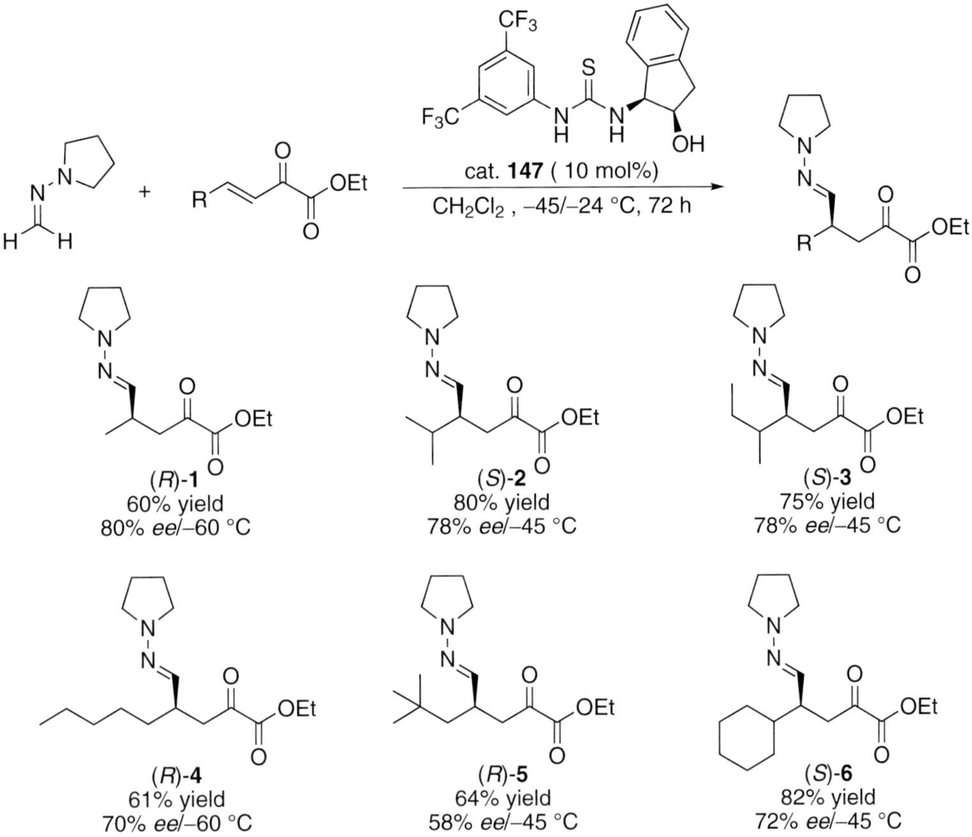

Scheme 6.152 Products of the asymmetric addition of 1-methyleneaminopyrrolidine to aliphatic γ,β-unsaturated α-keto ethylesters in the presence of catalyst **147**.

out to be a suitable formyl anion equivalent (umpoled formaldehyde) that underwent the symmetric nucleophilic addition to γ,β-unsaturated α-keto ethylesters without side reactions.

Mechanistically, the authors proposed a stereochemical model in which the hydroxy functionality of bifunctional catalyst **147** controlled the approach of the hydrazone through single hydrogen bonding such that the triple hydrogen-bonded unsaturated ketoester was preferably attacked from the *Re* face resulting in observed (*R*)-configured adducts (Scheme 6.153). The synthetic utility of the prepared adducts was demonstrated by straightforward transformations using magnesium monoperoxyphthalate hexahydrate (MMPP·6H$_2$O) for the racemization-free oxidative cleavage of the hydrazone moiety into the corresponding 4-cyano α-keto ethylesters; the ozonolytic cleavage, subsequent oxidation and

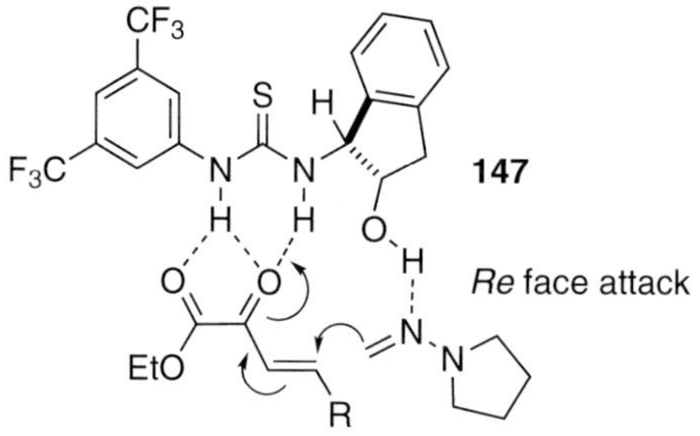

Scheme 6.153 Proposed stereochemical model for the asymmetric addition of 1-methyleneaminopyrrolidine to aliphatic γ,β-unsaturated α-keto ethylesters in the presence of catalyst **147**.

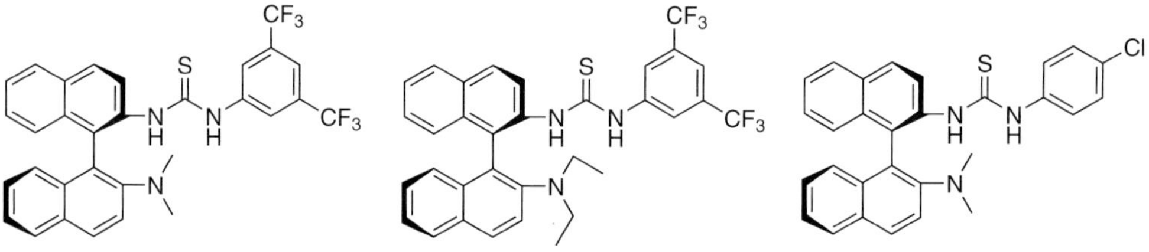

Figure 6.49 Binaphthyl-derived tertiary amine-functionalized bifunctional thiourea derivatives screened in the MBH reaction between 2-cyclohexen-1-one and 3-phenylpropionaldehyde at rt in dichloromethane.

148
83% yield
71% ee

149
56% yield
73% ee

150
18% yield
ee not det.

treatment with $SOCl_2$/MeOH allowed access to the respective succinate derivatives resulting from deoxidative decarboxylation [311].

6.2.2.4 Binaphthyl-Based (Thio)urea Derivatives

In 2005, Wang and co-workers utilized the catalytically "privileged" [119] axially chiral binaphthyl framework to design a new class of bifunctional hydrogen-bonding thiourea derivatives **148–150** readily prepared from commercially available (R)-binaphthyl diamine (Figure 6.49) [313]. This bifunctional catalyst design rationale has been already successfully realized in Takemoto's tertiary amine-functionalized thiourea catalyst **12** derived from (R,R)-1,2-diaminocyclohexane [129]. Screening studies regarding catalyst, solvent, and temperature effects on the enantioselective MBH reaction [176, 177] of 2-cyclohexen-1-one with 3-phenylpropionaldehyde identified thiourea **148** as the most active binaphthyl organocatalyst in terms of reaction yield (83%) and enantioselectivity (71% ee). While **149** was found to show similar asymmetric induction (73% ee), the yield of the model MBH product was lower (**149**: 56% yield; **150**: 18% yield) after 48 h at rt (Figure 6.49).

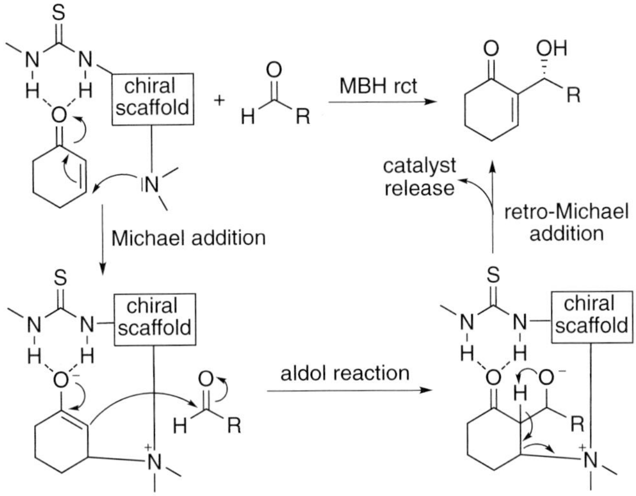

Scheme 6.154 Proposed catalytic cycle for the binaphthyl amine thiourea-promoted MBH reaction of aldehydes with 2-cyclohexen-1-one revealing the bifunctional mode of action of catalyst **148**, **149**, and **150**.

The significant difference in catalyst activity was in line with the known effect of the electron-withdrawing 3,5-bis(trifluoromethyl) phenyl functionality on the hydrogen-bonding strength of (thio)urea catalysts [1, 116], which led to a stronger binding of 2-cyclohexene-1-one and activation through double hydrogen bonding. This interaction facilitated the initial Michael addition of the tertiary amine to the β-position and initiated the proposed catalytic cycle (Scheme 6.154). A sequence of aldol/retro-Michael reactions produced synthetically valuable chiral allylic alcohols **1–6** in moderate to good yields (55–75%) and enantioselectivities (60–94% ee). The optimized protocol was reported to operate under mild conditions at rather low catalyst loadings and practical reaction times for aliphatic, aromatic, and sterically demanding aldehydes (Scheme 6.155) [313].

Almost simultaneously, the Wang group published an additional application of binaphthyl thiourea catalyst **148** [314]. Asymmetric Michael reactions [149–152] of 2,4-pentanedione with a series of *trans*-β-nitrostyrenes having electron-withdrawing and electron-donating substituents were performed at room temperature in the presence of **148** (1 mol% loading) producing the desired Michael adducts **1–5** in high yields (86–92%) and ee values (83–97%) in reasonable reaction times (Scheme 6.156).

Thiourea derivative **148** was expected to coordinate and stabilize the enolized 2,4-pentanedione through the basic amine function to facilitate the nucleophilic attack at the activated β-position of the hydrogen-bonded *trans*-nitroalkene resulting in the observed products (Scheme 6.157). Using Michael adduct **1** depicted in Scheme 6.156 as the starting material, the authors developed a synthetically useful

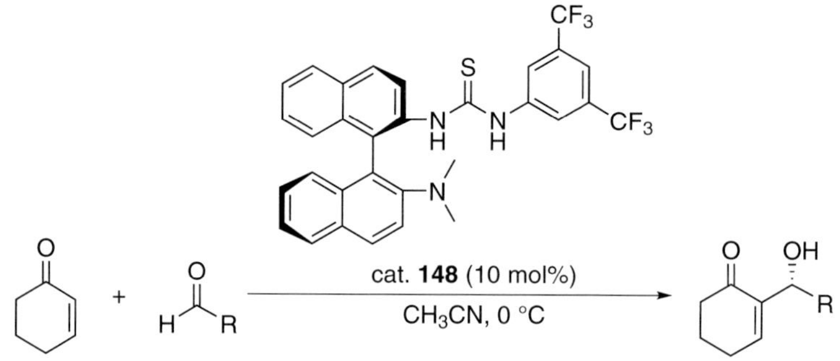

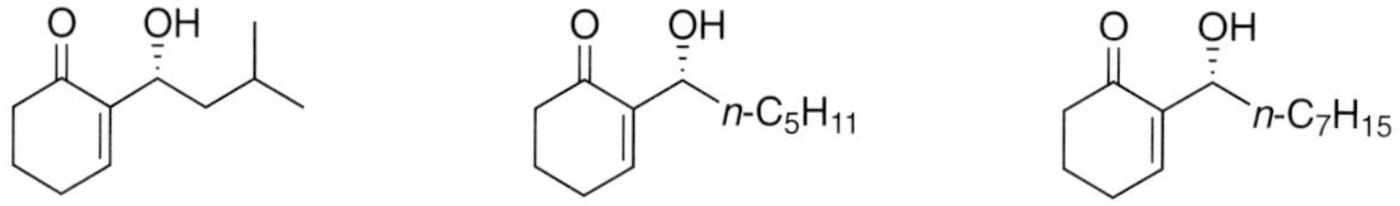

(R)-**1**
72% yield/82 h
80% *ee*

(R)-**2**
75% yield/60 h
81% *ee*

(R)-**3**
74% yield/72 h
82% *ee*

(R)-**4**
63% yield/72 h
94% *ee*

(R)-**5**
67% yield/120 h
92% *ee*

(S)-**6**
55% yield/108 h
60% *ee*

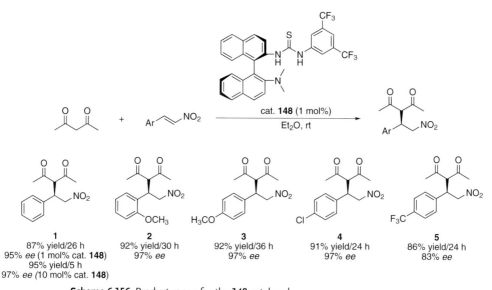

Scheme 6.155 Representative products resulting from enantioselective MBH reactions catalyzed by binaphthyl thiourea **148**.

1
87% yield/26 h
95% *ee* (1 mol% cat. **148**)
95% yield/5 h
97% *ee* (10 mol% cat. **148**)

2
92% yield/30 h
97% *ee*

3
92% yield/36 h
97% *ee*

4
91% yield/24 h
97% *ee*

5
86% yield/24 h
83% *ee*

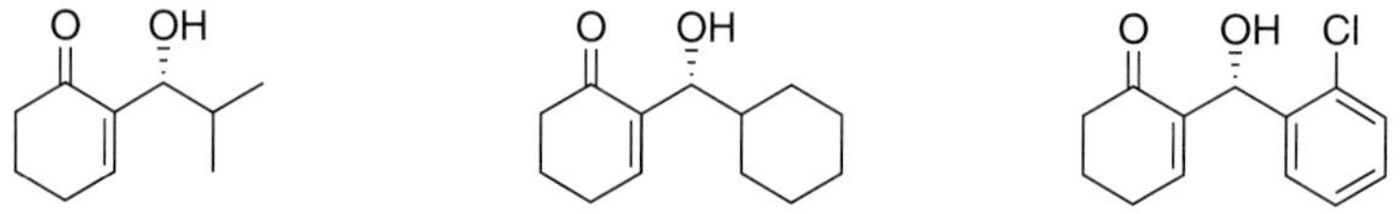

Scheme 6.156 Product range for the **148**-catalyzed asymmetric Michael addition of 2,4-pentandione to *trans*-β-nitrostyrenes.

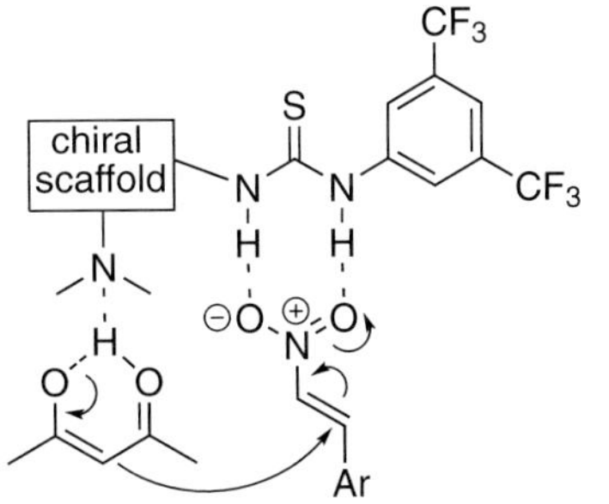

Scheme 6.157 Mechanistic proposal for the bifunctional coordination and simultaneous activation of 2,4-pentandione and *trans*-β-nitrostyrene through thiourea catalyst **148** leading to chiral Michael adducts.

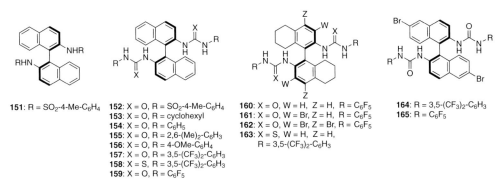

151: R = SO$_2$-4-Me-C$_6$H$_4$

152: X = O, R = SO$_2$-4-Me-C$_6$H$_4$
153: X = O, R = cyclohexyl
154: X = O, R = C$_6$H$_5$
155: X = O, R = 2,6-(Me)$_2$-C$_6$H$_3$
156: X = O, R = 4-OMe-C$_6$H$_4$
157: X = O, R = 3,5-(CF$_3$)$_2$-C$_6$H$_3$
158: X = S, R = 3,5-(CF$_3$)$_2$-C$_6$H$_3$
159: X = O, R = C$_6$F$_5$

160: X = O, W = H, Z = H, R = C$_6$F$_5$
161: X = O, W = Br, Z = H, R = C$_6$F$_5$
162: X = O, W = Br, Z = Br, R = C$_6$F$_5$
163: X = S, W = H, Z = H, R = 3,5-(CF$_3$)$_2$-C$_6$H$_3$

164: R = 3,5-(CF$_3$)$_2$-C$_6$H$_3$
165: R = C$_6$F$_5$

Figure 6.50 (*R*)-bis-*N*-tosyl-BINAM-derivative **151** and axially chiral bis(thio)ureas **152–165** screened for catalytic efficiency in the asymmetric addition of indole and *N*-methylindole to nitroalkenes.

method for the synthesis of the corresponding α-substituted- β-amino building block (38% yield) [314].

Connon and co-workers synthesized a small library of novel axially chiral binaphthyl-derived bis(thio)ureas **152–165** and elucidated the influence of the steric and electronic characteristics of both the chiral backbone and the achiral *N*-aryl(alkyl) substituents on catalyst efficiency and stereodifferentiation in the FC type additions of indole and *N*-methylindole to nitroalkenes (Figure 6.50) [315].

A screening of (*R*)-bis-*N*-tosyl-BINAM **151** and axially chiral (thio)urea derivatives **152–162** (10 mol% loading; 0.36 M catalyst concentration incorporating the *N*-aryl(alkyl) structural motif was performed at various reaction temperatures in d_1-chloroform using the asymmetric FC addition of *N*-methylindole to *trans*-β-nitrostyrene as model reaction (product **1**; Scheme 6.158). The structure of bis(3,5-bistrifluoromethyl) phenyl functionalized binaphthyl bisthiourea **158** was identified

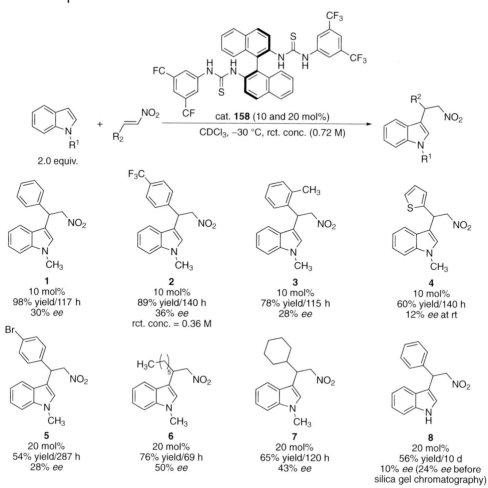

Scheme 6.158 Product spectrum of the **158**-catalyzed asymmetric FC addition of indole and *N*-methylindole to various nitroakenes. The product configurations were not determined.

to represent the best practical balance of both catalytic activity and stereoinduction. For (*R*)-bis-*N*-tosyl-BINAM **151** no catalytic effect was detected (adduct **1**: <2% conv.; no *ee* after 72 h at rt) comparable to the uncatalyzed reference experiment producing only trace levels of FC adduct **1** after the same reaction time. The importance of the (thio)urea motif for the catalytic efficiency became evident from the screening results obtained for BINAM-derivative **151** in comparison to the accelerating effect of tosyl-urea **152** (adduct **1**: 80% conv./20 h; 100% conv./166 h, no *ee*). Urea derivative structures **153–156** having either aliphatic (cyclohexyl) or electron-rich/sterically hindered aromatic substituents produced model FC adduct

1 in very low enantioselectivities (7–10%), and exhibited low conversions (2–14%) even after 113 h reaction time. In contrast, the structural analogs **157–159** incorporating electron-deficient aromatic substituents displayed a higher catalytic efficiencies ranging from 100% conv. (113 h) and 11% *ee* for urea structure **157** and 100% conv. after 160 h (12% *ee*) for thiourea counterpart **158** to 93% conv. (166 h) to 15% *ee* for decafluoro urea **159**. At −20 °C thiourea **159** revealed a loss of catalytic activity, but an increase of stereoinduction (adduct **1**: 42% conv./167 h; 34% *ee*), while catalyst **158** showed at −30 °C an increase of the *ee* value (30%) without a remarkable decrease of catalytic activity (88%/64 h). A higher catalyst concentration enhanced the conversion (e.g., **159** at −20 °C: 80% conv.; 28% *ee* after 70 h at 0.76 M conc.). The comparability and interpretation of the experimental screening data, however, were affected by different reaction times and conditions chosen by the authors. The modification of the axially chiral backbone represented by the (thio)ureas **160–163** led to no appreciable improvements in both rate enhancement and stereoinduction (no *ee* to <30% *ee*). The octahydro-analogs of **158** and **159** (**160** and **163**, respectively) were significantly less efficient in the formation of model adduct **1** (**160**: 23% conv./65 h; 20% *ee*/rt; **163**: 15% conv./113 h; 28% *ee*/−30 °C). Mono- and di-bromo derivatives of **160** (**161** and **162**, respectively) gave the racemic product. The catalytic activity of urea-based catalyst **157** could be substantially increased through the incorporation of bromo substituents at C-6 and C-6′ position (urea **164**: 100% conv./22 h; 5% *ee* at rt). Under optimized reaction conditions, thiourea **158** (10 and 20 mol%) catalyzed the asymmetric addition of indole and *N*-methylindole to aliphatic, and predominantly electron-rich as well as electron-poor aromatic *trans*-β-nitroalkenes. The respective FC adducts **1–8** were obtained in moderate to excellent yields (54–98%) and in low enantioselectivities (12–50% *ee*) in long reaction times (69–287 h), as shown in Scheme 6.158. For understanding the mode of action of catalysts **158** and the origin of stereoinduction, the authors performed a crystal structure analysis of catalyst **158**, which revealed the *s-trans, cis* conformation of the thiourea moiety instead of the expected *s-cis, cis* alternative [1, 116] and a long distance of the most suitably oriented thiourea amide hydrogen atoms of 3.53 Å (O–O distance in nitroolefin was considerably shorter: approx. 2.15 Å). Assuming that the conformation of the crystalline thiourea catalyst **158** represented the catalytically active conformation, the authors suggested a nitroalkene activation through single hydrogen bonding of the nitro functionality similar to primary amine-functionalized thiourea catalysts **99** (Scheme 6.102) and **100** (Scheme 6.104).

M. Shi and Y.-L. Shi reported the synthesis and application of new bifunctional axially chiral (thio)urea-phosphine organocatalysts in the asymmetric aza-Morita–Baylis–Hillman (MBH) reaction [176, 177] of *N*-sulfonated imines with methyl vinyl ketone (MVK), phenyl vinyl ketone (PVK), ethyl vinyl ketone (EVK) or acrolein [316]. The design of the catalyst structure is based on axially chiral BINOL-derived phosphines [317, 318] that have already been successfully utilized as bifunctional catalysts in asymmetric aza-MBH reactions. The formal replacement of the hydrogen-bonding phenol group with a (thio)urea functionality led to catalysts **166–168** (Figure 6.51).

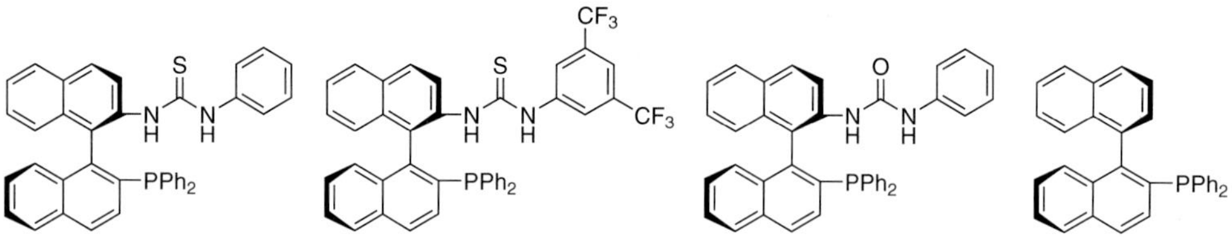

166 (70% yield)
97% yield/10 h
91% ee at rt

167 (94% yield)
50% yield/10 h
70% ee at rt

168 (78% yield)
63% yield/10 h
64% ee at rt

169
62% yield/17 h
13% ee at rt

Figure 6.51 Bifunctional (thio)urea-phoshine catalysts **166–168** prepared from (R)-2′-diphenylphosphanyl-[1,1′]binaphthalenyl-2-ylamine and the corresponding iso(thio)cyanate; the yield of the catalysts is given in parentheses. Catalyst screening results of the model product (S)-**1** (Scheme 6.159) formation in the presence of 5 mol % benzoic acid are given below and identifies **166** as the most active catalyst. **169** lacking the thiourea group was significantly less active.

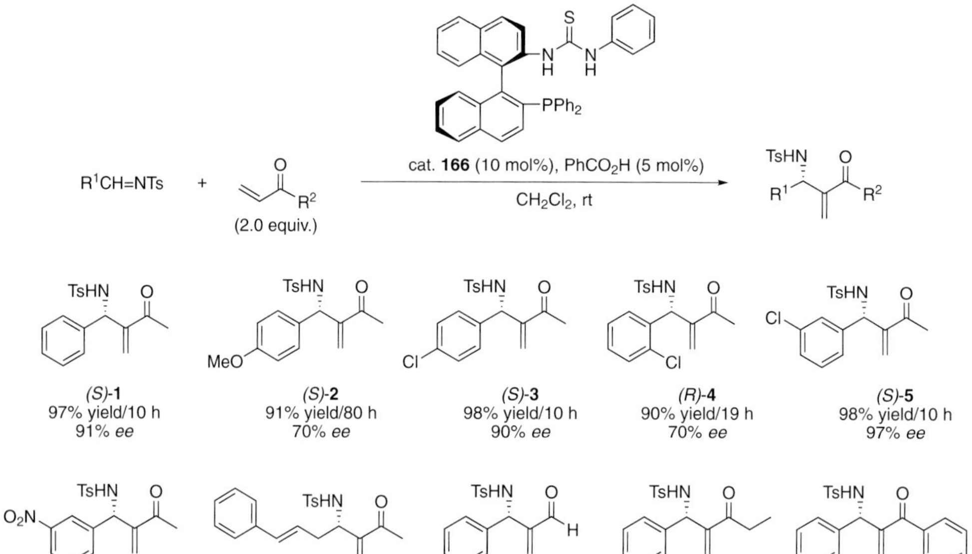

$R^1CH=NTs$ + (alkene with R^2, 2.0 equiv.) →[cat. **166** (10 mol%), PhCO$_2$H (5 mol%); CH$_2$Cl$_2$, rt] product

(S)-**1**
97% yield/10 h
91% ee

(S)-**2**
91% yield/80 h
70% ee

(S)-**3**
98% yield/10 h
90% ee

(R)-**4**
90% yield/19 h
70% ee

(S)-**5**
98% yield/10 h
97% ee

(S)-**6**
95% yield/5 h
88% ee

(S)-**7**
61% yield/10 h
67% ee

(S)-**8**
81% yield/3 h
67% ee

(S)-**9**
69% yield/56 h
77% ee

(S)-**10**
80% yield/5 h
73% ee

Scheme 6.159 Representative products obtained from the **166**-catalyzed asymmetric aza-MBH reaction between N-sulfonated imines α,β-unsaturated ketones and acrolein.

The aza-MBH model reaction of N-benzylidene-4-methylbenzenesulfonamide and methyl vinyl ketone in the presence of catalyst **166** (10 mol%) furnished different yields and ee values of the resulting adduct (S)-**1** (Scheme 6.159), when using long-stored (adduct **1**: 78% yield; 70% ee) or freshly prepared (adduct **1**: 12% yield; 16% ee) sulfonamide starting material under otherwise identical conditions. ^{1}H

NMR analysis of the long-stored N-benzylidene-4-methylbenzenesulfonamide showed 4-methylbenzenesulfonamide and benzoic acid as impurities formed by decomposition of the starting material. The addition of benzoic acid (10 mol%) to the model MBH reaction of freshly prepared N-benzylidene-4-methylbenzenesulfonamide with MVK indicated that this additive accelerated the conversion (96% yield/10 h) and increased the *ee* value (87%/rt/CH$_2$Cl$_2$) of adduct **1**, while the addition of 4-methylbenzenesulfonamide turned out to be without impact on the progress and outcome of this aza-MBH reaction. The optimization of the reactions conditions including benzoic acid loading, acidic additives, solvent, temperature, and the catalytic efficiency of **166–168** were carried out in the formation of model product **1** (Scheme 6.159). Variable loadings of benzoic acid in the range of 2.0 mol% (adduct **1**: 81% yield/10 h; 87% *ee*/rt) to 50 mol% (adduct **1**: 11% yield/10 h; 74% *ee*/rt) to demonstrate 5 mol% (adduct **1**: 97%, 91% *ee*/rt) loading to be the best. Various acidic additives with different steric hindrances and acididities were explored; these experiments revealed that additives with weaker (e.g., *p*-nitrophenol, pK_a = 7.15: yield 32%/10 h; 17% *ee*) or stronger (e.g., *o*-iodobenzoic acid, pK_a = 2.86, 41% yield/10 h 76% *ee*/rt) acidity resulted in lower yields and *ee* values of adduct **1**. Only those additives with similar acidity to benzoic acid (pK_a = 4.20) gave satisfactory results (e.g., naphthalene-2-carboxylic acid, pK_a 4.16, 95% yield/10 h, 88% *ee*/rt); benzoic acid produced the highest *ee* value (91%) and yield (97%) of the aza-MBH adduct **1** in 10 h reaction time. A longer reaction time and a lower reaction temperature (0 °C) had no significant effect on yield or enantioselectivity (adduct **1**: 95% yield; 92% *ee* after 72 h; 10 mol% **166**, 5 mol% benzoic acid). Thiourea **166** proved to be the most active thiourea catalyst (Figure 6.51) and was utilized at 10 mol% loading under optimized conditions to organocatalyze aza-MBH reactions of various *o*-, *m*-, and *p*-substituted aromatic N-sulfonated imines with simple α,β-unsaturated carbonyl compounds affording the adducts **1–10** in moderate to excellent yields (61–98%) and enantioselectivities (67–97% *ee*) (Scheme 6.159). The thiourea functionality appeared to be essential for the accelerating effect as well for the stereoinduction as shown by the application of **159** under optimized conditions and longer reaction time (Figure 6.51). The mechanism proposed for the **166**-catalyzed aza-MBH reaction and the function of benzoic acid based upon ^{31}P NMR spectroscopic investigations (Scheme 6.160). Thiourea **166** was suggested to operate as a bifunctional organocatalyst such that the phosphine group serves as the nucleophile to initiate the transformation by a Michael addition step, while the thiourea group serves as a hydrogen-bonding donor to stabilize the *in situ* generated enolate intermediate **A**. A subsequent Mannich reaction furnishes the sterically favored intermediate **C** instead of disfavored **D**, and elimination gives the observed (*S*)-configured product **1** in Scheme 6.159. Benzoic acid serves as a proton source leading to rate enhancement and improved asymmetric induction. The protonation of **A** gives intermediate **B** that is additionally stabilized through benzoate acting as a hydrogen-bond acceptor; benzoic acid, however, protonated the negatively charged nitrogen in **C** to give (*S*)-configured adduct **1** and suppressed reversible steps that could led to decreased enantioselectivities. Too much benzoic acid or addition of stronger acids, however, disfavored

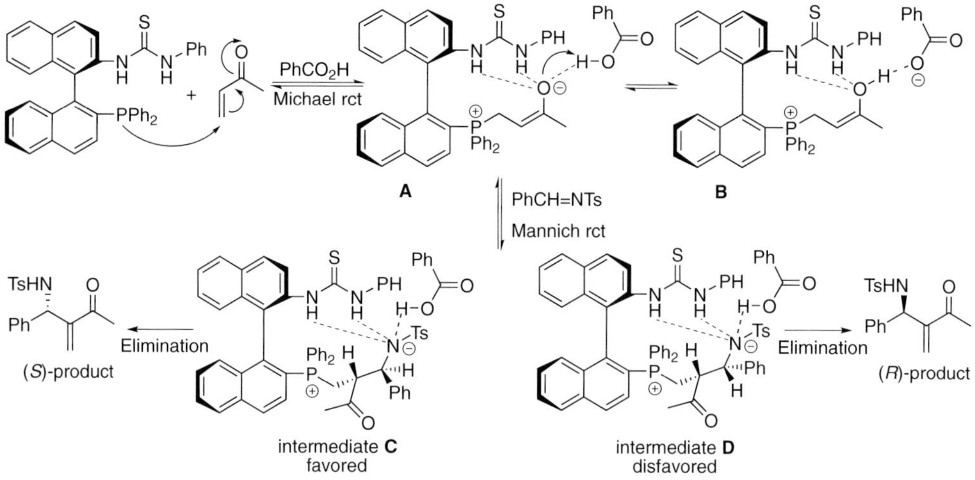

Scheme 6.160 Mechanistic proposal for the **166**-catalyzed aza-MBH reaction using benzoic acid as additive.

the formation of intermediate **A** in its rapid equilibrium with intermediate **B** (Scheme 6.160). In contrast, weaker acids proved to be less efficient as proton sources to generate the enol intermediate and to accelerate the reaction rate. This methodology represents a rare example for dual catalysis between a Brønsted acid and a hydrogen-bonding thiourea derivative similar to the cooperative Brønsted-acid type organocatalysis of the styrene oxides alkoholysis developed by the Schreiner group (Scheme 6.28) [173].

M. Shi and co-workers utilized axially chiral bis(arylurea)- and bis(arylthiourea)-based organocatalysts already introduced by Connon et al. [315]. Figure 6.50 in the enantioselective Henry addition [224] of nitromethane to arylaldehydes [319]. The synthesis of these thiourea organocatalysts is based on axially chiral (R)-(+)-binaphthalenediamine (BINAM) as the starting material to obtain successively the hydrogenated and the 3,3'-disubstituted BINAM derivatives that were subsequently condensded with 3,5-bis(trifluoromethyl) phenyl iso(thio)cyanate providing binaphthyl (thio)ureas **157**, **158**, **163**, and **170–175** in yields ranging from 58% (**170**) to 99% (**172**) (Figure 6.52).

The catalyst screening experiments were performed in the asymmetric Henry addition of nitromethane (10 equiv.) to 4-nitrobenzaldehyde in the presence of DABCO (20 mol %) as the base and (thio)ureas **157**, **158**, **163**, and **170–175** (each 10 mol% loading). After 12 h in reaction time at room temperature and in THF as the solvent, the corresponding Henry adduct was obtained in excellent yields (99%) but with very low *ee* values (7–17%) nearly independently of the sterical hindrance of the axially chiral backbone skeleton (e.g., **172**: and **174**: each 99% yield; 11% *ee*). Thioureas appeared slightly more enantioselective (e.g., **163**: 83% yield, 33% *ee*; **171**: 99% yield, 15% *ee*) than their urea counterparts probably due

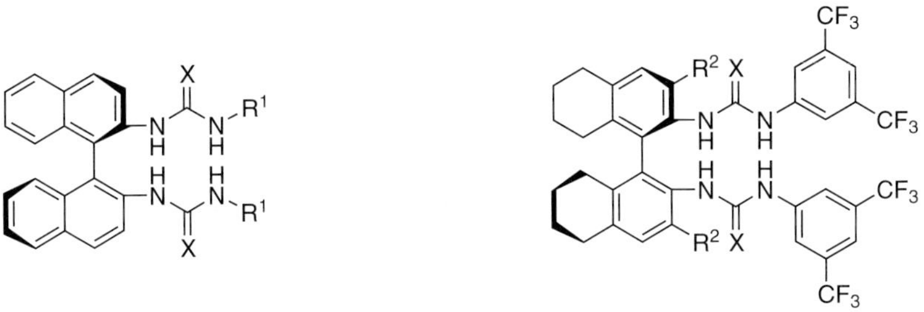

170: X = S, R¹ = Ph
157: X = O, R¹ = 3,5-(CF$_3$)$_2$C$_6$H$_3$
158: X = S, R¹ = 3,5-(CF$_3$)$_2$C$_6$H$_3$

171: X = O, R² = H
163: X = S, R² = H
172: X = O, R² = Ph

173: X = S, R² = Ph
174: X = S, R² = 4-MeC$_6$H$_4$
175: X = S, R² = 3,5-Me$_2$C$_6$H$_3$

Figure 6.52 Axially chiral bis(thio)ureas prepared from BINAM (**157**, **158**, and **170**), (R)-(+)5,5′,6,6′,7,7′,8,8′-octahydro-1,1′-binaphthyl-2,2′-diamine (**163** and **171**), and (R)-(+)3,3′-disubstituted-5,6,7,8,5′,6′,7′,8′-octahydro-[1,1′]binaphthalenyl-2,2′-diamines (**172–175**), respectively, for organocatalysis of Henry reactions between aromatic aldehydes and nitromethane.

to more efficient substrate complexation through stronger hydrogen-bonding interactions. Thiourea **163** derived from (R)-(+) 5,5′,6,6′,7,7′,8,8′-octahydro-1,1′-binaphthyl-2,2′-diamine was identified to be the most active catalyst and was employed for the Henry reactions of electron-rich and electron-poor aromatic aldehydes with nitromethane under solvent-, temperature-, and base additive optimized conditions. The protocol worked under mild conditions to furnish the corresponding nitroalcohols **1–8** in good to excellent yields (75–99%) and moderate to good enantioselectivities (50–75% *ee*) as depicted in Scheme 6.161.

The authors interpreted the outcome of these Henry reactions with an activation of the aldehyde component through double hydrogen-bonding interactions with the thiourea moiety facilitating the product-forming nucleophilic attack of the *in situ* generated nitronate (Scheme 6.162) [319].

In 2008, M. Shi and Liu reported the synthesis of novel axially chiral bis(thio)ureas from the (R)-(−)5,5′,6,6′,7,7′,8,8′-octahydro-1,1′-binaphthyl-2,2′-diamine (H$_8$-BINAM) building block and the application of these (thio)ureas as hydrogen-bonding catalysts in the enantioselective Morita–Baylis–Hillman reaction [176, 177] between predominantly 2-cyclohexen-1-one and a wide range of electron-deficient and electron-rich benzaldehydes in the presence of DABCO (20 mol%) [320]. Employing the optimized reaction parameters (toluene, rt, 3 d reaction time), the most efficient axially chiral bisthiourea **176** (20 mol%) incorporating the 3,5-bis(trifluoromethyl)phenyl moiety at the 3,3′-position furnished the desired MBH adducts **1–8** in moderate (50%) to excellent yields (99%) and in good enantioselectivities (62–88%) as shown in Scheme 6.163. The MBH reaction between 2-cyclopenten-1-one and 4-nitrobenzaldehyde was described to proceed under

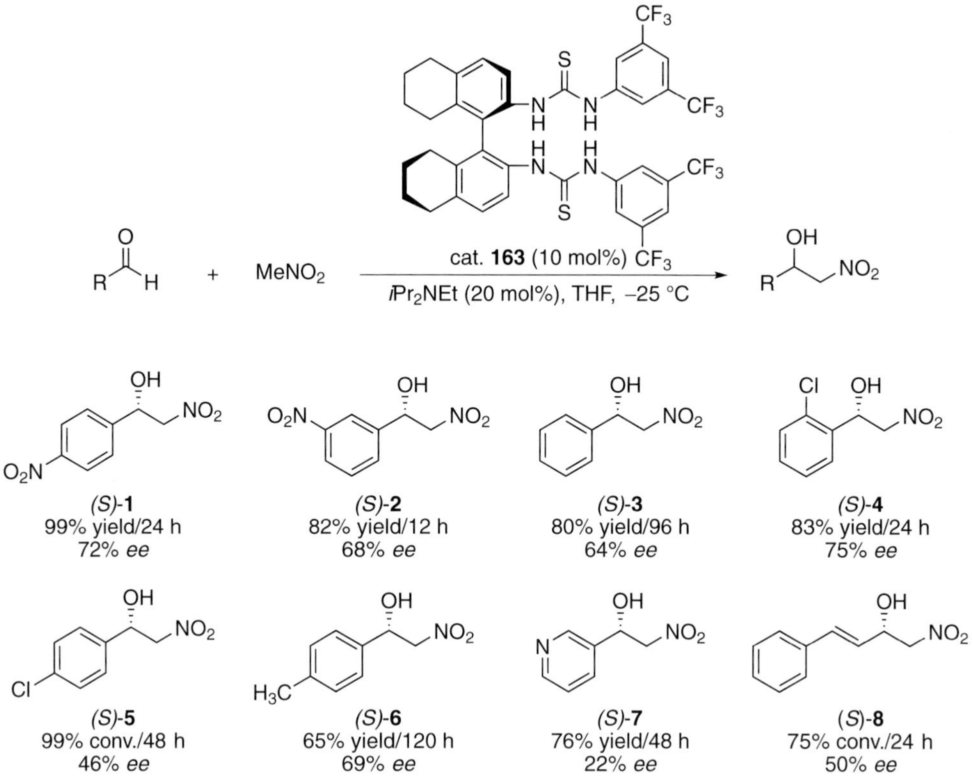

Scheme 6.161 Product range for the **163**-catalyzed enantioselective Henry reaction of arylaldehydes with nitromethane.

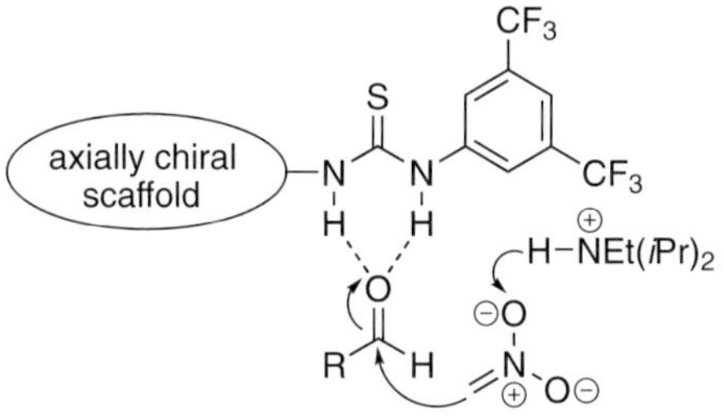

(S)-**1** 99% yield/24 h 72% ee

(S)-**2** 82% yield/12 h 68% ee

(S)-**3** 80% yield/96 h 64% ee

(S)-**4** 83% yield/24 h 75% ee

(S)-**5** 99% conv./48 h 46% ee

(S)-**6** 65% yield/120 h 69% ee

(S)-**7** 76% yield/48 h 22% ee

(S)-**8** 75% conv./24 h 50% ee

Scheme 6.162 Mechanistic proposal for the Henry reaction catalyzed by bifunctional double hydrogen-bonding thiourea derivative **163**.

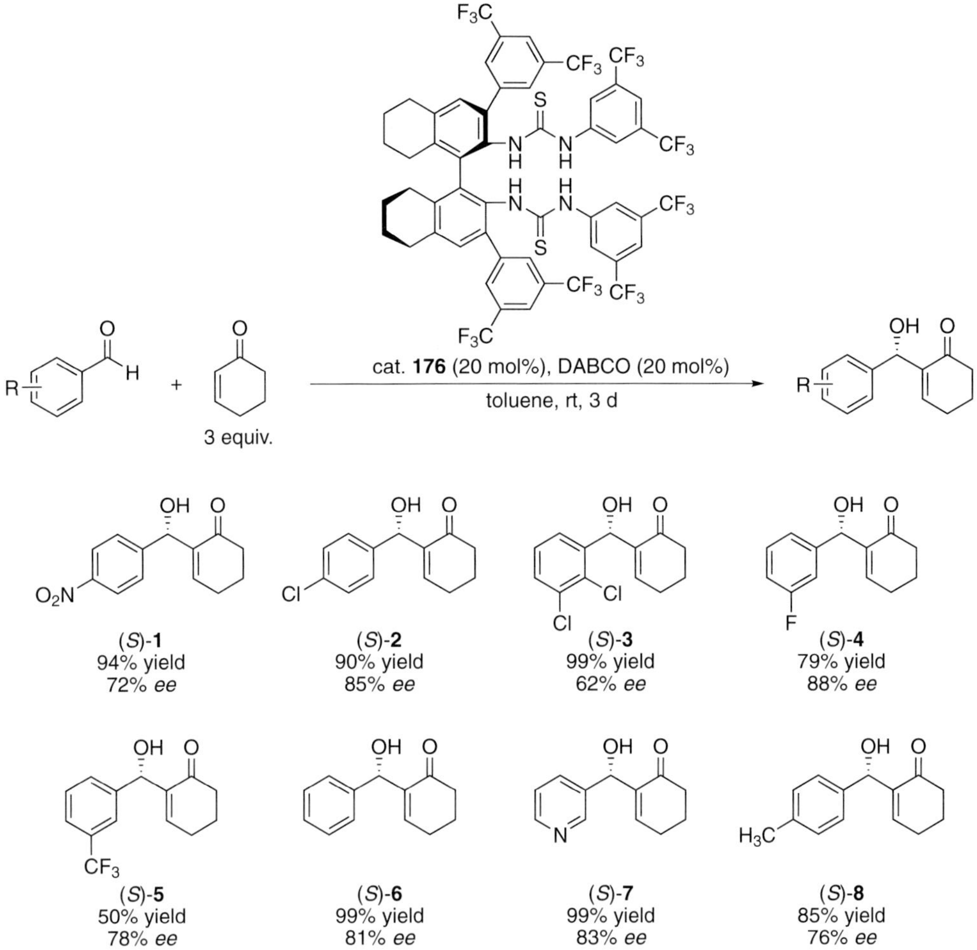

Scheme 6.163 Product range of the asymmetric MBH reaction catalyzed by bisthiourea **176** in the presence of DABCO.

standard conditions producing the corresponding adduct in 43% yield and 60% ee. Notably, this protocol turned out to favor the formation of the (S)-configured MBH adducts.

6.2.2.5 Guanidine-Based Thiourea Derivatives

Guanidines and guanidinium salts have been widely used in organic synthesis [321] and due to their hydrogen-bond donor ability they are utilized for anion recognition [111]. In 1996, guanidine-functionalized diketopiperazine **177** was introduced and suggested to be an efficient catalyst for the asymmetric Strecker reaction of various aldimines (80–97% yield/up to 99% ee) [322]. In the

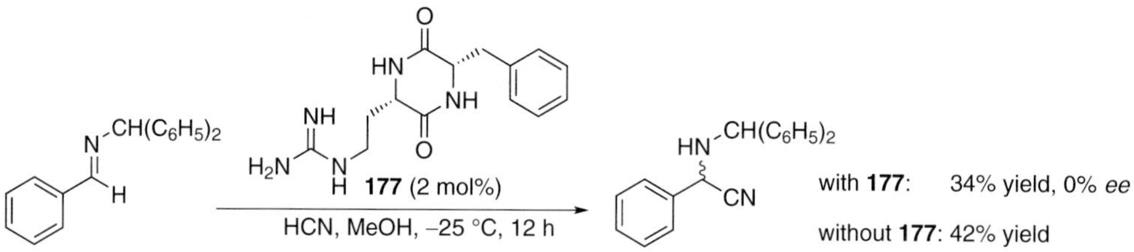

Scheme 6.164 Strecker reaction for examination of the catalytic efficiency of guanidine-functionalized diketopiperazine **177**.

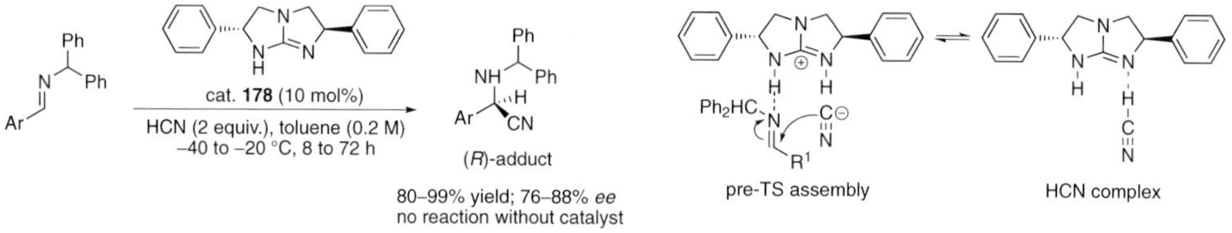

with **177**: 34% yield, 0% ee
without **177**: 42% yield

Scheme 6.165 Enantioselective Strecker reactions catalyzed by bifunctional hydrogen-bonding guanidine organocatalyst **178**. Catalytic action of **178**: HCN hydrogen bonds to **178** and generates a guanidinium cyanide complex after protonation, which activates the aldimine through single hydrogen bonding and facilitates stereoselective cyanide attack and product formation.

following decade of growing interest in organocatalysis, this publication has been cited frequently as the first example of an organocatalyzed enantioselective Strecker reaction. During their investigations toward the development of a metal-free protocol for the Strecker reaction the Kunz group could not, however, reproduce the reported synthesis of guanidine-diketopiperazine **177** nor the reported catalytic effect [323]. Structure **177** was prepared by a different procedure and structurally confirmed by X-ray analysis; this compound has proven not to be an enantioselective organocatalyst for the Strecker reaction at 0 °C, −25 °C, or −70 °C, neither as the hydroacetate nor as the hydronitrate, and without **177** the Strecker reaction was found to proceed even slightly faster (Scheme 6.164) [323].

The first organocatalyzed enantioselective Strecker reaction thus has been reported by Corey and Grogan in 1999 [324]. The C_2-symmetric chiral bicyclic guanidine **178** was described to operate as a bifunctional single hydrogen-bonding organocatalyst, which activated the aldimine substrate and formed the corresponding Strecker adducts in good to excellent yields (80–99%) and good enantioselectivities (76–88%) (Scheme 6.165). This guanidine catalyst was found to be recoverable (80–90%) and reusable in these asymmetric Strecker reactions.

In 2005, Nagasawa and co-workers developed a new catalyst design concept leading to novel hydrogen-bonding bifunctional organocatalysts, which incorpo-

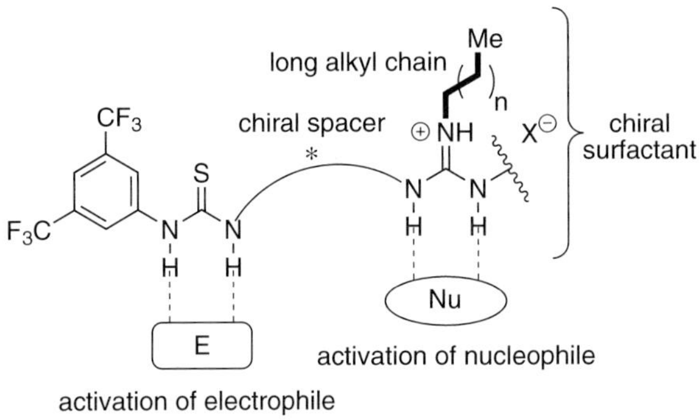

Figure 6.53 Structural concept for the design of guanidine-based bifunctional thiourea organocatalysts **179–189** and mode of bifunctional substrate activation.

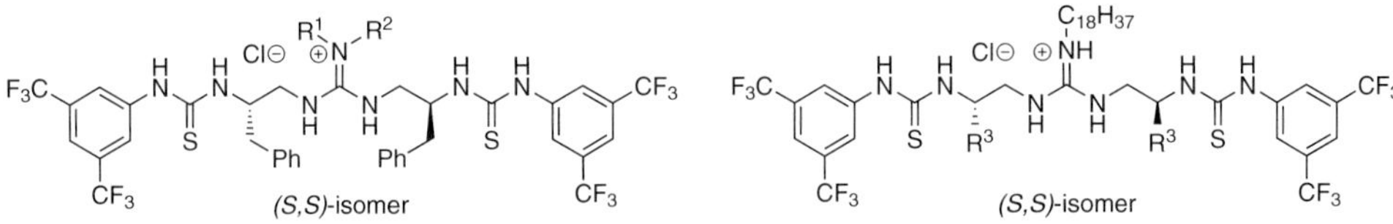

179: $R^1 = C_4H_9$, $R^2 = H$
180: $R^1 = R^2 = C_4H_9$
181: $R^1 = R^2 = (CH_2)_4$
182: $R^1 = C_8H_{17}$, $R^2 = H$
183: $R^1 = C_{18}H_{37}$, $R^2 = H$
184: $R^1 = 3,5\text{-}(CF_3)_2\text{-}C_6H_4$, $R^2 = H$
185: $R^1 = 3,5\text{-}(OMe)_2\text{-}C_6H_4$, $R^2 = H$
186: $R^1 = (R,R)$-isomer of **183**
187: $R^3 = Me$
188: $R^3 = iPr$
189: $R^3 = tBu$

Figure 6.54 Structures of guanidine-based thiourea derivatives screened in the Henry reaction of nitromethane with cyclohexane carboxaldehyde under phase-transfer conditions.

rate both a guanidinium alkyl chain moiety and two electron-poor thiourea functional groups linked by a modular chiral spacer (Figure 6.53) [325–327]. Based on this concept, a series of C_2-symmetric guanidine-thiourea derivatives **179–189** having various substituents were prepared via multistep synthesis starting from enantiomerically pure amino acids (Figure 6.54). These were screened for their catalytic efficiency in the enantioselective Henry (nitroaldol) reaction [224] between cyclohexane carboxaldehyde and nitromethane under toluene/aqueous potassium hydroxide biphasic conditions at 0 °C and 5 mol% catalyst loading [325].

Guanidine-thiourea **183** incorporating an octadecyl substituent (R^1) at the guanidine moiety and benzyl groups as chiral spacers (R^3) revealed the highest efficiency in the formation of the model Henry adduct considering both reaction rate (91% yield/24 h) and asymmetric induction (43% *ee*) while catalyst candidates **179–182**, **184–185**, and **187–189** gave poorer yields (24–89%) and *ee* values (8–36%) [325]. Time course studies with catalyst **183** for the above-screening reaction indicated that the retro-nitroaldol reaction occurred under the chosen biphasic reaction conditions resulting in the (*R*)-configured adduct and gradual decrease of enantioselectivity (after 1 h: 52% yield/52% *ee*; 12 h: 90% yield/47% *ee*; 24 h: 91% yield/43% *ee*) [327]. Since counter anions of not only ammonium salts but also guanidinium salts have been reported to influence catalytic activity and stereoinduction the authors examined various alkali bromides, iodides, and tetrafluoroborates as additives (50 mol%) in the standard screening reaction. A hard counter anion furnished no improvement of the *ee* value (KBF$_4$: 77% yield; 47% *ee*/24 h) while soft KI was identified to serve as an effective inhibitor of the retro-mode reaction (KI: 88% yield; 74% *ee*/24 h). Utilizing catalyst **183** with iodide instead of chloride Henry adduct (*R*)-1-cyclohexyl-2-nitro-ethanol was formed with 85% yield and 72% *ee*. This result suggested that under these biphasic conditions the chloride counter anion of **183** was replaced with iodide *in situ*. Under optimized reactions conditions concerning the choice of the organic solvent, ratio of organic solvent and water, the amount of base, and the additive the enantioselective and *syn*-diastereoselective Henry reactions of nitromethane [325] and prochiral nitroalkanes [326] with aliphatic cyclic and α-branched aldehydes were performed (Scheme 6.166). The resulting nitroalcohol products (**1–10**) were isolated in moderate to good yields (51–88%) with moderate to excellent enantioselectivities (51–99% *ee*) (Scheme 6.166). In all cases, the addition reactions proceeded with high

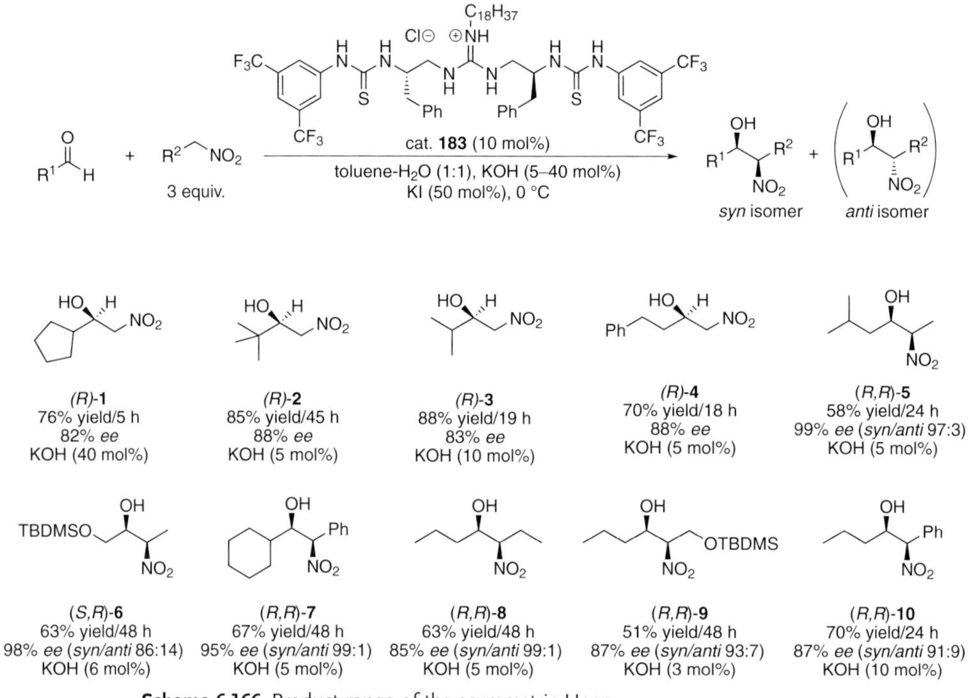

Scheme 6.166 Product range of the asymmetric Henry (nitroaldol) reaction of aldehydes with various nitroalkanes in the presence of (*S,S*)-configured catalyst **183**.

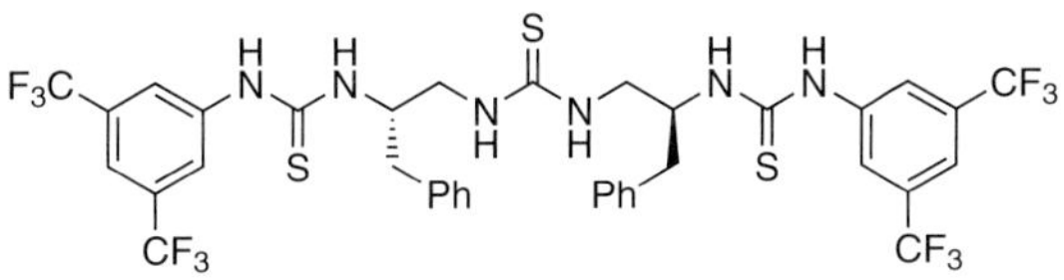

190
0% yield

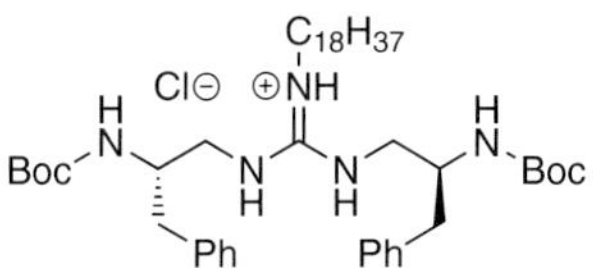

191
1% yield, 6% ee

Figure 6.55 Uncharged tris-thiourea **190** without guanidinium moiety and charged guanidinium structure **191** without a thiourea group appeared catalytically inactive in the Henry reaction of cyclohexane carboxaldehyde with nitromethane, while guanidine-thiourea **183** gave 99% yield and 95% ee under identical (optimized) conditions.

syn-diastereoselectivity (up to 99 : 1 dr) [325–327]. The derivatives of catalyst **183** without a guanidine moiety (**190**) or without a thiourea functionality (**191**) revealed no or only a very poor catalytic efficiency under optimized reaction conditions (Figure 6.55) [327].

To gain insight into the structure–activity relationships various substituent patterns of (S,S)-catalyst **183** were studied utilizing the standard screening reaction. The substituent on the guanidinium moiety influenced the catalytic efficiency resulting in a poor yield when using an alkyl chain shorter than 12 carbon atoms (e.g., **182** C_8H_{17}: 8% yield/86% ee, but **183** $C_{18}H_{37}$: 91% yield/92% ee, in 24 h; Figure 6.54). The introduction of an aromatic group at the terminal position of a short alkyl chain (e.g., catalysts **184** and **185**; Figure 6.54) also enhanced the reaction rate and the asymmetric induction (e.g., **184**: 88% yield/82% ee; **185**: 82% yield/97% ee in 24 h) in comparison to an unsubstituted short alkyl chain (e.g., **179**: 8% yield/87% ee in 24 h; Figure 6.54). The substituent R^3 at the chiral spacer was found to be critical for both the reaction rate and the selectivity. The ee values decreased as the bulk of the R^3 group was increased (structure **187**: R^3 = Me: 3% yield/76% ee; **189**: R^3 = tBu: 65 yield/6% ee after 24 h; Figure 6.54). Anionic (SDS, AOT) and nonionic (Triton® X-100)[8] surfactants (20 mol %) completely inhibited the catalytic activity of **183** in the nitroaldol screening reaction, while the cationic surfactant CTAB at 20 mol% loading lowered the catalytic activity to obtain 53% yield and 80% ee, but suppressed the reaction entirely at 50 mol% loading. The authors assumed that the cationic surfactant properties under biphasic conditions were important for the reactivity as well as for the selectivity, which were controlled through self-aggregation. The alkyl chain of the guanidinium moiety was suggested to contribute to the reactivity by controlling the proximity of the substrates due to hydrophobic interactions, and the benzyl group (= R^3) linked to the

---

8) Triton X-100 (octyl phenol ethoxylate) is a registered trademark of Union Carbide and is commonly used as a detergent in biochemistry laboratories; formula $C_{14}H_{22}O(C_2H_4O)n$ (n = 9, 10), [CAS-RN 9002-93-1].

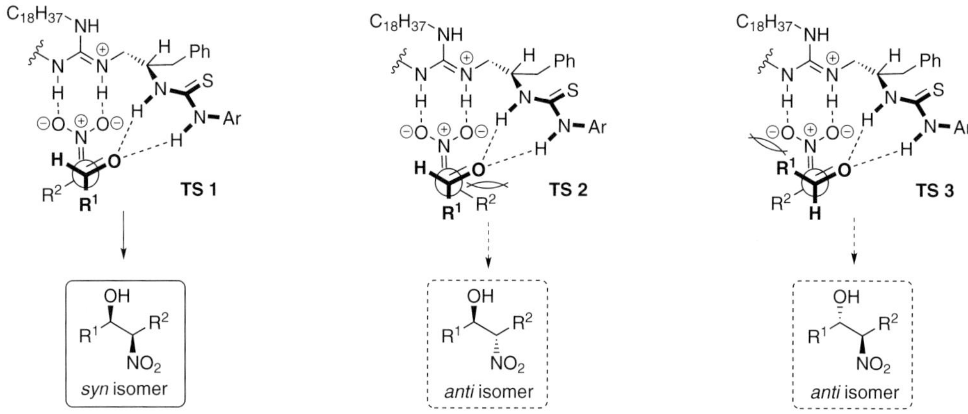

Scheme 6.167 Proposed transition-state models for the enantioselective Henry (nitroaldol) reaction in the presence of (S,S)-configured catalyst **183**: **TS 1**: *anti, anti* conformation; **TS 2**: gauche-*anti* conformation; **TS 3**: gauche-*anti* conformation.

chiral spacer fix the high-order asymmetric structure through intramolecular π–π stacking interactions [327]. On the basis of these experimental findings, the authors explained the observed selectivities with cooperative effects of the guanidinium moiety and the thiourea functionality for a chemoselective coordination and dual activation of the substrates through ionic as well as hydrogen-bonding interactions as demonstrated in the transition-state model visualized in Scheme 6.167 [325–327].

The guanidinium cation and the thiourea group were suggested to coordinate selectively to the *in situ* formed nitronate and the carbonyl oxygen of the aldehyde, respectively, and substituents in both aldehyde (R^1) and nitronate (R^2) favor an *anti* relationship to minimize steric repulsion (Scheme 6.167: **TS 1**: *anti, anti* conformation). This geometry induced the observed high *ee* and *dr* values of the nitroaldol products. The enantiodifferentiation in the case of prochiral ($R^2 \neq H$) was found to be higher than those with nitromethane ($R^2 = H$); this finding was in line with the model, because the less sterically hindered **TS 1** conformation was more favorable than the gauche, *anti* conformations of **TS 2** and **TS 3**, respectively (Scheme 6.167). The synthetic utility of this *syn*-selective enantioselective nitroaldol reaction was demonstrated by the multistep synthesis of 4-*epi*-cytoxazone and cytoxazone [327]. (S,S)-configured catalyst **183** and its (R,R)-enantiomer **186** catalyzed the initial asymmetric key steps of these syntheses, in which two stereogenic centers were constructed (Scheme 6.168) [326, 327].

Since α-branched aldehydes gave rather higher asymmetric induction (Scheme 6.166), Nagasawa *et al.* extended the biphasic strategy to the diastereoselective Henry reaction of nitromethane with enantiomerically pure (S)-configured N,N'-dibenzyl protected α-amino aldehydes and α-hydroxy aldehydes protected as silyl ethers. The screening reaction (Scheme 6.169) demonstrated a match/mismatch

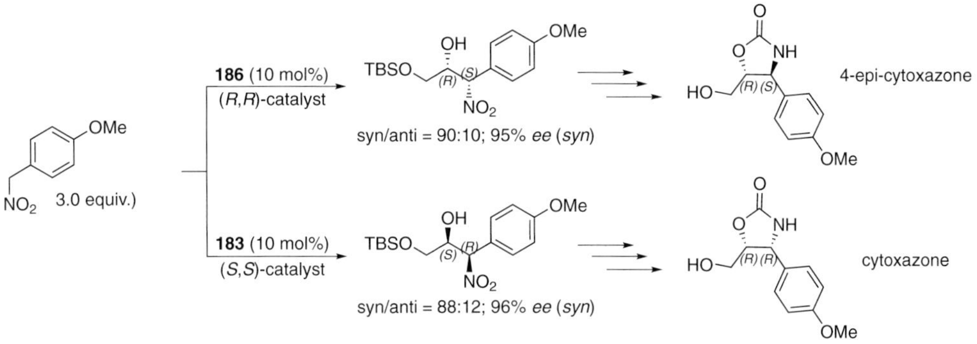

Scheme 6.168 Syntheses of 4-*epi*-cytoxazone and cytoxazone utilizing guanidine-thioureas **183** and **186** for the initial asymmetric Henry reaction step.

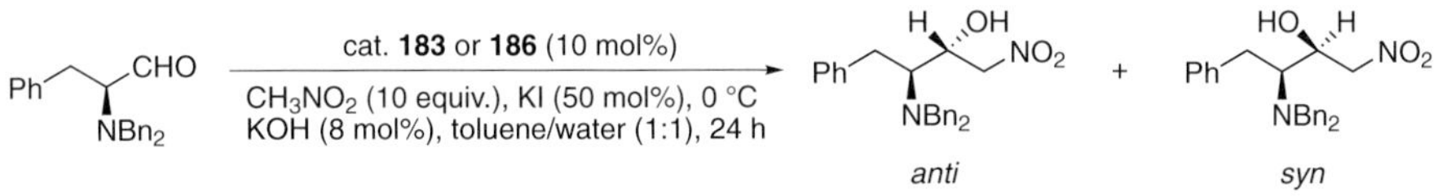

(S,S)-catalyst **183**: 8% yield, *anti/syn* 64:36, 80% *ee* (*anti*), 62% *ee* (*syn*)
(R,R)-catalyst **186**: 75% yield, *anti/syn* 95:5, 99% *ee* (*anti*), 85% *ee* (*syn*)

Scheme 6.169 Screening reaction to identify (R,R)-configured guanidine-thiourea **186** as matching catalyst for the *anti*-diastereoselective and enantioselective Henry reaction of (S)-α-amino aldehydes with nitromethane.

relationship between the guanidine catalysts **183** and **186** and (S)-α-amino aldehyde substrate.

High *anti*-diastereoselectivity (95 : 5 *dr*) and enantioselectivity of the major isomer (99% *ee*) were obtained when utilizing the combination of (R,R)-catalyst and (S)-aldehyde. This stereochemical outcome (Scheme 6.169) was explained in terms of the Cram rule proposed transition-state model. The substituent on the aldehyde would be located in an *anti*-relationship to the nitronate. As the largest substituent (R_L) should be in an *anti* position to the carbonyl group of the carbonyl substrate, the combination of (R,R)-catalyst **186** and (S)-substrate (**TS 1**) was favored rather than that of (S,S)-catalyst **183** and (S)-substrate (**TS 2**) because of the steric repulsion between R_S (smallest substituent) and nitronate (Scheme 6.170).

Utilizing 10 mol% of (R,R)-guanidine-thiourea catalyst **186** under optimized biphasic conditions for the Henry reaction [224] of (S)-α-amino aldehydes with nitromethane furnished the corresponding nitroalcohols **1–6** in yields ranging from 33 to 82% and with excellent diastereoselectivities (up to 99 : 1 *anti*/*syn*) and enantioselectivities of the major isomer (95–99% *ee*) (Scheme 6.171) [328].

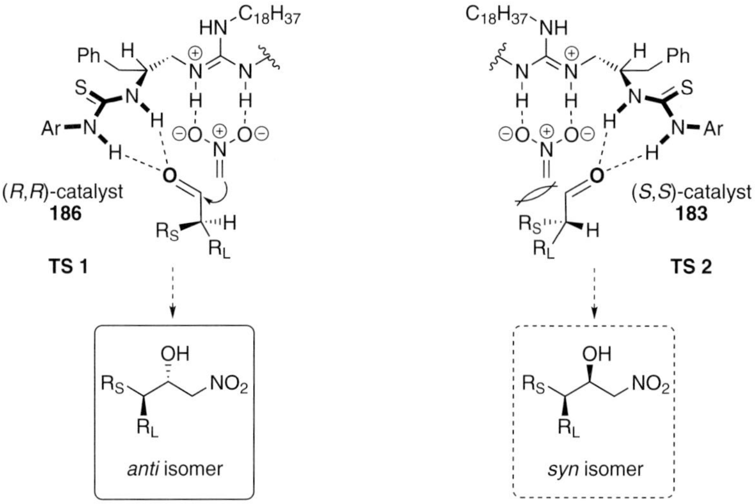

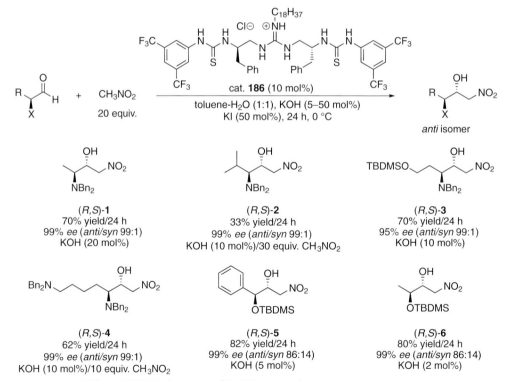

Scheme 6.170 Suggested transitions states for the *anti*-diastereoselective Henry (nitroaldol) reaction promoted by (R,R)-catalyst **186** (**TS 1**) and its (S,S)-isomer **183** (**TS 2**) to demonstrate the match/mismatch relationship between guanidine–thiourea catalyst and (S)-α-aldehyde.

Scheme 6.171 Product range of the **186**-catalyzed *anti*-diastereoselective Henry (nitroaldol) reaction of α-chiral aldehydes with nitromethane.

Boc- and Cbz-protective groups turned out to be not tolerable in this protocol as indicated by epimerization of Boc- or Cbz-protected α-amino aldehydes substrates (e.g., *N*-Boc-protected 2-amino-3-phenyl-propionaldehyde: 70% yield/24 h, *anti/syn* ratio 50:50, 20% *ee* with **186**).

The Nagasawa group modified the guanidine-thiourea catalyzed nitroaldol reaction [224] of nitroalkanes to aldehydes [325, 326, 328] proceeding under biphasic conditions and at 0 °C and published in 2008 a new protocol working at subzero temperatures (−20 to −35 °C) for α-ketoesters reacting with chiral tertiary nitroalcohols [329]. The optimized protocol utilized organocatalysts **183** and **186** (10 mol%) under biphasic conditions in the presence of KOH (10 mol%) as base additive, and KI (50 mol%) to reduce the retro-mode of the reaction at a minimized toluene–water ratio (10:1 and 5:1); employing cyclic, branched-type, and linear α-ketoesters as well as nitroalkanes as substrates, the protocol produced the desired *tert*-nitroaldol alcohol adducts **1–8** in moderate to excellent yields (35–99%) with moderate to very good *ee* values (35–93%) as depicted in Scheme 6.172. An exemplified aromatic α-ketoester (R^1 = Ph) gave only a moderate yield (56%) and a very low *ee* value (5%) (adduct **4**; Scheme 6.172). The addition of nitroethane and 1-nitropropane was reported to proceed under *syn*-diastereoselectivity (e.g., adduct **7**, Scheme 6.172: *syn/anti* 92:8) und afforded the (*S,S*)-adducts when using (*R,R*)-catalysts **186** and the enantiomeric (*R,R*)-adducts in the presence (10 mol%) of the (*S,S*)-catalyst **183**. Performing the **183**-catalyzed (10 mol%) formation of nitroaldol adduct **1** (Scheme 6.172) under water-free conditions in pure toluene (−20 °C, 5 mol% KOH, 50 mol% KI), the respective adduct was obtained in only 8% yield and 15% *ee*, while the aqueous protocol (toluene/water 5:1) under otherwise identical conditions gave the same adduct with 99% yield and 68% *ee* (Scheme 6.172); this emphasized the key role of water in having both the accelerating effect and the asymmetric induction through guanidine-thiourea derivatives such as **183** and **186**, respectively. The stereochemical outcome of this nitroaldol reaction was ascribed to a transition state comparable to that proposed for aldehyde substrates (Schemes 6.167 and 6.170) and was based on the hydrogen-bonding mediated chemoselective dual activation of the reactants. To minimize steric repulsion, the larger substituent of the R^1 group of the α-ketoester and the R^2 group of the nitroalkane was considered to favor the relative *anti* geometry with respect to the nitroalkane and the ketone producing the (*R,R*)-configured *syn*-isomeric adduct **6** under catalysis with (*S,S*)-configured catalyst **183** (Schemes 6.172 and 6.173) [329].

6.2.2.6 Saccharide-Based (Thio)urea Derivatives

Carbohydrates are configurationally stable, easily available in enantiopure forms from the chiral pool, and they show a high density of chiral information per molecular unit. Their polyfunctionality and structural diversity facilitate their tailor-made modification, derivatization, and structural optimization for a broad spectrum of synthetic applications. While derivatives of various saccharides have already been utilized as versatile starting materials and building blocks for chiral auxiliaries, ligands, and reagents [330] their obvious role as precursors for the

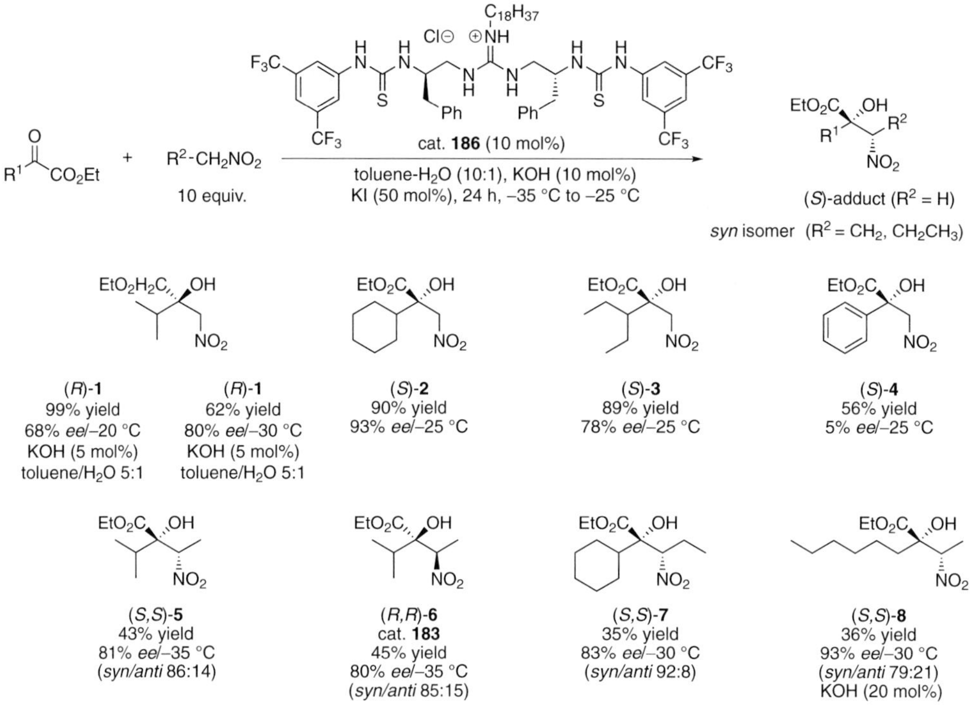

Scheme 6.172 Typical chiral adducts obtained from the **186**-catalyzed nitroaldol reaction between α-ketoesters and nitroalkanes.

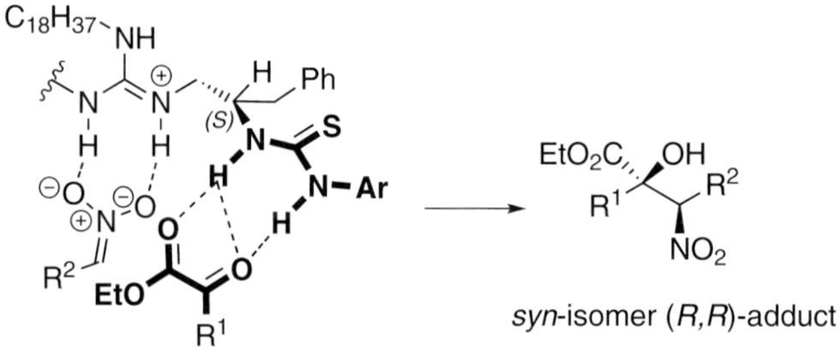

Scheme 6.173 Proposed transitions-state geometry for the nitroaldol reaction of nitroalkanes with α-ketoesters in the presence of (S,S)-configured guanidine thiourea **183**.

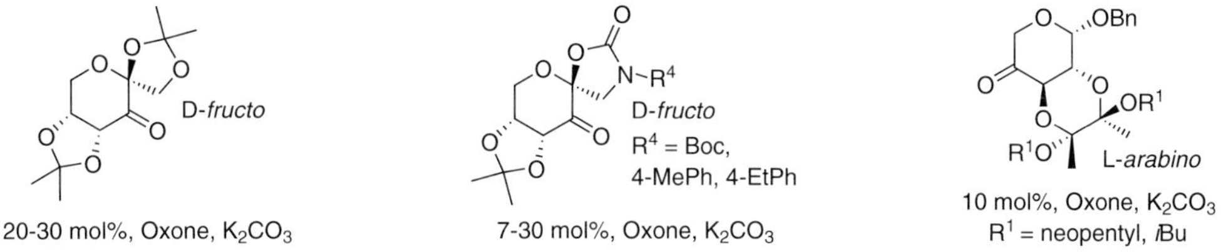

6.2 Synthetic Applications of Hydrogen-Bonding (Thio)urea Organocatalysts | 317

192
20-30 mol%, Oxone, K$_2$CO$_3$
D-fructo

193
7-30 mol%, Oxone, K$_2$CO$_3$
D-fructo
R^4 = Boc, 4-MePh, 4-EtPh

194
10 mol%, Oxone, K$_2$CO$_3$
R^1 = neopentyl, iBu
L-arabino

Figure 6.56 Saccharide-based epoxidizing organocatalysts derived from D-fructose (**192** and **193**) and L-arabinose (**194**).

synthesis of organocatalysts is underdeveloped. Shi and co-workers introduced D-fructose-based ketone **192** and its L-enantiomer as epoxidizing organocatalysts (Figure 6.56). In combination with oxone®[9], a wide range of unsaturated compounds such as hydroxy alkenes [331], enynes [332], and esters [333] were highly stereoselectively epoxidized to the corresponding oxiranes. Replacing the spiroketal functionality of **192** with a spiro-oxazolidinone moiety to obtain organocatalyst **193** supplemented the substrate scope of **192** to terminal and 1,2-*cis*-configured alkenes as well as styrenes [334]; Catalysts **192** and **193** have also been successfully applied to the oxidation of disulfides to chiral thiosulfinates [335, 336]. The Shing group prepared ketone derivatives of L-arabinose such as catalyst **194** to achieve good to excellent results for the epoxidation of aromatic *trans*- and trisubstituted alkenes (Figure 6.56) [337].

Series of various mono,- bi-, and poly-(thio)urea-functionalized (poly)saccharides have already been synthesized and studied for molecular recognition of, e.g., dimethyl and phenylphosphate as model compounds for monoanionic and polyanionic phosphate esters, respectively [111]. Thiourea derivatives such as **195–197** were analytically identified to provide double hydrogen bonding mediated host–guest complexes of well-defined dimension and orientations and were also reported to serve as phosphate binders even in the hydrogen bonding environment of water (Figure 6.57) [111].

Kunz and co-workers for the first time realized the potential of saccharides to facilitate synthetic access to a novel class of hydrogen-bonding organocatalysts such as urea derivative **198** [338]. The rational design of these urea-functionalized monosaccharide is based on the structure of Jacobsen's *trans*-1,2-diaminocyclohexane-derived urea Schiff base catalyst **42** (Figure 6.14) that had been described as being highly efficient (up to 99% yield and *ee*) in the enantioselective hydrocyanation of a broad spectrum of aldimines (Strecker reaction) (Scheme 6.41;

---

9) Oxone is a registered trademark of DuPont with potassium peroxymonosulfate KHSO$_5$ (potassium monopersulfate) as oxidizing ingredient of a triple salt with the formula 2KHSO$_5$·KHSO$_4$·K$_2$SO$_4$ (potassium hydrogen peroxymonosulfate sulfate), [CAS-RN 70693-62-8].

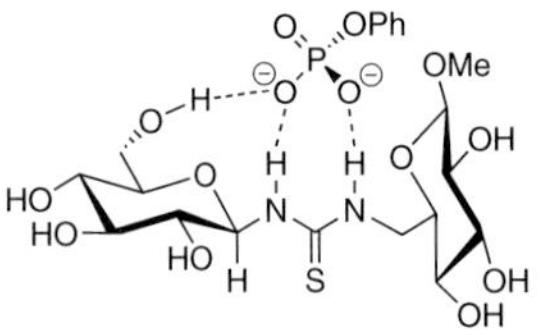

195

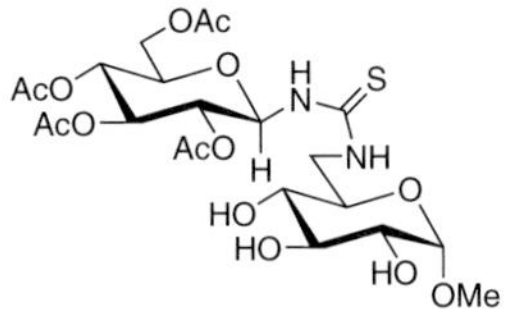

196

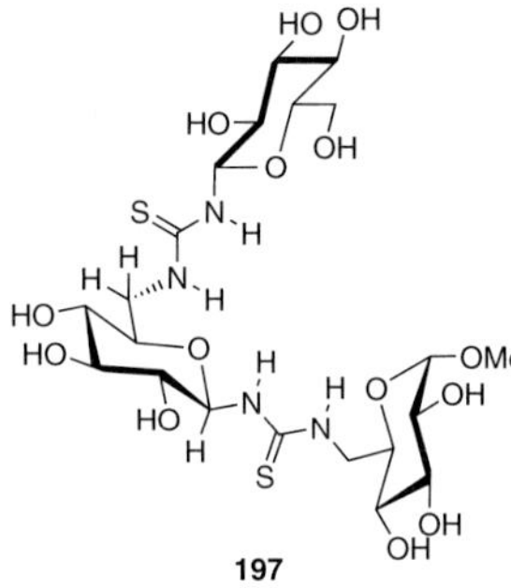

197

Figure 6.57 Saccharide-based (thio)ureas that coordinate phosphate anions through hydrogen bonding even in water.

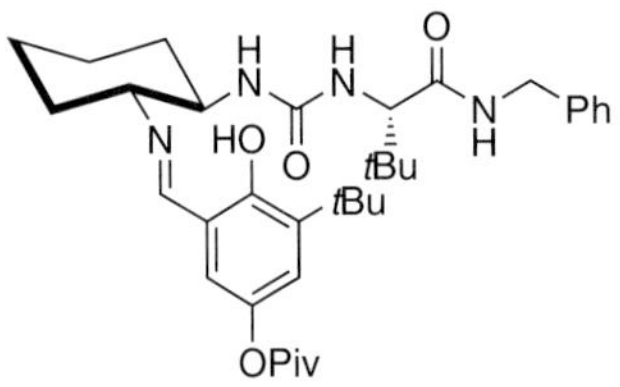

42

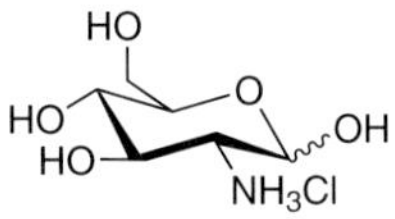

glucosamine hydrochloride

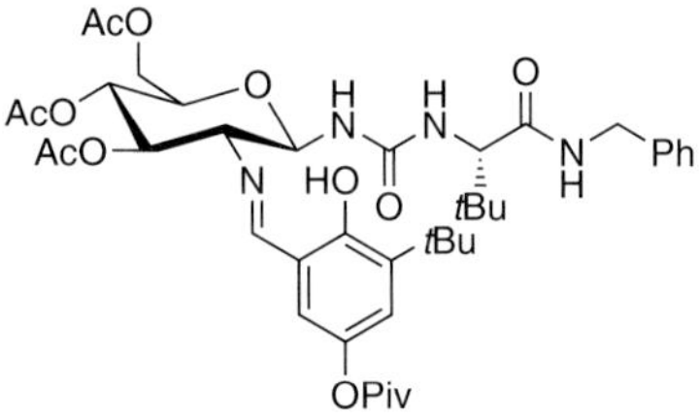

198

Figure 6.58 Schiff base catalyst **42** and glucosamine hydrochloride as starting material for the synthesis of saccharide-based catalyst **198**.

45) [122, 196]. Urea **198** was prepared from enantiomerically pure polyfunctional glucoseamine hydrochloride, which is readily accessible from chitin as a component of the natural chiral pool; it appeared to be an alternative backbone structure supplanting the *trans*-1,2-diamocyclohexane of Schiff base catalyst **42** (Figure 6.58).

Glucosaminylurea derivative **198** proved to be an efficient catalyst for the asymmetric hydrocyanation of a broad range of aromatic aldimines (Scheme 6.174) producing the respective Strecker adducts (*S*)-**1**–(*S*)-**12** in high yields (up to 98%) and *ee* values (up to 84%) at −50 °C, while the protocol developed by the Jacobsen group was reported to operate at −70 °C providing 74% yield (95% *ee*). Performing the urea **198**-catalyzed Strecker reaction of benzaldimine at −70 °C under otherwise identical conditions increased the yield (86%) and enantioselectivity (95%) of Strecker adduct (*S*)-**12** (Scheme 6.174). Aliphatic aldimines or ketimines furnished only products with moderate yields and enantioselectivities (adducts (*S*)-**9** and (*S*)-**11**) and nitrophenylglycinonitrile (adduct (*S*)-**6**) underwent racemization during the Strecker reaction (Scheme 6.174). In the case of the 2-furfuryl derivatived aldimine, the noncatalyzed transformation took place to a perceptible extent resulting in a moderate *ee* value of only 50% (adduct (*S*)-**7**).

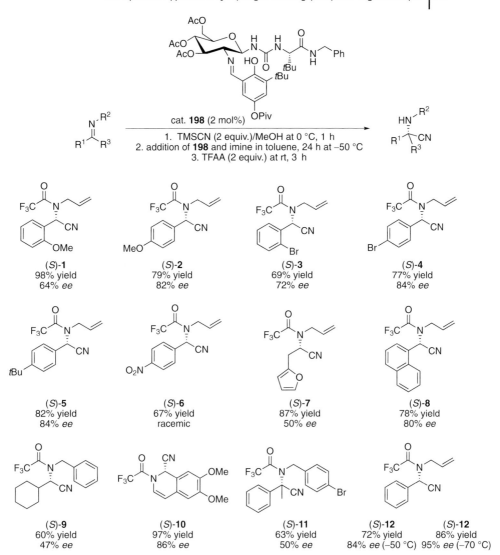

Scheme 6.174 Product range obtained from asymmetric Strecker reactions catalyzed by saccharide urea **198**.

Structural modifications of catalyst **198** concerned the positions of the urea and Schiff base side-chains (**199–201**; Figure 6.59), the alterations of the protective group pattern of the D-glucosamine scaffold (**202–205**; Figure 6.59), as well as immobilization of the catalyst on polystyrene (**202**, **203**, and **205**) were prepared to perform structure–efficiency relationships for the asymmetric model Strecker reaction of benzaldimine providing adduct **12** at −50/−70 °C and 2–8 mol% catalyst loadings. The experimental results visualized in Figure 6.59 suggested that the saccharide backbone was not just a surrogate of the cyclohexane ring incorporated

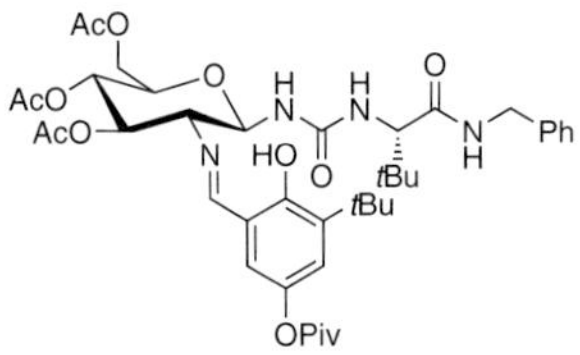

198
(2 mol%)/−70 °C
86% yield
95% ee (S)-**12**

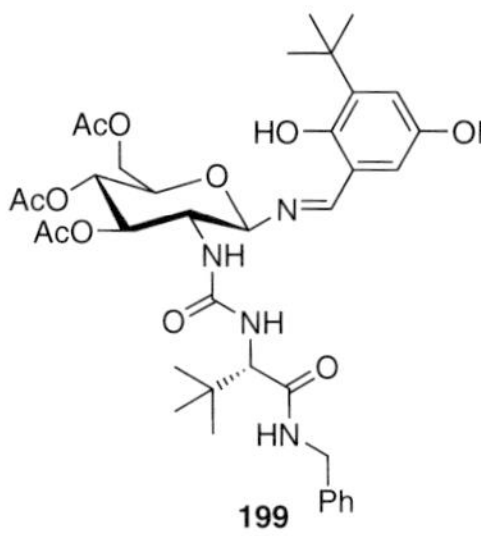

199
(5 mol%)/−70 °C
100% yield (by ¹H NMR)
15% ee (S)-**12**

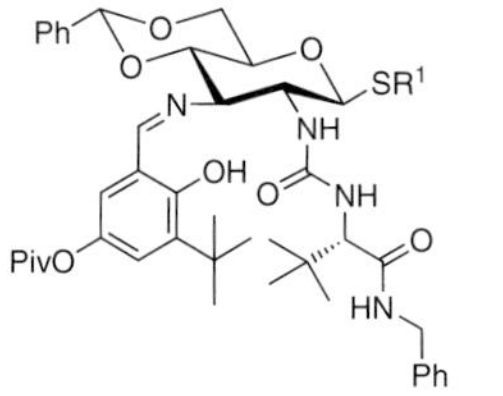

200
(5 mol%)/−50 °C
64% yield
30% ee (S)-**12**

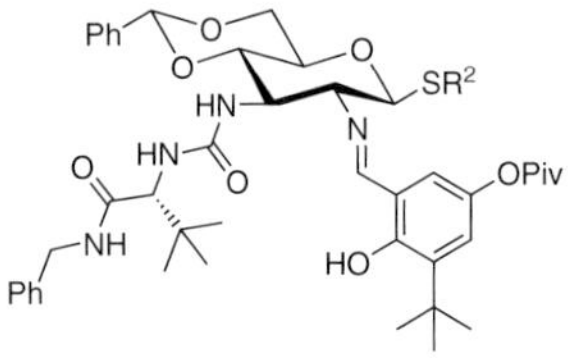

201
(2 mol%)/−50 °C
85% yield
23% ee (R)-**12**

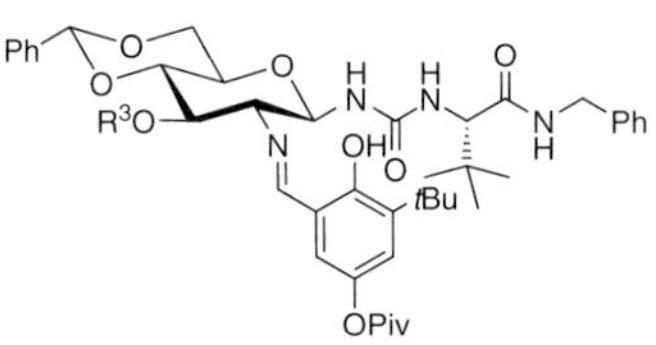

202
(8 mol%)/−50 °C
35% yield
12% ee (R)-**12**

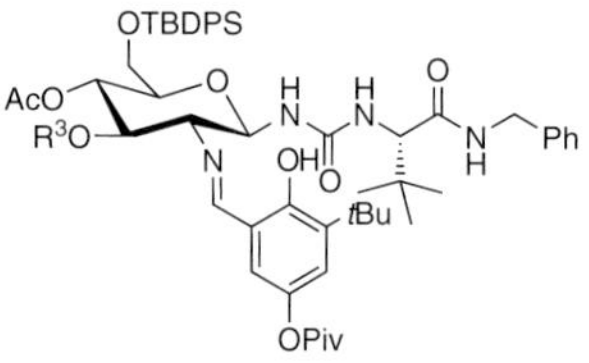

203
(4 mol%)/−50 °C
30% yield
15% ee (S)-**12**

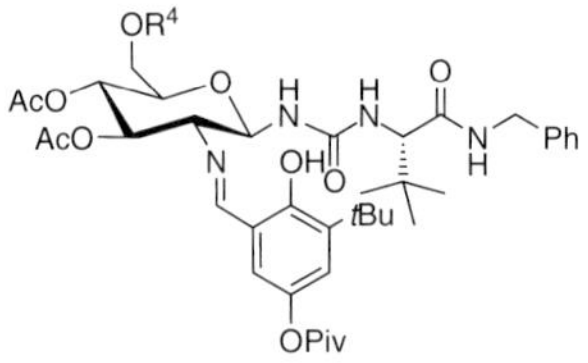

204
(2 mol%)/−70 °C
45% yield
36% ee (S)-**12**

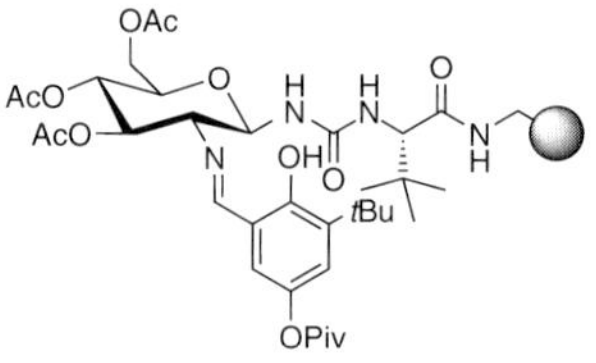

205 (2 mol%)/−50 °C
54% yield
4% ee (S)-**12**

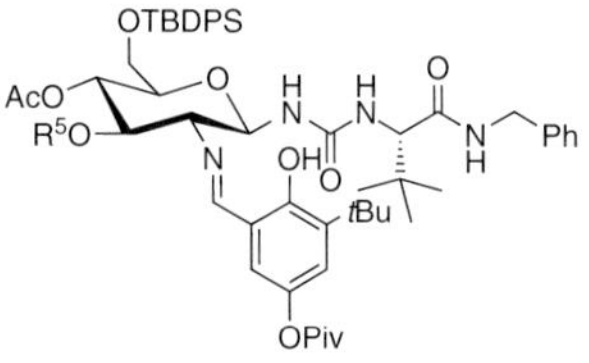

206
(2 mol%)/−70 °C
18% yield
10% ee (S)-**12**

R^1 = Et; R^2 = -(CH$_2$)$_4$COOEt; R^3 = -(CO)(CH$_2$)$_2$(CO)O(CH$_2$)$_8$O(CO)Ps; R^4 = -(CO)(CH$_2$)$_2$(CO)O(CH$_2$)$_8$OTBDS; R^5 = -(CO)(CH$_2$)$_9$OTBDS

Figure 6.59 Structural modifications of glucosamine-derived urea catalysts; screening in the formation of Strecker adduct **12** (Scheme 6.174).

in catalyst **42**, but decreased drastically the yields and enantioselectivities (4–36% ee). While the *exo*-anomeric effect, that is, the π-electron delocalization of the nitrogen substituent in the anomeric position, enhanced the NH-acidity in catalyst **198** and made it a better hydrogen-bond donor; in catalyst **199**, however, the electron density of the imine nitrogen was reduced and thus its ability to form a intramolecular hydrogen-bond to the phenol hydroxy group stabilizing the salen structure. Catalysts **200** and **201** are 4,6-*O*-benzylidene acetal protected and contain reversed positions of the urea and the salen side-chain, but gave low selectivities potentially due to the rigid heterodecaline framework. The influence of substituents in the carbohydrate moiety even of those quite remote from the catalytic center on the enantiodifferentiating potency of the catalysts is illustrated in

Strecker reactions catalyzed by **202–206** (Figure 6.59). The use of catalysts with larger protecting groups and the introduction of linkers (catalysts **204** and **206**) instead of the *O*-acetyl groups resulted in a decrease of enantioinduction. Immobilization on polystyrene either via the carbohydrate (ureas **202** and **203**) or via the amino acid amide (urea **205**) produced catalysts that also displayed low enantioselectivities (Figure 6.59). The authors suggested that the modifications in the protecting group pattern forced the carbohydrate scaffold to adopt a less favorable conformation. None of the examined catalysts (**199–206**) appeared to be competitive to Schiff base catalyst **42**. Only urea catalyst **198** showed an attractive catalytic efficiency obviously due to the enhanced hydrogen-bonding donor capacity, which may compensate the reduced conformational flexibility compared to diaminocyclohexane-derived urea catalyst **42**. Additionally, the high density of polar functions within the saccharide backbone offers too many basic centers for interaction with HCN and decreases enantiodifferentiation, while in the structure of the more effective Jacobson-type catalysts **42** such an interaction is located close to the catalytic center [338].

Glucosaminylurea derivative **198** was also applied to catalyze the enantioselective Mannich addition [72] of a silyl ketene acetal to the *N*-Boc-protected imine naphthalene-2-carbaldehyde resulting in the desired β-amino acid ester (Scheme 6.175). Under optimized reaction conditions with regard to temperature, solvent, equivalents of the ketene acetal, and concentration of the reaction solution good yields (up to 76%) and moderate *ee* values (about 50%) were reached (Scheme 6.175). In contrast to the results obtained by Jacobsen *et al.* when utilizing Schiff base catalyst **42**, the decrease of reaction temperature to −40 °C reduced the yield as well as enantioselectivity of the resulting Mannich adduct (Scheme 6.175) [201]. Catalyst **198** found to be less effective in the Mannich reaction in terms of yield and enantiomeric induction due to reduced basicity of the *N*-acylamine and weaker hydrogen-bonding interactions compared to the more basic Strecker substrates (Scheme 6.174).

The Ma group introduced bifunctional primary amine-thiourea organocatalysts incorporating the *trans*-diaminocyclohexane as well as a saccharide backbone for enantioselective Michael addition reactions of aromatic ketones to a range of nitroalkenes [339]. The saccharide-based catalysts **207–210** were prepared in moderate to good yields from the simple addition reaction of racemic *trans*-diaminocyclohexane to the corresponding saccharide-derived isothiocyanates (Scheme 6.176); **207–210** were screened at 15 mol% loading under various conditions in the addition reactions of acetophenones to nitroalkenes. In dichlorometh-

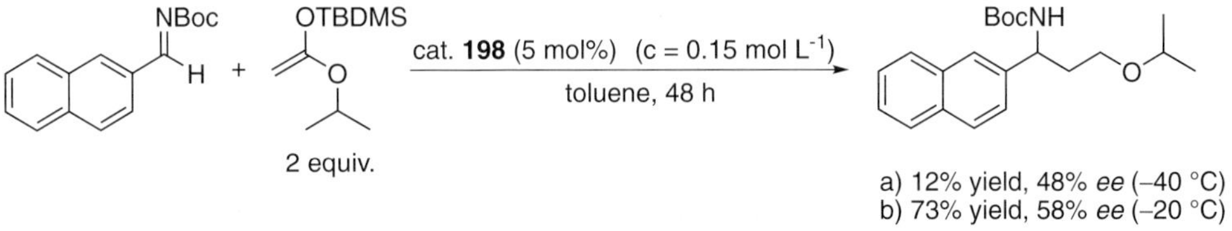

Scheme 6.175 Mannich reaction catalyzed by glucosaminylurea derivative **198**.

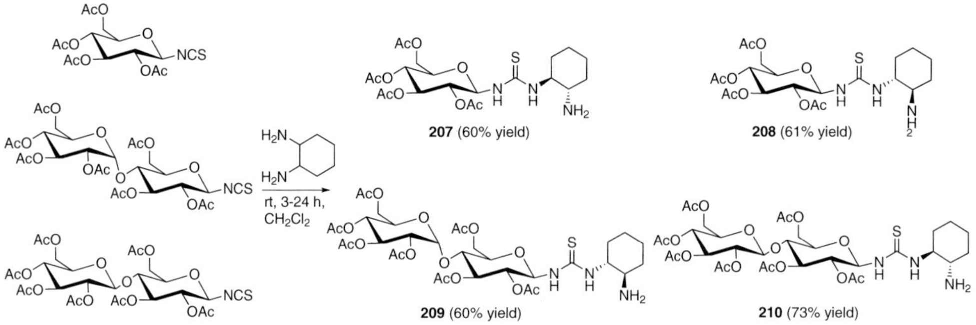

Scheme 6.176 Synthesis of bifunctional saccharide-based amine thioureas **207–210** from the corresponding glucose-, maltose, and lactose-isothiocyanates, respectively.

ane at room temperature, (S,S)-configured catalyst **207** furnished adduct (R)-**3** (Scheme 6.176) with good enantioselectivity (87% *ee*) and 46% yield, while (S,S)-configured catalyst **208** inducing the opposite sense of asymmetric induction producing product (S)-**3** with 97% *ee* in a moderate 60% yield; maltose-(**209**) and lactose-based (**210**) catalyst also bearing the *trans*-1,2-diaminocyclohexane moiety gave product (S)-**3a** in very low yields, but high enantioselectivities (**209**: <10% yield, 93% *ee*; **210**: 17% yield, 96% *ee*). These results indicated that the (R,R)-configuration of *trans*-1,2-diaminocyclohexane matched the β-D-glucopyranose scaffold and thus enhanced the stereochemical control. The choice of solvent showed no significant effect on the asymmetric reduction but reduced the yield of Michael adduct (S)-**3** (catalyst **208**, e.g., 62% yield, 96% *ee* in CHCl$_3$; 23% yield, 95% *ee* in diethyl ether).

Asymmetric Michael additions of various aromatic and aliphatic nitroalkenes and acetophenones utilizing catalyst **208** revealed that aromatic- and heteroaromatic-substituted nitroalkenes gave excellent yields (65–99%) and high enantioselectivities (94–98% *ee*) of the corresponding Michael adducts **1–6** and **8–10**, while the ethyl-substituted nitroalkene substrate polymerized resulting in a poor yield (20%) with a high *ee* value (94%) of adduct **7** (Scheme 6.177). **208**-catalyzed conjugate addition processes were also applicable to various aromatic methyl ketones in moderate to high yields (42–92%) and excellent enantioselectivities (95–97% *ee*) as visualized in Scheme 6.177. This method was described to operate under mild conditions and tolerated a broad range of substituents, independent of the substituent pattern or the electronic properties of these substituents and without the loss of stereocontrol. The authors suggested a bifunctional catalytic mechanism in which the thiourea moiety interacted through hydrogen bonding with the nitro functionality of the nitroalkenes to increase electrophilicity, while the neighboring primary amine group activated the ketone through formation of an enamine intermediate [55, 58, 77]. The primary amine seemed to play a key role, since dimethylation of catalyst **208** to the tertiary amine saccharide-thiourea **211** (Figure 6.60), an analogue of Takemoto's bifunctional thiourea catalyst

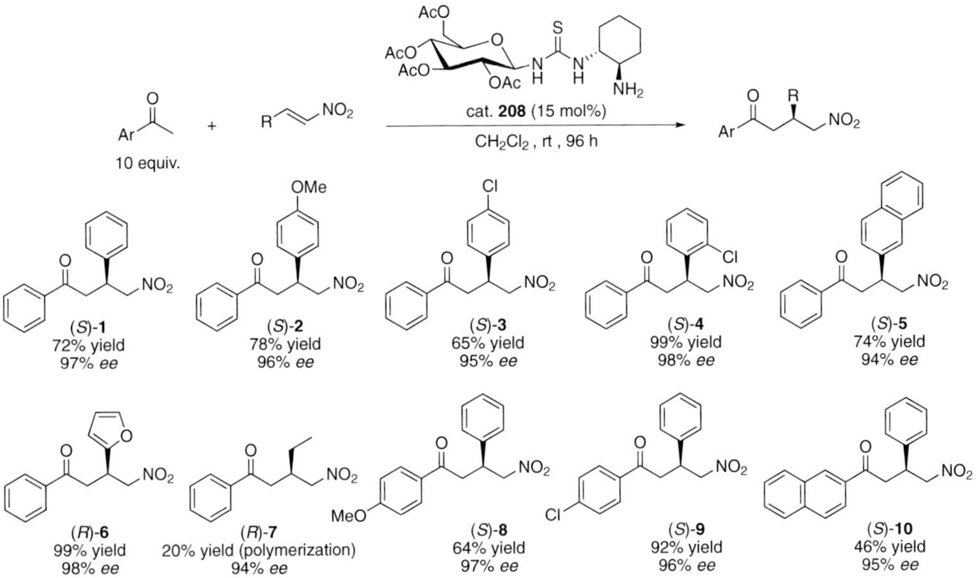

6.2 Synthetic Applications of Hydrogen-Bonding (Thio)urea Organocatalysts | 323

Scheme 6.177 Typical products obtained from the **208**-catalyzed asymmetric Michael addition of ketones to nitroalkenes.

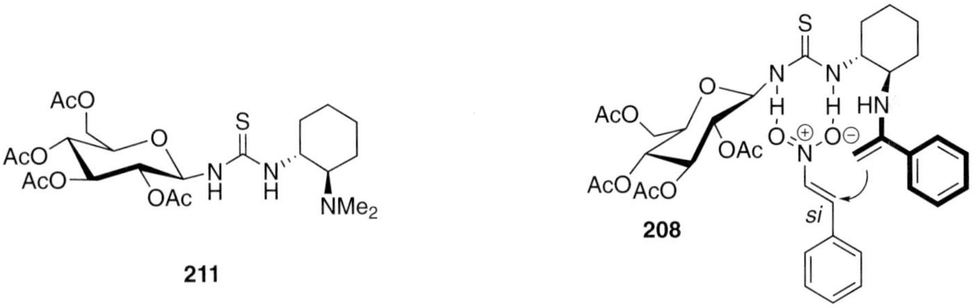

Figure 6.60 Tertiary amine saccharide-thiourea **211** and proposed transition-state model to explain the stereodifferentiation induced by **208**.

12 [129], eliminated any catalytic efficiency. The absolute configuration (S) of the conjugate adduct was explained by the proposed transition-state assembly in which the *si*-face of the nitroalkene was predominantly approached by the enamine intermediate generated from the ketone and the primary amine group of the bifunctional catalyst. The attack of the enamine to the *re*-face of the nitroalkene was restricted by the bulky cyclohexyl scaffold of the catalyst (Figure 6.60) [339].

In 2008, Tang and co-workers reported the utilization of tertiary amine-functionalized saccharide-thiourea **211** as a bifunctional hydrogen-bonding catalyst for the enantioselective aza-Henry [224] (nitro-Mannich) addition [72] of

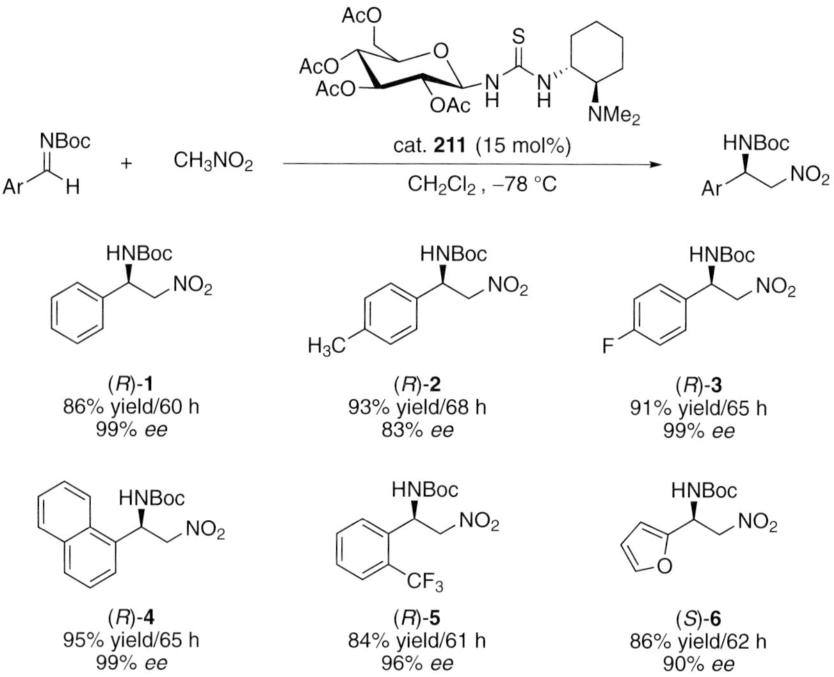

Scheme 6.178 Typical products provided from the asymmetric aza-Henry addition of nitromethane to N-Boc-protected aldimines in the presence of saccharide thiourea **211** as bifunctional hydrogen-bonding catalyst.

nitromethane to electron-rich and electron-deficient (hetero)aromatic N-Boc-protected aldimines [340]. The optimized protocol (15 mol% loading of **211**; CH$_2$Cl$_2$ as the solvent; −78 °C) furnished predominantly (R)-configured aza-Henry adducts such as **1–6** in synthetically useful yields (84–95%) and enantioselectivities (83–99%) as shown in Scheme 6.178.

6.2.2.7 Miscellaneous Stereoselective (Thio)urea Derivatives

This section considers the applications of bifunctional hydrogen-bonding (thio) urea derivatives that have been designed and utilized for asymmetric organocatalysis, but cannot clearly be assigned to one of the structural classifications mentioned above or are the catalysts of choice in only one publication that can mark the basis of further research efforts.

Inspired by the excellent catalytic efficiency of the bifunctional organocatalyst proline and its derivatives in enantioselective organic transformations [55, 63, 80, 81, 341], the Tang group in 2006 introduced new (thio)urea derivatives **212** and **213** incorporating a pyrrolidine ring as chiral scaffold and evaluated their potential as bifunctional organocatalysts in the Michael reaction between cyclohexanone and trans-nitrostyrene (Figure 6.61) [150, 342].

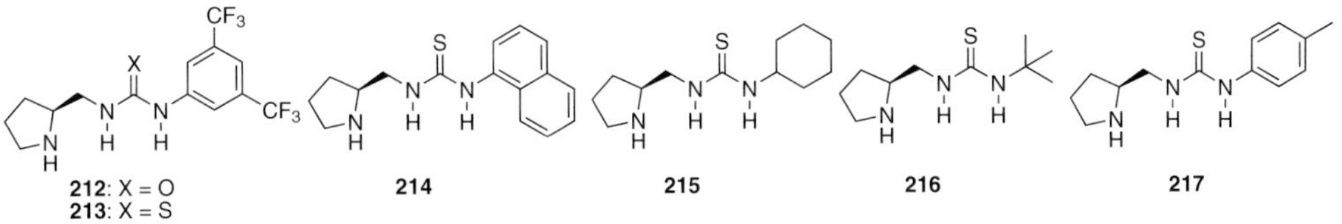

6.2 Synthetic Applications of Hydrogen-Bonding (Thio)urea Organocatalysts | 325

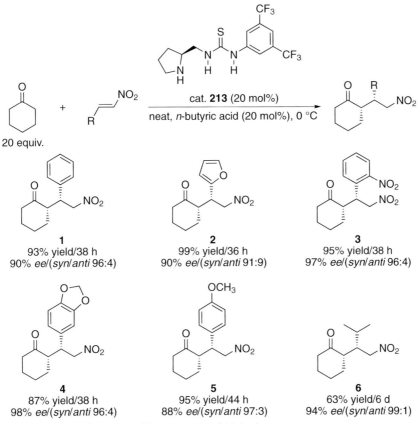

Figure 6.61 Bifunctional hydrogen-bonding pyrrolidine-(thio)ureas utilized for Michael reactions of ketones with nitroalkenes.

Scheme 6.179 Product range of the **213**-catalyzed Michael reaction of cyclohexanone with various nitroalkenes.

At room temperature and 20 mol% (thio)urea loading polar solvents such as MeOH or THF gave only trace amounts of the model adduct **1** (Scheme 6.179), while aprotic, nonpolar solvents such as n-hexane (73% conv.; 80% ee in 4 d) and benzene (68% conv.; 67% ee in 2 d) found to be much more suitable. The addition of an organic acid (10 mol%) increased the reaction rate without loss of stereoin-

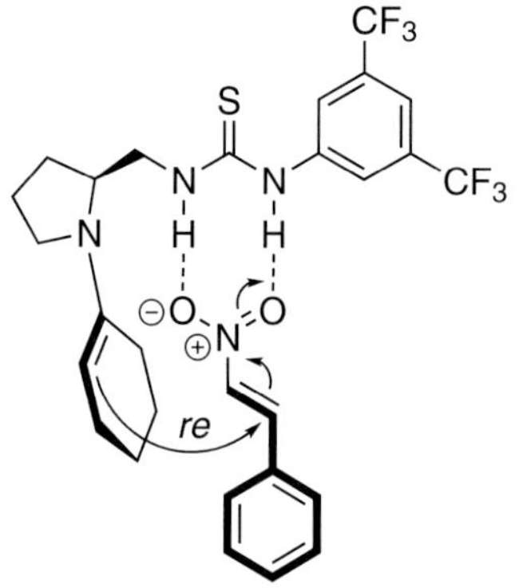

Scheme 6.180 Mechanistic proposal for the Michael reaction of cyclohexanone with *trans*-nitrostyrene catalyzed by **213**.

duction (e.g., *n*-butyric acid: 100% conv., 80% *ee* in 12 h). Under solvent-free conditions and at 0 °C the *ee* value was improved to 87% and, additionally, when using thiourea **213** instead of **212** the *ee* value of adduct **1** reached 90% at 100% conversion in 1.5 h. This result indicated that thiourea **213** was slightly more efficient than its urea analog **212**.

Scheme 6.179 visualizes the typical product range of the **213**-catalyzed Michael reactions of cyclohexanone with various aromatic *trans*-nitroalkenes, when running the reaction under optimized conditions. The adducts **1–6** were formed in yields ranging from 87 to 99%, *ee* values from 88 to 98%, and in *dr* up to 97 : 3. As exemplified for the synthesis of product **6** (63% yield; 94% *ee*) an aliphatic nitroalkene was also successfully converted, but required a longer reaction time (6 d). Pyrrolidine-thiourea **213** (20 mol%) could also organocatalyze the Michael reaction between *trans*-nitrostyrene and other ketones; with acetone the corresponding adduct was formed in 80% yield and 48% *ee*, while cyclohexanone gave its adduct in only 27% yield and 71% *ee*. The catalytic efficiency of catalyst **213** was proposed to be similar to a proline-like substrate activation mode [55, 63, 343], whereby the pyrrolidine moiety activated the ketone through formation of a nucleophilic enamine [55, 58, 77], which attacked the hydrogen-bonded nitroalkene from the *re*-face producing the desired adduct. The organic acid additive may facilitate the generation of the enamine intermediate and thus may lead to an additional accelerating effect (Scheme 6.180).

The Michael reactions [149–152] between cyclohexanone and *trans*-nitroalkenes were also explored by Xiao and co-workers utilizing bifunctional pyrrolidine-thiourea **213** and the pyrrolidine-thioureas **214–217** (Figure 6.61) [344]. The model Michael reaction between cyclohexanone and *trans*-nitrostyrene identified water as the best solvent and **217** to be the most efficient catalysts concerning the activity and asymmetric induction (90% yield; 96% *ee*; *dr* 98 : 2 in 12 h at 35 °C) in the presence of benzoic acid (10 mol%) as additive. The optimized catalytic system allowed the formation of a broad spectrum of Michael adducts such as **1–6** resulting from

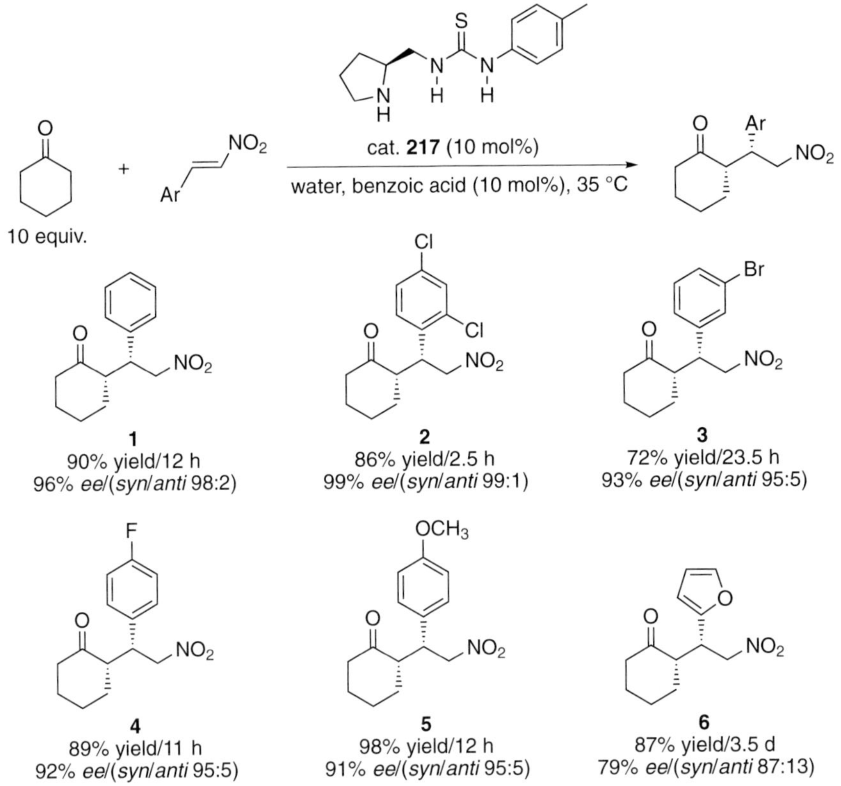

Scheme 6.181 Typical products of the **217**-catalyzed Michael reaction between cyclohexanone and various nitroalkenes.

nitroalkenes with both electron-donating and electron-withdrawing substituents at the aryl group; the adducts were obtained in yields ranging from 79 to 98%, in ee values between 79 and 99%, and dr values up to 99 : 1 (Scheme 6.181). The Michael adduct of acetone and *trans*-nitrostyrene was prepared in 65% yield (57% ee) in 6 d.

Xue et al., in 2007, presented a novel class of thiourea organocatalysts incorporating a chiral oxazoline-ring [345], which is widely utilized for enantiocontrolling ligands in asymmetric organometallic catalysis [141]. Oxazoline-thioureas **218–222** (Figure 6.62) were obtained from the high-yielding (78–92%) addition of the aminomethylene-functionalized phenyl-oxazoline to the corresponding isothiocyante and were screened for their efficiency in the enantioselective formation of aza-Henry adduct 1 in Scheme 6.182.

Electron-deficient oxazoline thiourea **222** turned out to be the most effective catalyst concerning activity (93% yield/48 h/THF) and asymmetric induction (88% ee/rt) in contrast to **218–221**, which gave poor results (Figure 6.62). The solvent screening revealed aprotic THF to be the solvent of choice, while polar protic

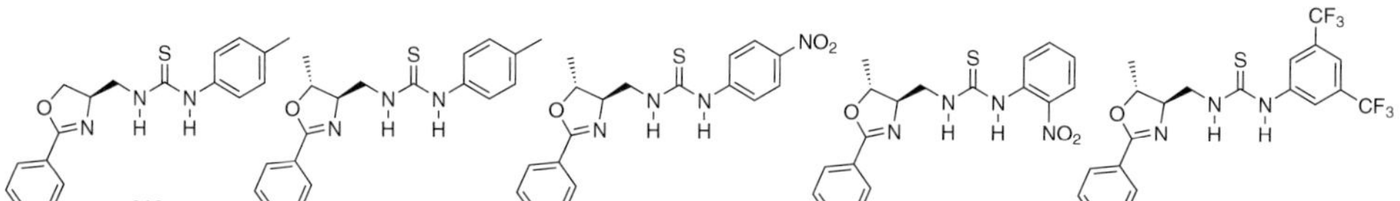

Figure 6.62 Oxazoline-thioureas screened in the aza-Henry reaction of N-Boc-protected benzaldimine with nitromethane.

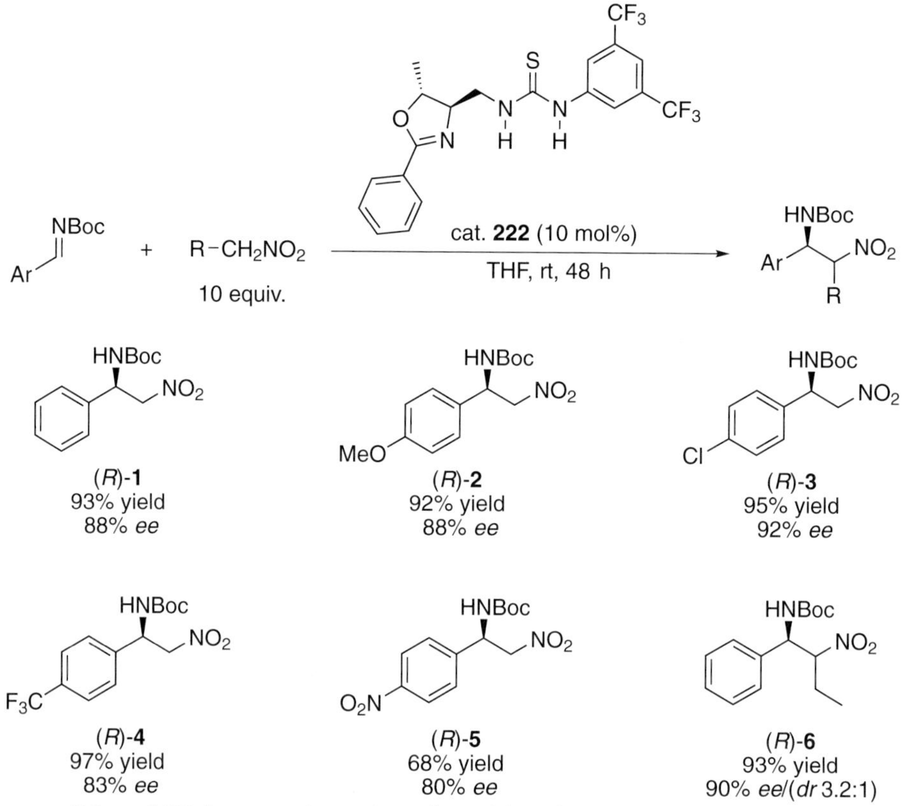

218 89% yield/48 44% ee/rt/THF

219 92% yield/48 38% ee/rt/THF

220 86% yield/48 70% ee/rt/THF

221 90% yield/48 82% ee/rt/THF

222 93% yield/48 88% ee/rt/THF

222 64% yield/48 31% ee/rt/MeOH

Scheme 6.182 Representative products obtained from the **222**-catalyzed aza-Henry reaction of N-Boc-protected aromatic imines with nitromethane and nitroethane.

(R)-**1** 93% yield 88% ee

(R)-**2** 92% yield 88% ee

(R)-**3** 95% yield 92% ee

(R)-**4** 97% yield 83% ee

(R)-**5** 68% yield 80% ee

(R)-**6** 93% yield 90% ee/(dr 3.2:1)

solvents such as methanol reduce the yield (64%/48 h) and the *ee* value (31% at rt) potentially due to the destruction of explicit hydrogen-bonding interactions between catalyst and reactants. Lowering the reaction temperature from rt to 0 °C did not provide a higher *ee* value for the model adduct **1** (72% yield; 89% *ee*/48 h/THF). Under optimized reaction conditions oxazoline-thiourea **222** (10 mol%) promoted the asymmetric aza-Henry reaction [224] of nitromethane and in one

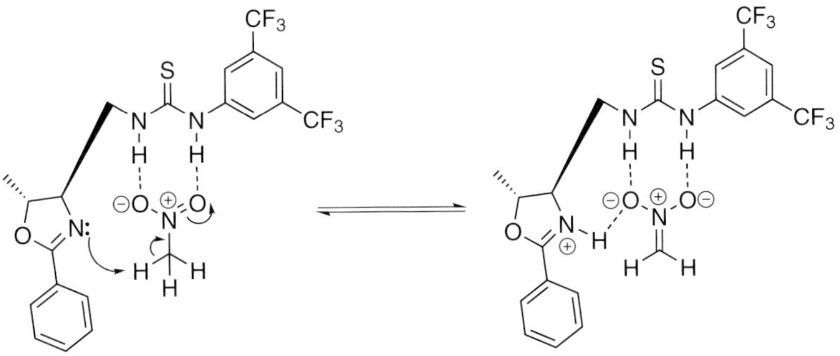

Scheme 6.183 Proposed bifunctional activation mode of oxazoline-thiourea catalyst **222**.

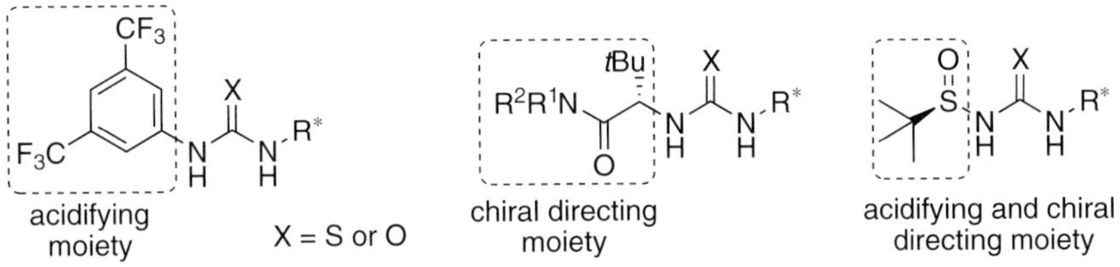

Figure 6.63 Design concept for the development of bifunctional N-sulfinyl (thio)urea derivatives.

- acidifying moiety   X = S or O
- chiral directing moiety
- acidifying and chiral directing moiety

example of nitropropane with various N-protected phenyl imines to furnish the desired adducts **1–6** in 48 h at room temperature in good to excellent yields (68–97%) and enantioselectivities (80–92%) as depicted in Scheme 6.182.

The authors suggested that **222** operates in a bifunctional mode by hydrogen-bonding activation of the nitroalkane and subsequent α-deprotonation through the basic oxazoline nitrogen providing a nucleophilic nitronate, which attacks the imine and give the observed aza-Henry adduct (Scheme 6.183) [345].

The most popular classes of (thio)urea-based hydrogen-bonding organocatalysts either incorporate an acidifying group such as the privileged 3,5-bis(trifluoromethyl) phenyl moiety directly adjacent to an amide nitrogen [1, 107, 112, 116], or a chiral-directing group first developed by Jacobsen (Figure 6.14) [122, 196]. Ellman and co-workers, in 2007, constituted a novel class of hydrogen-bonding organocatalysts, which are characterized by the N-sulfinyl substituent combining both the amide-proton acidification (2–3 pK_a units more acidifying than the 3,5-bis(trifluoromethyl)phenyl group) and the chiral directing function due to its directly contiguous stereogenic center (Figure 6.63) [346].

The N-sulfinyl (thio)ureas are modular and easily accessible in one step by condensing *tert*-butanesulfinamide with the appropriate isocyanate or isothiocyanate. Figure 6.64 shows a representative selection of the prepared N-sulfinyl (thio)ureas evaluated for their catalytic activity in the aza-Henry (nitro-Mannich) reaction of N-Boc-protected benzaldimine and nitroethane producing adduct **1**.

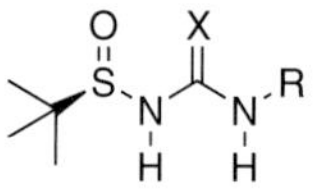

223: X = O, R = C$_6$H$_5$
224: X = S, R = C$_6$H$_5$

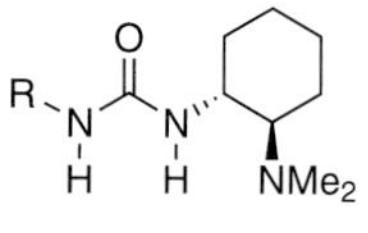

225: R = (R)-S(O)tBu
226: R = (S)-S(O)tBu

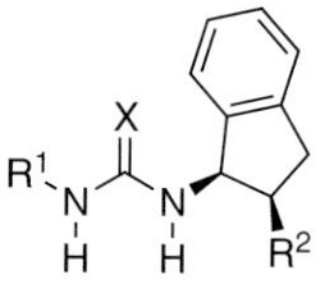

227: X = O, R^1 = (R)-S(O)tBu, R^2 = OH
228: X = S, R^1 = (S)-S(O)tBu, R^2 = OH
229: X = O, R^1 = (R)-S(O)tBu, R^2 = OTBS
230: X = O, R^1 = SO$_2$tBu, R^2 = H
231: X = O, R^1 = 3,5-(CF$_3$)$_2$C$_6$H$_3$, R^2 = OH

Figure 6.64 Representative N-sulfinyl (thio)ureas evaluated for catalytic activity in the asymmetric aza-Henry reaction of N-Boc-protected benzaldimine with nitroethane affording model product **1**.

The catalyst screening was performed in dichloromethane at −40 °C, which revealed 59% conv. and 30% ee for urea **223**, while the sulfur analog **224** gave only 17% conv. and a racemic product mixture. Diaminocyclohexane derivative **225** and cis-amino indanol derivative **227** having additional chirality and functionality were identified to be the most effective catalysts for the investigated screening reaction (**225**: 85% conv.; 89% ee; **227**: 82% conv.; 90% ee). Optimization of the reaction conditions including solvent and stoichiometry revealed high selectivity in acetonitrile with 0.5 equiv. of iPr$_2$NEt. Since thiourea derivative **228** (42% conv.; 58% ee) and the more acidic sulfonyl derivative **230** (35% conv.; 16% ee) turned out to be less efficient than **227**, acidity seems not to be the leading factor responsible for the reaction rate; a catalyst too acidic may be deactivated by deprotonation through the base additive. For high asymmetric induction, the OH-group of **227** is essential as demonstrated by the results obtained with silyl-protected **229** (73% conv.; 7% ee) and the derivative of **227** (99% conv.; racemic) lacking the coordinating OH-functionality [307]. Replacing the chiral sulfinyl-group with the benchmark 3,5-bis(trifluoromethyl)phenyl moiety resulted in urea **231**, which exhibited reduced stereoinduction (80% ee) and conversion (82%) for the screening reaction. The protocol utilizing N-sulfinyl urea derivative **227** was found to be applicable not only to electron-rich and electron-deficient aromatic N-Boc-protected imines, but also to aliphatic imines and furnished the corresponding products **1–6** in yields ranging from 62 to 92%, in very good enantioselectivities (92–96%), and good diastereoselectivities (up to dr 93:7) as depicted in Scheme 6.184 [346].

6.3
Summary and Outlook

This chapter reviews the last decade of intensive research toward explicit double hydrogen-bonding (thio)urea organocatalysts, summarizes the development and

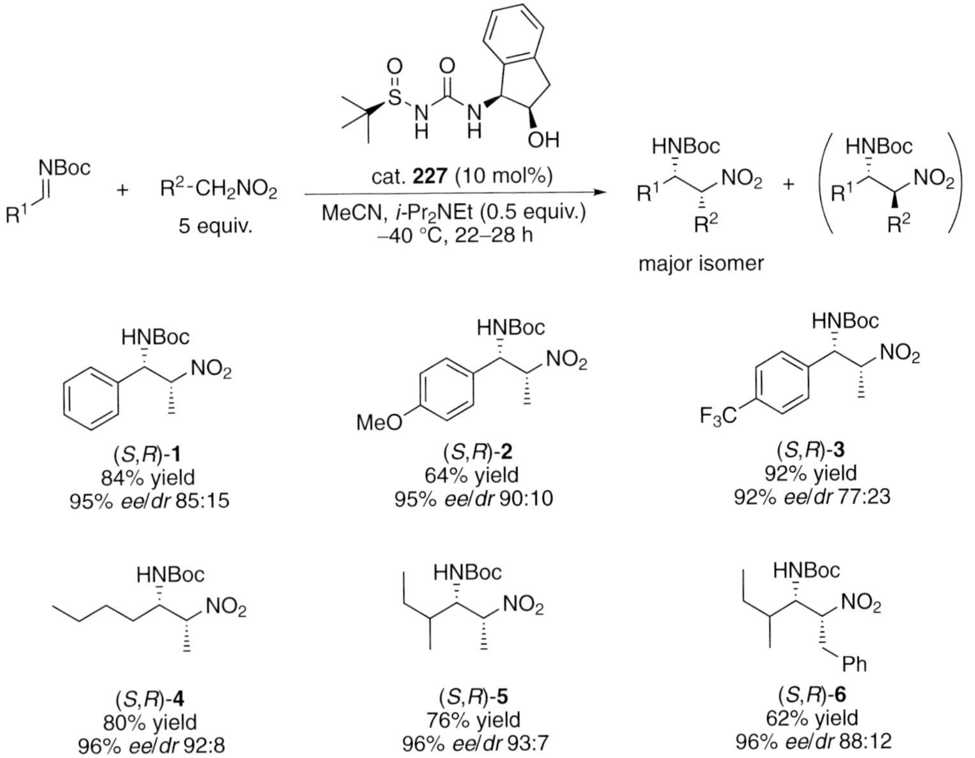

Scheme 6.184 Product range of the aza-Henry reaction of N-Boc-protected aliphatic and aromatic imines catalyzed by N-sulfinyl urea **227**.

synthetic applications of nonstereoselective and stereoselective (thio)urea derivatives in this period, describes catalyst design principles (e.g., bifunctional catalysts, impact of various chiral backbones), and presents mechanistic proposals to explain the observed (stereochemical) outcome of each catalytic procedure. Nearly 150 journal articles including four review articles exclusively revolving around (thio)urea organocatalysts (118 publications from 2005 to date) impressively confirm the success story of these catalysts, and emphasize that the scientific community – after initial ignorance and scepticism in the early years – has recognized the potential and importance of (thio)urea derivatives for catalysis. The growing interest in this research field in the end has not been surprising, since the milestone achievements and seminal guidelines for (thio)urea catalyst design as discussed in the introduction of this chapter have paved the avenue for further research efforts in this new field and underline the evident advantages of (thio)urea organocatalysts in comparison to traditional Brønsted acid and metal(-ion) containing Lewis acid catalysis. Hydrogen-bonding (thio)urea derivatives are readily synthesized from inexpensive raw materials, are easy to handle, water compatible, air stable, non- or low-toxic and thus environmentally benign, easily modulated from both electronic and steric standpoints, can be immobilized on solid support (polymer-bound catalysts), show no product inhibition due to weak enthalpic binding, operate under (nearly) neutral conditions tolerating acid-labile substrates, and are highly catalytically efficient. These features have been utilized for the development of various structurally diverse hydrogen-bonding mono- and bifunctional (thio)urea organocatalysts to catalyze and stereochemically alter predominantly addition reactions such as Michael, (aza)-Henry, and Mannich reactions that have been preferably used for catalyst evaluation and model reactions in many examples. With a critical and constructive view of the achievements and research results of the last decade some important conceptual and practical points have to be emphasized: (a) several reactions that require the activation of strong bonds, e.g., in Heck, Suzuki, Stille reactions are still only feasible with metal catalysts; (b) uncatalyzed reference reactions should always be reported and catalyzed reactions should be expressed in terms of TON or better TOF values; both would make the catalyst performance more transparent; (c) procedures utilizing (thio)urea organocatalysts are typically performed on very small (0.1 to 0.2 mmol) scale model experiments. Up-scaled (5–10 mmol) experiments should be reported as well. Upscaled experiments are also crucial to leave the "proof-of-concept" phase toward the challenging phase in which research should focus on broader, even large-scale applications. (d) Bifunctional (thio)urea organocatalysts are able to activate simultaneously the electrophile and the nucleophile resulting in high catalytic efficiencies. However, the key principle of bifunctional catalysis is usually interpreted in terms of a triple-collision scenario, although this mechanistic model appears to be entropically disfavored; refined models have to be developed on the basis of experimental as well as computational techniques. The research on hydrogen-bonding (thio)urea organocatalysts can help in enzyme profiling and provides closer insights to enzyme catalysis and biochemical processes on a molecular level. The advantages of (thio)urea organocatalysts mentioned above have not been entirely realized and utilized to date. For instance, methods in water, which are attractive for pharmaceutical applications, polymer-bound (thio)urea catalysts easily that are separable from reaction mixtures and are reusable, tandem reactions, multistep syntheses, the activation of new (not the standard small molecule) starting materials including acid-labile substrates, and further reduction of the catalyst loading still offer, along with novel analytical as well as computational methods, numerous perspectives and challenges, both experimentally and theoretically.

Acknowledgment

The author Mike Kotke heartfelt thanks Eva Kotke for untiring support, interest, and helpful advice making the realization of this manuscript possible. Mike Kotke and Peter R. Schreiner thank the Deutschen Forschungsgemeinschaft (DFG) for the support (SPP 1179).

Abbreviations and Acronyms

9-AECN	9-amino(9-deoxy) epicinchonine
Ac	acetyl
AOT	sodium bis(2-ethylhexyl) sulphosuccinate
approx.	approximately
aq	aqueous
Ar	aromatic substituent
asym.	asymmetric
BINAM	(*R*)-(+)-binaphthalenediamine
BINOL	1,1′-bi-2,2′-naphthol
Bn	benzyl
Boc	*tert*-butoxycarbonyl
Bs	brosyl (*p*-bromobenzenesulfonyl)
Bu	butyl
Bz	benzoyl
cat.	catalyst
Cbz	benzyloxycarbonyl
CD	cinchonidine
CN	cinchonine
comp.	compound(s)
conc.	concentration
config. not det.	configuration not determined
conv.	conversion
COSY	correlation spectroscopy
Cp	cyclopentadiene
CPD	cupreidine
CPN	cupreine
CTAB	cetyltrimethylammonium bromide (cetyl: 1-hexadecanol)
DABCO	1,4-diazabicyclo [2.2.2]octane
DBU	1,8-diazabicyclo [5.4.0]undec-7-ene
DCC	dicyclohexylcarbodiimide
DCM	dichloromethane
de	diastereomeric excess
decomp.	decomposition
DFT	density functional theory
DHP	3,4-dihydro-2*H*-pyran
DHQD	dihydroquinidine
DHQN	dihydroquinine
DIPEA	diisopropylethylamine (Hünig's base)
DKR	dynamic kinetic resolution
DME	1,2-dimethoxyethane
DMSO	dimethylsulfoxide
DOSY	diffusion ordered spectroscopy
DPP	diphenylphosphinoyl

dr	diastereomeric ratio
ee	enantiomeric excess
epi	epimeric
equiv.	equivalent
er	enantiomeric ratio
FC	Friedel–Crafts
fw	formula weight (g mol^{-1})
GC/MS	gas chromatography/mass spectrometry
ΔG_{rot}	Gibbs energy for rotation; rotational barrier (kcal mol^{-1})
HMPA	hexamethylphosphoramide
HOMO	highest unoccupied molecular orbital
HPLC	high-performance liquid chromatography
HTS	high-throughput screening
i	iso-
*i*Bu	*iso*-butyl
*i*Pr	isopropyl
IR	infrared
k_{obs}	observed rate constant
KR	kinetic resolution
k_{rel}	relative rate constant (relative to uncatalyzed rct.)
LUMO	lowest unoccupied molecular orbital
M	molarity (mol l^{-1}); molecular mass (g mol^{-1})
MBH	Morita–Baylis–Hillman
Me	methyl
MOP	2-methoxypropene; 2-methoxypropenyl
MS	molecular sieve; mass spectrometry
Ms	mesyl (methanesulfonyl)
MTBE	methyl *tert*-butyl ether
MVK	methyl vinyl ketone
MW	microwave
M_w	number-average molecular weight
NADH	dihydronicotinamide adenine dinucleotide
NADPH	nicotinamide adenine dinucleotide phosphate
*n*Bu	*n*-butyl
NCS	*N*-chlorosuccinimide
nd; not det.	not determined
NMR	nuclear magnetic resonance
NOE	nuclear overhauser effect
NOESY	nuclear overhauser enhancement spectroscopy
Nu	nucleophile
oxid.	oxidation
oxone®	potassium hydrogen peroxymonosulfate sulfate, triple salt 2KHSO$_5$·KHSO$_4$·K$_2$SO$_4$
PDI	polydispersity index = weigth-average molecular weight (M_w)/ number-average molecular weight (M_n) ≤ 1.07

PEG	poly(ethylene glycol)
PG	protective group
Ph	phenyl
Phthal	phthalimide
Piv	pivaloyl- (= pivalyl; trimethylacetyl)
PMP	*p*-methoxyphenyl
ppm	parts per million
Pr	propyl
Ps	polystyrene
PTMC	poly(trimethylene carbonate)
QD	quinidine
QN	quinine
quant.	quantitative
R*	chiral organic residue
rac	racemic
Ra–Ni	Raney nickel
rct.	reaction
recryst.	recrystallization
rel.	relative
ROESY	rotational frame nuclear overhauser effect spectroscopy
ROP	ring-opening polymerization
rr	regioisomer ratio
rt	room temperature
SDS	sodium dodecyl sulfate
t	*tert*-; tertiary
TBAA	tetrabutylammonium acetate
TBDMS (TBS)	*tert*-butyldimethylsilyl
TBDPS	*tert*-butyldiphenylsilyl
TBME	*tert*-butylmethyl ether
TBSCN	*tert*-butylsilyl cyanide
*t*Bu	*tert*-butyl
TEA	triethylamine
tert	tertiary
TES	triethylsilyl
Tf	trifluoromethanesulfonyl
TFAA	trifluoroacetic acid
THF	tetrahydrofuran
THP	2-tetrahydropyranyl
TMC	trimethylene carbonate
TMG	1,1,3,3-tetramethylguanidine
TMS	trimethylsilyl
TMSCN	trimethylsilyl cyanide
TMSOF	2-trimethylsilyloxyfuran
TOF	turnover frequency (h^{-1})
TON	turnover number

TPhP	tetraphenylphthalimide
Triton® X-100	octyl phenol ethoxylate
Troc	2,2,2-trichloroethoxycarbonyl
TrocCl	2,2,2-trichloroethyl chloroformate
Ts	tosyl (*p*-toluenesulfonyl)
UV/Vis	ultraviolet-visible
yl.	yield

References

1 Schreiner, P.R. (2003) *Chem. Soc. Rev.*, **32**, 289–296.
2 Pihko, P.M. (2004) *Angew. Chem. Int. Ed.*, **43**, 2062–2064.
3 Takemoto, Y. (2005) *Org. Biomol. Chem.*, **3**, 4299–4306.
4 Connon, S.J. (2006) *Chem. Eur. J.*, **12**, 5418–5427.
5 Taylor, M.S. and Jacobsen, E.N. (2006) *Angew. Chem. Int. Ed.*, **45**, 1520–1543.
6 Akiyama, T., Itoh, J. and Fuchibe, K. (2006) *Adv. Synth. Catal.*, **348**, 999–1010.
7 Doyle, A.G. and Jacobsen, E.N. (2007) *Chem. Rev.*, **107**, 5713–5743.
8 Bernardi, L., Fini, F., Herrera, R.P., Ricci, A. and Sgarzani, V. (2007) *Tetrahedron*, **62**, 375–380.
9 Special issue on enzyme catalysis: Schramm, V.L. (ed.) (2006) Princpiles of enzymatic catalysis. *Chem. Rev.*, **106**, 3029–3496.
10 Carey, P.R. (2006) *Chem. Rev.*, **106**, 3043–3054.
11 Boehr, D.D., Dyson, H.J. and Wright, P.E. (2006) *Chem. Rev.*, **106**, 3055–3079.
12 Antoniou, D., Basner, J., Núñez, S. and Schwartz, S.D. (2006) *Chem. Rev.*, **106**, 3170–3187.
13 Bruice, T.C. (2006) *Chem. Rev.*, **106**, 3119–3139.
14 Goa, J., Ma, S., Major, D.T., Nam, K., Pu, J. and Truhlar, D.G. (2006) *Chem. Rev.*, **106**, 3188–3209.
15 Pauling, L. (1968) *The Nature of the Chemical Bond*, 3rd edn, Cornell University Press, Ithaca.
16 Pimentel, G.C. and McClellan, A.L. (1960) *The Hydrogen Bond*, Freeman, San Francisco.
17 Jeffrey, G.A. (1997) *An Introduction to Hydrogen Bonding*, Oxford University Press, New York.
18 Prins, L.J., Reinhoudt, D.N. and Timmerman, P. (2001) *Angew. Chem. Int. Ed.*, **40**, 2382–2426.
19 Steiner, T. (2002) *Angew. Chem. Int. Ed.*, **41**, 48–76.
20 Hunter, C.A. (2004) *Angew. Chem. Int. Ed.*, **43**, 5310–5324.
21 Kim, S.P., Leach, A.G. and Houk, K.N. (2002) *J. Org. Chem.*, **67**, 4250–4260.
22 Breslow, R. (1991) *Acc. Chem. Res.*, **24**, 159–163.
23 Pratt, L.R. and Pohorille, A. (2002) *Chem. Soc. Rev.*, **102**, 2671–2692.
24 Meyer, E.A., Castellano, R.K. and Diederich, F. (2003) *Angew. Chem. Int. Ed.*, **42**, 1210–1250.
25 Dessent, C.E.H. and Müller-Dethlefs, K. (2000) *Chem. Rev.*, **100**, 3999–4021.
26 Wear, M.A., Kan, D., Rabu, A. and Walkinshaw, M.D. (2000) *Angew. Chem. Int. Ed.*, **46**, 6453–6456.
27 Kraut, D.A., Sigala, P.A., Pybus, B., Liu, C.W., Ringe, D., Petsko, G.A. and Herschlag, D. (2006) *PLoS Biol.*, **4**, 501–519.
28 Warshel, A., Sharma, P.K., Kato, M., Xiang, Y., Liu, H. and Olsson, M.H.M. (2006) *Chem. Rev.*, **106**, 3210–3235.
29 Bruice, T.C. and Benkovic, S.J.J. (2000) *Biochemistry*, **39**, 6267–6274.
30 Williams, D.H., Stephens, E., O'Brien, D.P. and Zhou, M. (2004) *Angew. Chem. Int. Ed.*, **43**, 6596–6616.
31 Evans, M.J. and Cravatt, B.F. (2006) *Chem. Rev.*, **106**, 3279–3301.
32 de Vries, E.J. and Janssen, D.B. (2003) *Curr. Opin. Biotechnol.*, **14**, 414–420.

33 Lau, E.Y., Newby, Z.E. and Bruice, T.C. (2001) *J. Am. Chem. Soc.*, **123**, 3350–3357.
34 Hopmann, K.H. and Himo, F. (2006) *J. Phys. Chem. B*, **110**, 21299–21310.
35 Hopmann, K.H. and Fahmi, H. (2006) *Chem. Eur. J.*, **12**, 6898–6909.
36 Wharton, C.W. (1998) *Comprehensive Biological Catalysis*, Vol. **1**, Academic Press, London.
37 Product inhibition and substrate inhibition are effects also known in enzyme catalysis that can reduce catalytic efficiency. Generally, catalytic systems (natural or artificial) based on covalent interactions are more sensitive towards inhibitions than non-covalent systems utilizing weak interactions: Garcia-Junceda, E. (2008) *Multi-Step Enzyme Catalysis*, Wiley-VCH Verlag GmbH, Weinheim, Germany.
38 Blackmond, D.G., Armstrong, A., Coombe, V. and Wells, A. (2007) *Angew. Chem. Int. Ed.*, **46**, 3798–3800.
39 Sheldon, R.A. (2008) *Chem. Commun.*, 3352–3365.
40 Breslow, R. (1982) *Science*, **218**, 532–537.
41 Breslow, R. (2005) *Artificial Enzymes*, Wiley-VCH Verlag GmbH, Weinheim.
42 Pascal, R. (2003) *Eur. J. Org. Chem.*, 1813–1824.
43 Pauling, L. (1946) *Chem. Eng. News*, **24**, 1375–1377.
44 Pauling, L. (1948) *Nature*, **161**, 707–709.
45 Zhang, X. and Houk, K.N. (2005) *Acc. Chem. Res.*, **38**, 379–385.
46 Kohen, A. and Klinman, J.P. (1998) *Acc. Chem. Res.*, **31**, 397–404.
47 Akiyama, T. (2007) *Chem. Rev.*, **107**, 5744–5758.
48 Jencks, W.P. (1976) *Acc. Chem. Res.*, **9**, 425–432.
49 Jencks, W.P. (1980) *Acc. Chem. Res.*, **13**, 161–169.
50 Michaelis, L. and Menten, M.L. (1913) *Biochem. Z.*, **49**, 333–369.
51 Bruice, T.C. (2002) *Acc. Chem. Res.*, **35**, 139–148.
52 Dalko, P.I. and Moisan, L. (2001) *Angew. Chem. Int. Ed.*, **40**, 3726–3748.
53 Jarvo, E.R. and Miller, S.J. (2002) *Tetrahedron*, **58**, 2481–2495.
54 Dalko, P.I. and Moisan, L. (2004) *Angew. Chem. Int. Ed.*, **43**, 5138–5175.
55 List, B. (2004) *Acc. Chem. Res.*, **37**, 548–557.
56 Fonseca, M.H. and List, B. (2004) *Curr. Opin. Chem. Biol.*, **8**, 319–326.
57 Schreiner, P.R. and Fokin, A.A. (2004) *Chem. Rec.*, **3**, 247–257.
58 Notz, W., Tanaka, F. and Barbas, C.F. III (2004) *Acc. Chem. Res.*, **37**, 580–591.
59 Berkessel, A. and Gröger, H. (2005) *Asymmetric Organocatalysis*, Wiley-VCH Verlag GmbH, Weinheim, Germany.
60 Gaunt, M.J., Johansson, C.C.C., McNally, A. and Vo, N.T. (2006) *Drug Discov. Today*, **12**, 8–27.
61 Cozzi, F. (2006) *Adv. Synth. Catal.*, **348**, 1367–1390.
62 Benaglia, M. (2006) *New J. Chem.*, **30**, 1525–1533.
63 List, B. (2006) *Chem. Commun.*, 819–824.
64 Marcelli, T., van Maarseveen, J.H. and Hiemstra, H. (2006) *Angew. Chem. Int. Ed.*, **45**, 7496–7504.
65 Pellissier, H. (2007) *Tetrahedron*, **63**, 9267–9331.
66 Dalko, P.I. (2007) *Enantioselective Organocatalysis – Reactions and Experimental Procedures*, Wiley-VCH Verlag GmbH, Weinheim, Germany.
67 Davie, E.A.C., Mennen, S.M., Xu, Y. and Miller, S.J. (2007) *Chem. Rev.*, **107**, 5759–5812.
68 Enders, D., Niemeier, O. and Henseler, A. (2007) *Chem. Rev.*, **107**, 5606–5655.
69 de Figueiredo, R.M. and Christmann, M. (2007) *Eur. J. Org. Chem.*, 2575–2600.
70 Enders, D., Grondal, C. and Hüttl, M.R.M. (2007) *Angew. Chem. Int. Ed.*, **46**, 1570–1581.
71 Revell, J.D. and Wennemers, H. (2007) *Curr. Opin. Chem. Biol.*, **11**, 269–278.
72 Ting, A. and Schaus, S.E. (2007) *Eur. J. Org. Chem.*, 5797–5815.
73 Walsh, P.J., Li, H. and de Parrodi, C.A. (2007) *Chem. Rev.*, **107**, 2503–2545.
74 Guillena, G., Nájera, C. and Ramón, D.J. (2007) *Tetrahedron Asymm.*, **18**, 2249–2293.
75 Kamber, N.E., Jeong, W., Waymouth, R.M., Pratt, R.C., Lohmeijer, B.G.G. and Hedrick, J.L. (2007) *Chem. Rev.*, **107**, 5817–5840.
76 Erkkilä, A., Majander, I. and Pihko, P.M. (2007) *Chem. Rev.*, **107**, 5416–5470.

77 Mukherjee, S., Yang, J.W., Hoffmann, S. and List, B. (2007) *Chem. Rev.*, **12**, 5471–5569.
78 Verkade, J.M.M., van Hemert, L.J.C., Quaedflieg, P.J.L.M. and Rutjes, F.P.J.T. (2008) *Chem. Soc. Rev.*, **37**, 29–41.
79 Yua, X. and Wang, W. (2008) *Org. Biomol. Chem.*, **6**, 2021–2216.
80 Xua, L.-W. and Lu, Y. (2008) *Org. Biomol. Chem.*, **6**, 2047–2053.
81 Dondoni, A. and Massi, A. (2008) *Angew. Chem. Int. Ed.*, **47**, 4638–4660.
82 MacMillan, D.W.C. (2008) *Nature*, **455**, 304–308.
83 Special journal issues on organocatalysis: (a) Houk, K.N. and List, B. (eds) (2004) Asymmetric organocatalysis. *Acc. Chem. Res.*, **37** (8), 487–631.
(b) Various authors (2004) Organic catalysis. *Adv. Synth. Catal.*, **346** (9–10), 1007–1249.
(c) Kočovský, P. and Malkov, A.V. (eds) (2006) Organocatalysis in organic synthesis. *Tetrahedron*, **62** (2–3), 243–250.
(d) List, B. (ed.) (2007) Organocatalysis. *Chem. Rev.*, **107** (12), 5413–5883.
(e) Alexakis, A. (ed.) (2007) Organocatalysis. *Chimia*, **61** (5), 212–281.
84 Hine, J., Ahn, K., Gallucci, J.C. and Linden, S.-M. (1984) *J. Am. Chem. Soc.*, **106**, 7980–7981.
85 A crystal structure of an unstable N,N'-[bis-(á-tosylbenzyl)urea-acetone hydrogen-bonded adduct had been reported earlier: Tel, R M. and Engberts, J.B.F.N. (1976). *J. Chem. Soc. Perkin Trans. 2*, 483–488.
86 Hine, J., Linden, S.M. and Kanagasabapathy, V.M. (1985) *J. Am. Chem. Soc.*, **107**, 1082–1083.
87 Hine, J., Hahn, S., Miles, D.E. and Ahn, K. (1985) *J. Org. Chem.*, **50**, 5092–5096.
88 Hine, J., Hahn, S. and Miles, D.E. (1986) *J. Org. Chem.*, **51**, 577–584.
89 Hine, J. and Ahn, K. (1987) *J. Org. Chem.*, **52**, 2089–2091.
90 Hine, J. and Ahn, K. (1987) *J. Org. Chem.*, **52**, 2083–2086.
91 Kelly, T.R., Meghani, P. and Ekkundi, V.S. (1990) *Tetrahedron Lett.*, **31**, 3381–3384.
92 Kelly, T.R. and Kim, M.H. (1994) *J. Am. Chem. Soc.*, **116**, 7072–7080.
93 Blake, J.F. and Jorgensen, W.L. (1991) *J. Am. Chem. Soc.*, **113**, 7430–7432.
94 Blake, J.F., Lim, D. and Jorgensen, W.L. (1994) *J. Org. Chem.*, **59**, 803–805.
95 Severance, D.L. and Jorgensen, W.L. (1992) *J. Am. Chem. Soc.*, **114**, 10966–10968.
96 Etter, M.C. and Baures, P.W. (1988) *J. Am. Chem. Soc.*, **110**, 639–640.
97 Etter, M.C. and P.T.W. (1988) *J. Am. Chem. Soc.*, **110**, 5896–5897.
98 Etter, M.C. (1991) *J. Phys. Chem.*, **95**, 4601–4610.
99 Etter, M.C. (1990) *Acc. Chem. Res.*, **23**, 120–126.
100 Etter, M.C. and Reutzel, S.M. (1991) *J. Am. Chem. Soc.*, **113**, 2586–2589.
101 Wolf, K.L., Frahm, H. and Harms, H. (1937) *Z. Phys. Chem. Abt. B*, **36**, 17.
102 Smith, P.J., Reddington, M.V. and Wilcox, C.S. (1992) *Tetrahedron Lett.*, **33**, 6085–6088.
103 Wilcox, C.S., James, J., Adrian, C., Webb, T.H. and Zawacki, F.J. (1992) *J. Am. Chem. Soc.*, **114**, 10189–10197.
104 Wilcox, C.S., Kim, E., Romano, D., Kuo, L.H., Burt, A.L. and Curran, D.P. (1995) *Tetrahedron Lett.*, **51**, 621–634.
105 Curran, D.P. and Kuo, L.H. (1994) *J. Org. Chem.*, **59**, 3259–3261.
106 Curran, D.P. and Kuo, L.H. (1995) *Tetrahedron Lett.*, **36**, 6647–6650.
107 Wittkopp, A. (1997) Diploma Thesis. University of Göttingen, Germany.
108 Fan, E., Arman, S.A.V., Kincaid, S. and Hamilton, A.D. (1993) *J. Am. Chem. Soc.*, **115**, 369–370.
109 Gómez, D.E., Fabbrizzi, L., Licchelli, M. and Monzani, E. (2005) *Org. Biomol. Chem.*, **3**, 1495–1500.
110 Bonizzoni, M., Fabbrizzi, L., Taglietti, A. and Tiengo, F. (2006) *Eur. J. Org. Chem.*, 3567–3574.
111 Blanco, J.L.J., Bootello, P., Benito, J.M., Mellet, C.O. and Ferna'ndez, J.M.G. (2006) *J. Org. Chem.*, **71**, 5136–5143.
112 Wittkopp, A. and Schreiner, P.R. (2000) *The Chemistry of Dienes and Polyenes*, Vol. 2 (ed. Z. Rappoport), John Wiley & Sons, Ltd, Chichester, pp. 1029–1088.

113 Ostwald, W. (1910) *Chem. Z.*, **34**, 397–399.
114 Wittkopp, A. and Schreiner, P.R. (2003) *Chem. Eur. J.*, **9**, 407–414.
115 Evans, D.A., Chapman, K.T. and Bisaha, J. (1984) *J. Am. Chem. Soc.*, **106**, 4261.
116 Schreiner, P.R. and Wittkopp, A. (2002) *Org. Lett.*, **4**, 217–220.
117 Custelcean, R., Gorbunova, M.G. and Bonnesen, P.V. (2005) *Chem. Eur. J.*, **11**, 1459–1466.
118 Kotke, M. (forthcoming) Dissertation, University Giessen, University of Giessen, Germany; online access via the Electronic Library.
119 Yoon, T.P. and Jacobsen, E.N. (2003) *Science*, **299**, 1691–1693.
120 Schuster, T., Kurtz, M. and Göbel, M.W. (2000) *J. Org. Chem.*, **65**, 1697–1701.
121 Schuster, T., Bauch, M., Dürner, G. and Göbel, M.W. (2000) *Org. Lett.*, **2**, 179–181.
122 Sigman, M.S. and Jacobsen, E.N. (1998) *J. Am. Chem. Soc.*, **120**, 4901–4902.
123 Wittkopp, A. (2001) Dissertation, University of Göttingen, Germany, published online.
124 Vachal, P. and Jacobsen, E.N. (2002) *J. Am. Chem. Soc.*, **124**, 10012–10014.
125 Anastas, P.T. and Warner, J.C. (1998) *Green Chemistry: Theory and Practice*, Oxford University Press, Oxford, UK.
126 Anastas, P.T. and Zimmerman, J.B. (2003) Green chemistry. *Environ. Sci. Technol.*, A95–A101.
127 Lindström, U.M. (2002) *Chem. Rev.*, **102**, 2751–2772.
128 Manabe, K. and Kobayashi, S. (2002) *Chem. Eur. J.*, **8**, 4094–4101.
129 Okino, T., Hoashi, Y. and Takemoto, Y. (2003) *J. Am. Chem. Soc.*, **125**, 12672–12673.
130 Ahrendt, K.A., Borths, C.J. and MacMillan, D.W.C. (2000) *J. Am. Chem. Soc.*, **122**, 4243–4244.
131 Liebig, J.v. (1860) *Liebigs Ann.*, **113**, 246.
132 Liebig, J.v. (1870) *Liebigs Ann.*, **153**, 9.
133 Ostwald, W. (1900) *Z. Physikalische Chem.*, **34**, 510.
134 (a) Langenbeck, W. (1927) *Ber. Dtsch. Chem. Ges.*, **60**, 930.
(b) Langenbeck, W. (1928) *Angew. Chem.*, **41**, 740–745.
135 Langenbeck, W. (1932) *Angew. Chem.*, **45**, 97–99.
136 (a) Langenbeck, W. (1949) *Die organischen Katalysatoren und ihre Beziehungen zu den Fermenten (Organic Catalysts and their Relation to the Enzymes)*, 2nd edn, Springer, Berlin.
(b) Langenbeck, W. (1958) *Tetrahedron*, **3**, 185–196.
137 Bennani, Y.L. and Hanessian, S. (1997) *Chem. Rev.*, **97**, 3161–3195.
138 Fache, F., Schulz, E., Tommasino, M.L. and Lemaire, M. (2000) *Chem. Rev.*, **100**, 2159–2231.
139 Hoffmann, H.M.R. and Frackenpohl, J. (2004) *Eur. J. Org. Chem.*, 4293–4312.
140 Chen, Y., Yekta, S. and Yudin, A.K. (2003) *Chem. Rev.*, **103**, 3155–3211.
141 McManus, H.A. and Guiry, P.J. (2004) *Chem. Rev.*, **104**, 4151–4202.
142 Desimoni, G., Faita, G. and Jørgensen, K.A. (2006) *Chem. Rev.*, **106**, 3561–3651.
143 Mellah, M., Voituriez, A. and Schulz, E. (2007) *Chem. Rev.*, **107**, 5133–5209.
144 Pellissier, H. (2007) *Tetrahedron*, **63**, 1297–1330.
145 Kotke, M. and Schreiner, P.R. (2007) *Synthesis*, 779–790.
146 Kotke, M. and Schreiner, P.R. (2006) *Tetrahedron*, **62**, 434–439.
147 Okino, T., Hoashi, Y. and Takemoto, Y. (2003) *Tetrahedron Lett.*, **44**, 2817–2821.
148 Dessole, G., Herrera, R.P. and Ricci, A. (2004) *Synlett*, 2374–2378.
149 Enders, D., Luttgen, K. and Narine, A.A. (2007) *Synthesis*, 959–980.
150 Sulzer-Mossé, S. and Alexakis, A. (2007) *Chem. Commun.*, 3123–3135.
151 Vicario, J.L., Badía, D. and Carrillo, L. (2007) *Synthesis*, **14**, 2065–2092.
152 Tsogoeva, S.B. (2007) *Eur. J. Org. Chem.*, 1701–1716.
153 Kleiner, C.M. and Schreiner, P.R. (2006) *Chem. Commun.*, 4315–4317.
154 Franks, F. (2000) *Water–A Matrix of Life*, 2nd edn, Royal Society of Chemistry, Cambridge, UK.
155 Engberts, J.B.F.N. (1995) *Pure Appl. Chem.*, **67**, 823–828.

156 Pettersen, D., Herrera, R.P., Bernardi, L., Fini, F., Sgarzani, V., Fernandez, R., Lassaletta, J.M. and Ricci, A. (2006) *Synlett*, 239–242.
157 Gröger, H. (2003) *Chem. Rev.*, **103**, 2795–2827.
158 Pan, S.C., Zhou, J. and List, B. (2006) *Synlett*, **19**, 3275–3276.
159 Dornow, A. and Lüpfert, S. (1956) *Chem. Ber.*, **89**, 2718.
160 Dornow, A. and Lüpfert, S. (1957) *Chem. Ber.*, **90**, 1780.
161 Pan, S.C., Zhou, J. and List, B. (2007) *Angew. Chem. Int. Ed.*, **46**, 2971–2971.
162 Pan, S.C. and List, B. (2007) *Synlett*, 318–320.
163 Muñiz, F.M., Montero, V.A., de Arriba, A.L.F., Simón, L., Raposo, C. and Morán, J.R. (2008) *Tetrahedron Lett.*, **49**, 5050–5052.
164 Kirsten, M., Rehbein, J., Hiersemann, M. and Strassner, T. (2007) *J. Org. Chem.*, **72**, 4001–4011.
165 Huang, Y. (2007) *Synlett*, **14**, 2304–2305.
166 Zhang, Z. and Schreiner, P.R. (2007) *Synlett*, 1455–1457.
167 Kawai, Y., Inaba, Y., Hayash, M. and Tokitoh, N. (2001) *Tetrahedron Lett.*, **42**, 3367.
168 Kawai, Y., Inaba, Y. and Tokitoh, N. (2001) *Tetrahedron Asymm.*, **12**, 309.
169 Zhang, Z. and Schreiner, P.R. (2007) *Synthesis*, **16**, 2559–2564.
170 Korn, M., Wodarz, R., Schoknecht, W., Weichardt, H. and Bayer, E. (1984) *Arch. Toxicol.*, **55**, 59–63.
171 Kanai, M., Kato, N., Ichikawa, E. and Shibasaki, M. (2005) *Synlett*, **10**, 1491–1508.
172 Park, Y.J., Park, J.-W. and Jun, J.-H. (2008) *Acc. Chem. Res.*, **41**, 222–234.
173 Weil, T., Kotke, M., Kleiner, C.M. and Schreiner, P.R. (2008) *Org. Lett.*, **10**, 1513–1516.
174 List, B. and Reisinger, C. (2008) *Synfacts*, **6**, 644.
175 McGilvra, J.D., Unni, A.K., Modi, K. and Rawal, V.H. (2006) *Angew. Chem. Int. Ed.*, **45**, 6130–6133.
176 Shi, Y.-L. and Shi, M. (2007) *Eur. J. Org. Chem.*, 2905–2916.
177 Singh, V. and Batra, S. (2008) *Tetrahedron*, **64**, 4511–4574.
178 Maher, D.J. and Connon, S.J. (2004) *Tetrahedron Lett.*, **45**, 1301–1305.
179 Kavanagh, S.A., Piccinini, A., Fleming, E.M. and Connon, S.J. (2008) *Org. Biomol. Chem.*, **6**, 1339–1343.
180 Meinwald, J., Labana, S.S. and Chadha, S.S. (1963) *J. Am. Chem. Soc.*, **85**, 582.
181 Rosa, M. De, Citro, L. and Soriente, A. (2006) *Tetrahedron Lett.*, **47**, 8507–8510.
182 Costero, A.M., Rodríguez-Muñiz, G.M., Gaviña, P., Gil, S. and Domenech, A. (2007) *Tetrahedron Lett.*, **48**, 6992–6995.
183 Procuranti, B. and Connon, S.J. (2007) *Chem. Commun.*, 1421–1423.
184 Fleming, E.M., Quigley, C., Rozas, I. and Connon, S.J. (2008) *J. Org. Chem.*, **73**, 948–956.
185 Ema, T., Fujii, T., Ozaki, M., Korenaga, T. and Sakai, T. (2005) *Chem. Commun.*, 4650–4651.
186 Otera, J. (1993) *Chem. Rev.*, **93**, 1449–1470.
187 Ishihara, K., Kosugi, Y. and Akakura, M. (2004) *J. Am. Chem. Soc.*, **126**, 12212–12213.
188 Ema, T., Tanida, D., Matsukawa, T. and Sakai, T. (2008) *Chem. Commun.*, 957–959.
189 Dove, A.P., Pratt, R.C., Lohmeijer, B.G.G., Waymouth, R.M. and Hedrick, J.L. (2005) *J. Am. Chem. Soc.*, **127**, 13798–13799.
190 Pratt, R.C., Lohmeijer, B.G.G., Long, D.A., Lundberg, P.N.P., Dove, A.P., Li, H.B., Wade, C.G., Waymouth, R.M. and Hedrick, J.L. (2006) *Macromolecules*, **39**, 7863–7871.
191 Lohmeijer, B.G.G., Pratt, R.C., Leibfarth, F., Logan, J.W., Long, D.A., Dove, A.P., Nederberg, F., Choi, J., Wade, C., Waymouth, R.M. and Hedrick, J.L. (2006) *Macromolecules*, **39**, 8574–8583.
192 Nederberg, F., Lohmeijer, B.G.G., Leibfarth, F., Pratt, R.C., Choi, J., Dove, A.P., Waymouth, R.M. and Hedrick, J.L. (2007) *Biomacromolecules*, **8**, 153–160.
193 Wieland, H., Schlichtung, O. and Langsdorf, W.V. (1926) *Z. Phys. Chem.*, **161**, 74.

194 Glasbol, F., Steenbol, P. and Sorenson, S.B. (1980) *Acta Chem. Scand.*, **26**, 3605.

195 Gennari, C. and Piarulli, U. (2003) *Chem. Rev.*, **103**, 3071–3100.

196 Sigman, M.S., Vachal, P. and Jacobsen, E.N. (2000) *Angew. Chem. Int. Ed.*, **39**, 1279–1281.

197 HCN was generated by the method of Ziegler. Alternatively, the generation of HCN *in situ* by reaction of TMSCN and MeOH afforded identical results. Ziegler, K. (1932) *Organic Synthesis Collective*, Volume 1 (eds H. Gilman and A.H. Blatt), John Wiley & Sons, Inc., New York, p. 314.

198 Vachal, P. and Jacobsen, E.N. (2000) *Org. Lett.*, **2**, 867–870.

199 Harriman, D.J., Deleavey, G.F., Lambropoulos, A. and Deslongchamps, G. (2007) *Tetrahedron*, **63**, 13032–13038.

200 Joly, G.D. and Jacobsen, E.N. (2004) *J. Am. Chem. Soc.*, **126**, 4102–4103.

201 Wenzel, A.G. and Jacobsen, E.N. (2002) *J. Am. Chem. Soc.*, **124**, 12964–12965.

202 Taylor, M.S. and Jacobsen, E.N. (2004) *J. Am. Chem. Soc.*, **126**, 10558–10559.

203 Bentley, K.W. (2004) *Nat. Prod. Rep.*, **21**, 395–424.

204 Taylor, M.S., Tokunaga, T. and Jacobsen, E.N. (2005) *Angew. Chem. Int. Ed.*, **44**, 6700–6704.

205 Raheem, I.T., Thiara, P.S., Peterson, E.A. and Jacobsen, E.N. (2007) *J. Am. Chem. Soc.*, **129**, 13404–13405.

206 Martin, N.J.A., Ozores, L. and List, B. (2007) *J. Am. Chem. Soc.*, **129**, 8976–8977.

207 Okino, T., Hoashi, Y., Furukawa, T., Xu, X. and Takemoto, Y. (2005) *J. Am. Chem. Soc.*, **127**, 119–125.

208 Hoashi, Y., Yabuta, T. and Takemoto, Y. (2004) *Tetrahedron Lett.*, **45**, 9185–9188.

209 Hoashi, Y., Yabuta, T., Yuan, P., Miyabe, H. and Takemoto, Y. (2006) *Tetrahedron*, **62**, 365–374.

210 Zhu, R.X., Zhang, D.J., Wu, J. and Liu, C.B. (2006) *Tetrahedron Asymm.*, **17**, 1611–1616.

211 Hamza, A., Schubert, G., Soos, T. and Papai, I. (2006) *J. Am. Chem. Soc.*, **128**, 13151–13160.

212 Miyabe, H., Tuchida, S., Yamauchi, M. and Takemoto, Y. (2006) *Synthesis*, 3295–3300.

213 Li, H., Wang, J., Zu, L. and Wang, W. (2006) *Tetrahedron Lett.*, **47**, 2585–2589.

214 Li, H., Wang, H., Zu, L.S. and Wang, W. (2006) *Tetrahedron Lett.*, **47**, 3145–3148.

215 Liu, T.Y., Long, J., Li, B.J., Jiang, L., Li, R., Wu, Y., Ding, L.S. and Chen, Y.C. (2006) *Org. Biomol. Chem.*, **4**, 2097–2099.

216 Liu, T.Y., Li, R., Chai, Q., Long, J., Li, B.J., Wu, Y., Ding, L.S. and Chen, Y.C. (2007) *Chem. Eur. J.*, **13**, 319–327.

217 Hoashi, Y., Okino, T. and Takemoto, Y. (2005) *Angew. Chem. Int. Ed.*, **44**, 4032–4035.

218 Zhang, D., Wang, G. and Zhu, R.X. (2008) *Tetrahedron Asymm.*, **19**, 568–576.

219 Wodrich, M.D., Corminboeuf, C., Schreiner, P.R., Fokin, A.A. and Schleyer, P.v.R. (2007) *Org. Lett.*, **9**, 1851–1854.

220 Jose, D.A., Singh, A., Das, A. and Ganguly, B. (2007) *Tetrahedron Lett.*, **48**, 3695–3698.

221 Fu, A. and Thiel, W. (2006) *J. Mol. Struct.*, **765**, 45–52.

222 Inokuma, T., Hoashi, Y. and Takemoto, Y. (2006) *J. Am. Chem. Soc.*, **128**, 9413–9419.

223 Zu, L., Xie, H., Li, H., Wang, J., Jiang, W. and Wang, W. (2007) *Adv. Synth. Catal.*, **349**, 1882–1886.

224 Palomo, C., Oiarbide, M. and Laso, A. (2007) *Eur. J. Org. Chem.*, 2561–2574.

225 Okino, T., Nakamura, S., Furukawa, T. and Takemoto, Y. (2004) *Org. Lett.*, **6**, 625–627.

226 Xu, X., Furukawa, T., Okino, T., Miyabe, H. and Takemoto, Y. (2006) *Chem. Eur. J.*, **12**, 466–476.

227 Buffat, M.G.P. (2004) *Tetrahedron*, **60**, 1701–1729.

228 Atobe, M., Yamazaki, N. and Kibayashi, C. (2004) *J. Org. Chem.*, **69**, 5595–5607.

229 Yamaoka, Y., Miyabe, H., Yasui, Y. and Takemoto, Y. (2007) *Synthesis*, **16**, 2571–2575.

230 Liu, T.Y., Cui, H.L., Long, J., Li, B.J., Wu, Y., Ding, L.S. and Chen, Y.C. (2007) *J. Am. Chem. Soc.*, **129**, 1878–1879.

231 Berkessel, A., Mukherjee, S., Cleemann, F., Müller, T.N. and Lex, J. (2005) *Chem. Commun.*, 1898–1900.

232 Berkessel, A., Mukherjee, S., Müller, T.N., Cleemann, F., Roland, K., Brandenburg, M., Neudörfl, J.M. and Lex, J. (2006) *Org. Biomol. Chem.*, **4**, 4319–4330.
233 Berkessel, A., Cleemann, F. and Mukherjee, S. (2005) *Angew. Chem. Int. Ed.*, **44**, 7466–7469.
234 Xu, X.N., Yabuta, T., Yuan, P. and Takemoto, Y. (2006) *Synlett*, 137–140.
235 Pihko, P.M. and Pohjakallio, A. (2004) *Synlett*, 2115–2118.
236 Saaby, S., Bella, M. and Jørgensen, K.A. (2004) *J. Am. Chem. Soc.*, **126**, 8120–8121.
237 Cativiela, C. and Diaz-de-Villegas, M.D. (2000) *Tetrahedron Asymm.*, **11**, 645.
238 Watts, J., Benn, A., Flinn, N., Monk, T., Ramjee, M., Ray, P., Wang, Y. and Quibell, M. (2004) *Bioorg. Med. Chem.*, **12**, 2903–2925.
239 Yamaoka, Y., Miyabe, H. and Takemoto, Y. (2007) *J. Am. Chem. Soc.*, **129**, 6686–6687.
240 Fuerst, D.E. and Jacobsen, E.N. (2005) *J. Am. Chem. Soc.*, **127**, 8964–8965.
241 Gregory, R.J. (1999) *Chem. Rev.*, **99**, 3649–3682.
242 Zuend, S.J. and Jacobsen, E.N. (2007) *J. Am. Chem. Soc.*, **129**, 15872–15883.
243 Fang, Y.-Q. and Jacobsen, E.N. (2008) *J. Am. Chem. Soc.*, **130**, 5660–5661.
244 Both enantiomers of *trans*-1,2-aminocyclohexanephosphine were readily accessed on a multi-gram scale using a modification of the literature procedure: Caiazzo, A., Dalili, S. and Yudin, A.K. (2002) *Org. Lett.*, **4**, 2597–2600.
245 Wenzel, A.G., Lalonde, M.P. and J.E.N. (2003) *Synlett*, **12**, 1919–1922.
246 Berkessel, A., Cleemann, F., Mukherjee, S., Müller, T.N. and Lex, J. (2005) *Angew. Chem. Int. Ed.*, **44**, 807–811.
247 Faigl, F., Fogassy, E., Nógrádi, M., Pálovics, E. and Schindler, J. (2008) *Tetrahedron Asymm.*, **19**, 515–536.
248 Pellissier, H. (2008) *Tetrahedron*, **64**, 1563–1601.
249 Carter, H.E. (1946) *Org. React.*, **3**, 198–239.
250 Tan, C.Y.K. and Weaver, D.F. (2002) *Tetrahedron*, **58**, 7449–7461.
251 Chen, C.S., Fujimoto, Y., Girdaukas, G. and Sih, C.J. (1982) *J. Am. Chem. Soc.*, **104**, 7294–7299.
252 Wang, C.-J., Zhang, Z.-H., Dong, X.-Q. and Wu, X.-J. (2008) *Chem. Commun.*, 1431–1433.
253 Dixon, D.J. and Richardson, R.D. (2006) *Synlett*, 81–85.
254 Ballini, R., Bosica, G., Fiorini, D., Palmieri, A. and Petrini, M. (2005) *Chem. Rev.*, **105**, 933–971.
255 Steele, R.M., Monti, C., Gennari, C., Piarulli, U., Andreoli, F., Vanthuyne, N. and Roussel, C. (2006) *Tetrahedron Asymm.*, **17**, 999.
256 Yoon, T.P. and Jacobsen, E.N. (2005) *Angew. Chem. Int. Ed.*, **44**, 466–468.
257 Tan, K.L. and Jacobsen, E.N. (2007) *Angew. Chem. Int. Ed.*, **46**, 1315–1317.
258 Tsogoeva, S.B., Hateley, M.J., Yalalov, D.A., Meindl, K., Weckbecker, C. and Huthmacher, K. (2005) *Bioorg. Med. Chem.*, **13**, 5680–5685.
259 Tsogoeva, S.B., Yalalov, D.A., Hateley, M.J., Weckbecker, C. and Huthmacher, K. (2005) *Eur. J. Org. Chem.*, 4995–5000.
260 Yalalov, D.A., Tsogoeva, S.B. and Schmatz, S. (2006) *Adv. Synth. Catal.*, **348**, 826–832.
261 Tsogoeva, S.B. and Wei, S.W. (2006) *Chem. Commun.*, 1451–1453.
262 Lalonde, M.P., Chen, Y.G. and Jacobsen, E.N. (2006) *Angew. Chem. Int. Ed.*, **45**, 6366–6370.
263 Huang, H.B. and Jacobsen, E.N. (2006) *J. Am. Chem. Soc.*, **128**, 7170–7171.
264 Sohtome, Y., Tanatani, A., Hashimoto, Y. and Nagasawa, K. (2004) *Tetrahedron Lett.*, **45**, 5589–5592.
265 Nemoto, T., Fukuyama, T., Yamamoto, E., Tamura, S., Fukuda, T., Matsumoto, T., Akimoto, Y. and Hamada, Y. (2007) *Org. Lett.*, **9**, 927.
266 Zhang, Y., Liu, Y.-K., Kang, T.-R., Hu, Z.-K. and Chen, Y.-C. (2008) *J. Am. Chem. Soc.*, **130**, 2456–2457.
267 Berkessel, A., Roland, K. and Neudörfl, J.M. (2006) *Org. Lett.*, **8**, 4195–4198.
268 Tillack, J., Schmalstieg, L., Puetz, W. and Ruttmann, G. (2001) German patent WO 002001009215A1.

269 Chen, Y.-C. (2008) *Synlett*, **13**, 1919–1930.
270 Bredig, G. and Fiske, P.S. (1913) *Biochem. Z.*, **46**, 7–23.
271 Pracejus, H. (1960) *Liebigs Ann.*, **634**, 9–22.
272 Hiemstra, H. and Wynberg, H. (1981) *J. Am. Chem. Soc.*, **103**, 417–430.
273 Li, B.J., Jiang, L., Liu, M., Chen, Y.C., Ding, L.S. and Wu, Y. (2005) *Synlett*, 603–606.
274 Ye, J.X., Dixon, D.J. and Hynes, P.S. (2005) *Chem. Commun.*, 4481–4483.
275 Tillman, A.L., Ye, J.X. and Dixon, D.J. (2006) *Chem. Commun.*, 1191–1193.
276 Hynes, P.S., Stranges, D., Stupple, P.A., Guarna, A. and Dixon, D.J. (2007) *Org. Lett.*, **9**, 2107–2110.
277 Liu, T.-Y., Cui, H.-L., Chai, Q., Long, J., Li, B.-J., Wu, Y., Ding, L.-S. and Chen, Y.-C. (2007) *Chem. Commun.*, 2228–2230.
278 Vakulya, B., Varga, S., Csampai, A. and Soos, T. (2005) *Org. Lett.*, **7**, 1967–1969.
279 McCooey, S.H. and Connon, S.J. (2005) *Angew. Chem. Int. Ed.*, **44**, 6367–6370.
280 McCooey, S.H., McCabe, T. and Connon, S.J. (2006) *J. Org. Chem.*, **71**, 7494–7497.
281 Bode, C.A., Ting, A. and Schaus, S.E. (2006) *Tetrahedron*, **62**, 11499–11505.
282 Wang, J., Li, H., Zu, L.S., Jiang, W., Xie, H.X., Duan, W.H. and Wang, W. (2006) *J. Am. Chem. Soc.*, **128**, 12652–12653.
283 Gu, C.L., Liu, L., Sui, Y., Zhao, J.L., Wang, D. and Chen, Y.J. (2007) *Tetrahedron Asymm.*, **18**, 455–463.
284 Song, J., Wang, Y. and Deng, L. (2006) *J. Am. Chem. Soc.*, **128**, 6048–6049.
285 Wang, Y.Q., Song, J., Hong, R., Li, H.M. and Deng, L. (2006) *J. Am. Chem. Soc.*, **128**, 8156–8157.
286 Wang, B.M., Wu, F.H., Wang, Y., Liu, X.F. and Deng, L. (2007) *J. Am. Chem. Soc.*, **129**, 768–769.
287 Zu, L., Wang, J., Li, H., Xie, H., Jiang, W. and Wang, W. (2007) *J. Am. Chem. Soc.*, **129**, 1036–1037.
288 Song, J., Shih, H.W. and Deng, L. (2007) *Org. Lett.*, **9**, 603–606.
289 Wang, Y., Li, H., Wang, Y.-Q., Liu, Y., Foxman, B.M. and Deng, L. (2007) *J. Am. Chem. Soc.*, **129**, 6364–6365.
290 Amere, M., Lasne, M.-C. and Rouden, J. (2007) *Org. Lett.*, **9**, 2621–2624.
291 Zu, L., Li, H., Xie, H. and Wang, W. (2007) *Synthesis*, **16**, 2576–2580.
292 Dinér, P., Nielsen, M., Bertelsen, S., Niess, B. and Jørgensen, K.A. (2007) *Chem. Commun.*, 3646–3648.
293 Pettersen, D., Piana, F., Bernardi, L., Fini, F., Fochi, M., Sgarzani, V. and Ricci, A. (2007) *Tetrahedron Lett.*, **48**, 7805–7808.
294 Li, D.R., Murugan, A. and Falck, J.R. (2008) *J. Am. Chem. Soc.*, **130**, 46–48.
295 Dollt, H., Hammann, P. and Blechert, S. (1999) *Helv. Chim. Acta*, **82**, 1111–1122 and references cited therein.
296 Dodda, R., Mandal, T. and Zhao, C.-G. (2008) *Tetrahedron Lett.*, **49**, 1899–1902.
297 Rho, H.S., Oh, S.H., Lee, J.W., Lee, J.Y., Chin, J. and Song, C.E. (2008) *Chem. Commun.*, 1208–1210.
298 Marcelli, T., van der Haas, R.N.S., van Maarseveen, J.H. and Hiemstra, H. (2006) *Angew. Chem. Int. Ed.*, **45**, 929–931.
299 Tian, S.-K., Chen, Y., Hang, J., Tang, L., McDaid, P. and Deng, L. (2004) *Acc. Chem. Res.*, **37**, 621–631.
300 Liu, X., Li, H. and Deng, L. (2005) *Org. Lett.*, **7**, 167–169.
301 Li, H., Song, J., Liu, X. and Deng, L. (2005) *J. Am. Chem. Soc.*, **127**, 13774–13775.
302 Marcelli, T., van der Haas, R.N.S., van Maarseveen, J.H. and Hiemstra, H. (2005) *Synlett*, **18**, 2817–2819.
303 Hammar, P., Marcelli, T., Hiemstra, H. and Himo, F. (2007) *Adv. Synth. Catal.*, **349**, 2537–2548.
304 Zhong, G.F., Fana, J. and Barbas, C.F. (2004) *Tetrahedron Lett.*, **45**, 5681–5684.
305 Almasi, D., Alonso, D.A. and Nájera, C. (2006) *Tetrahedron Asymm.*, **17**, 2064–2068.
306 Xu, X.-Y., Wang, Y.-Z. and Gong, L.-Z. (2007) *Org. Lett.*, **9**, 4247–4249.
307 Herrera, R.P., Sgarzani, V., Bernardi, L. and Ricci, A. (2005) *Angew. Chem. Int. Ed.*, **44**, 6576–6579.

308 Sibi, M.P. and Itoh, K. (2007) *J. Am. Chem. Soc.*, **129**, 8064–8065.
309 Sibi, M.P., Shay, J.J., Liu, M. and Jasperse, C.P. (1998) *J. Am. Chem. Soc.*, **120**, 6615–6616.
310 Lattanzi, A. (2007) *Synlett*, 2106–2110.
311 Herrera, R.P., Monge, D., Martín-Zamora, E., Fernández, R. and Lassaletta, J.M. (2007) *Org. Lett.*, **9**, 3303–3306.
312 Matsumoto, T., Yamamoto, H. and Inoue, S. (1984) *J. Am. Chem. Soc.*, **106**, 4829–4832.
313 Wang, J., Li, H., Yu, X.H., Zu, L.S. and Wang, W. (2005) *Org. Lett.*, **7**, 4293–4296.
314 Wang, J., Li, H., Duan, W.H., Zu, L.S. and Wang, W. (2005) *Org. Lett.*, **7**, 4713–4716.
315 Fleming, E.M., McCabe, T. and Connon, S.J. (2006) *Tetrahedron Lett.*, **47**, 7037–7042.
316 Shi, Y.-L. and Shi, M. (2007) *Adv. Synth. Catal.*, **349**, 2129–2135.
317 Shi, M. and Chen, L.H. (2003) *Chem. Commun.*, 1310–1311.
318 Qi, M.-J., Ai, T., Shi, M. and Li, G. (2008) *Tetrahedron*, **64**, 1181–1186.
319 Liu, X.-G., Jianga, J.-J. and Shi, M. (2007) *Tetrahedron Asymm.*, **18**, 2773–2781.
320 Shi, M. and Liu, X.-G. (2008) *Org. Lett.*, **10**, 1043–1046.
321 Ishikawa, T. and Kumamoto, T. (2006) *Synthesis*, **5**, 737–752.
322 Iyer, M.S., Gigstad, K.M., Namdev, N.D. and Lipton, M. (1996) *J. Am. Chem. Soc.*, **118**, 4910–4911.
323 Becker, C., Hoben, C., Schollmeyer, D., Scherr, G. and Kunz, H. (2005) *Eur. J. Org. Chem.*, 1497–1499.
324 Corey, E.J. and Grogan, M.J. (1999) *Org. Lett.*, **1**, 157–160.
325 Sohtome, Y., Hashimoto, Y. and Nagasawa, K. (2005) *Adv. Synth. Catal.*, **347**, 1643–1648.
326 Sohtome, Y., Hashimoto, Y. and Nagasawa, K. (2006) *Eur. J. Org. Chem.*, 2894–2897.
327 Sohtome, Y., Takemura, N., Takada, K., Takagi, R., Iguchi, T. and Nagasawa, K. (2007) *Chem. Asian J.*, **2**, 1150–1160.
328 Sohtome, Y., Takemura, N., Iguchi, T., Hashimoto, Y. and Nagasawa, K. (2006) *Synlett*, 144–146.
329 Takada, K., Takemura, N., Cho, K., Sohtome, Y. and Nagasawa, K. (2008) *Tetrahedron Lett.*, **49**, 1623–1626.
330 Boysen, M.M.K. (2007) *Chem. Eur. J.*, **13**, 8648–8659.
331 Wang, Z.-X. and Shi, Y. (1998) *J. Org. Chem.*, **63**, 3099–3104.
332 Wang, Z.-X., Cao, G.-A. and Shi, Y. (1999) *J. Org. Chem.*, **64**, 7646–7650.
333 Zhu, Y., Tu, Y., Yu, H. and Shi, Y. (1998) *Tetrahedron*, **39**, 7819–7822.
334 Tian, H., She, X., Yu, H., Shu, L. and Shi, Y. (2002) *J. Org. Chem.*, **67**, 2435–2446.
335 Colonna, S., Pironti, V., Drabowicz, J., Brebion, F., Fensterbank, L. and Malacria, M. (2005) *Eur. J. Org. Chem.*, 1727–1730.
336 Khiar, N., Mallouk, S., Valdivia, V., Bougrin, K., Soufiaoui, M. and Fernandez, I. (2007) *Org. Lett.*, **9**, 1255–1258.
337 Shing, T.K.M., Leung, G.Y.C. and Luk, T. (2005) *J. Org. Chem.*, 70.
338 Becker, C., Hoben, C. and Kunz, H. (2007) *Adv. Synth. Catal.*, **349**, 417–424.
339 Liu, K., Cui, H.F., Nie, J., Dong, K.Y., Li, X.J. and Ma, J.A. (2007) *Org. Lett.*, **9**, 923–925.
340 Wang, C., Zhou, Z. and Tang, C. (2008) *Org. Lett.*, **10**, 1707–1710.
341 Klussmann, M., Iwamura, H., Mathew, S.P., Wells, D.H. Jr., Pandya, U., Armstrong, A. and Blackmond, D.G. (2006) *Nature*, 441.
342 Cao, C.L., Ye, M.C., Sun, X.L. and Tang, Y. (2006) *Org. Lett.*, **8**, 2901–2904.
343 Allemann, C., Gordillo, R., Clemente, F.R., Cheong, P.H.-Y. and Houk, K.N. (2004) *Acc. Chem. Res.*, **37**, 558–569.
344 Cao, Y.J., Lai, Y.Y., Wang, X., Li, Y.H. and Xiao, W.J. (2007) *Tetrahedron Lett.*, **48**, 21–24.
345 Chang, Y.-W., Yang, J.-J., Dang, J.-N. and Xue, Y.-X. (2007) *Synlett*, **14**, 2283–2285.
346 Robak, M.T., Trincado, M. and Ellman, J.A. (2007) *J. Am. Chem. Soc.*, **129**, 15110–15111.

Appendix: Structure Index

This survey illustrates all (thio)urea derivatives discussed in the chapter "(Thio)urea Organocatalysts."

346 | *6 (Thio)urea Organocatalysts*

Appendix: Structure Index

348 | *6 (Thio)urea Organocatalysts*

Appendix: Structure Index | 349

350 | *6 (Thio)urea Organocatalysts*

Appendix: Structure Index | 351

$R^2 = -(CH_2)_4COOEt; R^3 = -(CO)(CH_2)_2(CO)O(CH_2)_8O(CO)Ps$

$R^5 = -(CO)(CH_2)_9OTBDS$

7
Highlights of Hydrogen Bonding in Total Synthesis
Mitsuru Shoji and Yujiro Hayashi

7.1
Introduction

In the total synthesis of natural products, hydrogen bonding causes several effects, such as change of stereoselectivity and reaction rate. Intramolecular hydrogen bonding acts for fixing a conformation of molecules. In some cases, the desired stereoisomer predominates owing to the hydrogen bonding, which overrides the stabilization of anomeric effect. Hayes and Heathcock converted spiroacetal **1** to the desired **2** under acidic condition in the synthesis of the AB-ring segment of altohyrtin A (Scheme 7.1) [1]. In the gas phase, the stabilization energy of neutral O–H···O-hydrogen bonding is calculated as 5–7 kcal/mol [2], which is greater than that of anomeric effect (1.5–2 kcal/mol). The utilization of the hydrogen bonding and the suppression of undesired hydrogen bonding interaction would also be important issues in the natural product synthesis. For instance, selectivity of epoxidation of 2-cyclohexen-1-ol (**3**) using *m*-chloroperoxybenzoic acid (*m*CPBA) varies depending on protective groups of the hyrdroxyl group (Scheme 7.2) [3]. This kind of diastereoselective epoxidation is efficiently utilized in a number of total syntheses of natural products. Furthermore, intermolecular hydrogen bonding approximates two molecules and accelerates reaction rates and/or change stereoselectivities. In this chapter, intra- and intermolecular hydrogen bondings in total syntheses are described.

7.2
Intramolecular Hydrogen Bonding in Total Syntheses

7.2.1
Thermodynamic Control of Stereochemistry

7.2.1.1 Pinnatoxin A
Pinnatoxin A (**5**) (Figure 7.1) was isolated by Uemura and co-workers in 1995, which is one of the major toxic principles responsible for outbreaks of *Pinna*

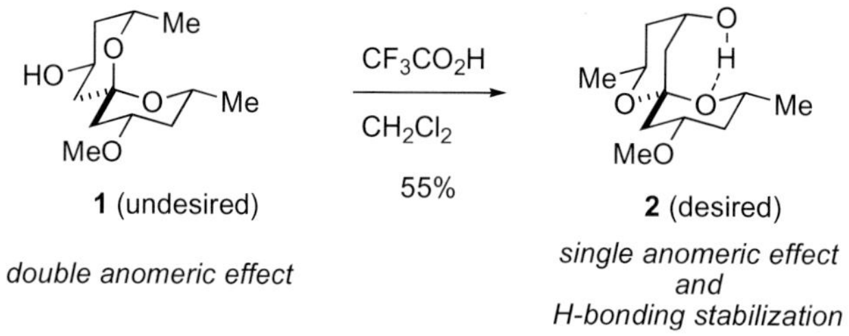
Me
HO
O
O
Me
MeO
1 (undesired)
double anomeric effect
CF3CO2H
CH2Cl2
55%
O
H
Me
O
O
Me
MeO
2 (desired)
single anomeric effect
and
H-bonding stabilization

Scheme 7.1

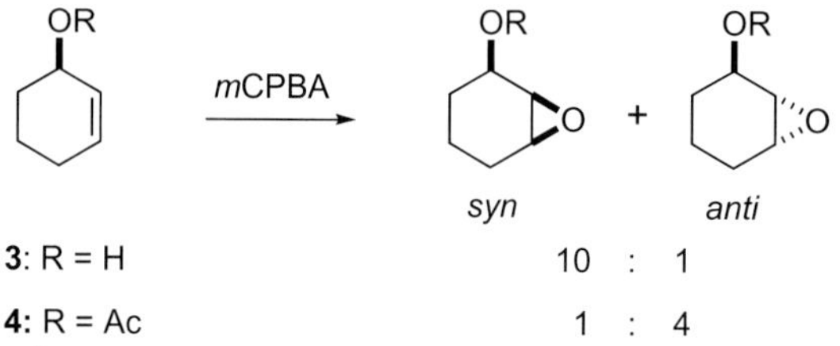
OR
mCPBA
OR
O
+
OR
O
syn
anti
3: R = H
10 : 1
4: R = Ac
1 : 4

Scheme 7.2

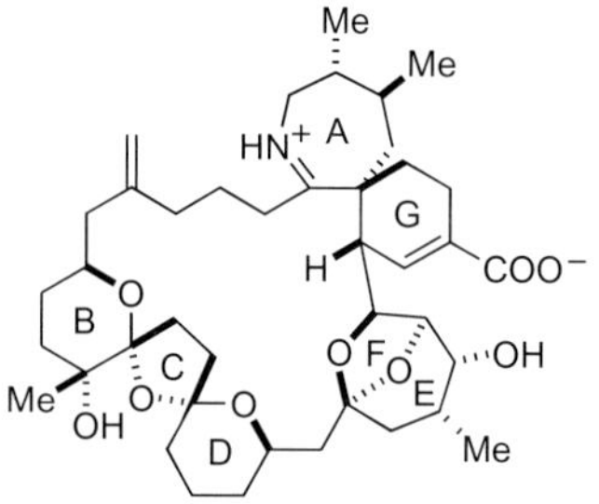
Me
Me
HN+ A
G
H
COO−
O
B
C
Me
OH
O
O
D
O
F
O
E
OH
Me

Pinnatoxin A (**5**)

Figure 7.1 Structure of Pinnatoxin A.

shellfish poisonings in China and Japan [4]. In 1998, Kishi and co-workers accomplished the first total synthesis of **5** and reported interconversion of bis-spiroketals **6** and **7** [5]. Undesired bis-spiroketal **6** was transformed to desired bis-spiroketal **7** under silylation condition, whereas **7** was converted to **6** in the presence of magnesium bromide with removal of two TBS groups (Scheme 7.3). These results suggested that intramolecular hydrogen bonding of **6** stabilizes its conformation.

In contrast, Hirama and co-workers stereoselectively constructed BCD ring moiety **8**, utilizing intramolecular hydrogen bonding between the primary and

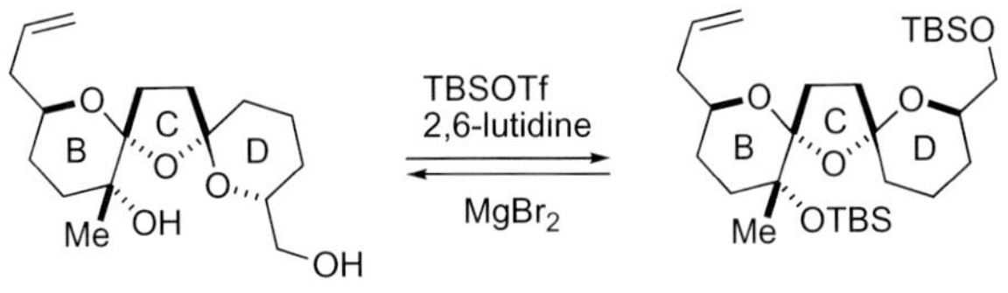

Scheme 7.3

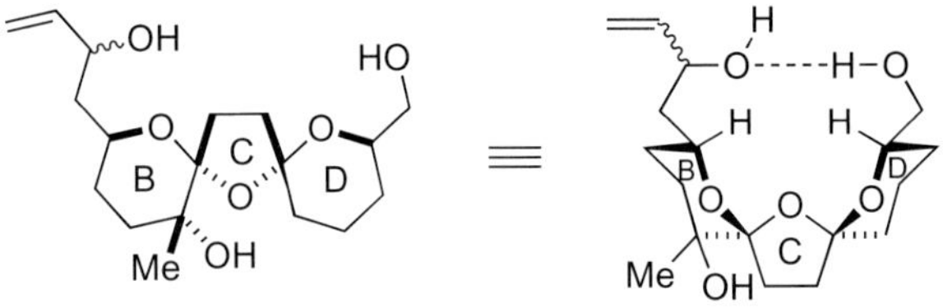

Figure 7.2 Intramolecular hydrogen bonding of diol **8**.

secondary hydroxyl groups [6] (Figure 7.2). Interestingly, the secondary hydroxyl group contributes to the formation of the desired stereoisomer **8**.

7.2.1.2 Azaspiracid-1

Azaspiracid-1 (**9**) (Figure 7.3), which causes human illness with diarrhetic shellfish poisoning-like symptoms, was isolated by Yasumoto, Satake, and co-workers, from mussels *Mytilus edulis* of Killary Harbour, Ireland [7]. Although the relative structure of azaspiracid-1 was proposed as **10**, Nicolaou and co-workers revised it to **9** [8]. In the synthesis of **10**, they controlled C13 stereochemistry utilizing intramolecular hydrogen bonding, as shown in Scheme 7.4. Researchers suggest that bis-spiroketal **12** is thermodynamically stable because of intramolecular hydrogen bonding between C9 hydroxyl group and two oxygens on B and C rings, overcoming double anomeric effect in **11**.

Zezschwitz and co-workers controlled the stereochemistry of hydroxyl spiro [4.5] acetal skeleton of **13** and **14** utilizing intramolecular hydrogen bonding between the secondary hydroxyl group and pyran oxygen (Scheme 7.5) [9].

7.2.2
Kinetic Control Stereochemistry

7.2.2.1 Pancratistatin

Pancratistatin (**15**) (Figure 7.4) was isolated by Pettit and co-workers from the roots of *Pancratium littorale*, which exhibits anticancer activity against murine P-5076 ovarian sarcoma and P-388 lymphocytic leukemia [10]. Rigby and co-workers performed photocyclization as a key step in the total synthesis of **15**,

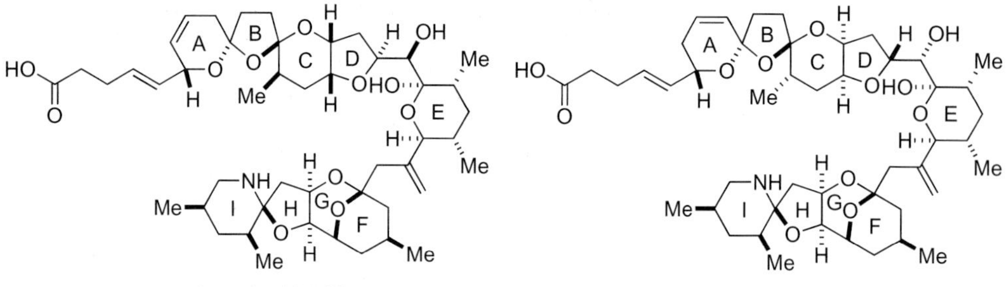

Azaspiracid-1 (9)

originally proposed structure 10

Figure 7.3 Structures of Azaspiracid-1 and its original.

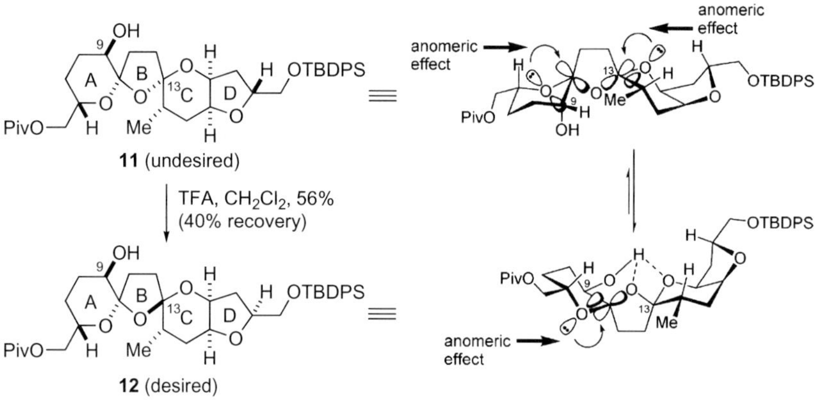

11 (undesired)

TFA, CH$_2$Cl$_2$, 56%
(40% recovery)

12 (desired)

anomeric effect

Scheme 7.4

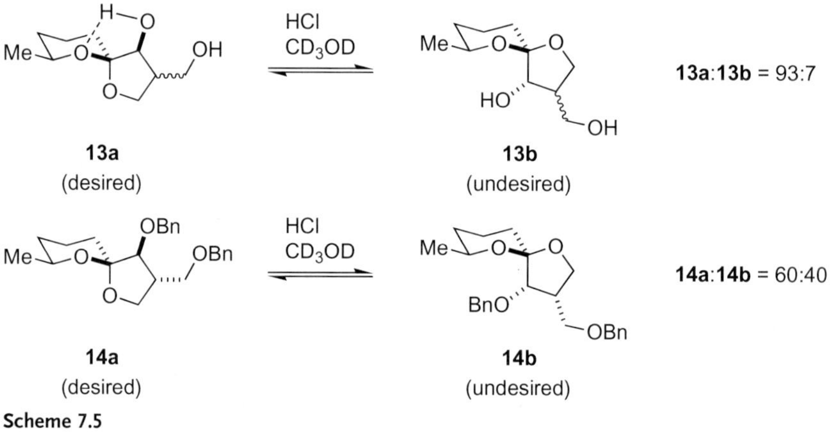

13a (desired) ⇌ HCl, CD$_3$OD ⇌ **13b** (undesired)   **13a:13b** = 93:7

14a (desired) ⇌ HCl, CD$_3$OD ⇌ **14b** (undesired)   **14a:14b** = 60:40

Scheme 7.5

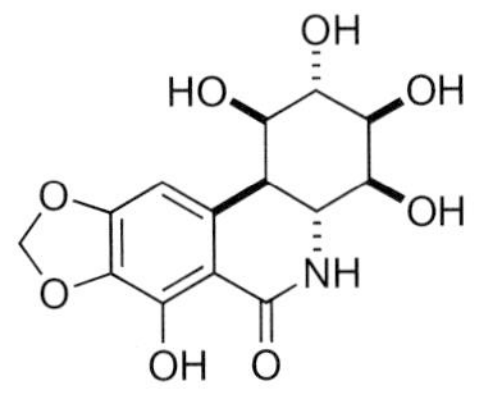

Pancratistatin (**15**)   **Figure 7.4** Structure of Pancratistatin.

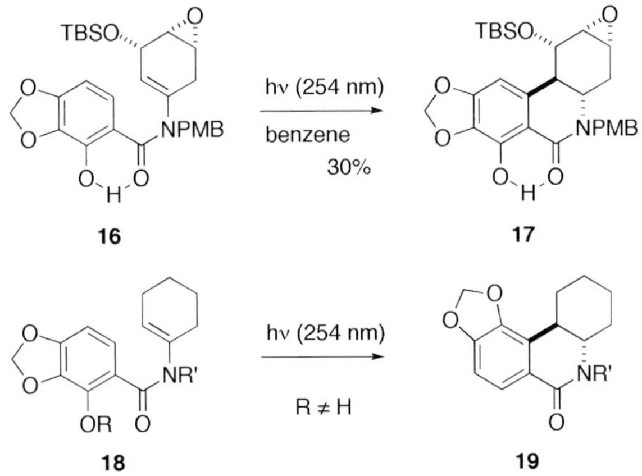

Scheme 7.6

utilizing intramolecular hydrogen bonding between the phenolic hydroxyl group and the amide carbonyl oxygen to fix the conformation of precursor **16**, affording *trans* isomer **17** exclusively (Scheme 7.6) [11]. Interestingly, photocyclization of phenol-protected **18** exclusively gave ipso cyclization product **19**, which means intramolecular hydrogen bonding is essential for the photocyclization.

7.2.2.2 Tunicamycins

In the total synthesis of tunicamycins, Myers and co-workers utilized intramolecular hydrogen bonding in stereoselective radical cyclization [12]. When **20** was treated with Et$_3$B and ^{n}Bu$_3$SnH, undesired 5′S diastereomer **21** via transition state **22** was obtained exclusively (Scheme 7.7). Another transition state **23** toward desired 5′R product has steric repulsion between TBS group and acetamide group. On the other hand, the radical cyclization of **24** proceeded to produce desired **26** as a major product (5′R:5′S = 7.5:1) via transition state **25**, in which intramolecular hydrogen bonding between the hydroxyl group and the amide carbonyl oxygen would predominate the formation of **26**. The reaction conducted in MeOH gave a 1.6:1 mixture of 5′R and 5′S diastereomers, which suggests the hydrogen bonding would be disrupted in the protic solvent.

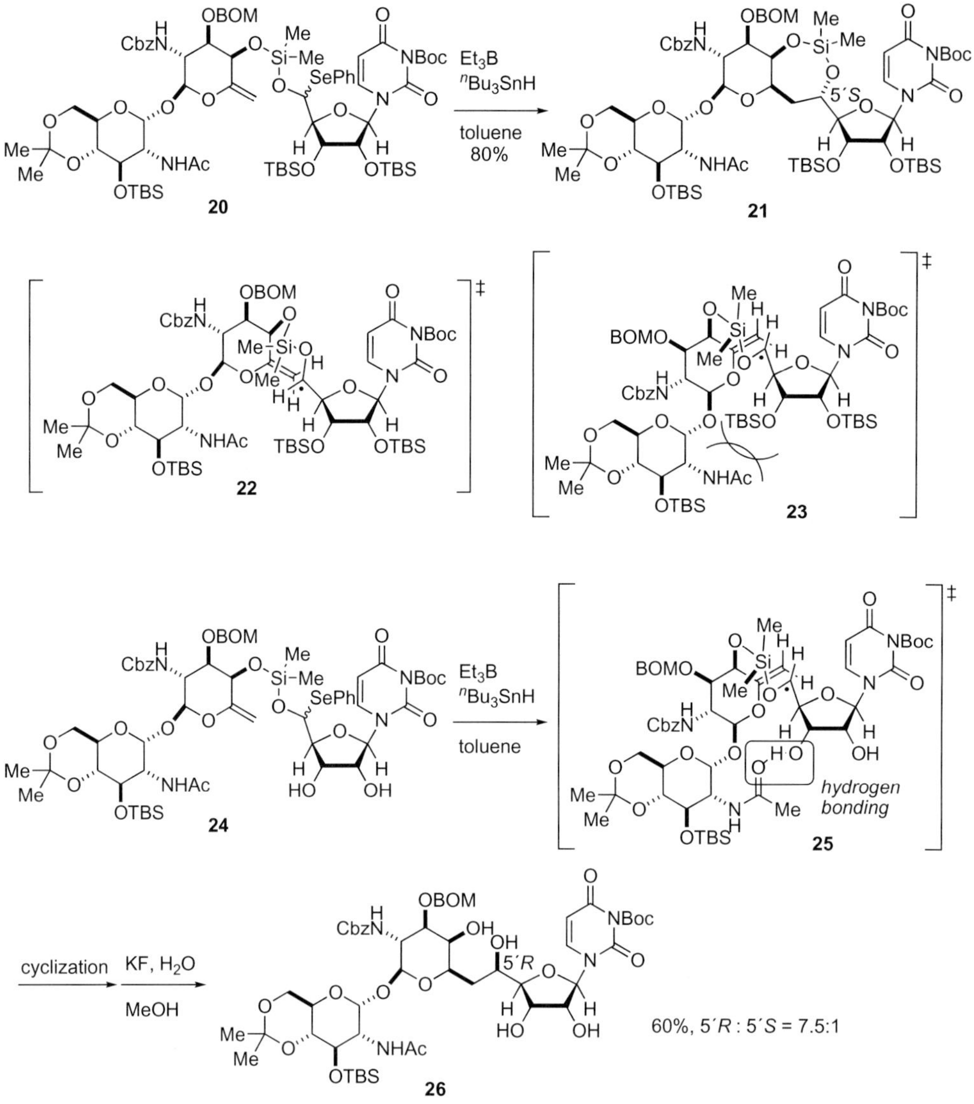

Scheme 7.7

7.2.2.3 Callystatin

Callystatin A (**27**) was isolated by Kobayashi and co-workers from the sponge *Callyspongia truncata* in 1997; it shows anticancer activity against KB tumor cells and L1210 cells [13]. During the total synthesis of **27**, Marshall and co-workers met interesting phenomena in Wittig reaction [14]. Homologation of *syn,syn,syn,syn* aldehydes **28** and **29** proceeded to give α,β-unsaturated esters **30** and **31** in high yield (Scheme 7.8), whereas that of *anti,anti,syn,syn* aldehydes **32** and **33** afforded

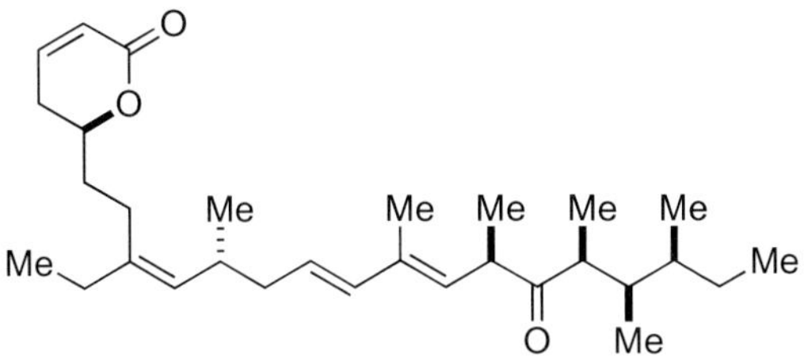

Callystatin (**27**)

Figure 7.5 Structure of Callystatin.

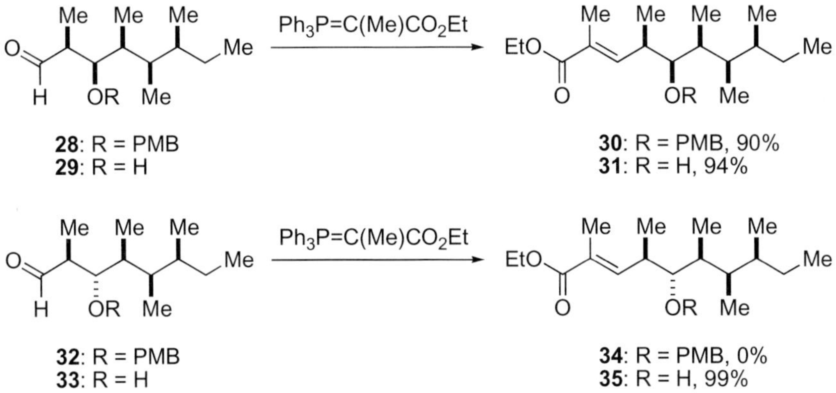

28: R = PMB
29: R = H

30: R = PMB, 90%
31: R = H, 94%

32: R = PMB
33: R = H

34: R = PMB, 0%
35: R = H, 99%

Scheme 7.8

quite different results (0% of **34** and 99% of **35**, respectively). Marshall and co-workers explained these contrasting results using theoretical calculations, which indicate that carbony groups are not blocked from nucleophilic attack in aldehydes **28** and **29** (Figure 7.6). On the other hand, both faces of aldehydes **32** are crowded and the C–C bond could not be formed, whereas hydrogen bonding between β-hydroxyl group and carbonyl oxygen of **33** make the nucleophile easier to approach to aldehyde.

7.2.2.4 Resorcylides

Trans- and *cis-*Resorcylides (**36** and **37**) (Figure 7.7) are both plant-growth inhibitors, isolated independently from different *Penicillium* species [15]. In the first total syntheses of **36** and **37** accomplished by Couladouros and co-workers, they managed the stereochemistry of alkenes of **36** and **37** in the ring-closing olefin metathesis with catalyst **39**, altering the conformation of diene precursors **38** and **41** through the control of the intramolecular hydrogen bonding (Scheme 7.9) [16]. Related results of *E/Z* selectivity change at ring-closing metathesis were also reported by Fürstner's and Winssinger's groups in the total synthesis of

28 29 32 33

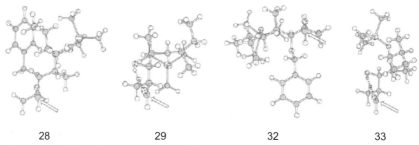

Figure 7.6 Calculated conformations of aldehydes **28**, **29**, **32**, and **33**. Dot lines on **29** and **33** indicate hydrogen bondings and arrows show the approach trajectory of nucleophile.

trans-Resorcylide (**36**) *cis*-Resorcylide (**37**)

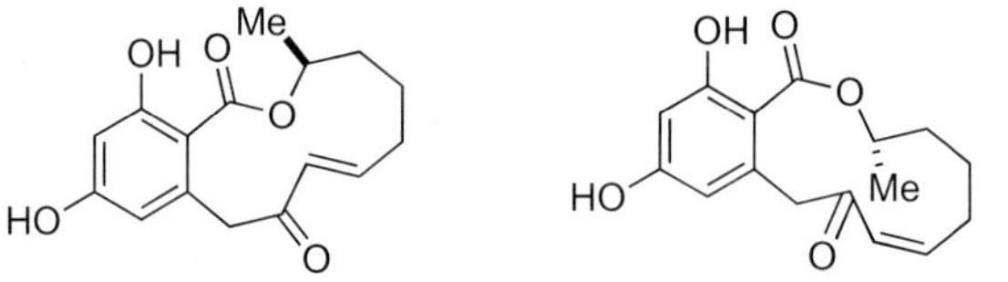

Figure 7.7 Structures of *trans*- and *cis*-Resorcylide.

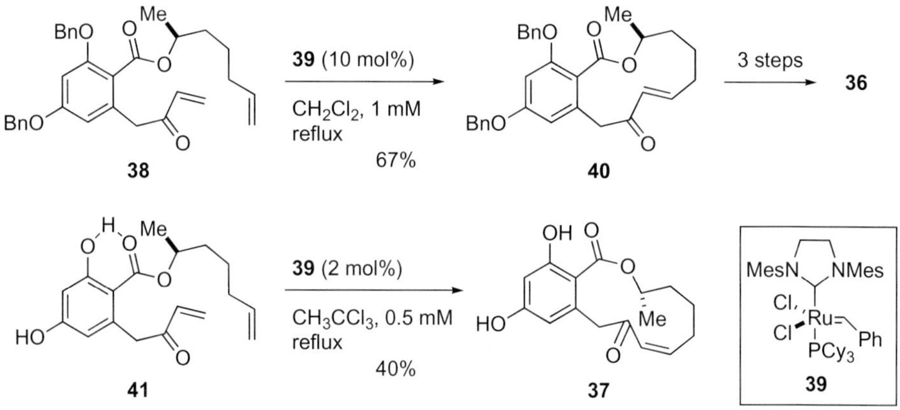

Scheme 7.9

salicylihalamide [17] and radicicol/pochonin C [18], respectively. In the total synthesis of kendomycin, Smith and co-workers described the difference of reactivity in ring-closing metathesis between two diastereomers, both of them have intramolecular hydrogen bonding [19].

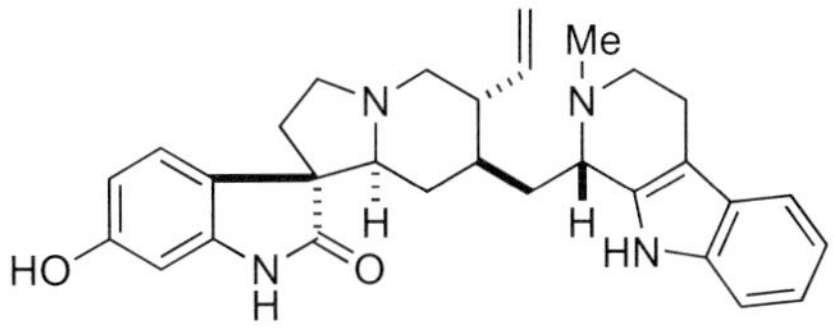

Strychnofoline (42)

Figure 7.8 Structure of Strychnofoline.

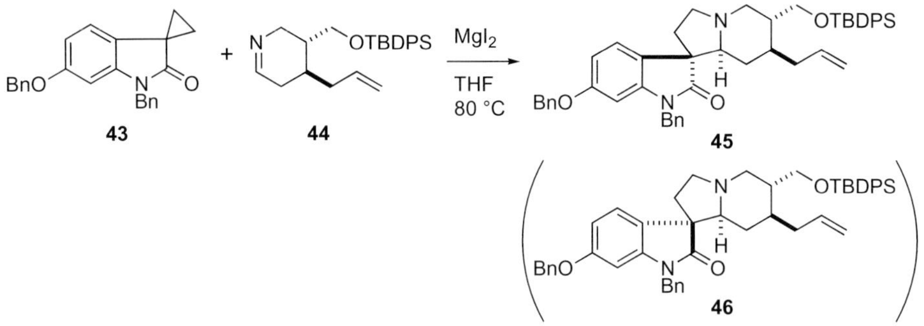

Scheme 7.10

7.2.2.5 Strychnofoline

Strychnofoline (**42**) (Figure 7.8) was isolated by Angenot and co-workers from the leaves of *Strychnos usambarensis* [20], and **42** was disclosed to inhibit mitosis in a number of cell lines, such as mouse melanoma B16, Ehrlich, and Hepatom HW165 [21]. In the total synthesis of **42**, Carreira and co-workers reported an MgI$_2$-mediated ring-expansion reaction of cyclopropyl oxyindole **43** and aldimine **44**, affording **45** diastereoselectively (Scheme 7.10) [22]. In contrast, **46** dominates in acidic medium. Oxindoles **45** and **46** are known to be interconverted via an intramolecular retro-Mannich/Mannich reaction sequence. As shown in Figure 7.9, **46**-type A is thermodynamically stable in the presence of acid because of intramolecular hydorogen bonding, whereas electrostatic repulsion of **46**-type B predominates the formation of **45**-type B under basic condition.

7.2.2.6 Asialo GM$_1$

Danishefsky and co-workers reported that the selectivity of glycosylation depended on the substituent of C4-position [23]. When hydroxyl-free glycosyl donor **47** was coupled with acceptor **49**, desired β-anomer was obtained predominantly (β:α = 13:1, Scheme 7.11). On the other hand, hydroxyl-protected donor **48** provided α-anomer as a major product (β:α = 1:10). This β-selectivity of **47** is explained by intramolecular hydrogen bonding between axial hydroxyl group and pyranose ring oxygen; preventing the easy cleavage of the axial EtS group by

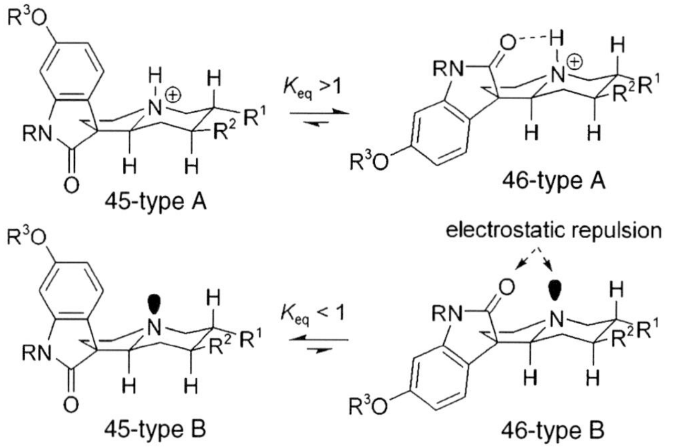

Figure 7.9 The effect of acidity of the medium on the diastereomer population.

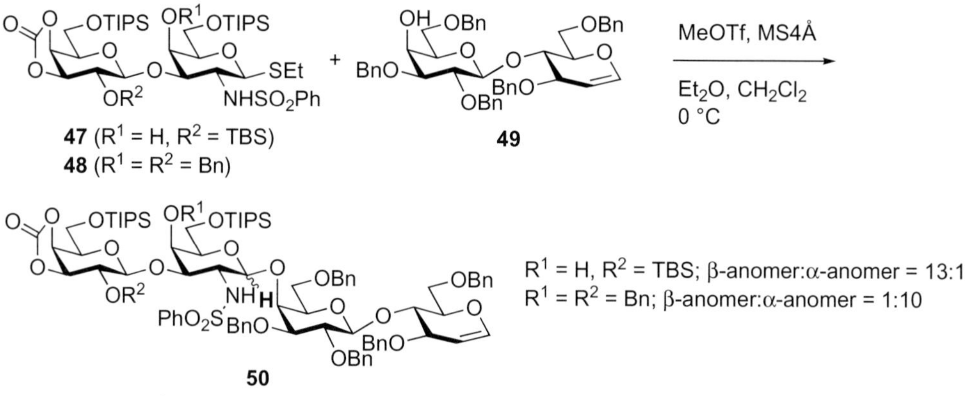

Scheme 7.11

stabilizing the conformer where the EtS leaving group is equatorial (Scheme 7.12). This facilitates neighboring group participation by the sulfonamide group and allows the β-anomer to be formed.

7.2.3
Activation/Deactivation of Reactions

7.2.3.1 Rishirilide B
Rishirilide B (**57**) (Figure 7.10) was isolated by Iwaki and co-workers from *Streptomyces rishiriensis* OFR-1056 [24], and this compound exhibits antithrombotic activity through selective α_2-macroglobulin inhibition. Allen and Danishefsky synthesized **57** utilizing regioselective Diels–Alder reaction of the quinodimethide

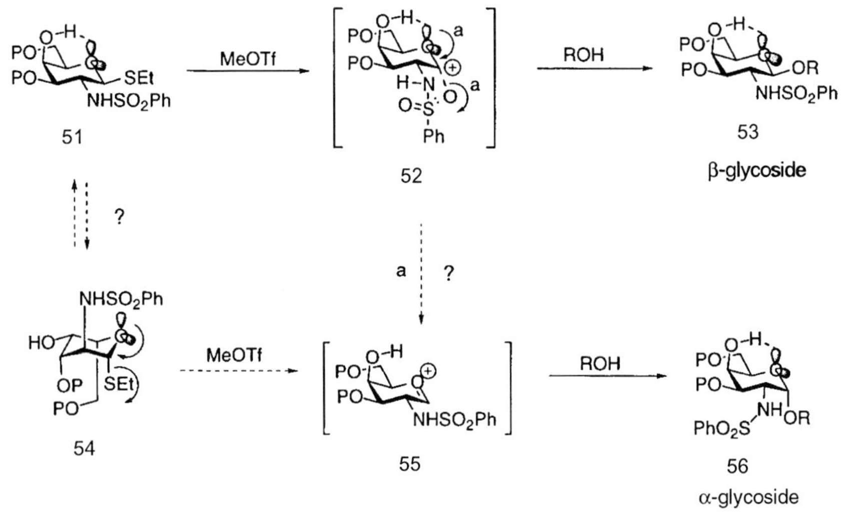

Scheme 7.12

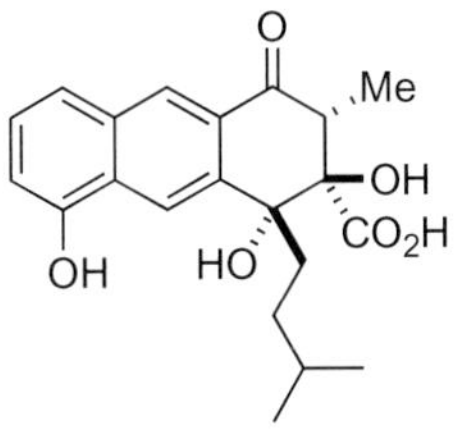

Rishirilide B (**57**)

Figure 7.10 Structure of Rishirilide.

derived from **58** and enedione **59** (Scheme 7.13) [25]. Intramoleculer hydrogen bonding of **59** activates one of the two carbonyl groups, which directs the formation of **60** regioselectively.

7.2.3.2 2-Desoxystemodione

In the total synthesis of 2-desoxystemodione, White and co-workers reported that intramolecular hydrogen bonding in α-hydroxyl aldehyde **61** activated the carbonyl group and constrained it in a reactive conformation of ene reaction to afford **62** in excellent yield (Scheme 7.14) [26]. In this ene reaction, one out of three allylic positions was involved.

7.2.3.3 Leucascandrolide A

Carreira and co-workers reported a hydrogen bonding network between hydroxyl groups and cyclic ether oxygens that might prevent macrolactonization in the total

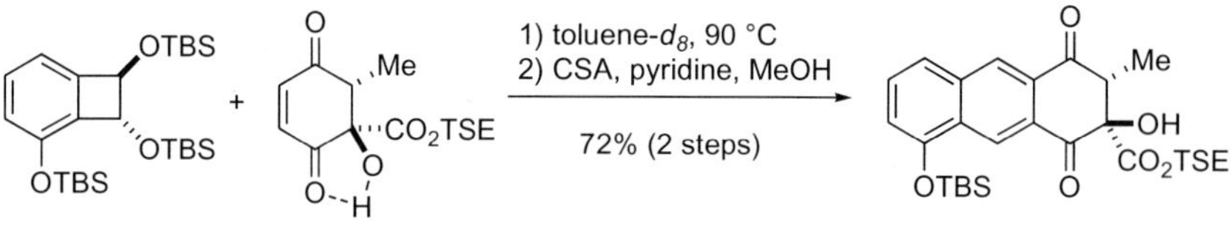

Scheme 7.13

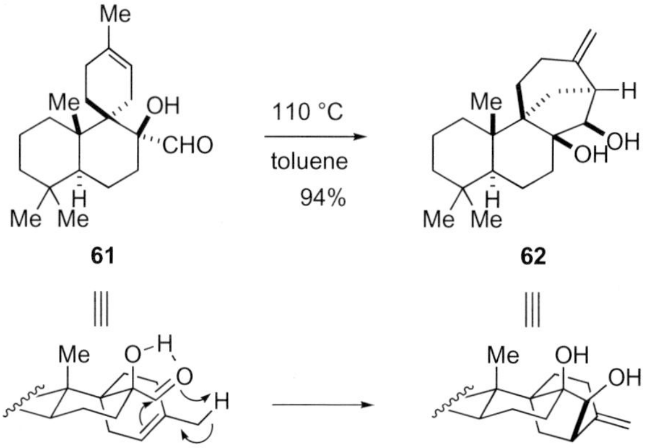

Scheme 7.14

synthesis of leucascandrolide A [27]. That is, Yamaguchi macrolactonization of **63** did not proceed in benzene, toluene, xylene, and THF, whereas the reaction in THF/DMF afforded the desired macrolactone **64** in good yield. A similar seco acid possessing 9-OMe substituent formed the cyclized product in high yield (Scheme 7.15). These results indicate that DMF disrupt intramolecular hydrogen-bonding network of **63** to assist the formation of **64**. Hunter and co-workers have studied these solvent effects extensively [28].

7.2.3.4 Azaspirene

Azaspirene (**65**) (Figure 7.11) was isolated by Kakeya, Osada, and co-workers in 2002, from the fungus *Neosartrya* sp., as an angiogenesis inhibitor [29]. It has a highly oxygenated azaspiro[4.4]nonenedione skeleton and two hydroxyl groups at C8 and C9 positions. Hayashi and co-workers accomplished the first total synthesis of azaspirene (**65**) and reported the importance of the order of the last two transformations [30]. It implies that removal of TIPS group of optically pure **66**, followed by hydration of the enamide moiety afforded racemic azaspirene, whereas hydration followed by deprotection gave optically pure azaspirene (Scheme 7.16). The racemization would proceed via retro-aldol reaction, but it did not occur during the sequence of hydration and deprotection. Based on these results, the

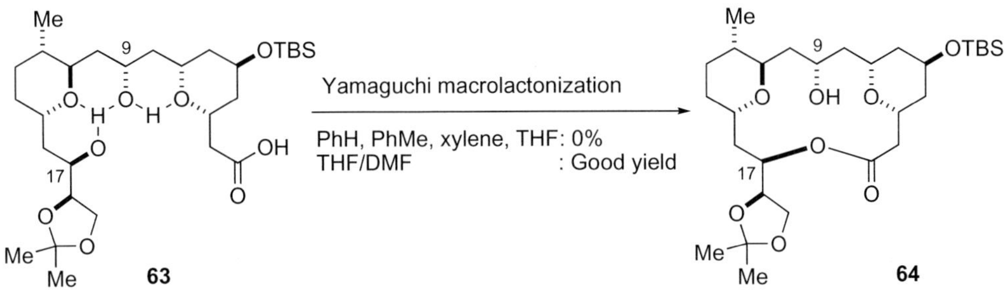

7.3 Intermolecular Hydrogen Bondings in Total Syntheses

Scheme 7.15

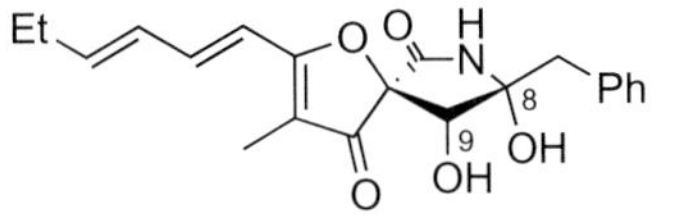

Yamaguchi macrolactonization
PhH, PhMe, xylene, THF: 0%
THF/DMF : Good yield

63 → 64

Figure 7.11 Structure of Azaspirene.

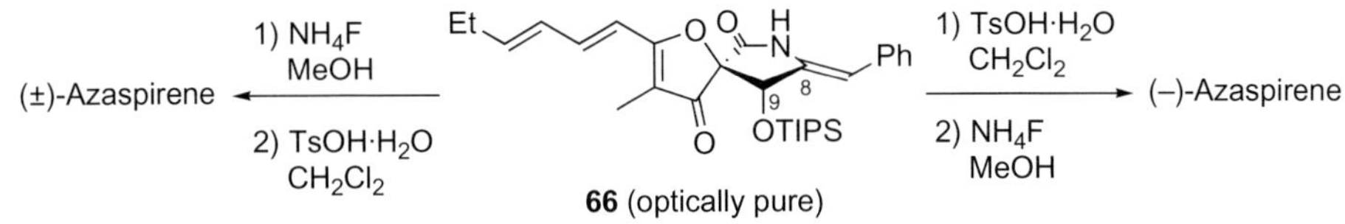

Azaspirene (**65**)

Scheme 7.16

(±)-Azaspirene ← 1) NH$_4$F MeOH; 2) TsOH·H$_2$O CH$_2$Cl$_2$ — **66** (optically pure) → 1) TsOH·H$_2$O CH$_2$Cl$_2$; 2) NH$_4$F MeOH → (−)-Azaspirene

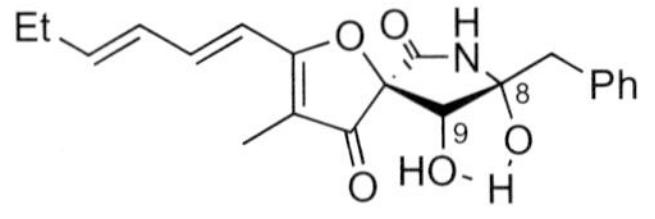

Figure 7.12 Intramolecular hydrogen bonding of Azaspirene.

researchers suggested that an intramolecular hydrogen bond between the C8 and C9 hydroxyl groups would be responsible for preventing the unwanted retro-aldol reaction and subsequent racemization (Figure 7.12).

7.3 Intermolecular Hydrogen Bondings in Total Syntheses

7.3.1 Henbest Epoxidation

In 1957, Henbest and Wilson reported hydroxyl-directed epoxidation using m-chloroperoxybenzoic acid (mCPBA) [3]. In the synthetic studies of juvenile

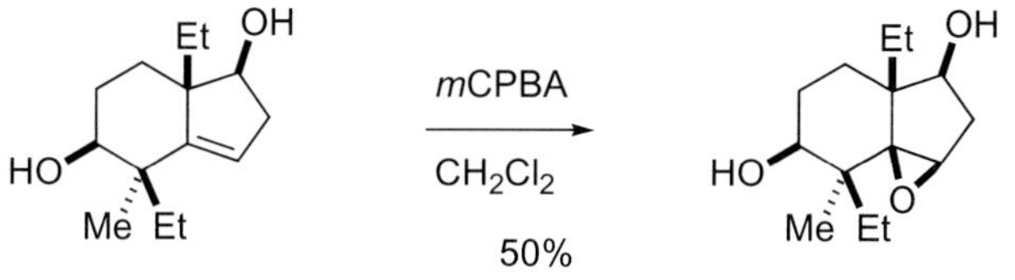

Scheme 7.17

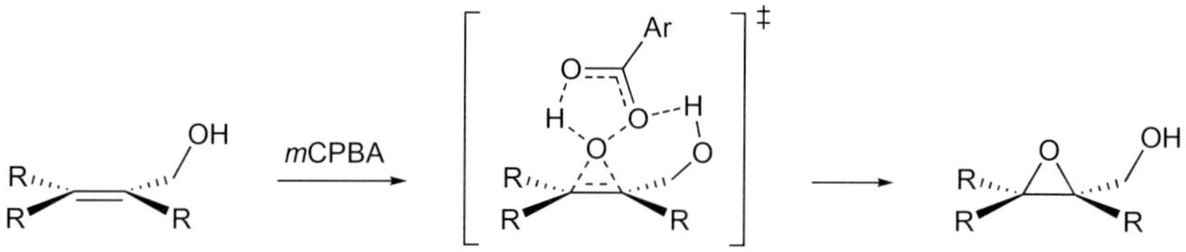

Scheme 7.18

hormones, Zurflüh and co-workers obtained β-epoxide **68** from homoallylic alcohol **67** by mCPBA epoxidation in CH$_2$Cl$_2$, in contrast, α-epoxide was exclusively formed in Et$_2$O (Scheme 7.17) [31]. The difference of diastereoselectivities would be explained with the transition state of epoxidation, which involves intermolecular hydrogen bonding between the hydroxyl group and mCPBA (Scheme 7.18). That is, the solvent effect in Scheme 7.17 indicates that CH$_2$Cl$_2$ allows hydroxyl-directed epoxidation, whereas Et$_2$O cleaves the intermolecular hydrogen bonding.

7.3.2
Epoxyquinols

Epoxyquinols A (**69**) and B (**70**), and epoxytwinol A (**71**) (Figure 7.13) were isolated by Kakeya, Osada, and co-workers from unidentified fungus, which were discovered to be strong angiogenesis inhibitors [32]. Shoji, Hayashi, and co-workers accomplished the first total synthesis of **69** and **70**, using biomimetic oxidative dimerization of the monomer **72**, that is, regioselective oxidation of the primary allylic alcohol **72**, followed by spontaneous 6π-electrocyclization of **73** gave 2H-pyrans **74** and **75** (Scheme 7.19) [33]. Intermolecular Diels–Alder reaction of the 2H-pyrans furnished **69** and **70**. Interestingly, stereoselectivity of the Diels–Alder dimerization varied dependent on solvent. For instance, dimerization in neat form provided **69** and **70** in 40 and 25% yield, whereas that in toluene afforded **69** and **70** in 25 and 45% yield, respectively. Hayashi and co-workers disclosed that intermolecular hydrogen bondings between the 2H-pyrans affect the stereoselectivity in the dimerization with performing theoretical calculations [34].

Inter- and intramolecular hydrogen bondings are also important for the formation of epoxytwinol A (**71**), which would be produced by the quite rare formal [4 + 4] cycloaddition reaction. Theoretical calculations suggest that preassociation

Epoxyquinol A (**69**) Epoxyquinol B (**70**) Epoxytwinol A (**71**)

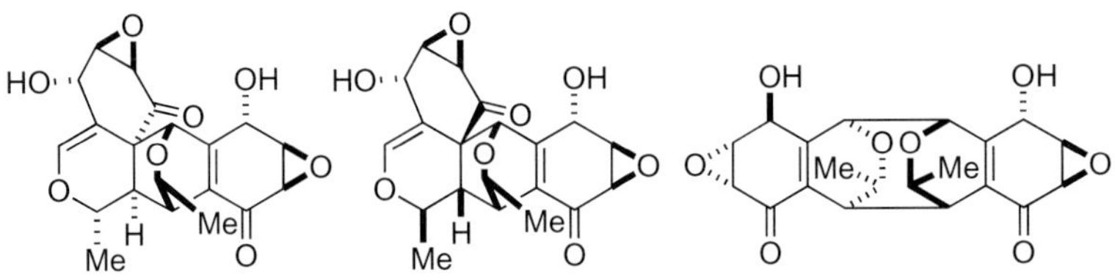

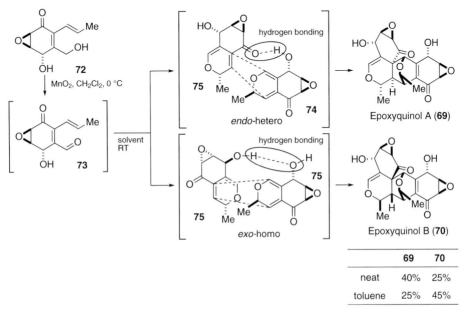

Figure 7.13 Structures of Epoxyquinols A and B, and Epoxytwinol A.

	69	70
neat	40%	25%
toluene	25%	45%

Scheme 7.19

of two 2*H*-pyrans **74** forms complex **76** stabilized by intermolecular hydrogen bonding (Scheme 7.20) [35]. Successive C8–C8′ bond formation, C8–C8′ bond rotation, and C1–C1′ bond formation would produce epoxytwinol A (**71**), in which intramolecular hydrogen bondings stabilize intermediates **77** and **78**.

7.3.3
Epoxide-Opening Cascades

Brevetoxin B (**79**) is one of the most representative polyether compounds, which was isolated by Nakanishi, Clardy, and co-workers, from *Gymnodynium breve* Davis [36]. Nakanishi and co-workers postulated the biosynthetic pathway of **79** as shown in Scheme 7.21, that is, stereoselective oxidation of polyene **80** would give

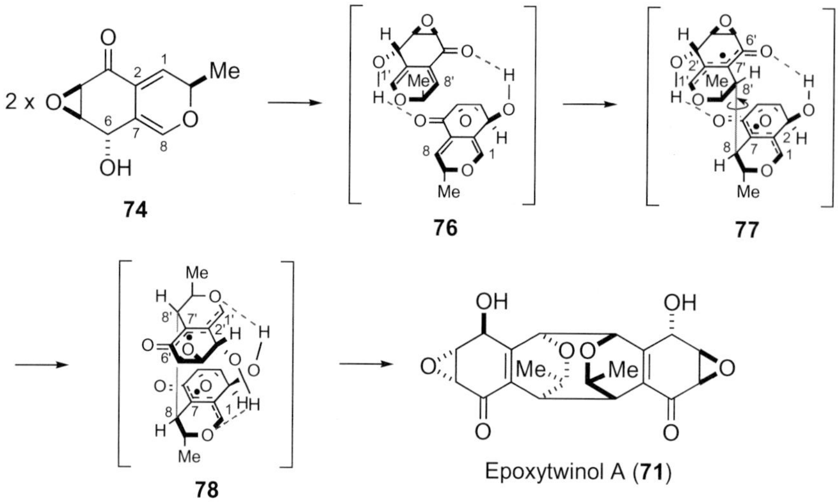

368 | *7 Highlights of Hydrogen Bonding in Total Synthesis*

Epoxytwinol A (**71**)

Scheme 7.20

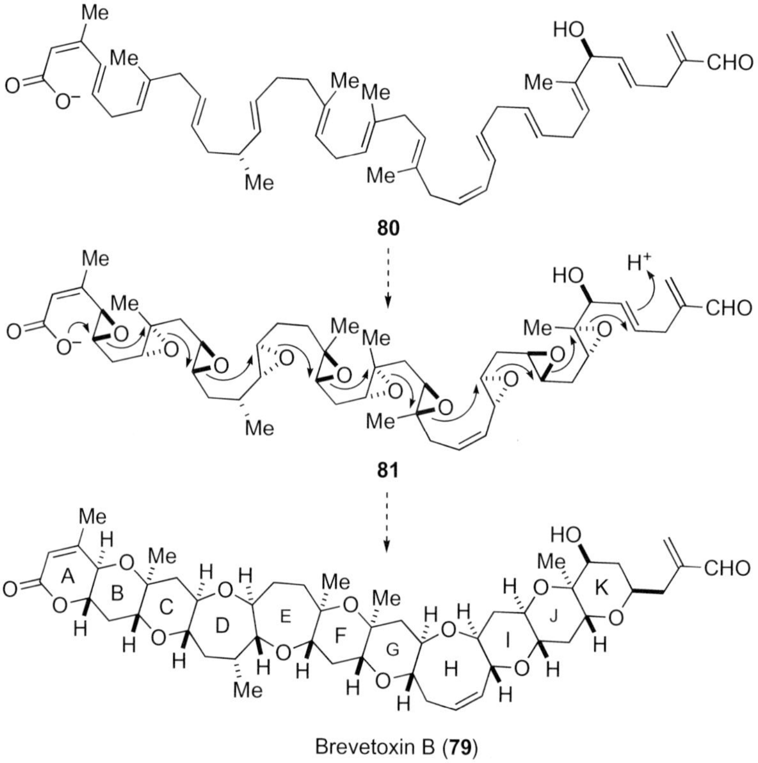

80

81

Brevetoxin B (**79**)

Scheme 7.21

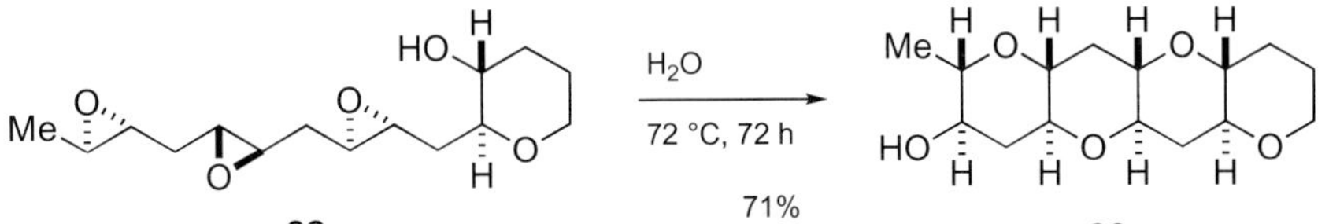

Scheme 7.22

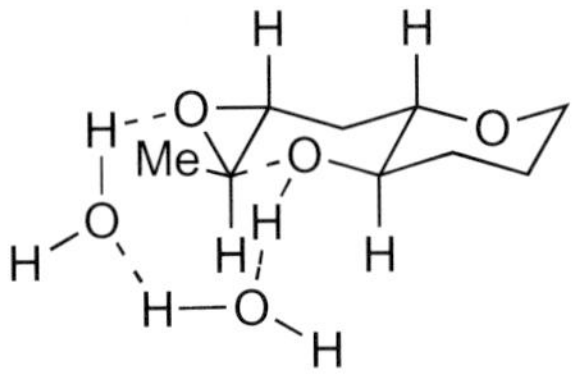

Figure 7.14 A model of epoxide-opening reaction promoted by water.

polyepoxide precursor **81**, followed by polyether formation cascade would furnish polycyclic ether **79** [37]. According to this hypothesis, Vilotijevic and Jamison accomplished the water-promoting epoxide-opening cascade of **82**, providing tetracyclic ether **83** in high yield with regio- and diastereoselectivity (Scheme 7.22) [38]. In general, the undesired 5-*exo* cyclization predominates the desired 6-*endo* [39]. For accounting this reactivity and stereoselectivity, they made a cooperative network of hydrogen bonding model **84** as shown in Figure 7.14. Two water molecules activate the hydroxyl group and the epoxide with hydrogen bonding, which would induce to produce the tetracyclic compound **83**.

7.4 Conclusions

Natural organisms produce many compounds, with high efficiency and selectivity, utilizing intra- and intermolecular hydrogen bondings. Hydrogen bondings between acidic protons and basic heteroatoms often affect reactivity and stereoselectivity in the total synthesis of natural products. Although it is not easy to predict how the hydrogen bonding works, it can be utilized for more efficient synthesis of many compounds. If we could control hydrogen bondings at will, such as nature, a lot of natural compounds would be synthesized much more efficiently.

References

1 Hayes, C.J. and Heathcock, C.H. (1997) *J. Org. Chem.*, **62**, 2678–2679.

2 Weiner, S.J., Kollman, P.A., Case, D.A., Singh, U.C., Ghio, C., Alagona, G.,

Profeta, S. and Weiner, P. (1984) *J. Am. Chem. Soc.*, **106**, 765–784.

3 Henbest, H.B. and Wilson, R.A.L. (1957) *J. Chem. Soc.*, 1958–1965.

4 Uemura, D., Chuo, T., Haino, T., Nagatsu, A., Fukuzawa, S., Zheng, S. and Chen, H. (1995) *J. Am. Chem. Soc.*, **117**, 1155–1156.

5 McCauley, J.A., Nagasawa, K., Lander, P.A., Mischke, S.G., Semones, M.A. and Kishi, Y. (1998) *J. Am. Chem. Soc.*, **120**, 7647–7648.

6 (a) Noda, T., Ishiwata, A., Uemura, S., Sakamoto, S. and Hirama, M. (1998) *Synlett*, 298–300.
(b) Sakamoto, S., Sakazaki, H., Hagiwara, K., Kamada, K., Ishii, K., Noda, T., Inoue, M. and Hirama, M. (2004) *Angew. Chem. Int. Ed.*, **43**, 6505–6510.

7 Satake, M., Ofuji, K., Naoki, H., James, K.J., Furey, A., McMahon, T., Silke, J. and Yasumoto, T. (1998) *J. Am. Chem. Soc.*, **120**, 9967–9968.

8 (a) Nicolaou, K. C., Qian, W., Bernal, F., Uesaka, N., Pihko, P.M. and Hinrichs, J. (2001) *Angew. Chem. Int. Ed.*, **40**, 4068–4071.
(b) Nicolaou, K.C., Pihko, P.M., Bernal, F., Frederick, M.O., Qian, W., Uesaka, N., Diedrichs, N., Hinrichs, J., Koftis, T.V., Loizidou, E., Petrovic, G., Rodriquez, M., Sarlah, D. and Zou, N. (2006) *J. Am. Chem. Soc.*, **128**, 2244–2257.

9 Bender, T., Schuhmann, T., Magull, J., Grond, S. and von Zezschwitz, P. (2006) *J. Org. Chem.*, **71**, 7125–7132.

10 (a) Pettit, G. R., Gaddamidi, V., Cragg, G.M., Herald, D.L. and Sagawa, Y. (1984) *J. Chem. Soc., Chem. Commun.*, 1693–1694.
(b) Pettit, G. R., Gaddamidi, V. and Cragg, G.M. (1984) *J. Nat. Prod.*, **47**, 1018–1020.
(c) Pettit, G.R., Gaddamidi, V., Herald, D.L., Singh, S.B., Cragg, G.M., Schmidt, J.M., Boettner, F.E., Williams, M. and Sagawa, Y. (1986) *J. Nat. Prod.*, **49**, 995–1002.

11 Rigby, J.H., Maharoof, U.S.M. and Mateo, M.E. (2000) *J. Am. Chem. Soc.*, **122**, 6624–6628.

12 Myers, A.G., Gin, D.Y. and Rogers, D.H. (1994) *J. Am. Chem. Soc.*, **116**, 4697–4718.

13 (a) Kobayashi, M., Higuchi, K., Murakami, N., Tajima, H. and Aoki, S. (1997) *Tetrahedron Lett.*, **38**, 2859–2862.
(b) Murakami, N., Wang, W., Aoki, M., Tsutsui, Y., Sugimoto, M. and Kobayashi, M. (1998) *Tetrahedron Lett.*, **39**, 2349–2352.

14 Marshall, J.A. and Bourbeau, M.P. (2002) *J. Org. Chem.*, **67**, 2751–2754.

15 (a) Oyama, H., Sassa, T. and Ikeda, M. (1978) *Agric. Biol. Chem.*, **42**, 2407–2409.
(b) Barrow, C.J. (1997) *J. Nat. Prod.*, **60**, 1023–1025.

16 Couladouros, E.A., Mihou, A.P. and Bouzas, E.A. (2004) *Org. Lett.*, **6**, 977–980.

17 Fürstner, A., Dierkes, T., Thiel, O.R. and Blanda, G. (2001) *Chem. Eur. J.*, **7**, 5286–5298.

18 Barluenga, S., Moulin, E., Lopez, P. and Winssinger, N. (2005) *Chem. Eur. J.*, **11**, 4935–4952.

19 Smith, A.B., Mesaros, E.F. III and Meyer, E. (2006) *J. Am. Chem. Soc.*, **128**, 5292–5299.

20 (a) Angenot, L. (1978) *Plant. Med. Phytother.*, **12**, 123–129.
(b) Angenot, L., Coune, C. and Tits, M. (1978) *J. Pharm. Belg.*, **33**, 11–23.

21 Bassleer, R., Depauw-Gillet, M., Massart, B., Marnette, J., Wiliquet, P., Caprasse, M. and Angenot, L. (1982) *Planta Med.*, **45**, 123–126.

22 Lerchner, A. and Carreira, E.M. (2006) *Chem. Eur. J.*, **12**, 8208–8219.

23 Kwon, O. and Danishefsky, S.J. (1998) *J. Am. Chem. Soc.*, **120**, 1588–1599.

24 Iwaki, H., Nakayama, Y., Takahashi, M., Uetsuki, S., Kido, M. and Fukuyama, Y. (1984) *J. Antibiot.*, **37**, 1091–1093.

25 Allen, J.G. and Danishefsky, S.J. (2001) *J. Am. Chem. Soc.*, **123**, 351–352.

26 White, J.D. and Somers, T.C. (1994) *J. Am. Chem. Soc.*, **116**, 9912–9920.

27 Fettes, A. and Carreira, E.M. (2003) *J. Org. Chem.*, **68**, 9274–9283.

28 Cook, J.L., Hunter, C.A., Low, C.M.R., Perez-Velasco, A. and Vinter, J.G. (2007) *Angew. Chem. Int. Ed.*, **46**, 3706–3709 and references cited therein.

29 Asami, Y., Kakeya, H., Onose, R., Yoshida, A., Matsuzaki, H. and Osada, H. (2002) *Org. Lett.*, **4**, 2845–2848.

30 (a) Hayashi, Y., Shoji, M., Yamaguchi, J., Sato, K., Yamaguchi, S., Mukaiyama, T., Sakai, K., Asami, Y., Kakeya, H. and Osada, H. (2002) *J. Am. Chem. Soc.*, **124**, 12078–12079.
(b) Hayashi, Y., Shoji, M., Mukaiyama, T., Gotoh, H., Yamaguchi, S., Nakata, M., Kakeya, H. and Osada, H. (2005) *J. Org. Chem.*, **70**, 5643–5654.

31 Zurflüh, R., Wall, E.N., Siddall, J.B. and Edwards, J.A. (1968) *J. Am. Chem. Soc.*, **90**, 6224–6225.

32 (a) Kakeya, H., Onose, R., Koshino, H., Yoshida, A., Kobayashi, K., Kageyama, S.-I. and Osada, H. (2002) *J. Am. Chem. Soc.*, **124**, 3496–3497.
(b) Kakeya, H., Onose, R., Yoshida, A., Koshino, H. and Osada, H. (2002) *J. Antibiot.*, **55**, 829–831.
(c) Kakeya, H., Onose, R., Koshino, H. and Osada, H. (2005) *Chem. Commun.*, 2575–2577.

33 (a) Shoji, M., Yamaguchi, J., Kakeya, H., Osada, H. and Hayashi, Y. (2002) *Angew. Chem. Int. Ed.*, **41**, 3192–3194.
(b) Shoji, M. and Hayashi, Y. (2007) *Eur. J. Org. Chem.*, 3783–3800.
(c) Shoji, M. (2007) *Bull. Chem. Soc. Jpn.*, **80**, 1672–1690.

34 (a) Shoji, M., Kishida, S., Kodera, Y., Shiina, I., Kakeya, H., Osada, H. and Hayashi, Y. (2003) *Tetrahedron Lett.*, **44**, 7205–7207.
(b) Shoji, M., Imai, H., Shiina, I., Kakeya, H., Osada, H. and Hayashi, Y. (2004) *J. Org. Chem.*, **69**, 1548–1556.

35 Shiina, I., Uchimaru, T., Shoji, M., Kakeya, H., Osada, H. and Hayashi, Y. (2006) *Org. Lett*, **8**, 1041–1044.

36 Lin, Y.-Y., Risk, M., Ray, S.M., Engen, D.V., Clardy, J., Golik, J., James, J.C. and Nakanishi, K. (1981) *J. Am. Chem. Soc.*, **103**, 6773–6775.

37 Lee, M.S., Nakanishi, G.-W., Qin, K. and Zagorski, M.G. (1989) *J. Am. Chem. Soc.*, **111**, 6234–6241.

38 Vilotijevic, I. and Jamison, T.F. (2007) *Science*, **317**, 1189–1192.

39 (a) Baldwin, J. E. (1976) *J. Chem. Soc., Chem. Commun.*, 734–735.
(b) Baldwin, J.E., Cutting, J., DuPont, W., Kruse, L., Silberman, L. and Thomas, R.C. (1976) *J. Chem. Soc., Chem. Commun.*, 736–737.
(c) Baldwin, J. E. (1976) *J. Chem. Soc., Chem. Commun.*, 738–739.

Index

a

acetalizations   12, 156–159, 201
acetamide thioureas   240
acetylacetone (2,4-pentanedione)   25–27, 297–299
acroleins   112, 127, 176, 301, 302
activation/deactivation of reactions   18, 362–365
activation energies   10, 20, 23, 39, 40
activation enthalpy   35, 36
activation entropy   35, 36, 44
activation modes, electrophilic   15, 16
addition reactions. see Henry; Mannich; Michael reactions
aggregation-induced enhancement   32
alcohols
– β-alkoxy-   173, 174, 183
– THP protection   162, 163, 164
aldehydes
– α-branched   91, 312, 313, 314
– homologation via epoxide ring-opening   177
aldimines
– aza-Henry reactions   219, 220, 241, 265, 267, 270
– BNP catalyzed Diels-Alder addition   83
– BNP catalyzed Friedel-Crafts addition   79, 80
– Brønsted acid catalysis   110
– cyclopropyl oxindole ring expansion   361
– guanidine catalyzed Diels-Alder addition   102
– N-phospinoyl   218
– (thio)urea catalyzed cyanation   161, 162, 189–192, 195, 196, 318, 319
– thiourea catalyzed Mannich addition   221–223, 267, 269
– thiourea catalyzed transfer hydrogenation   169, 170

aldol reactions
– see also Henry (nitroaldol) reaction
– γ-butenolides from TMSOF   179
– enzymes and   59–61, 66
– Mukaiyama aldol reaction   39, 151
– nitroso-aldol reaction   116–120
– proline- and prolinamide catalysis   12, 15, 17, 288
– retro-aldol reaction   364, 365
– Strecker reaction   308
– thio-Michael addition and   272
alkene epoxidations   29–37, 317
allylation of hydrazones   242
altohyrtin A   353
amidinium salts   106
amines, (thio)ureas from   185–235
amino acids
– α-, from hemimalonates   277
– enantiomerically pure α-   191, 192
– epimerization/stereo inversion   64, 66, 233
– optically active β-   270
– oxazolidinone derivatives   223
– precursors of β-   291
– synthesis of γ-   206
amino alcohol (thio)urea derivatives   288–296
ammonium salt catalysis   106–108
anhydride desymmetrization   285
anomeric effects
– exo-anomeric effect   320
– hydrogen bonding opposition   353–355
– nonanomeric spiroketals   3
arginine   99
artificial enzymes. see enzyme mimetics
Asialo GM_1 361, 362
atropoisomerism   238–240
aza-Darzens reaction   83

aza-Diels-Alder reaction 83, 85
aza-Henry reaction. *see* nitro-Mannich reaction
aza-hetero-Diels-Alder reaction 83
aza-Michael addition 83, 279, 280
aza-Morita-Baylis-Hillman (aza-MBH) reaction 115, 116, 250, 301–305
azaspiracid-1 355
azaspirene 364, 365
azlactone DKR catalysis 19–25, 221, 231–233
azodicarboxylates 105, 223, 224

b

B3LYP. *see* Becke's three-parameter method
baclofen 206
BAMOL. *see* biaryl-2,2′-dimethanol
Baylis-Hillman reactions. *see* Morita-Baylis-Hillman
Becke's half-and-half method (BHandHLYP) 10
Becke's three-parameter method (B3LYP) 17, 28, 29, 34, 36, 37
benzaldehyde
– HDA reaction with Danishefsky's diene 37–39
– reaction with acetals 151
benzamides as Michael acceptors 215, 216
benzoins 180–182
1H-benzotriazole 277–279
O-benzylhydroxylamide 292, 294
BHandHLYP. *see* Becke's half-and-half method
1,1′-biaryl-2,2′-dimethanol (BAMOL) 112–114
bifunctionality 332
– cinchona alkaloids 255
– proline 12
– thiourea organocatalysts 25–29, 202, 203, 289, 296, 302, 324–330
Biginelli reaction 86
binaphthalenediamine (BINAM) 304, 305
– bis-N-tosyl- 299, 300
2,2′-bi-1-naphthol, 3,3′-diphenyl- (VANOL) 93, 113
1,1′-bi-2-naphthol (BINOL) 5, 6
– as a Brønsted acid 115, 116, 120
– cooperative hydrogen bonding 18
– derived phosphines 301
– Diels-Alder reactions 117
– vinylogous Mukaiyama aldol reaction 113
binaphthyl guanidines 105
binaphthyl phosphoric acid (BNP) derivatives 75–91

binaphthyl thiourea derivatives 296–307
BINOL. *see* bi-2-naphthol
biomimetics *171*
– *see also* enzyme mimetics
4-biphenanthrol, 2,2′-diphenyl- (VAPOL) 93, 113
1,8-biphenylenediol catalysts 142–144
bis-thioureas 250–253
BNP. *see* binaphthyl phosphoric acid
bond lengths
– hydrogen bond flexibility and 18, 21, 26, 28
– interatomic distances 2, 6, 11, 12
– intramolecular hydrogen bonds 38
bond strengths (hydrogen bonds) 1, 2, 6, 7, 45, 47
boronates 124, 225, 226, 280–282
– *see also* oxazaborolidine
Brassard's diene 85, 112
brevetoxin 367, 368
Brønsted-acid-assisted Brønsted acid catalysis (BBA) 110–122
– diol activation of carbonyl electrophiles 111–116
– diol activation of other electrophiles 116–120
– N-triflyl phosphoramide 96
Brønsted-acid-assisted Lewis acid catalysis (BLA) 126–135
Brønsted acid catalysts 9, 10, *11*
– asymmetric counteranion-directed catalysis 98–99
– binaphthyl phosphoric acid derivatives 75–91
– cooperative catalysis 173, 174
– difference from hydrogen bond catalysts 6, 7–9, 12
– general Brønsted acid catalysts 15, 16, 19
– N-triflyl phosphoramide 95–98
– nonBINOL-based phosphoric acids 93, 95
– specific Brønsted acid catalysts 15, *16*
– super Brønsted acids 129
γ-butenolides 179

c

CALB. *see* Candida antarctica lipase B
calculated bond strengths 45
calculated reaction paths 3
– *see also* density functional theory calculations
callystatin 358, 359
Candida antarctica lipase B (CALB)
– enzymatic promiscuity 64, 66

carbohydrates. see saccharides
β-carbolines 198, 291
carbonyl groups/compounds
– see also aldehydes; ketones
– acid catalysis 7, 8
– activation modes 16
– diol activation of 111–116
– in enzymatic reactions 44, 49
– hydrogen bonding catalysis of nucleophilic additions 3, 4, 5
– intermediates in enzymatic additions 44
carboxylic acid donors 121
CAVEAT program 64
chalcones 120, 210–212, 261–263, 266–269, 279–281, 293
– aza-chalcones 147
chiral Brønsted-acid catalysts. see Brønsted acid catalysts
chiral proton catalysts 5
m-chloroperoxybenzoic acid 365, 366
chymotrypsin 49, 50, 52–54, 55
– see also serine proteases
cinchona alkaloids
– names and structures of 254
– stereoselective (thio)ureas organocatalysts from 253–288
cinchonidine 254, 256
cinchonine 254, 256–261, 262
cintrolellal 98
citral 98
citrate synthase 58–61
Claisen rearrangements 144, 145, 168, 169
CoA/ACP dependent enzymes 51
combined acid catalysis
– Brønsted-acid-assisted Brønsted acid catalysis (BBA) 110–122
– Brønsted-acid-assisted Lewis acid catalysis (BLA) 126–135
– four combinations 109–110
– Lewis-acid-assisted Brønsted acid catalysis (LBA) 122–126
computational studies
– see also density functional theory calculations
– alkene epoxidation in fluorinated alcohols 29–37
– DKR of azlactones 19–25
– HDA of Danishefsky's diene with benzaldehyde 37–39
– hydrogen bond catalysis 15–18
– thiourea catalyzed Michael additions 25–29
conformation and HFIP catalysis 32

conjugate additions
– see also Michael additions
– activation for 159
– asymmetry establishment and 98
– oxyanion holes and 64–66
cooperative catalysis 37, 115, 173, 174, 228
cooperative hydrogen bonds 18, 29, 34
Corey-Chaykovsky epoxidation 177, 178
counteranion-directed catalysis 98–99
covalent and non-covalent organocatalysis 15
CP-99,994 221
Cram rule 313
crotonase folds 58
crystalline structures 1, 2
crystallographic studies
– enzyme catalysis 47, 48, 54
– small molecule catalysis 100, 110, 111, 146, 242
cupreidine/cupreine 254, 286, 287
cutinase 47, 50–52, 54, 56, 57
cyanation 151, 152, 161, 162, 195
– see also Strecker reaction
cyanoacetates 211–213, 216, 235, 269, 270
cyanohydrins 101
cyanosilylation
– aldehydes 238
– ketones 227–229
– oxazaborolidines 133
cyclizations
– artificial cyclases 123, 124
– catalyzed by thiourea pyrroles 200, 201
– epoxyquinol synthesis 366
– intramolecular 86, 129, 198, 206
– Nazarov cyclization 96
– pancratistatin synthesis 355–357
– tunicamycin synthesis 357
cycloadditions
– see also Diels-Alder reactions
– [3,2] cycloadditions 229, 230
– [4,4] cycloadditions 366
– 1,3-dipolar 149, 150
– oxazaborolidines 133
cyclodextrins 61, 62
cyclohexane, trans-1,2-diamino-. see diaminocyclohexane
2-cyclohexen-1-ol 353
2-cyclohexen-1-one
– see also Diels-Alder reactions (as product); Michael additions (as reactant)
– 5,5-dimethyl- reaction with thiols 255
– Diels-Alder reactions of 83

- Morita-Baylis-Hillman reaction with aldehydes 115, 250–254, 293, 294, 296, 297, 306
cyclohexylamine, isophosphoronediamine [3-(aminomethyl)-3,5,5-trimethyl]- (IPDA) 252
cyclopentadienes
- Diels-Alder cyclization 127, 131, 144, 146, 147
cyclopentenones
- Nazarov cyclization 96
cyclopropanes, nitro- 264, *266*
cyclopropyl oxindoles 361
cysteine proteases 50

d

DABCO. see 1,4-diazabicyclo [2.2.2]octane
DACH. see cyclohexane-1,2-diamino-
Danishefsky's diene
- HDA reaction with benzaldehyde 37–39
- HDA reaction with glyoxylates 121, 122
- HDA reaction with N-aryl imines 83
density functional theory (DFT) calculations 10, 11
- alcoholytic ring opening of azlactones 19
- alkene epoxidation in fluorinated alcohols 29–37
- cyanosilylation 227
- epoxide alcoholysis 174
- epoxide aminolysis 156
- Hartree-Fock and 17
- Henry additions 287, 288
- TADDOL promoted HDA reaction 37–39
- thiourea-catalysed Claisen rearrangements 168
- thiourea-catalysed Michael additions 25, 208, 215
- thiourea-catalysed MOP protection 168
- tosylurea catalyst design 182
2-desoxystemodione 363
DFT. see density functional theory calculations
trans-1,2-diaminocyclohexane and derivatives
- Schiff base (thio)ureas from 190–195
- (thio)ureas from, as organocatalysts 185– 293, 322, 330
diastereoselectivity
- see also enantioselectivity
- cyanation 151
- epoxidation 353, 366
- Michael additions 206
- nitro-Mannich reaction 241, 265
- thio-Mannich reaction 274

1,4-diazabicyclo [2.2.2]octane (DABCO)
- see also Morita-Baylis-Hillman reaction
- in Michael additions 205, 281
Diels-Alder reactions
- see also hetero-Diels-Alder (HDA) reaction
- aza-Diels-Alder reaction 83
- binaphthyl phosphoric acid (BNP) derivatives 83, 84
- BINOL derivatives 117, 118
- Brønsted-acid-assisted Lewis acid catalysis (BLA) 126–132
- catalysis by 1,8-biphenylenediols 144
- catalysis by cinchona-related thioureas 274–277
- guanidine catalyzed 101
- intramolecular 129
- synthesis of epoxyquinols 366
- synthesis of rishirilide B 362–365
- transannular Diels-Alder reactions 129
diffuse functions 17
dihydroquinidine/dihydroquinine *254,* 261–264
diisopropylethylamine (DIPEA, Hünig's base) 240, 241
diol activation
- of carbonyl electrophiles 111–116
- of other electrophiles 116–120
diol catalysts
- 1,8-biphenylenediols 142–144
1,3-dioxolane-4,5-dimethanol, α,α,α',α'-diphenyl- (TADDOL). see diphenylmethanol
DIPEA. see diisopropylethylamine
diphenylmethanol, (*trans*-α,α'-(dimethyl-1,3-dioxolane-4,5-diyl)-*bis*- (TADDOL) 5, 6
- enantioselective HDA reaction 37–39, 111–114
- as a hydrogen bond catalyst 9, 11, 18, 111–114
- N-nitroso aldol reaction 116
diphenylphosphinoyl (DPP) protection 229, 230
disulfides, oxidation 317
DKR. see dynamic kinetic resolution
DPP. see diphenylphosphinoyl-
dynamic kinetic resolution (DKR)
- α-branched aldehydes 91
- of azlactones, thiourea catalysis 19–25, 221, 231–233
- of oxazinones 234

e

electrophilic activation modes 15, *16*
- see also activation
enamides
- azaspirene synthesis 364

- BNP catalyzed Friedel–Crafts addition 79
- BNP reaction with imines 86
enamines
- conjugate additions 237–239
- iminium intermediate favored over 83
- Michael reaction intermediates 245–248, 250, 322, 323, 326
- nitroso-aldol reaction of 116–118
- proline catalysis and 12, 15, 17
- transfer hydrogenation intermediates 91
enantioselective Brønsted acid catalysts. see Brønsted acid catalysts
enantioselectivity
- see also diastereoselectivity
- alcoholytic ring opening of azlactones 23–25
- aza-Henry reactions 265
- BNP catalyzed Friedel-Crafts addition 79
- Brønsted-acid-assisted Brønsted acid catalysis (BBA) 110
- cyanosilylation 229
- Diels-Alder reactions 132
- Henry reactions 313
- hetero-Diels-Alder reactions 37–39
- Mannich-type reactions 9, 223, 269, 321
- Michael additions 26, 29, 205, 206, 210
- Petasis reaction 225
- Pictet-Spengler reactions 200
- retro-nitroaldol reaction 310
- Strecker reaction 308, 320, 321
enecarbamates 79, 85
energy profile
- HFIP epoxidation of alkenes 31
- Michael additions 29
- proton transfer reactions 10
enolate intermediates 49, 51, 56–61
enolization, enzymatic 44, 46
enoyl-CoA isomerase 58–60
entropy. see activation entropy
enzymatic catalysis
- contribution of oxyanion holes 44–47, 49–52
- nitroalkene reductase mimetics 169, 170
enzymatic promiscuity 64–67
enzyme mimetics 142, 170, 180, 183, 203
epibatidine 207
epimerization
- amino acids 64, 66
- nitronates 219, 220
epoxidations
- alkenes 29–37, 317
- α,β-unsaturated aldehydes 98

- Corey–Chaykovsky epoxidation 177, 178
- diastereoselective 353
- enzymatic promiscuity and 66, 67
- Henbest epoxidation 365, 366
- hydrolysis 171, 172
- N-tosylurea catalyzed ring opening 183
epoxide aminolysis 155–157
epoxide hydrolase 141
epoxide-opening cascades 367–369
epoxyquinols and epoxytwinols 366, 367
esterase oxyanion holes 50

f
fluorinated alcohols 29–37
formate dehydrogenase 141, 142
Friedel-Crafts alkylation
- amino alcohol thiourea catalysis 288–290, 293
- binaphthylthiourea catalyzed 299, 300
- BNP catalyzed 79–83
- cinchona alkaloid derivatives and 261, 262, 270, 271
- intramolecular cyclizations 198
- N,N'-bisarylthiourea catalyzed 153–155, 159

g
general Brønsted acid catalysts 15, 16, 19
glyoxylates 121, 122
guanidine-based thiourea derivatives 307–315
guanidine organic base catalysis 99–105

h
haloalcohol dehydrogenase 141, 142
Hantzsch esters 89, 90, 169–171, 202, 223
Hartree-Fock theory 17
Henbest epoxidation 365, 366
Henry (nitroaldol) reaction
- see also nitro-Mannich (aza-Henry) reaction
- aminocyclohexylthiourea catalysts and 206
- bisaryl(thio)urea catalysts and 304–306
- cinchona-derived thiourea catalyzed 25, 286–288
- guanidine-based thiourea catalysts and 309–316
- N-H hydrogen bond catalysts and 108
hetero-Diels-Alder (HDA) reaction
- aza-hetero-Diels-Alder reaction 83
- carbonyl compounds 11, 12, 111, 112
- Danishefsky's diene with benzaldehyde 37–40

– Danishefsky's diene with glyoxylates 121, 122
1,1,1,3,3,3-hexafluoro-2-propanol (HFIP) 29–36
H_3O^+ ion 1
Hünig's base (diisopropylethylamine, DIPEA) 240, 241
hybrid density functional theory. *see* density functional theory calculations
hydratases 58
hydrazones 159–161, 242, 295
hydrocyanation 89, 187, 190, 192–195, 317, 318
hydrogen bond catalysis 11–13
– catalytic functions performed 18
– computational studies 15–18
– difference from Brønsted acid catalysis 6, 7–9, 12
– N-H hydrogen bond catalysis 99–108
– nucleophilic additions to CO groups 3, 4
– proline-catalyzed reactions as 12
hydrogen bond flexibility 18
hydrogen bonds described 1, 6
hydrogen bonds synthetic uses 3
hydrogen peroxide 29–37
hydrophobic amplification/hydration 155, 156, 172
hydrophosphonylation
– binaphthyl phosphoric acid (BNP) derivatives 76–80
– mechanism 242
– Schiff base thioureas 196, *198*, 242
hydropyridine 89, 169, 181
γ- and δ-hydroxy-α,β-enones 282, 283
α-hydroxyester cyclization 98
hydroxylactams 200–202

i
ice, hydrogen bonding in 6
imines
– *see also* aldimines; ketimines
– ammonium salt catalyzed HCN addition 108
– BNP catalyzed Diels-Alder reactions 83
– BNP catalyzed Friedel-Crafts additions 79, 80, 83
– BNP catalyzed hydrophosphonylation 78–79
– BNP catalyzed Mannich reactions 75–78
– equilibrium with enamines 247
– Povarov reaction with dihydrofuran 243
iminium ions
– intermediates in Brønsted acid catalysis 9, 10, 80, 87, 89, 91, 99

– intermediates in Petasis reactions 224, 225
– intermediates in Pictet-Spengler reactions 197, 200–202
– intermediates in Povarov reactions 243
– intermediates in Strecker reactions 161
– modes of electrophilic activation 15, 16
– use in acetylization 12, 161
– use in epoxidation 67
indanol (thio)urea derivatives 294–296, 330
indoles
– additions to styrene oxides 182
– binaphthyl thiourea catalyzed Friedel–Crafts addition 299, 300
– BNP catalyzed Friedel–Crafts addition 79–83
– thioureas derived from 154, 155, 270, 288–294
interatomic distances 2, 6, 11, 12
– *see also* bond lengths
intermediates
– *compare also* transition states
– alcoholytic ring opening of azlactones 22, 23
– enamines in Michael additions 245–247
– enzymatic addition to carbonyls 44, 49
– enzymes with oxyanion holes 49
– tetrahedral intermediates 3, 22, 44, 47–50, 52–60
intermolecular aldol reactions 12
intermolecular hydrogen bonding 2, 18, 144, 146, 148
– involving TADDOL 38, 39, 112, 114
– total synthesis of natural products 365–369
intermolecular Michael additions 206
intermolecular nuclear overhauser effect 233
intramolecular cyclizations 86, 129, 198, 206
intramolecular hydride donation 181
intramolecular hydrogen bonding 3, 18, 38, 39
– in 2-methoxybenzamides 216
– activation/deactivation of reactions 362–365
– in bifunctional thioureas 205
– Brønsted acid catalysis 112, 114, 116, 120, 122
– in diols 114
– in epoxide hydrolysis 174
– kinetic control 355–362
– in oxyanion hole mimics 63
– thermodynamic control 353–355

– (thio)urea catalysis 144, 174, 204, 205, 216, 242, 243, 320
– total synthesis of natural products 353–367
intramolecular Michael additions 83, 206, 280–283
ion pairs, Brønsted acid catalysis 8
IPDA. see cyclohexylamine
isomerases
– enolate intermediates 58
– ketosteroid isomerase 48
– oxyanion hole geometry 51
isoquinolines 190, 194, 200, 201

k

kendomycin 360
ketimines 80, 169, 193–195
α-keto esters 315, 316
β-keto esters 223, 224
ketones
– BNP catalyzed Mannich reactions 77
– as oxyanion hole mimics 62, 63
ketosteroid isomerase (KSI) 48
kinetic control of stereochemistry 355–362
kinetic use of hydrogen bonds 3
KSI. see ketosteroid isomerase

l

L-lactide 184
β-lactones 135
Lee-Yang-Parr (LYP) correlation functional 10, 17, 28, 29, 34, 36, 37
leucascandrolide A 363, 364
Lewis-acid-assisted Brønsted acid catalysis (LBA) 122–126
Lewis acid catalysis
– Brønsted-acid-assisted (BLA) 126–135
Lewis/Brønsted bifunctionality 202, 203
life, dependence on hydrogen bonding 2
lipases 64–67, 183
low-barrier hydrogen bonds (LBHB) 18, 34, 47, 48
lowest unoccupied molecular orbital (LUMO) 5, 32, 111, 147, 171
LYP. see Lee-Yang-Parr (LYP) correlation functional

m

macrolactonization 364
Mannich-type reactions
– see also nitro-Mannich reactions
– bifunctional thiourea catalyzed 221
– binaphthyl phosphoric acid (BNP) derivatives 75–78

– BINOL derivatives 118, 120
– bis-thiourea catalyzed 252
– carboxylic acids 121
– cinchona alkaloids catalysis of 257, 258, 267, 269, 274
– phosphoric acid catalyzed 9, 10, 11
– pyrrole thiourea catalyzed 196–201
– retro-Mannich reaction 361
– saccharide-based thiourea catalyzed 321
– Schiff base thiourea catalyzed 196–198
MBH. see Morita-Baylis-Hillman reaction
mCPBA. see chloroperoxybenzoic acid
menthol 98
methanolysis of anhydrides 285, 286
2-methoxypropene (MOP) 167
Michael acceptors
– benzamides 215, 216
– nitroalkenes 153, 154, 159
Michael additions
– aza-Michael addition 279–280
– DABCO attack 176
– enzymatic promiscuity and 66
– guanidine catalysis of 104
– hetero-Michael reactions 117–120
– intramolecular 280–283
– mechanism of 208
– oxazaborolidine catalysis of 134
– oxy-Michael addition 280, 282
– thio-Michael additions 272–275
– (thio)urea catalysis
– – amino alcohol derivatives 291, 294
– – aminocyclohexyl (thio)ureas 25, 26, 203–217, 235–237, 245–247
– – cinchona alkaloid derivatives 255–257, 260–264, 268, 269, 271, 277–279
– – pyrrolidine thioureas 324–326
– – saccharide-based thioureas 321, 322
Michael-aldol reaction 217, 218, 272–275
Michael-Knoevanagel reaction 284
molecular recognition 317
molecular sieves 161
MOP. see methoxypropene
Morita–Baylis–Hillman (MBH) reaction 115, 174–177, 250–254, 293–298, 305, 307
– see also aza-Morita-Baylis-Hillman
Mukaiyama aldol reaction 151, 153
mutagenesis studies 46, 47

n

N-H hydrogen bond catalysis 99–109
– by ammonium salts 106–108
– by guanidine organic bases 99–106
– by tetraaminophosphonium salts 109
NADH/NADPH 169, 170, 180

natural bond orbital (NBO) charges 20, *21*
natural product total synthesis
– intermolecular hydrogen bonding
  365–369
– intramolecular hydrogen bonding
  353–365
Nazarov cyclization 96
neurokinin-1 (NK-1) receptor antagonist (–)-
  CP-99,994 221
nitro-aldol reaction. *see* Henry reaction
nitro-Mannich (aza-Henry) reaction
– amidinium salt catalysis 106
– aminocyclohexylthiourea catalysis
  218–221
– cinchona-derived (thio)urea catalysis 265,
  267, 270, 272
– oxazoline thioureas 327–329
– saccharide (thio)urea catalysis 323, 324
– *N*-sulfinyl-thioureas 329–331
nitro-Michael reactions 26, 244
nitroalkanes 170, 171
– *see also* aza-Henry reaction
nitroalkenes. *see* Michael additions
nitronates
– in aza-Henry reactions 219, 329
– in Henry reactions 109, 288, 305, 311,
  312
– in Michael additions 27, 208
– in nitro-Mannich reactions 241
nitrones 150–152
nitroso-aldol reaction 116–120
nitroso-Diels-Alder reaction 117
non-covalent catalysts 15, 73
nuclear overhauser effect (NOE) 194, 233
nucleophilic additions to CO groups 3, 4, 5

o

O-H distance. *see* interatomic distances
old yellow enzyme (OYE) 169
olefins
– epoxidations 29–37, 317
– ring-closing olefin metathesis 359
ONIOM (Our owN n-layered Integrated
  molecular Orbital + molecular mechanics
  Method) 37, *38*
organic acid catalysts 5
organic synthesis, uses of hydrogen bonds
  3
oxazaborolidene BLA catalysis 126–135
oxazinone dynamic resolution 234
oxazolidinones 146, 223, 317
oxazoline thioureas 327–329
oxy-Michael additions 280, 282
oxyanion hole mimics 61–67, 183

oxyanion holes 44
– contribution to enzymatic catalysis
  44–47, 49–52
– description of the two classes 49–52
– with enolate intermediates 56–61
– enzymatic promiscuity and 64–67
– geometry of 50–52, 54–56
– properties of hydrogen bonds 47, 48
– with tetrahedral intermediates 52–56
– thiourea catalysis of azlactone DKR and
  19–25

p

pancratistatin 355–357
2,4-pentanedione (acetylacetone) 25–27,
  297–299
peptide inhibitors 54, *55*
peptide synthesis 223
Petasis reaction 224–226
phase transfer catalysts (PTC)
– ammonium compounds 120
– guanidine containing 104
phosphine oxides 134
phosphinothioures 229, 230, 301
phosphoric acid catalysts 5, 6, 9, 10, 89
phosphorus ylides 252
phthalimides 237, 238
Pictet–Spengler reaction 81, 197–202
pinnatoxin A 353, 354
piperazine, diketo- 307, 308
polarization activation 18
polyether formation cascades 369
polymer-bound (thio)urea catalysts 165,
  166, 187, 190, 209
post-Hartree-Fock methods 17
Povarov reaction 243
proline-based thioureas 244, 324
proline-catalyzed reactions 12, 15
proline-derived catalysts 288
proline substitution effects 46
2-propanol, 1,1,1,3,3,3,-hexafluoro-. *see*
  hexafluoro-2-propanol
proton transfer reactions
– barrier-free, cascade-like 35
– in Brønsted acid catalysts 15
– energy profile 10
– hydrogen bonds as incipient 7
protonated chiral catalysts 5
pseudo-Lewis acids 15, 20
pseudoephedrine-derived catalyst 25
PTC. *see* phase transfer catalysts
pyrazole, 2,4-dimethyl- 284, 292
pyrrole, 2-aryl-2,5-dihydro- 229, 230
pyrrole thioureas 200–203

pyrrolidine (thio)ureas 325, 326
pyrrolidines, methyleneamino- 118, 294–296
pyrrolidinones 213, 214, 217

q
quantum mechanics 17
quinidine
– stereoselective (thio)ureas organocatalysts from 261, 262, 268–271, 274–277
– structure of 254
quinine
– O-acetyl 255
– stereoselective (thio)ureas organocatalysts from 261, 262, 268–286
– structure of 254
quinoline cascade hydrogenation 89
quinolinium ions 225, 226

r
radicicol/pochonin C 360
RB3LYP. *see* Becke's three-parameter functional
reactant activation. *see* activation
reactant spatial preorganization 18, 46
resorcylides 359, 360
ring-closing metathesis 359, 360
ring-expansion 361
ring-opening aldehyde homologation 177
ring-opening polymerization (ROP) 184, 185
rishirilide B 362–365

s
saccharide-based (thio)urea catalysts 315–324
salicylihalamide 360
Schiff base thioureas 143, 187, 188, 190, 192, 195–199
Schiff base ureas 190–195, 317, 318, 321
serine hydrolases 183
serine proteases
– oxyanion hole geometry and formation 50, 52, 54–56
– (thio)ureas as enzyme mimetics 141, 142
σ-cooperativity principle 110
6-31G and 6-31+G (split-valence basis set) 17, *28, 29, 34, 36*
small molecule oxyanion hole mimics 61–67
solvation of H_3O^+ 1
solvent effects
– Diels-Alder cyclizations 144, 366
– HDA kinetics 111
– hydrogen bond strengths 2
– intramolecular hydrogen bonding 364
– Michael additions 278
spatial preorganization of reactants 18, 46
specific Brønsted acid catalysts 15, *16*
spiro-acetals 353
spiro-oxazolidinones 317
spiroketals 3, 317, 354, 355
starting complexes, azlactones ring opening 19, 20, *24, 25*
stationary-point structures *34, 36*
steric effects
– azlactone ring opening 23
– γ-butenolide formation 179
– cyanosilylation 227
– Diels-Alder reactions 131, 132
– favoring D-amino acids 64, 65
– Friedel-Crafts additions 299, 300
– Henry reaction 312, 313
– Mannich reaction 77
– Michael additions 208, 213, 214, 218, 219, 284, 285
– nitrone additions 95
– protection of hindered alcohols 163, 164
– Strecker reaction 195
– Wittig reaction 359
Stetter reaction 294
Strecker reaction 101, 186–197, 317–321
– *see also* cyanation
– asymmetric 147, 186–193, 195, 196, 242–244, 307, 308, 319
streptenols 282
strychnofoline 361
styrene oxide. *see* epoxidations
subtilisin 50, 52
N-sulfinyl thioureas 329, 330, *331*
sulfuric acid solvation 1
super Brønsted acids 129
surfactants 311
synthetic uses of hydrogen bonds 3
"synzymes." *see* enzyme mimetics

t
TADDOL. *see* diphenylmethanol, (dimethyl-1,3-dioxolane-4,5-diyl)*bis*-
tertiary amine-functionalized (thio)ureas 231, 234, 235
tetraaminophosphonium salts 109
tetrahedral intermediates 3, 22, 44, 47–50, 52–60
tetrahydropyran (THP) 162–166, 168
tetrakis-(3,5-bis(trifluoromethyl)phenyl) borate) (TFPB) 106

1,1,3,3-tetramethylguanidine (TMG)   99, 206, 207
TFPB. see tetrakis(3,5-bis(trifluoromethyl)phenyl)borate
thermodynamic control of stereochemistry   353–355
thermodynamic use of hydrogen bonds   3
thiochromanes   217, 218, 221, 272, 273, 282, 284
thiolase enolate intermediates   58
thiourea organocatalysts   5, 9, 12
– advantages over ureas   145
– background and enzyme mimetic activity   141–149
– bifunctional thioureas   202, 203, 209, 221, 222
– catalysis of azlactone DKR   19–25, 221
– catalysis of Michael additions   25–29
– double hydrogen bonding catalysts   142, 143
– growth of literature on   148, 149, 331
– indolyl thioureas   289–293
– N,N′-diaryl-   143, 174
– – electron deficient   146
– – N,N′-bis[3,5-(trifluoromethyl)phenyl]-   146–174, 204
– – N,N′-diphenyl-   147
– non-stereoselective   149–185
– oxyanion hole mimics   64, 183
– phosphinothioures   229, 230, 301
– polymer-bound   165, 166, 209
– pyrrole thioureas   200–203
– saccharide-based   315–324
– Schiff base thioureas   143, 187, 188, 190, 192
– stereoselective 1,2-diaminocyclohexane derivatives   185–293
– – acetamide thioureas   240
– – bifunctional   202–204, 324–330
– – bis-thioureas   250–253, 304
– – guanidine-based   307–315
– – Michael addition catalysts   203–223
– – phthalimide containing   237, 238
– – primary amine-functionalized   244–250
– – tertiary amine-functionalized   231, 235
– stereoselective amino alcohol derivatives   288–296
– stereoselective binaphthyl derivatives   296–307
– stereoselective cinchona alkaloid derivatives   253–288
– stereoselective N-sulfinyl derivatives   329, 330

– stereoselective oxazoline derivatives   327–329
– structures of (Chapter 6 Appendix)   345–351
THP. see tetrahydropyran
TMG. see tetramethylguanidine
TMSOF. see trimethylsilyloxyfuran
trans-cyclohexane-1,2-diamine (DACH)   20, 25
transannular Diels-Alder (TADA) reactions   129
transesterification reactions   183, 184
transfer hydrogenation
– of aldehydes and ketones   98
– of aldimines   169, 170
– binaphthyl phosphoric acid (BNP) derivatives   89–91
– of nitroalkanes   202
– VANOL-derived phosphoric acids   93
transition state analogs   56
transition states
– alcoholytic ring opening of azlactones   20, 22, 23, 25
– γ-butenolides from TMSOF   179, 180
– Claisen rearrangements   169
– cyanosilylation   227, 228
– epoxidations   30, 31, 366
– epoxide aminolysis   156
– Henry (nitroaldol) reactions   312, 314, 316
– hetero-Diels-Alder reactions   39, 40
– Mannich-type reactions   11
– stabilization mechanisms   16, 18
– thiourea-catalysed Michael additions   26, 272, 323
triethylsilyl ether (TES)   282
N-triflyl phosphoramide   95–98
2-trimethylsilyloxyfuran (TMSOF)   178–180
trypsin   50, 52, 54, 55
– see also serine proteases
tunicamycins   357
turnover frequency (TOF)   142, 157, 163, 165, 195
turnover number (TON)   157, 163, 195

u

umpolung (polarity inversion)   294, 295
urea organocatalysts   5, 6
– see also thiourea organocatalysts
– background and enzyme mimetic activity   141–149
– N,N′-diaryl   144, 145, 153, 177
– saccharide-based   315–324
– Schiff base urea catalysts   190–195

– stereoselective 1,2-diaminocyclohexane derivatives 185–293
– – tertiary amine-functionalized 231
– stereoselective amino alcohol derivatives 288–296
– stereoselective binaphthyl derivatives 300, 301, 303
– stereoselective cinchona alkaloid derivatives 253–288
– structures of (Chapter 6 Appendix) 345–351
– thiourea advantages over 145

v
VANOL. *see* bi-1-naphthol, 3,3′-diphenyl-
VAPOL. *see* biphenanthrol, 2,2′-diphenyl-
vinylogous Mannich reaction 221–223
vinylogous Mukaiyama aldol (VMA) reaction 113

w
water
– activation entropy of enzymatic additions 44, 46
– in Diels-Alder reactions 144
– as HDA proton source 122
– hydrophobic amplification/hydration 155, 156
– as oxyanion hole hydrogen bond donor 57, 58, 60
Wittig reactions 358

x
X-ray structures 31
xanthone oxyanion hole mimics 64, 65

z
Zimmerman-Traxler transition states 176